D0087062

STRUCTURAL ANALYSIS AND BEHAVIOR

F. Arbabi

*Professor of Civil Engineering
and Adjunct Professor
of Engineering Mechanics
Michigan Technological University*

McGraw-Hill, Inc.

New York   St. Louis   San Francisco   Auckland   Bogotá   Caracas
Hamburg   Lisbon   London   Madrid   Mexico   Milan   Montreal
New Delhi   Paris   San Juan   São Paulo   Singapore
Sydney   Tokyo   Toronto

To my wife, Sharon,
AND MY DAUGHTERS,
Niku and Kassia

This book was set in Times Roman.
The editors were B. J. Clark and John M. Morriss;
the production supervisor was Denise L. Puryear.
The cover was designed by Joseph Gillians.
Project supervision was done by The Universities Press (Belfast) Ltd.
R. R. Donnelley & Sons Company was printer and binder.

STRUCTURAL ANALYSIS AND BEHAVIOR

Copyright © 1991 by McGraw-Hill, Inc. All rights reserved. Printed in the United States of America. Except as permitted under the United States Copyright Act of 1976, no part of this publication may be reproduced or distributed in any form or by any means, or stored in a data-base or retrieval system, without the prior written permission of the publisher.

234567890   DOC DOC   954321

ISBN   0-07-002143-0

Library of Congress Cataloging-in-Publication Data

Arbabi, F. (Freydoon)
    Structural analysis and behavior / by F. Arbabi.
        p.   cm.
    Includes bibliographical references.
    ISBN 0-07-002143-0
    1. Structural analysis (Engineering).   I. Title.
TA645.A73   1991
    624.1'71—dc20   89-77498

ABOUT THE AUTHOR

F. Arbabi is Professor of Civil Engineering and Adjunct Professor of Engineering Mechanics at Michigan Technological University. He received his B.S. degree from the Tehran Polytechnic Institute in 1965, his M.S. from the University of California–Berkeley in 1969, and his Ph.D. degree from the Illinois Institute of Technology in 1975. He also studied at the School of Bridges and Highways in Paris, France. He taught in the Civil Engineering Department at the Illinois Institute of Technology in 1975–1976, and was a visiting research professor at the University of Waterloo in 1984.

Professor Arbabi has performed research on the nonlinear and dynamic response of railroad tracks, creep deformations of shell structures, and macroelements. He is the author of a number of papers and research reports. He has developed a reduced-scale models laboratory and a number of experiments for teaching structural behavior. In 1986 he received the Inland Steel–Ryerson teaching innovation award, and was the co-recipient of the ASEE Elgin–Wickendon award for the best paper in engineering education. From 1969 to 1974 he worked as a stress analyst and development engineer dealing with shell structures. Prior to that he had worked for three years for a consulting engineering firm as a design engineer. He has also served as a consultant to a number of companies and government agencies. He is a member of the American Society of Civil Engineers, American Concrete Institute, American Railroad Engineering Association, and American Academy of Mechanics, and is a registered professional engineer in Michigan.

CONTENTS

6   Energy Methods

7   Method of Consistent Deformations

PREFACE

This text presents the methods of structural analysis with some emphasis on the physical behavior of structures. Classical methods of analysis provide useful means of understanding the concepts and the behavior of structures. Modern matrix methods, on the other hand, present systematic general solution procedures. Therefore, an attempt has been made to strike a balance between the classical and the modern methods. The direct stiffness method has emerged as the most common method of solution because of its generality and ease of programming. This is the matrix method used throughout the book. However, an introduction to the flexibility method is presented in Chapter 13.

Additional features of the text include small-scale model experiments, at the end of each chapter, for observing the behavior and verifying the principles; introduction of the singularity method; in-depth discussion of cables, arches, buckling, torsion, and three-dimensional frames, with both classical and matrix methods.

The laboratory experiments may be conducted by the students, demonstrated by the instructor, or simply used as reading material to supplement the text. Reduced-scale structural engineering laboratories have been used at a number of engineering schools, for example, Cornell University and MIT. Some of the experiments included here have been used at the above institutions. Undergraduate structural engineering laboratories have also been recommended by the American Board of Engineering and Technology, to enhance the students' intuition and knowledge of structural behavior.

In Chapters 1 to 10 the emphasis is on the classical methods. Chapters 11 to 13 and 18 emphasize the stiffness method. Chapters 14 to 16 and 17 include both classical and matrix methods.

Chapter 1 is a general introduction to the field of structural engineering. Since throughout the text problems are presented both in metric and US customary units, the conversion between these units is discussed at the end of Chapter 1. Chapters 2 and 3 discuss reactions and trusses. Chapters 4 and 5 present the internal forces and deflections of beams and frames. Calculation of deflections by energy methods is covered in Chapter 6. Chapter 7 discusses the method of consistent deformations. Chapter 8 presents influence lines. Chapter 9 presents the singularity method for beams under multiple loads, for variable sections, and for continuous beams. The singularity method also provides an elegant procedure for influence lines. Chapter 10 derives the force–displacement relations for beam elements, which are used for both classical and matrix methods, and covers the moment distribution and slope–deflection methods.

Chapter 11 derives the procedure for the direct stiffness method and describes its application to beams. Chapter 12 presents application of the stiffness method to two-dimensional trusses and frames. Chapter 13 presents additional topics in matrix methods. These include support settlement, internal hinges, elastic supports, flexible connections, variable section elements, and thermal stresses. The last section of this chapter gives an introduction to the flexibility method. Chapters 14 and 15 discuss cables and arches. Chapter 16 describes beam–columns and stability of beams and frames. Chapter 17 discusses torsion as well as grid systems and bow-girders. Chapter 18 discusses the methods of solution for three-dimensional (space) trusses and frames, and Chapter 19 discusses application of computers in structural analysis and programming the stiffness method.

The text is primarily intended for undergraduate courses. However, the more advanced topics may be used in graduate courses. Chapters 1 to 5, 7, 8, 10, and 11 provide the material for a first course in structural analysis. It is possible to treat the beam deflection problem by a single method. The virtual work method with visual integration, discussed in Chapter 6, would be suitable for this purpose. This is a semi-graphical method similar to the moment-area method. However, it applies to all problems of beams and frames in a straightforward manner. For calculating the deflection at a variable distance (or for finding the maximum deflection) it may be preferable to integrate the virtual work equation analytically instead of by visual integration. Chapters 6, 9, and 12 to 18 can be used in an intermediate course in structural analysis, at the senior level, or in a first graduate course. Chapters 11 to 13, and the stiffness methods in Chapters 15, 16 and 19, provide sufficient material for a course on matrix methods of structural analysis. Finally, the text can be used as a reference by practicing engineers.

The example and problems at the end of the chapters are both the V.S. customary and SI units. Solutions to selected problems are given at the end of the book.

McGraw-Hill and the author would like to thank the following reviewers for their comments and suggestions: D. A. Gasperini, Case Western Reserve University; David Pecknold, University of Illinois; José Roesset, University of Texas at Austin; A. N. Sherbourne, University of Waterloo; Judith Stalnaker, University of Colorado; and Louis Tartaglione, University of Lowell.

F. Arbabi

LIST OF SYMBOLS

A	area, cross-sectional area, constant
A_i	partial area, cross-sectional area of truss bar i
A_o	enclosed area
A_s	shear area $= A/\kappa$
B	constants
$\mathbf{B}$	matrix $= \mathbf{KT}$, force transformation matrix
C	seismic coef., constant, carry over factor
C_e	exposure factor for snow and wind loads (different)
C_i	constants of integration
C_{ij}	carry over factor for variable section members, elements of transformation submatrix for 3-D beam elements
C_q	pressure coefficient
C_s	sloped roof factor for wind load
C_t	thermal factor
$\mathbf{C}$	transformation submatrix
D	constant, plane dimension in the direction of earthquake force, dead load
D_{ij}	distribution factor
E	modulus of elasticity, earthquake load
E_s, E_c, E_t	modulus of elasticity of steel, concrete and timber
F_i	applied lateral static load, force in truss bar i
F_{ij}	force in bar ij of truss, elements of the flexibility matrix
$FPS = USCS$ units, ft. lb. sec.	
F_t	force due to whiplash
F_u	ultimate strength of steel
F_v	force in axial spring
F_x	force at level x, force in the x direction
F_y	yield stress of steel, force in the y direction
$\mathbf{F}$	element flexibility matrix
$\mathbf{F}_c$	composite flexibility matrix
$\mathbf{F}_s$	structure flexibility matrix

$\mathbf{F}_{ij}$	submatrices of structure flexibility matrix
$F(j, k)$	fixity matrix
G, G_s	shear modulus, structure geometric stiffness matrix with $N = 1$
H	thrust of cable, height of the floor
$H(x - a) = H_a$	Heaviside unit step function
$H_i = H(x - a_i),\ H_\xi = H(x - \xi)$	unit step function
I	importance factor
I_i	moment of inertia of span i, moment of inertia of span to the right of support i, moment of inertia of segment i
IL	influence line
J	total number of terms in approximating function, torsion constant
JL	number of loaded joint
K	structural system coefficient, pressure coef., stiffness, Kelvin
K_{ij}	stiffness factor for member ij, terms of the stiffness matrix
K'_{ij}	reduced stiffness factor
$\mathbf{K}_e$	elastic stiffness matrix
$\mathbf{K}_{ES},\ \mathbf{K}_{GS}$	elastic and geometric structure stiffness matrices
$\mathbf{K}_G$	geometric stiffness matrix
$\mathbf{K}, \bar{\mathbf{K}}$	stiffness matrix in local and global coordinates
$\mathbf{K}_s$	structure stiffness matrix
$L(k)$	locator matrix
M	mass, moment at x, bending moment in torsional spring
M_0	applied concentrated moment
M_i	internal moment at point i, applied moment at i
$M_i, \bar{M}_i$	moment at end i in local and global coordinates
M_{ix}, M_{iz}	moments at point i about x and z coordinate axes
$\bar{M}_{iX}, \bar{M}_{iY}, \bar{M}_{iZ}$	moments at i about global coordinate axes X, Y, Z
M_i^j	moment at end i in element j
M_i^L, M_i^R	moment at left and right of point i
M_{ij}	moment in segment ij
M_{Li}	moment of the structure on left side of i calculated at i
M_L, M_R	moment at left and right of section
$M_{\max}$	maximum moment
M_u	unbalance moment
M^+, M^-	positive and negative moments
M_i^F	fixed-end moment at end i
MPa	mega pascal
M', M'_i	moments in the conjugate beam
N	number of stories, axial force, Newton, normal force in arches, axial force in beam columns
N_{cr}	critical (buckling) load
N_e	Euler buckling load
NI, NJ	first and second joints of a member

P, P_i	load, concentrated load
Pa	pascal
P_s, P_c, P_t	force in steel, concrete, timber
$\bar{P}$	resultant force
P_x, P_y, P_z	force components in x, y and z directions
Q	total weight, first moment, number of degrees of freedom
$Q(x)$	total axial load over x
$\mathbf{Q}$	matrix defined for arch elements, stability matrix
R	reduction factor for occupancy loads, reaction, radius of curvature of the centroidal axis
R_i	reaction at point i, element of force vector
R_{ix}, R_{iy}	reactional support i in the x and y directions
R_{ix}, R_{iy}	reactional support i in the x and y directions
R_i'	reaction at i, reaction of conjugate beam
R_i^0	reaction of primary structure at i
$\mathbf{R}$	structure load (force) vector
$\mathbf{R}_i$	submatrices of structure force vector
S	site structure resonance factor, snow load,
SI	system international (units)
$\mathbf{S}, \bar{\mathbf{S}}$	member end force vectors in local and global coordinates
$\mathbf{S}_c$	composite force vector
$\mathbf{S}^F$	Fixed end force vector
T	tension force in cable, fundamental (largest) period of vibration, temperature, torque, transpose (superscript)
T_i	torque at point i, applied concentrated torque at i
T_l	temperature at the lower fiber
T_{ij}	torque in segment ij
T_m	average temperature over depth
T_s	characteristic site period
T_u	temperature at the upper fiber
T_y	temperature at distance y
$\mathbf{T}$	Transformation matrix
U	normal force, elastic energy
U_a, U_b, U_s, U_t, U_S	axial, bending, shear, torsion and structure strain energy
U_i	normal force at point i
U_i^L, U_i^R	normal force at left and right of point i
$U_i, \bar{U}_i$	normal end forces in the local and global coordinates
V	shear force, wind velocity, man base shear
V_i	shear force at point i
V_0, V_1	shear without and due to sidesway
V', V_c'	shear in the conjugate beam
V_i^L, V_i^R	shear at left and right of point i
V_{ij}	shear in panel ij
V_i^+, V_i^-	shear at the right and left of support i
V_i^F	fixed end shear at end i

$V_i, \bar{V}_i$	end shear forces at point i in the local and global coordinates
W	weight, total structure dead load plus a portion of live load, wind load, work
W_i	weight of different floors, internal work, work
$W_i, \bar{W}_i$	shear force at end i in the local and global Z coordinates
W_e	external work
W_{ij}	partial work
W_x	weight of floor x
W^*	complementary elastic work
X, X_i	global X axis, X coordinate of point i
Y, Y_i	global Y axis, Y coordinate of point i
Z, Z_i	zone factor, global Z axis, Z coordinate of i
a	constant parameter, distance to load from the left end, base length for area calculation, constant in equation of a line
a_i	distance, unknown parameters
a_j	distance to force P_j from the left end, distance from the first load to another load
b	distance, distance to load from the right end, number of truss bars, width of section, constant in equation of a line
b_i	distance to the right of segment i
c	number of equations of condition, distance from the neutral axis to the extreme fiber, distance, direction cosine $= \cos \alpha$
c_i	direction cosines for 3-D truss members
cm	centimeter
d	differential sign, distance
d_i	parameters
d_{ij}	distance
d_s, d_S	length of fiber at neutral axis and at y from it
dv	differential of displacement
$d\theta$	change of angle (slope)
f	parameter, natural frequency of vibration (Hertz $=$ cycle/s)
f_i	force in truss bar i due to a unit applied load
f'_c	compressive strength of concrete at 28 days age
f_t	tensile stress
ft	foot
$f'_{c\perp}$	compressive stress perpendicular to grain of wood
$f_{c\parallel}$	compressive stress parallel to grain of wood
$f_{t\parallel}$	tensile stress parallel to grain of wood

f_v	shear stress
$f(x)$	function, moment due to loads and reactions of beam-columns (excluding the moment of the axial force)
$f_1(x)$	function
$f_{2i}(x)$	function in terms of step functions
g	gravitational acceleration, parameter
$g(x)$	function
$g_1(x)$	function
$g_{2i}(x)$	function in terms of step functions
h	sag of a cable, rise of an arch, depth, height for area calculation, height
h_1	distance
h_i	height of floor i
h_n	maximum height of building
h_x	height of floor x
$h_1(x)$	function
$h_{2i}(x)$	function in terms of step functions
in	inch
j	number of truss joints
kg	kilogram
k_i	rotational stiffness
kN	kilo Newton
k_v	stiffness of axial spring
k_θ	stiffness of torsional spring
l	length, span length
l_i	distance, length of span i, length of span on the right of support i, length of element i
l_{ijx}	projection of length of member ij in the x direction
l_x, l_y, l_z	projections of the length in X, Y, Z directions
lb	pound
m	meter, moment at x due to an applied unit load
$\bar{m}$	value of moment diagram for a unit load at the centroid of M/EI diagram
$m(x)$	moment due to the applied loads only
n	number of equations of statics, degree of parabola, number of concentrated loads, number of segments in a variable section member
p	snow load, design wind pressure, distributed lateral load
$p(x)$	distributed lateral load
q	shear flow $= \tau t$, ground snow load, water or soil pressure, distributed axial load, total load on span
q_s	basic wind pressure
$q(x)$	distributed axial load
r	radial polar coordinate, factor, radius, number of unknown reactions, radius of gyration, parameter

r_i	reaction to an applied unit load, distance, rotation of joint, displacement or rotation degree of freedom
$\bar{\mathbf{r}}, \bar{\mathbf{r}}_i$	vectors
$\mathbf{r}$	displacement vector
$\mathbf{r}_i$	submatrices of structure displacement vector
rad.	radian
s	sine of direction angle $= \sin \alpha$
$\mathbf{s}, \bar{\mathbf{s}}$	member-end displacement vectors in the local and global coordinates
$\mathbf{s}_c$	composite displacement vector
t	thickness
t_i	parameter
t_{ij}	tangent distance-distance from point i on the deflected beam to the tangent of point j
u	displacement in the x direction
u_i	displacement at point i in the x direction
$v, v(x)$	displacement, vertical displacement, displacement in the y direction
v_i	displacement, deflection at point i
v_{ij}	deflection at point i due to a force at point j
v_{io}	deflection at point i of primary structure
v_0	displacement
v_b, v_s	bending and shear deflections
v_{max}	maximum deflection
$v(x), v(\xi)$	deflection at x, and at ξ
vol	volume
w	unit weight of concrete, weight of surcharge, weight, load
w_0	maximum value of distributed triangular load
x	distance to an arbitrary section, local (member) axis
x_i	x of point i
$\bar{x}$	centroidal distance, location of resultant force
y	distance from the neutral axis, coordinate, local (member) coordinate, ordinate of an arch $y = f(x)$
z	local (member) z axis
Γ	perimeter
Δ	differential displacement (settlement)
	drift or lateral displacement of the floor
Δl	change of length
Δl_i	change of length of bar i
Δ_{jk}	displacement between points j and k along direction jk
ΔT	temperature change, temperature difference through depth
θ	polar coordinate
$\theta_x, \theta_y, \theta_z$	direction angles
$\theta_{xX}, \theta_{xY}, \theta_{xZ}$	angles of the local (member) x axis with X, Y, Z global coordinate axes

Π	total potential energy
Πs	total potential energy of a system or structure
Σ	summation sign
α	slope of roof, coef. of thermal expansion, angle, parameter, shear parameter, parameter angle, direction angle, beginning angle of arch, ratio of distance to load over span length, buckling parameter $= \sqrt{N/EI}$
$\alpha_i, \beta_i, \delta_i, \gamma_i$	parameters
β	angle, angle of arch, parameter, dimensionless parameter
β_1	parameter
γ	weight density, shear strain
δ	angle, variation symbol, virtual sign, deflection
δ_i	floor deflection due to applied static load, F_i
δ_{ij}	deflection at i due to a unit force at j
$\delta_{i\xi}$	displacement at i due to a unit load at ξ
$\delta_{\xi i}$	displacement at ξ due to a unit load at i
δ_{jk}	displacement between points j and k along direction jk due to unit forces at j and k
ε	strain, normal (axial) strain
$\varepsilon_1, \varepsilon_2$	strain at bottom and top fibers
$\varepsilon_s, \varepsilon_c, \varepsilon_t$	strain in steel, concrete and timber
θ	slope, rotation angle
θ_i	angle (slope) at point i, member-end rotation
θ_{ij}	change of slope between points i and j
θ_i^j	rotation at end i of element j
θ_i', θ_i''	rotation at support i calculated from the left and right members
$\theta_i'^j, \theta_i''^j$	rotation at i, calculated from the left and right members due to a unit moment at j
$\theta_i'^o, \theta_i''^o$	slopes at support i calculated from simply support spans under the applied loads (slopes at right of left span and left of right span)
κ	shape (form) factor for shear deflection
ν	Poisson's ratio
ξ	distance from the left end to the variable position of the load
π	number pi
σ	stress, normal stress
σ_a	allowable stress (in a cable)
σ_e	Euler buckling stress
σ_{cr}	critical stress (buckling)
τ	shear stress
$\tau_{\max}$	maximum shear stress
ϕ	internal pressure angle for soil, displacement (shape) function, slope of arch at a point, angular polar coordinate
ψ	rotation of the chord, angle of twist

ψ_i	member-end angles of twist
ω	circular frequency of free vibration rad./sec.
°C	degree Celsius
°F	degree Farenheit
$'$	ft, derivative with respect to x
$''$	inch, second derivative with respect to x
iv	fourth derivative with respect to x
$\langle\ \rangle$	Macaulay's brackets equal to argument if argument positive otherwise zero

CHAPTER
1

INTRODUCTION

1.1 ENGINEERING STRUCTURES

Engineering structures are systems built for a variety of purposes. The more common structures are buildings and bridges. Dams and other hydraulic structures, offshore oil platforms, water and gas tanks, piping systems, nuclear and chemical reactor vessels, ships, airplanes, buses and other transportation systems provide additional examples. Some examples of structures are shown in Fig. 1.1.

1.2 STRUCTURAL ENGINEERING

Structural engineering is the process of applying the laws and principles of science to the design and construction of structures. This involves several stages. Planning and conceptualization is the first stage of design. The concept must comply with the requirements of esthetics, functionality, and economy. The type of structure and the materials used are usually decided at this stage.

The tentative configurations established at the planning stage are analyzed for forces and displacements. These forces and displacements are used to determine the size of the members and finalize the dimensions of the members. If modifications to the original configuration become necessary, the

1

(b)

(a)

2

(c)

(d)

FIGURE 1.1
Examples of engineering structures. (a) First National Bank of Chicago. (b) Mackinack Bridge, Michigan. (c) Mossyrock Dam, Tacoma, Washington. (*Courtesy Harza Engineering*). (d) Maplewood Water Tank. (*Courtesy Chicago Bridge & Iron Co.*).

3

analysis and size-calculation cycle may be repeated to achieve compliance with safety, economy, and functionality. The dimensions thus obtained are used for developing the final drawings for use in construction and for calculating the amount and cost of the materials.

Structural engineers play a major role in all phases of construction projects, from the initial planning, through analysis, to inspection. The initial planning stage, where the type of structure and material are chosen, is a crucial phase and may significantly affect the cost and performance of the structure. The cost of a structure must be viewed as the total cost, that is, the initial cost plus maintenance costs. The design and analysis are done by structural engineers. Any changes to the original design must also be approved by the structural engineer to insure compliance with the design codes and to confirm that the safety of the structure is not compromised. During the course of the construction the structural engineer may be involved in frequent inspection to insure that the work is proceeding according to the design.

1.3 TYPES OF STRUCTURES

Structures may be composed of a variety of types and shapes. For analysis and construction purposes they are divided into a number of elements. Some common element types (Fig. 1.2) are beams, trusses, frames, arches, cables, plates (or slabs), and shells. Any of these types or combinations thereof may be used in buildings, bridges, and other structures. The choice of structure type depends on the materials used, loads to be carried and esthetic requirements. The choice of structure type is an important step in the construction process and is made early in the planning stage.

Beams are the commonest structural components. They carry their loads by developing bending moments and shear forces at different sections. Beams are used as single or continuous elements. Continuous beams are usually more economical. However, when there is a possibility of support settlement, either the settlement must be accurately predicted and included in the design, or single spans must be used. This also applies to other multispan or fixed-end structural components. The main beams supporting other beams are called girders. Small beams carried by other beams are called joists or rafters (in timber structures). The transverse beams in bridges are called floor beams.

Trusses carry their loads by developing axial forces in the bars. They are usually more economical than beams for moderately large spans. Their depth is larger than that of beams for the same span and load. They are common in factory buildings and bridges.

Frames are composed of beam members with different orientations. They are common in multistory buildings. They are rather easy to erect, but are not very strong in resisting large lateral loads of wind or earthquake. Therefore, frame buildings are often supplemented by lateral bracing or shear walls.

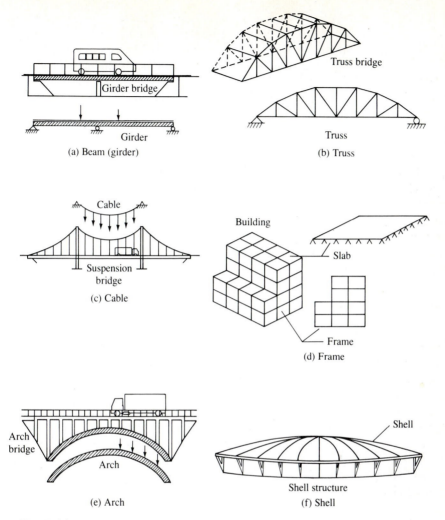

FIGURE 1.2

Arches are suitable for spanning deep openings such as gorges. Since arches with a significant curvature carry most of their loads by developing compressive stresses, they can be built with materials of high compressive strength and low tensile strength, such as masonry.

Cables are used for long-span structures such as suspension bridges and roofs to reduce the dead load of the structure. They carry their loads by developing axial tension.

Slabs are used to cover the floors in conjunction with beams. They may also be used as mat foundatons. They are common in buildings, bridges and elevated parking garages.

Shells have a curved surface. They are suitable for carrying distributed loads such as gas and water pressures encountered in industrial applications. Shells are more complex to build than other structures. For buildings they allow a variety of shapes that may be appealing if esthetics is the main concern.

Plates and shells are usually subjects of specialized texts and are not covered in this text. Most of the principles discussed in this book, however, apply to plates and shells as well.

1.4 MODELLING AND IDEALIZATION

For design and construction purposes a structure may be divided into a number of components. These components may be of different types, for example, frames and slabs, or of a number of similar components such as beams. Furthermore, if a mode of deformation is predominant, then the structure may be idealized to allow a simpler analysis. Many systems can be divided into components that can be idealized as two-dimensional structures. In the latter case all the members as well as forces and deformations are in one plane. Some simple structures carrying only axial forces and developing axial stresses can be categorized as one-dimensional. It is often possible to separate the different components of a structure from the rest and account for the effect of the rest of the structure by applying the loads or reactions that the latter may be exerting on the component under consideration. If a two-dimensional idealization cannot provide an accurate picture of the behavior of the structure, then a three-dimensional analysis must be performed. This may be the case when significant lateral loads act in two directions. However, more effort is needed for performing a three-dimensional analysis.

> **Example 1.1.** The roof of the building in Fig. 1.3a is composed of a number of trusses. These trusses are connected together by purlins. However, since the displacement of each truss occurs mainly in its own plane, each truss can be designed as a two-dimensional structure, Fig. 1.3b. The purlins, placed at the truss joints, serve to distribute the roof loads between trusses. Care must be exercised to insure that no failure occurs because of the loads acting normal to the plane of the trusses, such as wind loads.

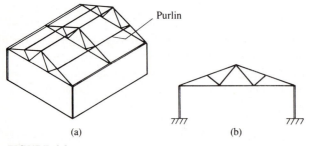

(a)   (b)

FIGURE 1.3

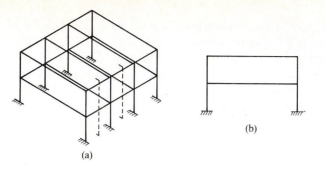

(a)

(b)

FIGURE 1.4

Example 1.2. The structure of Fig. 1.4a is a three-dimensional frame. For design purposes it can be divided into a number of two-dimensional frames as shown in Fig. 1.4b.

These frames deform in a plane. However, they must be braced laterally to resist the loads acting in the direction normal to their plane. Bracing may be provided by additional members or by masonry shear walls.

1.5 STRUCTURAL BEHAVIOR

The manner in which structures act to carry and transfer their loads and the deformation modes they assume is called *structural behavior*. Thus, study of structural behavior consists of establishing the primary modes of deformation and identifying the manner in which they carry their loads. Displacements may be caused by static or dynamic loads. Dynamic loads induce time-dependent responses (i.e., deformations and stresses) in structures. Another area of concern is instability, where one or more components of the structure buckle under compressive forces and lose their ability to carry loads. Since, unlike bending, buckling deformations occur abruptly, they may cause catastrophic failures and must therefore be prevented. Similarly, in reinforced-concrete structures a bending failure that starts by yielding of the reinforcement is ductile and accompanied by large displacements. This provides the occupants of the building with sufficient warning to evacuate the building before the failure occurs. For this reason the specifications of the American Concrete Institute, which is used as design code for concrete buildings in the United States, requires that bending members be under-reinforced. This means that the bending reinforcement must be sufficient for carrying the applied loads. However, they must yield under overload, in order to allow a ductile failure initiated in the reinforcement and avoid a brittle failure initiated in the compressed concrete.

For design purposes, structural components may be divided into different idealized members. However, the idealized structure must embody the behavior of the actual structure. For example, members of a truss are assumed to carry axial forces only. Therefore, the configuration of the truss and the manner in which the loads are applied must be such that this assumption holds true. If the truss members are long and slender and the loads are applied at the joints only, then the member forces are primarily axial, even when the members are welded at the joints. If, on the other hand, the loads are applied between the joints or the members are stocky, then significant bending is induced in the members. In the latter case the analysis and design procedures must include the bending, that is, the truss must be analyzed as a frame, so that the actual behavior of the structure is simulated.

Similarly, in a planar beam the applied loads are carried by shear (forces normal to the axis of the beam) and bending moments developed in the beam. Arches have curved configurations and carry their applied loads by developing shear and moment as well as axial force. Cables carry their loads by developing axial tension only. They do not provide any resistance against axial compression and bending. Furthermore, cables change their configuration significantly in order to carry the applied loads.

Frames are composed of beam elements. But, unlike continuous beams, their members can be oriented in any direction. However, like beams, their load-carrying capacity is due to shear forces and moments developed in the members.

1.6 MATERIALS

Many different types of materials are used in structures. The commonest ones used for load-carrying members are steel, concrete (reinforced or prestressed), masonry, and timber. Aluminum and plastics are also sometimes used. Typical stress–strain curves for steel, concrete, and wood are shown in Fig. 1.5. For stresses that are not very high the material remains elastic (the members resume their initial shape upon unloading) and the stress–strain curves are linear. In this book linear-elastic materials are assumed unless otherwise stated.

Stress–Strain Curves

Figure 1.5 shows the typical uniaxial test specimens for steel, concrete, and timber. For steel the basic test is a tension test, while for concrete and wood compression is the most common test.

In a testing machine, incremental force is applied to the specimen. The load and change of length of the specimen are recorded at the different stages of loading. Stress, σ, and strain, ε, are then calculated from

$$\sigma = \frac{P}{A} \tag{1.1}$$

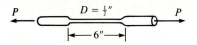

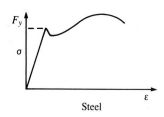

Steel

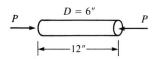

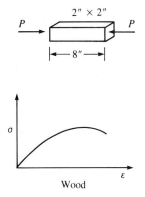

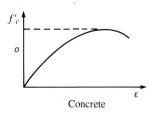

Concrete

Wood

FIGURE 1.5
Uniaxial tests and specimens.

and

$$\varepsilon = \frac{\Delta l}{l} \tag{1.2}$$

where P is the applied load and A is the area of the cross-section of the specimen. The change of the area of the cross-section is usually ignored and the original area is used. Δl and l are the change of length and the original length of the specimen. The stress–strain curves for different materials are shown in Fig. 1.5.

For an elastic analysis only the linear portion of the stress–strain curve is considered, or it is assumed that the curve is linear in the range of loading. The slope E of the stress–strain curve is called *Young's modulus* or the modulus of

TABLE 1.1
Examples of material strength

Material	Property	USCS units	SI units
Steel, A36	F_y	36 000 psi	250 MPa
	F_u*	58 000 psi	400 MPa
	E	29×10^6 psi	200 000 MPa
Concrete	f'_c	4 000 psi	28 MPa
$E = 33(w^{1.5})\sqrt{f'_c}$,			
for $f'_c = 4\,000$ psi,			
and $w = 145$ lb/ft^3**	E	$3\cdot644 \times 10^6$ psi	26×10^3 MPa
Timber	$f_t \parallel$ to grain	1 000–2 000 psi	7–14 MPa
(allowable	bending and		
stresses)	tension		
	$f_c \parallel$ to grain	600–1 800 psi	4–13 MPa
	$f_c \perp$ to grain	300–800 psi	2–6 MPa
	f_v	50–150 psi	0.3–1 MPa
	horizontal shear		
	E	$0.6–1.7 \times 10^6$ psi	$4–12 \times 10^3$ MPa

* F_u is the ultimate strength of steel.
** w is the weight density of concrete in lb/ft^3 or kg/m^3.

elasticity. Since

$$E = \frac{\sigma}{\varepsilon}$$

then

$$\sigma = E\varepsilon \qquad (1.3)$$

Equation (1.3) is Hooke's law for a uniaxial member. Stress is measured in pounds per square inch (psi) or Pascals ($1\,\text{Pa} = 1\,\text{newton/m}^2$) and strain is length/length and is therefore dimensionless.

Yield stresses F_y for steel of 36 ksi (248 MPa)[1] and 50 ksi (345 MPa) were common in the past. However, the trend is to use steels with higher strengths. E is the same for all grades of steel. It is equal 29 000 ksi (200 000 MPa) (see Table 1.1). Furthermore, for steel the compressive and tensile yield stresses are practically the same. However, long slender compressive members may buckle before yielding.

For concrete the reference stress is f'_c, the ultimate compressive stress at 28 days of age, when it has gained most of its strength. In the past, values of f'_c equal 4 to 6 ksi (27 to 41 MPa) were common for reinforced and prestressed concrete. Concretes with much higher strengths are finding general use. The strength of concrete in tension, f_t, is about 10% of its compressive strength. In reinforced concrete and masonry the steel reinforcement provided in the

---

[1] The systems of units are discussed at the end of this chapter in Sec. 1.9.

tension zone removes this deficiency. In prestressed concrete the tension region is compressed, usually by pretensioned cables buried inside the concrete.

The strength of a timber member not only depend on the species and grade of wood and the mode of loading (axial, bending, etc.), but also on whether the load is parallel or perpendicular to the grain of wood. The strength parallel to the grain or fibres of the wood is higher.

The commonest shape of cross-section for steel members is W (wide flange I shaped) section. Other shapes are channel [[) angle (L) and T-sections. For concrete and timber members the cross-sections are usually rectangular.

1.7 ANALYSIS AND DESIGN

Structural analysis consists of determining the stresses and deflections of a structural configuration with given dimensions and specified loads.

Design, on the other hand, consists of the complete process from conceptualization to sizing of the members, such that the safety, stability, and functionality of the structure is achieved without excessive cost. For simple structures, the sizing of the members may be done directly. For more complex structures, determination of member sizes may involve repeated analyses.

The allowable stress method was used for a long time. In this method the members are designed by using the appropriate allowable stresses to determine sections capable of carrying the applied loads. The applied loads were either calculated or estimated as discussed below. The allowable stresses were found by reducing the yield strength of steel or similar values for other materials by a reduction or safety factor. The problem with this method is that the yielding of the materials at some points of the streucture is not synonymous with failure or collapse of the structure.

The load and resistance factor design (LRFD) method, on the other hand, uses safety factors in relation to the ultimate strengths of the material. That is, the strengths are multiplied by factors less than 1, while the loads are multiplied by factors that are usually larger than 1. It further utilizes larger strength reduction factors for modes of failure that are harder to predict and larger load factors for harder to predict loads. this method has been in use for concrete structures for some time and is finding general application for other materials. For example, the bending strength of concrete is reduced by a factor 0.9, while a reduction factor of 0.85 is used for shear, which is more complex to predict accurately. Similar factors for other materials are given in the pertinent codes. The loads and their factors are discussed later in this chapter.

Codes and Specifications

The earliest building code document is that of Hammurabi, some 4 000 years ago. It specifies that, "if a builder constructs a building and it collapses and kills the owner, the builder shall be put to death. If it kills the son of the owner, the son of the builder shall be put to death. . ."

Current design codes do not have mitigating power, although they may be used in court to argue for or against a case. Design of most structures is done according to the pertinent codes. Current codes provide much useful information and procedures for design and construction. Some codes apply to a broad range of structures. The Uniform Building Code (UBC)[1], for example, specifies the levels of loading and allowable stresses for buildings made of a variety of materials. Similarly, American National Standards Institute (ANSI)[2] develops specifications for various loads applicable to many structures. American Association of Highway and Transportation Officials (AASHTO)[3] provides specifications for highway bridges of all types and materials.

More specific codes may pertain to a given material. Thus, the Specifications of the American Institute of Steel Construction (AISC)[4] applies to steel buildings, while that of the American Concrete Institute (ACI)[5] pertains to concrete buildings.

Acceptable levels of stress and deformation providing appropriate safety factors are specified by these codes. In addition, the specifications may include common design procedures.

Free Body Diagrams

In the analysis and design process it is often necessary to separate a portion of the structure from the rest or from the foundation. This is called a free body diagram and is obtained by applying the internal forces that the remainder of the structure effects on the portion under consideration. Furthermore, hypothetical sections can be taken through the structure in order to evaluate the internal forces needed for design.

One useful property of free bodies for analysis and design is that for a structure in equilibrium any free body thereof must also be in equilibrium.

1.8 LOADS

The loads applied to structures are the main causes of stress and deformation. Other causes may be support settlement and temperature change. Loads may be subdivided into dead loads and live loads.

---

[1] Uniform Building Code, 1988.

[2] American National Standards Institute, Specifications 1982.

[3] American Association of State Highway and Transportation Officials, Specifications.

[4] American Institute of Steel Structures, Manual of Steel construction, 9th ed., 1989.

[5] American Concrete Institute, Building Code Requirements for Reinforced Concrete, ACI 318–89.

Dead Loads

Dead loads consist of the weight of the basic frame, walls, and other structural components and the machinery that is permanently installed. Even though calculation of the dead load of a member with known dimensions is a routine task, computation of dead load may not be as easy during design. This is because at the design stage the final dimensions of the elements may not be known. In practice the designer starts out with a set of blueprints with tentative dimensions established on the basis of similar projects or estimated by an experienced designer. These dimensions may be used for dead load calculations. If in the course of design some of the dimensions are modified, then adjustments must be made to reflect these changes, unless the changes are insignificant.

In some cases member sizes may not be known at the design stage. If the weight of the member is not negligible then the designer must come up with an estimate. For example, the weight of reinforced concrete beams under a typical uniform load may be estimated at 30 to 40% of the load it is to carry. For steel members the weight is much smaller, about 5% or less of the applied load. In any event, the estimated weights of the members must be checked after the sizes of the members are determined. It should be noted that in calculating the total weight of sloped roofs the dimension along the slope must be used.

On the basis of the initial dimensions, the designer can calculate the dead loads using the unit weights of the materials. For steel sections the weight per unit length is given in the AISC Manual of Steel Construction. For some common structural materials approximate unit weights (densities) are given in Table 1.2. The unit weight of a reinforced concrete member depends on the amount of reinforcement in the section. This may vary from $145 \, \mathrm{lb/ft^3}$ ($22\,777 \, \mathrm{N/m^3}$) for moderately reinforced concrete to $150 \, \mathrm{lb/ft^3}$ ($23\,562 \, \mathrm{N/m^3}$) for concrete with a significant amount of reinforcement. The density of wood depends on its type and moisture content.

As mentioned before, if the dimensions of the member under consideration are not available, an estimate of the dead weight of the member can be used.

TABLE 1.2
Unit weight of materials

Material	Unit weight	
	lb/ft^3	N/m^3
Steel	490	76 969
Concrete	145	22 776
Wood (pine)	40	6 283

Live Loads

Unlike dead loads, live loads are movable or moving. Therefore, their magnitude and/or location may be varible. Live loads may be vertical, such as occupancy and snow loads, or lateral, such as earthquake, wind, soil, and water pressure. Wind loads and soil and water pressures usually act perpendicularly to the surface of the structure.

Snow loads are usually specified per unit horizontal projection of the area of the roof. Thus, the total snow load is obtained by multiplying the magnitude of the load by the horizontal projection of the loaded area. This is in contrast to the weight of pitched roofs, which is found by multiplying the length by the distance along the slope and the unit weight of the roof.

Occupancy loads. Occupancy loads may be due to people, furniture, etc. For design purposes their magnitudes are specified by the Uniform Building Code. some typical values are reproduced in Table 1.3.

For buildings, other than assembly halls, with a live load less than 100 psf, the occupancy load for members supporting an area A larger than $150\,\text{ft}^2$ may be reduced by a percentage $R = r(A - 150) \leq 23.1(1 + D/L)$. For floors, $r = 0.08$ (for roofs refer to the UBC). D and L are the dead and live loads per unit area supported by the member. R cannot exceed 40% for members carrying loads from one floor, and 60% for other members.

Snow loads. Roof snow loads depend on the geographic location and conditions of the building. The magnitudes of ground snow load, q, based on statistical values with a 50-year frequency are given in the snow map of the United States (Fig. 1.6) for different locations. Common values of ground snow load in the northern United States are given in lb/ft^2 (psf). They range from 20 psf (958 Pa) to 45 psf (2 155 Pa). The roof snow loads as specified by ANSI standard A58.1 (1982) are as follows

$$p = 0.7C_e C_t C_s Iq \tag{1.4}$$

C_e is the exposure factor, with values from 0.8 for very windy and exposed sites to 1.2 for highly sheltered locations.

C_t is the thermal factor. It is equal to 1 for heated structures, 1.1 for structures heated just above the freezing level, and 1.2 for unheated structures.

C_s is the sloped roof factor. It depends on whether the building is heated or not. For an unheated building a conservative value of 1 can be assumed. For other conditions the ANSI Standards can be consulted. For heated buildings, when $C_t = 1$, the following values are used. $C_s = 1$ for slippery surfaces with a slope $\alpha \geq 15°$ or other surfaces with $\alpha \leq 30°$. $C_s = 0$ for $\alpha \geq 70°$. For α between the above values, C_s is found by interpolation.

I is the importance factor. Its value is 1.2 for essential buildings such as hospitals, 1.1 for assembly halls, and 1 for other buildings, except for agricultural buildings, which is 0.8.

TABLE 1.3
Occupancy loads

Category	Magnitude	
	lb/ft²	Pa (N/m²)
Assembly areas, auditoriums and their balconies—fixed seating areas. Movable seating and other areas. Stage and en-closed platforms	50	2 394
Cornices, marquees and residential balconies	100	4 790
Garages		
General storage and repair	100	4 790
Private car storage	50	2 390
Hospital wards and rooms	40	1 915
Libraries		
Reading rooms	60	2 870
Stack rooms	125	5 985
Manufacturing		
Light	75	3 590
Heavy	125	5 985
Offices	50	2 390
Residential	40	1 915
School classrooms	40	1 915
Sidewalks and driveways, public access	250	11 970
Storage		
Light	125	5 985
Heavy	250	11 970
Stores		
Retail	75	3 590
Wholesale	100	4 790
Theaters		
Orchestra floor and Balcony	60	2 870
Stage	150	7 180

These values do not reflect the possibility of the snow drift onto portions of the roof. In many areas local codes specify design snow loads. Note that on sloped roofs snow loads apply to the horizontal projection of the roof. For snow loads larger than 20 psf and roof pitches more than 20° the snow load p may be reduced by an amount,

$$(\alpha - 20)\left(\frac{p}{40} - 0.5\right) \quad \text{psf} \tag{1.5}$$

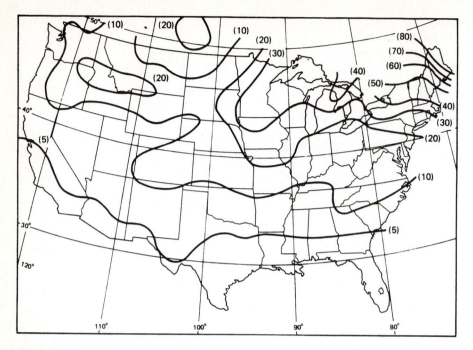

FIGURE 1.6
Minimum snow loads (in psf) in different areas of the United States. (Reproduced from ANSI Standards A58.1, with permission)

Wind loads. Wind loads are particularly significant on tall buildings. On small buildings the uplift caused by the wind may be severe enough to cause damage.

The basic wind pressure at a standard height of 30 ft is given by

$$q_s = 0.00256V^2 \tag{1.6}$$

where V is the maximum wind velocity (in miles per hour) and q_s is in psf. In the United States, V can be obtained from the basic wind speed map (Fig. 1.7). The wind speed ranges from 70 miles per hour (mph) (112 km/h) to 110 mph (177 km/h).

This magnitude must be modified on the basis of the type and condition of the building. The design pressure p (in psf) is given by the following equation

$$p = C_e C_q I q_s \tag{1.7}$$

where C_e is a factor for height, exposure and gust. It varies from 0.7 to 2.2 (the latter value for a height of 400 ft (122 m), or greater, and severe exposure), with an average value of 1.5. C_q is the pressure coefficient.

For gabled frame buildings the pressure is assumed normal to all the walls and roofs. The pressure coefficient C_q for this case is given in Table 1.4, where a negative value indicates an uplift or an outward pressure.

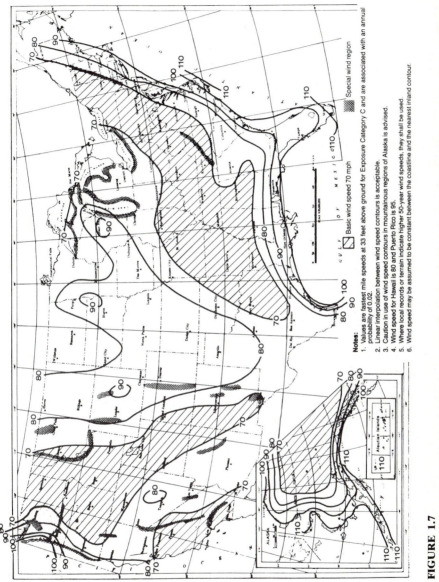

FIGURE 1.7

Wind map of the United States. (Reproduced from the 1988 edition of the Uniform Building Code, copyright © 1988, with the permission of the publishers, the International Conference of Building Officials.)

TABLE 1.4
Wind pressure coefficient C_q for gabled frames

Roof slope	Windward side	Leeward side
< 9:12	−0.7	−0.7
9:12 to 12:12	0.4	−0.7
> 12:12	0.7	−0.7

The value of C_q is 0.8 for windward walls and −0.5 for leeward walls. For buildings other than gabled frames with a height less than 200 ft (61 m) the values of C_q are as follows.

	C_q
For vertical projections:	
with height < 40 ft	±1.3 in any horizontal direction
and height > 40 ft	±1.4 in any horizontal direction
For horizontal projections	
and enclosed structures	−0.7
with more than 30% on any side open	1.2

I is the structure importance factor and is slightly different from the corresponding values for snow loads. It is equal to 1.15 for hospitals and other essential buildings and 1 for other buildings.

Example 1.3. Find the wind forces on the building of Fig. 1.8, which is constructed in an area with a maximum wind velocity 70 mph. The building is surrounded by trees.

Solution. Since the building is protected from wind we take $C_e = 0.7$, $I = 1$. The basic wind pressure is $q_s = 0.002\,56 \times (70)^2 = 12.54$ psf.

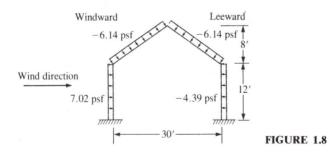

FIGURE 1.8

The slope of the roof is $8:15 = 6.4:12$, which gives a pressure coefficient C_q for both the windward and leeward sides of the sloped roof (Table 1.4) of $C_q = -0.7$. For the windward wall, $C_q = 0.8$ and for the leeward wall $C_q = -0.5$. Thus, the applied pressure on both the windward and leeward sides of the sloped roof is

$$p = 0.7 \times (-0.7) \times 12.54 = -6.14 \text{ psf}$$

That is, the roof is subjected to an uplift. For the windward wall the pressure is $0.7 \times 0.8 \times 12.54 = 7.02$ psf and for the leeward wall $0.7 \times (-0.5) \times 12.54 = -4.39$ psf. Note that the leeward wall is subjected to a suction. Figure 1.8 shows the wind pressure on the building components. A negative pressure indicates that the pressure is acting outward, that is, a suction. Note also that if the direction of the wind can change, then the maximum pressure for each member must be calculated considering the different wind directions.

Earthquake forces. Earthquake or seismic forces depend on the seismicity of the site and properties of the structure. The main forces are generally in the horizontal direction. In the past, a horizontal force equal to 10% of the weight was applied at the center of gravity of the system. This may be a reasonable approximation for massive elements such as retaining walls and small dams. For buildings, more detailed procedures are available for determining more accurate earthquake forces. The force at each level is calculated as a portion of the base shear, V, where

$$V = ZIKCSW \tag{1.8}$$

In Eq. (1.8), Z is the zone factor. Its values are, 3/16, 3/8, 3/4, and 1 for seismic zones 1, 2, 3 and 4, respectively. The zones can be found on the seismic map (Fig. 1.9). I is the importance factor, reflecting the hazards to human life that the building failure may cause. Its values are given in Table 1.5.

K is the structural system coefficient, with values for the different structural systems given in Table 1.6.

The Uniform Building Code imposes an additional requirement related to ductility that is not reflected in K. For all buildings in zones 3 and 4 and those in zone 2 having an importance factor greater than 1, all members in braced frames must be designed for lateral forces 25% greater than those found in the code provisions. Furthermore, connections must be designed either to develop the full capacity of the members or for the code forces without one-third increases in the allowable stresses, that are permitted when earthquake forces are included in the design.

C is the seismic coefficient,

$$C = \frac{1}{15\sqrt{T}} \le 0.12 \tag{1.9}$$

Thus, if C is found to be less than 0.12 it is taken as 0.12.

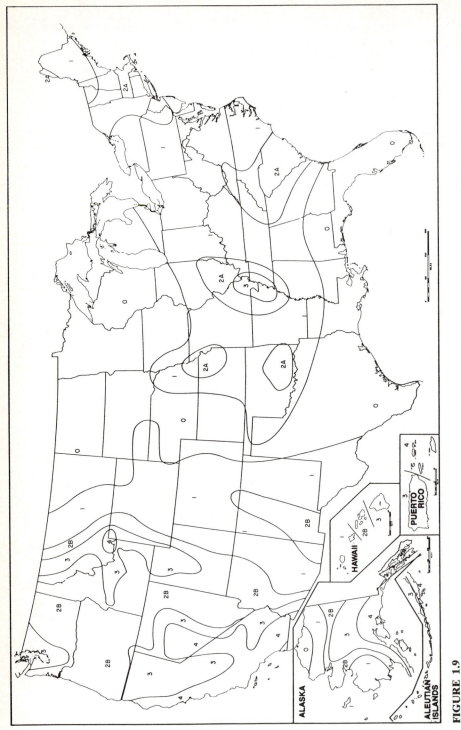

FIGURE 1.9
Seismic zones in the United States. (Reproduced from the 1988 edition of the Uniform Building Code, copyright © 1988, with the permission of the publishers, the International Conference of Building Officials.)

TABLE 1.5
Importance factor I

Type of occupancy	I
Essential facilities, including hospitals and other medical facilities having surgery or emergency treatment areas, fire and police stations, and municipal government disaster operation and communication centers deemed to be vital in emergencies	1.5
Any building where the primary occupancy is for assembly use for more than 300 persons in one room	1.25
All others	1

TABLE 1.6
Structural system coefficient K

Structural system	K
Buildings with box system: no complete vertical load-carrying space frame; lateral forces resisted by shear walls	1.33
Buildings with dual bracing system consisting of ductile moment-resisting space frame and shear walls, designed so that: (a) Frames and shear walls resist total lateral forces in accordance with their relative rigidities, considering the interaction of shear walls and frames, (b) Shear walls acting independently of space frame resist total required lateral force. (c) Ductile moment-resisting space frame has capacity to resist at least 25% of required lateral forces.	0.8
Buildings with ductile moment-resisting space frames designed to resist total required lateral forces	0.67
Other building framing systems	1.00
Elevated tanks, plus full contents, on four or more cross-braced legs and not supported by a building	2.5
Other structures	2.00

T is the fundamental period of vibration of the building in seconds. It can be found from the free vibration analysis of the building, or calculated from the following empirical formulas:

$$T = \begin{cases} 0.1\,N & \text{for ductile frames} \\ 0.05\,h_N/\sqrt{D} & \text{for other buildings (e.g., shear walls)} \end{cases} \qquad (1.10)$$

Where N is the number of stories, h_N is the maximum height of the building and D its plane dimension in the direction of the earthquake force.

If the loads have a distribution approximately equal to those discussed below, then a closer value of T can be calculated by

$$T = 2\pi \sqrt{\sum_{i=1}^{N} W_i \delta_i^2 \div \left(g \sum_{i=1}^{N} F_i \delta_i \right)} \qquad (1.11)$$

In the latter equation, W_i refers to the weight of different floors. δ_i are the floor deflections due to the applied static lateral loads F_i, and g is the gravitational acceleration.

S is the site-structure resonance factor, and can be computed from the ratio of the fundamental building period T and the characteristic site period T_s:

$$S = 1 + \frac{T}{T_s} - 0.5\left(\frac{T}{T_s}\right)^2 \qquad \text{for} \qquad \frac{T}{T_s} \le 1 \qquad (1.12a)$$

$$S = 1 + 0.6\frac{T}{T_s} - 0.3\left(\frac{T}{T_s}\right)^2 \qquad \text{for} \qquad \frac{T}{T_s} > 1 \qquad (1.12b)$$

But $S \ge 1$ in any case.

The UBC standards present procedures for determining the characteristic site period. In the absence of a geotechnical investigation for finding T_s, or when it has been shown by analysis that the building period T is larger than 2.5 s, S may be taken 1.5. If information is available about the soil profile then S may be estimated from Table 1.7.

In any case the product CS in Eq. (1.8) need not be taken larger than 0.14.

W is the total structure dead load plus a portion of the live load depending on duration of the live load. For example, for snow load, values as small as 25% of the snow load may be used.

The forces at each floor level are calculated by

$$F_x = \frac{W_x h_x}{\sum\limits_{i=1}^{N} W_i h_i} (V - F_t) \qquad (1.13)$$

where W_x and h_x are the weight and height at level x, for which the lateral load F_x is being calculated.

TABLE 1.7
Site-structure resonance coefficient S found from soil profile

Soil profile type		S
S_1	Rock of any characteristic, either shale-like or crystalline in nature (soil material may be characterized by a shear wave velocity greater than 2 500 ft/sec (762 m/sec); or stiff soil conditions, where the soil depth is less than 200 ft (61 m) and the soil types overlying rock are stable deposits of sands, gravels, or stiff clays	1.0
S_2	Deep cohesionless or stiff-clay soil conditions, including sites where the soil depth exceeds 200 ft (61 m) and the soil types overlying rock are stable deposits of sands, gravels, or stiff clays	1.2
S_3	Soft to medium-stiff clays and sands, characterized by 30 ft (9 m) or more of soft to medium-stiff clay with or without intervening layers of sand or other cohesionless soils.	1.5

In locations where the soil properties are not known in sufficient detail to determine the soil profile type or where the profile does not fit any of the three types, soil profile S_3 shall be used.

$F_t = 0.07TV$, not to exceed $0.25V$, is an additional force applied at the top level owing to whiplash if $T \ge 0.7$ sec. For $T < 0.7$, $F_t = 0$.

The drift at any level found by using the above lateral forces, unless it can be shown that larger drifts can be tolerated, is limited to

$$\Delta \le 0.005H \qquad K \ge 1 \qquad\qquad (1.14a)$$

$$\Delta \le 0.005KH \qquad K < 1 \qquad\qquad (1.14b)$$

where Δ is the lateral displacement within the story, and H is the height of the story.

Example 1.4. Find the earthquake forces for the frame structure of Fig. 1.10. The dead load for each floor is 200 lb/ft and the live load 400 lb/ft. Use the dead load plus 50% of the live load as the weight of each floor. The building is in zone 3.

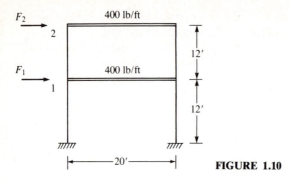

FIGURE 1.10

Solution. For zone 3, $Z = 3/4$ and for an ordinary building $I = 1$. Furthermore since the lateral loads are resisted by the frames only, $K = 1$.

$$T = 0.1 \times 2 = 0.2 \sec \qquad S = 1.5$$

$$C = \frac{1}{15\sqrt{0.2}} = 0.149 > 0.12, \qquad \text{use} \qquad C = 0.12$$

$$CS = 0.12 \times 1.5 = 0.20 > 0.14 \qquad \text{use} \qquad CS = 0.14$$

$$V = \tfrac{3}{4} \times 1 \times 1 \times 0.14 \times (2 \times 400 \times 20) = 1\,680 \text{ lb}$$

The horizontal forces acting at levels 1 and 2 are

$$F_1 = \frac{400 \times 20 \times 12}{400 \times 20 \times 12 + 400 \times 20 \times 24} \times 1\,680 = 560 \text{ lb}$$

$$F_2 = \frac{400 \times 20 \times 24}{400 \times 20 \times 12 + 400 \times 20 \times 24} \times 1\,680 = 1\,120 \text{ lb}$$

Water and soil pressures. Water and soil pressures are calculated with a linear distribution. The pressure, q, at depth h is

$$q = K\gamma h \qquad (1.15)$$

where γ is the weight density of water or soil. K is the pressure coefficient. It is equal to 1 for water, and for soils it can be calculated by methods of soil mechanics. For sand and gravel, γ is approximately 120 lb/ft^3 ($18\,850 \text{ N/m}^3$) and

$$K = \frac{1 - \sin \phi}{1 + \sin \phi} \qquad (1.16)$$

where ϕ is the angle of internal friction of the soil, obtained by shear tests in a laboratory. It is approximately equal to the natural slope of a heap of the soil on a horizontal plane. That is the angle that the side of the soil profile makes with horizontal. For sand and gravel, ϕ is approximately 30°. Any additional surcharge or weight of soil above the top of the retaining structure, w per unit area, produces an additional pressure $q = Kw$. When water can collect behind a retaining structure without being drained, then the water pressure must be added to soil pressure. However, the unit weight of soil can be reduced because it is assumed to be submerged. The weight density of water is $62.4 \, lb/ft^3$ ($9\,800 \, N/m^3$).

Thermal Loads

A member heated uniformly through its cross-section and without any restraints against expansion will undergo an increment in its length without inducing any stresses. This expansion is

$$\Delta l = \alpha l \, \Delta T \tag{1.17}$$

where α is the coefficient of thermal expansion, l is the member length, and ΔT is the change in temperature. Table 1.8 gives the coefficients of thermal expansion for some structural materials.

If the member is restrained against expansion, then a compressive stress

$$\sigma = E\alpha \, \Delta T \tag{1.18}$$

is developed. A reduction in temperature will result in a tensile stress or a shortening of the member depending on whether the member is restrained or not. A nonuniform temperature change will result in bending of the member.

Other loads include blast loads and those due to forced fitting of the members.

Load Combinations

The live loads specified by the code represent the maximum possible loads. The likelihood of these maximum loads occurring simultaneously is remote. Therefore, the following combinations are given by the Uniform Building Code for buildings. Let

D = dead load L = floor live load Lr = roof live load

W = wind load E = earthquake load S = snow load

H = load due to soil weight or pressure

T = temperature, creep and shrinkage effects

TABLE 1.8
Coefficient of thermal expansion, α

Material	in/in/°F	m/m/°C
Steel	6.5×10^{-6}	11.7×10^{-6}
Concrete	5.5×10^{-6}	9.9×10^{-6}

For the allowable stress design (ASD) the following combinations must be considered.

1. $D + L + [Lr$ or $S]$
2. $D + L + [W$ or $E]$
3. $D + L + W + S/2$
4. $D + L + S + W/2$
5. $D + L + E + [S$ if $S > 30$ psf]

Snow loads larger than 30 psf may be reduced by up to 75% with the approval of the building official.

For the load and resistance factor design (LRFD), the loads are multiplied by load factors and then combined. Load factors vary for the different load combintions and may be slightly different for concrete and steel structures. For example,

For concrete buildings

1. $1.4D + 1.7L$
2. $0.75(1.4D + 1.7L + 1.7W)$
3. $0.9D + 1.3W$
4. $1.4D + 1.7L + 1.7H$
5. $0.75(1.4D + 1.4T + 1.7L)$
6. $1.4(D + T)$

For steel buildings

1. $1.4D$
2. $1.2D + 1.6L + 0.5(Lr$ or $S)$
3. $1.2D + 0.5L$ (or $0.8W) + 1.6(Lr$ or $S)$
4. $1.2D + 0.5L + 0.5(Lr$ or $S) + 1.3W$
5. $1.2D + 0.5L$ (or $0.2S) + 1.5E$
6. $0.9D - 1.3W$ (or $1.5E$)

Design codes and texts give the load factors and resistance reduction factors for the pertinent materials. They also specify impact factors for members supporting moving loads, e.g. cranes and elevators.

Bridge Loads

For highway and railroad bridges design loads are given by the AASHTO and AREA specifications, respectively. Highway bridge loads must be chosen as the maximum of the alternate loading conditions—a design truck with specified

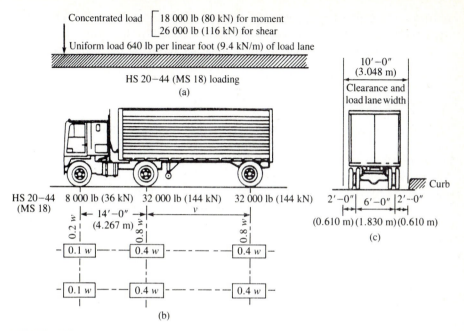

Concentrated load $\begin{bmatrix} 18\ 000 \text{ lb (80 kN) for moment} \\ 26\ 000 \text{ lb (116 kN) for shear} \end{bmatrix}$

Uniform load 640 lb per linear foot (9.4 kN/m) of load lane

HS 20−44 (MS 18) loading

(a)

10′−0″
(3.048 m)

Clearance and load lane width

HS 20−44 (MS 18)   8 000 lb (36 kN)   32 000 lb (144 kN)   32 000 lb (144 kN)

$\leftarrow$ 14′−0″ $\rightarrow$
(4.267 m)

v

0.2 w   0.8 w   0.8 w

Curb

2′−0″   6′−0″   2′--0″
(0.610 m) (1.830 m) (0.610 m)

(c)

0.1 w   0.4 w   0.4 w

0.1 w   0.4 w   0.4 w

(b)

FIGURE 1.11
HS 20 load.

axle loads and spacings (Fig. 1.11b, c), or a combination of uniform and concentrated loads (Fig. 1.11a). The latter is usually critical for long-span bridges.

For railroad bridges, locomotive axle loads are specified. Bridge codes also specify an impact factor that accounts for the dynamic action of the moving vehicles.

The HS-20 truck of Fig. 1.11b, c is used for design of bridges on main highways. For secondary roads, AASHTO specifies less severe loadings. It should be noted that this is not an actual truck but one that is assumed to represent the critical characteristics of common trucks. For any member under consideration, the loading must be so placed as to produce the largest stresses.

The impact factor includes the dynamic effect of the moving loads. The static loads are multiplied by the factor $(1 + I)$. For highway bridges, I is given in terms of the span length L, in feet, as

$$I = \frac{50}{L + 125} \leq 0.3 \tag{1.19}$$

Principle of Superposition

If a structure with linear elastic materials is subjected to more than one load or loading conditions, then the moment, shear, and deformation at any section can be found as the sum of the results for the different loading conditions. This law, known as the *principle of superposition,* makes it easier to analyze structures under complex loading conditions, by reducing the problem to one with a number of simpler load cases.

1.9 SYSTEMS OF UNITS

The United States customary system (USCS) is the legal standard of weights and measure in the United States. In this system Foot, Pound and Second (FPS) are the basic units. This system was originated and formerly used in England and was thus known as the English System. However, in most countries the System International (SI) is used.

The five basic units used in structural engineering are force, length, time, mass, and temperature. These units and their conversion factors from one system to another are given in Table 1.9. The factors for the units are given in Table 1.10 and Table 1.11 presents other units derived from the basic ones. Since the unit of time is second in both systems, it is eliminated from the table.

TABLE 1.9
Base weight and measure units

	USCS			SI		
Quantity	Unit	Abbreviation	Conversion to SI	Unit	Abbreviation	Conversion to USCS
Force	pound	lb	4.448 N	newton	N	0.2248 lb
Length	foot	ft	0.3048 m	meter	m	3.2808 ft
Mass*	slug	slug	0.4536 kg	kilogram	kg	2.2046 slug
Temperature	degree Fahrenheit	°F	(°F − 32)/1.8	degree Celsius**	°C	1.8 × °C + 32

* It should be noted that in addition to slug, pound mass (lbm) has also been used as a unit. 1 slug is the amount of mass that accelerates at the rate of 1 ft/s^2 under the action of 1 lbf. 1 lbm is the amount of matter which at earth's surface is drawn by gravity toward the earth by 1 lbf. Thus, 1 slug = 32.2 lbm.
** The SI unit for thermodynamic temperature is the kelvin (K): $T(K) = T(°C) + 273$.

TABLE 1.10
Factors of the USCS and SI units

USCS units		SI units		
Force	Length	Prefix	Symbol	Factor
kip = 1 000 lb	inch (in) = 1/12 ft	micro	μ	10^{-6}
ton = 2 000 lb	yard (yd) = 3 ft	milli	m	10^{-3}
		centi	c	10^{-2}
		kilo	k	10^{3}
		mega	M	10^{6}

TABLE 1.11
Derived units

Unit	USCS Abbreviation	USCS Conversion to SI	SI Abbreviation	SI Conversion to USCS
Area	in^2 ft^2	$6.452 \, cm^2$ $9.290 \times 10^{-2} \, m^2$	cm^2 m^2	$0.155 \, in^2$ $10.764 \, ft^2$
Volume	in^3 ft^3	$16.39 \, cm^3$ $2.8317 \times 10^{-2} \, m^3$	cm^3 m^3	$6.1 \times 10^{-2} \, in^3$ $35.31 \, ft^3$
Force/ unit length	lb/in lb/ft	$175.08 \, N/m$ $14.59 \, N/m$	N/m	$5.7 \times 10^{-3} \, lb/in$ $6.852 \times 10^{-2} \, lb/ft$
Bending moment	$lb \cdot in$ $lb \cdot ft$	$0.1130 \, N \cdot m$ $1.3558 \, N \cdot m$	$N \cdot m$	$8.851 \, lb \cdot in$ $0.7376 \, lb \cdot ft$
Stress	lb/in^2 (psi) lb/ft^2 (psf) kip/in^2 (ksi)	$6.895 \times 10^3 \, Pa$ $47.88 \, Pa$ $6.895 \, MPa$	Pa* (N/m^2) MPa	$0.1450 \times 10^{-3} \, lb/in^2$ $2.089 \times 10^{-2} \, lb/ft^2$ $0.1450 \, kip/in^2$
Weight density	lb/ft^3	$157.0796 \, N/m^3$	N/m^3	$6.366 \times 10^{-3} \, lb/ft^3$
Mass density	$slug/ft^3$	$16.0187 \, kg/m^3$	kg/m^3	$6.2427 \times 10^{-2} \, slug$

* Since the pascal is a very small unit, in structural engineering usually the megapascal (MPa) is used.

The metric system, which is sometimes used in structural engineering, is the precursor to SI. In that system the unit of weight is kilogram force, kgf, 1 kgf = 9.81 N, and 1 ton = 1 000 kgf.

For angles, units of radians, degrees, and grads are used in both systems. 1 rad = 57.295 78 degrees and 1 grad = 0.9 degree.

Another unit of length is the mile: 1 mile = 5 280 ft = 1.609 km. The unit for measuring land surfaces in the USCS is the acre and in SI the hectare. 1 acre = 43 560 ft^2 = 0.405 hectare; 1 hectare = 10 000 m^2. The measure of volume in the United States is gallon = 4 quarts (qt) = 4.545 liters, 1 qt = 32 ounces (oz). In SI, liter is used for measuring volume. 1 liter = 1 000 cm^3 = 0.880 qt.

The gravitational acceleration is 32.2 ft/sec^2 (386 in/sec^2) in the USCS and 9.81 m/sec^2 in SI units.

LABORATORY EXPERIMENTS

Uniaxial Test (Fig. L1.1)

This experiment is unnecessary if the students have already seen tension tests on steel specimens.

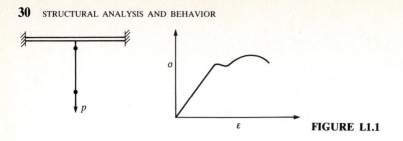

FIGURE L1.1

The purpose of the experiment is to obtain the stress–strain curve. A wire is fastened at its upper end and loaded by weight at its lower end. The deformation of a given length is measured at different load increments, from which the stresses and strains are obtained and plotted.

Use a 3–4 ft (0.8–1.1 m) hanger wire. Fasten directly or by an eye hook. A bucket with weights or sand may be used for loading. Place the bucket at about 8 in (20 cm) from the floor with a rag under to minimize impact in case of wire failure. Measure the wire diameter at several places with a caliper. Remove the kinks in the wire by preloading (~20 lb or 80 kN) and unloading. Increase the load gradually (15–20 lb, 50–80 kN increments) decrease load increments toward the end of the test. Record the loads (including the weight of bucket) and the elongation at the recorded loads. If tested to failure, examine the failure area (a magnifying glass may be used).

Wind Load (Fig. L1.2)

Wind loads cause uplift on the leeward side of gabled roofs and may cause pressure or uplift on the windward side depending on the roof slope.

A wooden or carboard box with a pitched roof, with light plastic roof covering, may be used. Leave the covering somewhat slack.

Place a fan at a short distance to apply wind load to one of the walls adjacent to the sloped roof. Observe roof pressure or suction by watching the plastic cover. Try this with roof slopes larger and smaller than 20°, as for small slopes both sides of the roof will have uplift.

Dead Loads and Snow Loads on Pitched Roofs

The total load due to the dead load of a roof depends on the weight of the roof (surface area) and not its slope. The snow load, on the other hand, depends on the slope, i.e. the projected surface of the roof (Fig. L1.3).

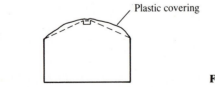

Plastic covering

Fan

FIGURE L1.2

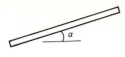

FIGURE L1.3

Use a solid board to represent the roofing material. Observe that the total weight of the roofing does not change by changing the slope. The total weight is the area times the weight per unit area.

On the other hand, the amount of snow collected on the roof depends on the horizontal projection of the roof. This may be shown by showering small particles (styrofoam) on the roof for different slopes. In the extreme cases the snow load is zero for a vertical side of a roof and maximum for a horizontal part.

Earthquake forces

Earthquake motions induce vibration of structures and dynamic loads. Dynamic loads are also produced by moving loads, such as trucks moving on bridges, moving cranes, as well as by wind. Dynamic loads are larger than the weight of the static trucks. Dynamic loads depend on a number of factors, including the dynamic characteristics of the structure and load.

Forces due to ground motion. To demonstrate how ground motions produce internal forces (stresses) in structures, use a one- or two-story one-bay frame made of some flexible material—spring steel, plastic, thin wood, or cardboard strips—with the base fixed to a board (Fig. L1.4). The length of the frame may be about 1 ft, the height 1 to 2 ft and the thickness of the strips 1/16 in or less.

Observe the deformation in the beams and columns (causing bending stresses) as the base is moved back and forth. By changing the frequency of the base motion observe that certain frequencies cause more motion. This also depends on the properties of the structure. One important parameter is the frequency of free vibration (number of oscillations per second that the structure makes when the structure, say the top of a building, is displaced some distance and then released); see the next experiment. In fact, when the frequency of the motion coincides with the frequency of free vibration of the structure (resonance), then the displacement keeps increasing.

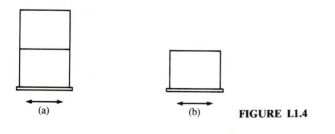

(a)   (b)   **FIGURE L1.4**

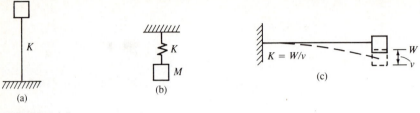

(a) (b) (c)

FIGURE L1.5

Free vibration. The frequency of free vibration depends on the mass and stiffness of a structure and is independent of the amplitude (the amount of initial displacement). To study these characteristics we can use the model of Fig. L1.5a, which may consist of an elastic strip (spring steel, plastic, wood strip, etc.) fixed at the base with a mass attached to the top. This model can represent a one-story frame structure carrying a slab roof. The slab causes the top of the frame to be highly rigid and carry most of the mass, while the columns provide most of the elasticity against the motion.

We can cause free vibration of the model in Fig. L1.5a by displacing the mass some distance and releasing it.

By displacing the mass different distances and observing the frequency (counting the number of oscillations for one minute and then calculating the number of oscillations per second) we see that the frequency of free vibration is independent of the amplitude (largest displacement) of vibration.

If we use a shorter strip for the model of Fig. L1.5a (higher stiffness) we observe that the frequency of free vibration increases. On the other hand, if we use a larger mass the frequency of free vibration decreases. The same observation can be made for the spring–mass model of Fig. L1.5b. In fact if the masses in Figs. L1.5a and L1.5b are the same and the stiffness of the spring is the same as that of the strip, then the frequencies will be the same.

The stiffness of a spring is the force required to produce a unit displacement. Similarly, the stiffness of the strip in Fig. L1.5a is the force required to produce a unit displacement at its top. If we know the modulus of elasticity of the strip, the stiffness can be found by calculating the lateral deflection of the strip v due to a lateral load P. Thus, $K = P/v = 3EI/l^3$ (see Chapter 5). If the modulus of elasticity of the strip is not known, we can find the stiffness by measuring the displacement v under a known weight W and calculating $K = W/v$.

For known stiffness and mass, the frequency of free vibration for a single degree of freedom system (one mass undergoing displacement) can be calculated by

$$ f = \frac{\omega}{2\pi} \quad \text{with} \quad \omega = \sqrt{\frac{K}{M}} $$

where f is called the *frequency* or natural frequency and is measured is cycles

FIGURE L1.6

per second (hertz); ω is called the *circular frequency* and is measured in radians per second.

Compare the calculated and measured frequencies for the model of Fig. L1.5a.

More elaborate experiments are possible (Fig. L1.6), by using actuators to induce specific base motions or simulate the motions of specific earthquakes.

PROBLEMS

1.1. (*a*) The stress–strain curve for a concrete specimen under compression is shown in Fig. P1.1a. The segment from zero stress to 5 000 psi is a straight line. Beyond that point the plot is a curve. Up to what total compressive stress in the section can the principle of superposition be applied?

(*b*) A timber member was tested as a beam by a mid-span loading. The plot of load versus deflection at the middle of the span is given in Fig. P1.1b. Up to a deflection of 0.2 in, the curve is a straight line. For what amount of total deflection can the principle of superposition be applied?

1.2. Find the loads described below.

(*a*) The weight per linear foot of a steel bar of 2 in diameter.

(*b*) The weight per unit length of a concrete beam with a 2 ft × 4 ft section.

(*c*) The live load on the balcony of a movie theater.

(*d*) Snow load for a residential building in Green Bay, Wisconsin, with a 30° pitched roof in a location surrounded by trees.

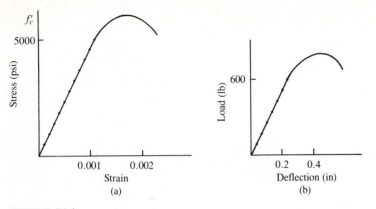

FIGURE P1.1

(e) Wind load per unit area of the roof in (d) if the low-rise building is located in Chicago, Illinois.

(f) Earthquake forces on a two-story frame building. The height of the floors is 12 ft. The width of the frame is 24 ft. The dead load plus 25% of the live load on each floor is to be used as weight, which equals 32 kips total on each floor. The frame alone is to resist the earthquake forces. The building is in zone 2 and no information is available about the soil.

1.3. A spherical storage tank of 10 ft radius is placed under water at a depth of 1 000 ft. What is the external pressure on the tank if the dimensions of the tank are assumed to be very small compared to the depth of the water.

1.4. A steel bridge with a main span of 3 000 ft has a temperature range of $-20°F$ to $+100°F$ during the year. What total change in length does this temperature difference produce? How is the length change accounted for in bridges? The coefficient of thermal expansion for steel is 0.000 0065/°F.

1.5. A 12-in concrete cylinder with diameter $= 6$ in fails in compression at a load $P = 100$ kips. What is the ultimate strength f'_c? Calculate the value of E for the concrete? Use $w = 145$ lb/ft^3 for unit weight of concrete.

1.6. Plain concrete and masonry are very weak in tension. Name some members made without the use of any reinforcement that these materials would be appropriate for.

1.7. Short columns with square sections, 1 in^2 area and 2 in length, made of steel, concrete and timber, are placed simultaneously between the parallel heads of a compression testing machine and loaded in compression. The location of the applied compression force is such that it does not produce any bending but only compression in the specimens. Each specimen undergoes the same strain, and all are kept within the elastic range of behavior. The moduli of elasticity of the three materials are as follows, $E_s = 29\ 000$ ksi, $E_c = 3\ 500$ ksi, and $E_t = 2\ 000$ ksi, with the subscripts referring to the materials. Calculate the proportions (P_s, P_c, P_t) of the applied load P carried by each specimen? What is the effectiveness of steel as compared with concrete. That is, how many square inches of concrete would be required to carry the load of 1 in^2 of steel? What is the effectiveness of concrete as compared with timber? *Hint:* $\varepsilon_s = \varepsilon_c = \varepsilon_t$, and $P = P_s + P_c + P_t$.

1.8. Convert the following USCS values to SI units, and the SI values to USCS units.

(*a*) 10 kip load; 10 ft height; 2 kip/ft load on a beam; 50 psf load on a slab; 29 000 ksi modulus of elasticity of steel; 4 000 ksi strength of concrete.

(*b*) 50 kN load; 3 m height; 6 kN/m load on a beam; 250 N/m^2 load on a slab; 20 000 MPa modulus of elasticity of concrete; 20 MPa strength of concrete.

CHAPTER
2

REACTIONS

The loads applied to a structure must be transferred to the ground by the supports, that is, the supports will generate reactions to counteract the action of the loads. The type and number of reactions depend on the support conditions. Analysis of structures usually requires the determination of the reactions. In this chapter we will discuss the reactions of two-dimensional or plane structures. Reactions of three-dimensional structures are discussed in Chapter 18.

2.1 SUPPORT CONDITIONS

A stable structure must be properly restrained in space. Restraints are provided by the supports, which may be rollers, hinges, or fixed ends.

Rollers

A roller (Fig. 2.1a) provides a restraint in one direction. For example, in Fig. 2.1d the right-hand support, 2, of the beam is a roller. Thus, support 2 can undergo horizontal movement and rotation without any reactions developed by the support against these movements. On the other hand, this end of the beam cannot move in the vertical direction. The roller support provides a reaction,

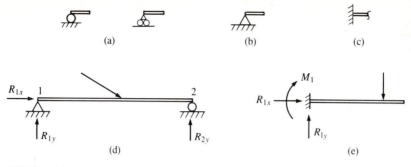

FIGURE 2.1
Support conditions.

R_{2y}, to prevent such a movement. Hence, a roller provides only one reaction. Note that a roller provides reaction against uplift as well.

Hinges

A hinge support (Fig. 2.1b) will allow only rotation but no displacements. The beam of Fig. 2.1d can freely rotate at support 1, but cannot move in either the vertical or horizontal direction. That is, the left-hand support develops reactions, R_{1x} and R_{1y} against such displacements. Thus, there may be two reactions at a hinged support.

Fixed Supports

A fixed support (Fig. 2c) allows neither rotation nor displacements. Consequently, there may be up to three reactions at the left-hand end of the beam in Fig. 2.1e—a moment M_1, a horizontal reaction R_{1x}, and a vertical reaction, R_{1y}.

A beam with a hinge and a roller support (as in Fig. 2.1d), is called a *simply supported beam*, whereas a beam with one end fixed and the other free is called a *cantilever beam*.

2.2   COMPUTATION OF REACTIONS

For most structures, deflections are small in comparison to the lengths of the members (less than 1%). In calculating the reactions of such structures the deflection can be neglected and the original configuration before loading can be used. Throughout this book (except for cables and beam columns) this assumption of small deflections is made.

The equations of statics of rigid bodies can thus be applied in computing the reactions of structures. For plane structures there are three equations of statics. These are

$$\sum F_x = 0 \tag{2.1a}$$

$$\sum F_y = 0 \tag{2.1b}$$

$$\sum M = 0 \tag{2.1c}$$

That is, the sum of all the forces in two perpendicular directions (usually horizontal and vertical), and the sum of moments with respect to an arbitrary point, must be zero. One or both of the force equations, (2.1a) and/or (2.1b), may be replaced by moment equations. Obviously the different moment equations are written with respect to different points. In applying each equilibrium equation a sign convention for the forces and moments must be chosen. It may be less confusing if the same sign convention is used throughout the problem (or even for all problems). The reactions are assumed to act in a given direction, generally the positive direction. If the value obtained for a reaction is negative, this means that the reaction acts in the opposite direction of the assumed reaction.

In the following examples the number of unknown reactions is equal to the number of equations of statics. If the number of unknown reactions is not equal to the number of equations then the structure is not statically determinate. Such structures are discussed in Section 2.4. For structures composed of multiple parts connected together by hinges, additional equations exist. This topic is discussed in Section 2.3.

Example 2.1. Find the reactions of the simply supported beam of Fig. 2.2, with a span length l under a concentrated load P, acting at a distance a from the left end.

Solution. First the supports are replaced by the reactions that they provide (Fig. 2.2b). Support 1 is a hinge and may thus have two reactions R_{1x} and R_{1y}. The right support is a roller, providing a vertical reaction only. The unknown

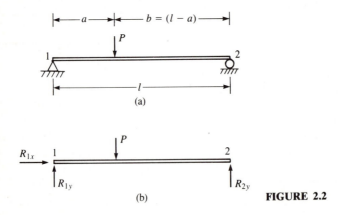

(a)

(b)

FIGURE 2.2

reactions are determined by applying the equations of static equilibrium (Eqs. 2.1).

Equation (2.1a): $\sum F_x = 0$ yields $R_{1x} = 0$.

Equation (2.1b): $\sum F_y = 0$ gives $R_{1y} - P + R_2y = 0$.

Equation (2.1c): $\sum M = 0$, with moments taken at support 2 results in

$$(R_{1y} \times l) - (P \times b) = 0 \qquad \text{or} \qquad R_{1y} = Pb/l$$

Substitution of R_{1y} in the second equation gives

$$R_{2y} = P - R_{1y} = P - \frac{Pb}{l} = P - \frac{P(l - a)}{l} = \frac{Pa}{l}$$

Notice that the reactions $R_{1y} = Pa/l$ and $R_{2y} = Pb/l$ of a simply supported beam under a single concentrated load are in proportion to the spacing of the load from the ends. The larger reaction is at the support closer to the load. Thus, the reaction at support 1 is equal to the ratio of the distance of the load from the right end to the span times the load.

Example 2.2. Find the reactions of the simply supported beam of Fig. 2.3 under a concentrated moment M_0.

Solution. Here again, as in Example 2.1, Eq. (2.1a) shows that the horizontal reaction of the left support is zero. Using Eqs. (2.1b) and (2.1c) we find the two vertical reactions.

Equation (2.1b) gives $R_1 + R_2 = 0$, and Eq. (2.1c) results in $(R_1 \times l) - M_0 = 0$. Thus, $R_1 = M_0/l$ and $R_2 = -M_0/l$.

It is seen that for a beam under a single concentrated moment the reactions are equal and opposite with a magnitude M_0/l. These reactions produce a couple $(M_0/l) \times l = M_0$. Which balances the applied moment.

Note that a roller can provide both upward and downward reactions.

Example 2.3. Find the reactions of the simply supported beam of Fig. 2.4 under a concentrated load of 40 kips acting at a distance 10 ft from the left-hand end and a concentrated moment of 200 kip · ft acting at 20 ft from the left-hand end. The span of the beam is 30 ft.

Solution. We will first find the reactions of this beam directly, using the equilibrium equations, and then by the principle of superposition. Referring to Fig. (2.4a), since the horizontal reaction is zero, we can use the equilibrium equations (2.1b) and (2.1c) to find the vertical reactions.

Equation (2.1b) gives $R_1 - 40 + R_2 = 0$.

Equation (2.1c) gives $R_1 \times 30 - 40 \times 20 + 200 = 0$.

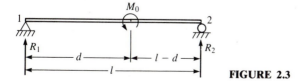

FIGURE 2.3

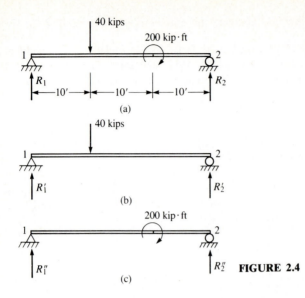

(a)

(b)

(c) FIGURE 2.4

The last equation gives $R_1 = 20$ kips. Substituting this value into the first equation results in $R_2 = 40 - R_1 = 40 - 20 = 20$ kips.

To find the reactions by superposition, we apply the load and the moment separately, as shown in Figs. 2.4b and 2.4c. The reactions for the load (Fig. 2.4b) can be found in a manner similar to Example 2.1. This gives

$$R_1' = 40 \times 20/30 = 26.67 \text{ kips} \qquad R_2' = 40 \times 10/30 = 13.33 \text{ kips}$$

The reactions due to the moment are found by a procedure similar to that of Example 2.2. In view of Fig. 2.4c, this results in

$$R_1'' = -200/30 = -6.67 \text{ kips} \qquad R_2'' = 200/30 = 6.67 \text{ kips}$$

Superimposing the reactions for the load and moment we have

$$R_1 = R_1' + R_1'' = 26.67 - 6.67 = 20 \text{ kips}$$
$$R_2 = R_2' + R_2'' = 13.33 + 6.67 = 20 \text{ kips}$$

As expected, the results obtained directly and those found by superposition are identical.

Example 2.4. Find the reactions of the cantilever beam of Fig. 2.5 with a 20 m span under the distributed uniform load of 200 N/m and a concentrated load of 800 N applied at a slope of 3/2.

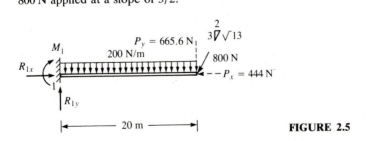

FIGURE 2.5

Solution. There are three reactions at the fixed end of the cantilever beam, a horizontal reaction, a vertical reaction and a moment. Since we will be summing up forces in the vertical and horizontal directions, we replace the inclined load by its horizontal and vertical components. The horizontal and vertical components of this load are

$$800 \times \frac{2}{\sqrt{13}} = 444 \text{ N} \quad \text{and} \quad 800 \times \frac{3}{\sqrt{13}} = 665.6 \text{ N}$$

For calculating the reactions, distributed loads can be replaced by their resultants. Summation of the horizontal forces gives

$$R_{1x} - 444 = 0 \quad \text{or} \quad R_{1x} = 444 \text{ N}$$

and summation of the vertical forces yields

$$R_{1y} - 200 \times 20 - 665.6 = 0$$

which gives

$$R_{1y} = 4\,665.6 \text{ N}$$

Finally setting the sum of moments with respect to the left-hand end equal to zero we have

$$M_1 + 200 \times 20 \times 20/2 + 665.6 \times 20 = 0 \quad \text{or} \quad M_1 = -53\,312.0 \text{ N m}$$

Example 2.5. Find the reactions of the beam of Fig. 2.6 with a hinged and an inclined support.

Solution. The roller at support 2 provides a reaction, R_2, normal to the inclined surface. We can replace this reaction by its horizontal component, $R_{2x} = \frac{3}{5}R_2$ and vertical component $R_{2y} = \frac{4}{5}R_2$. Thus,

$$R_{1y} - 3 \times 20 + R_{2y} = 0,$$

and

$$R_{1y} \times 60 - 3 \times 20 \times 40 = 0 \quad \text{or} \quad R_{1y} = 40 \text{ kips}$$

Then

$$R_{2y} = 60 - 40 = 20 \text{ kips}$$

and

$$R_{2x} = \tfrac{3}{4}R_{2y} = 15 \text{ kips}$$

$$R_{1x} = R_{2x} = 15 \text{ kips}$$

Example 2.6. Find the reactions of the simply supported beam of Fig. 2.7 with a 30 m span under a load varying linearly from 50 N/m at the left end to zero at the right end.

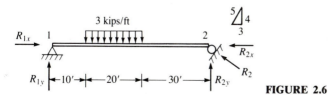

FIGURE 2.6

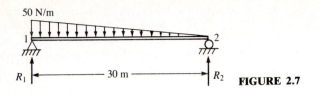

FIGURE 2.7

Solution. Using two moment equilibrium equations, by taking the sum of moments at the left-hand and right-hand supports we can find the two reactions directly. Thus, $\sum M_2 = 0$ gives

$$R_1 \times 30 - 50 \times \tfrac{30}{2} \times \tfrac{2}{3} \times 30 = 0 \qquad \text{or} \qquad R_1 = 500\,\text{N}$$

and $\sum M_1 = 0$ yields

$$50 \times \tfrac{30}{2} \times \tfrac{1}{3} \times 30 - R_2 \times 30 = 0 \qquad \text{or} \qquad R_2 = 250\,\text{N}$$

Example 2.7. Find the reactions of the truss of Fig. 2.8

Solution. The sum of the horizontal forces set equal to zero gives

$$R_{1x} + 2 + 4 + 4 + 4 + 2 = 0 \qquad \text{or} \qquad R_{1x} = -16\,\text{kips}$$

The sum of the vertical forces set equal to zero gives

$$R_{1y} - 5 - 10 - 10 - 10 - 10 - 10 - 10 - 10 - 5 + R_{2y} = 0$$

or

$$R_{1y} + R_{2y} = 80\,\text{kips}$$

The sum of moments with respect to the right end equal to zero results in

$$R_{1y} \times 120 - 5 \times 120 + 2 \times 10 - 10 \times 105 + 4 \times 20 - 10 \times 90 + 4 \times 30$$
$$- 10 \times 75 + 4 \times 40 - 10 \times 60 + 2 \times 50 - 10 \times 45 - 10 \times 30 - 10 \times 15 = 0$$

or $R_{1y} = 36$ kips, resulting in $R_{2y} = 80 - 36 = 44$ kips.

Example 2.8. Find the reactions of the frame of Fig. 2.9. The loads are a 200 lb/ft load due to snow over the horizontal projection, and a 100 lb/ft due to

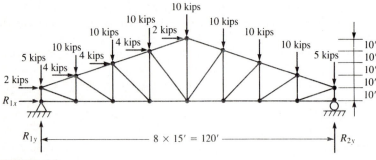

FIGURE 2.8

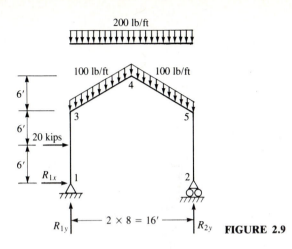

200 lb/ft

100 lb/ft   100 lb/ft

6'

6'

20 kips

6'

R_{1x} 1   2

R_{1y} — $2 \times 8 = 16'$ — R_{2y} **FIGURE 2.9**

the weight of the roofing. In addition a concentrated load of 20 kips is acting at the mid-height of the left column.

Solution. The horizontal reaction is found by setting the sum of the horizontal forces equal to zero,

$$R_{1x} + 20 = 0 \qquad \text{or} \qquad R_{1x} = -20 \text{ kips}$$

The length of the inclined member is $\sqrt{6^2 + 8^2} = 10$ ft.

To find the vertical reactions we set sum of the moments with respect to the left and right supports equal to zero,

$$20 \times 6 + 200 \times 16 \times 8 + 100 \times 10 \times 4 + 100 \times 10 \times 12 - R_{2y} \times 16 = 0$$

or

$$R_{2y} = 2\,607.5 \text{ kips}$$

and

$$R_{1y} \times 16 + 20 \times 6 - 200 \times 16 \times 8 - 100 \times 10 \times 12 - 100 \times 10 \times 4 = 0$$

which gives $R_{1y} = 2\,592.5$ kips.

2.3 EQUATIONS OF CONDITION

If a structure consists of two or more parts connected together by hinges, rollers, etc., then these connection conditions would provide additional equations that are needed for finding the unknown reactions. Such equations are called equations of condition.

Consider for example the structure of Fig. 2.10a. The left-hand part of the structure (part 12) is hinged at its left-hand end, point 1, and supported by the right-hand part through a roller at point 2. There are five unknown reactions, two at point 1 and three at point 4. Thus, the three equations of statics, Eqs. (2.1), which apply to the whole structure, are not sufficient for finding the end reactions. Two additional equations are needed, which can be obtained upon close examination of the structure.

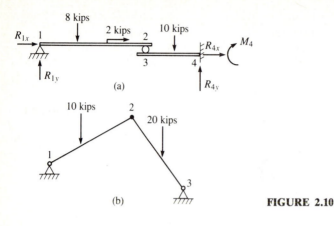

(a)

(b)

FIGURE 2.10

Since the roller of point 2 does not provide any resistance against horizontal forces, it is not difficult to see that the horizontal force of 2 kips acting on part 12, must be carried by the support at point 1. Thus, we can see that the sum of the horizontal forces on part 12, on the left of point 2, must add up to zero. This is a necessary condition. Otherwise the roller of point 2 cannot provide any resistance against a net horizontal force acting at point 2. In other words, a motion will result if such a net force exists at 2, and the structure will not be stable. Therefore, stability of the structure requires that part 12 be independently in equilibrium with respect to horizontal forces. This requirement will provide an equation, called the *equation of condition*. By a similar reasoning we can see that the total moment on part 12, calculated at point 2 must be zero. This results in another equation of condition. These two equations, along with the three equations of statics, are sufficient for finding all the reactions. Notice that the equations of statics, Eqs. (2.1), apply to the whole structure, while the equations of condition apply to a portion of the structure only.

Alternatively, using the principle of action–reaction, the two parts 12 and 34 can be separated upon application of the interaction forces to each part, at the roller. We see that part 12 exerts a downward force R to part 34, while part 34 provides an equal and opposite reaction R at point 3. The five unknown reactions at the two ends of the structure plus the unknown force R can be found by applying the three equations of statics to parts 12 and 34 separately, resulting in six equations for finding the six unknown reactions (including the interaction force at 2).

Similarly, in Fig. 2.10b the two parts 12 and 23 are hinged together at point 2. They are also hinged at supports 1 and 3. Here again there are four unknown reactions and three equations of statics and thus an additional equation must be sought for determination of the reactions.

Since a hinge rotates under any moment and cannot transfer the moment from one part of the structure to another, the forces on part 12 must produce no moment at point 2. That is, the sum of the moments of all the forces acting on part 12 at point 2 must be zero. This provides the necessary equation of condition for finding the four reactions.

Again this problem may be solved by separating the two parts 12 and 23 and applying the interaction forces at point 2. Since a hinge can transfer forces but not moments, equal and opposite horizontal and vertical forces are applied at point 2 to the two parts. Then the three equilibrium equations, applied to the two parts separately, provide six equations that are sufficient for finding the four unknown reactions and two interaction forces.

Example 2.9. Find the reactions at points 1 and 3 of the structure of Fig. 2.11a. The two parts are hinged at point 2.

Solution. The equations of statics, Eqs. (2.1),

$$\sum F_x = 0 \qquad \sum F_y = 0 \qquad \sum M = 0$$

apply to the whole structure. Since the structure consists of two rigid parts, hinged together at point 2, an equation of condition is also available. The latter equation is found by setting the sum of the forces acting on the left part of the structure with respect to the hinge (point 2) equal to zero. Thus,

$$\sum F_x = 0 \qquad \text{gives} \qquad R_{1x} + R_{3x} = 0$$

$$\sum F_y = 0 \qquad \text{or} \qquad R_{1y} + R_{3y} - 25 = 0$$

$$\sum M_3 = 0 \qquad \text{or} \qquad R_{1x} \times 20 + R_{1y} \times 120 - 25 \times 60 = 0$$

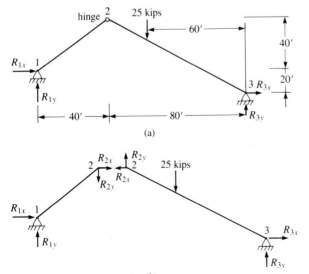

(a)

(b)

FIGURE 2.11

and the equation of condition, that is the sum of the moments on the left-hand part of the structure taken with respect to point 2 is

$$\sum M_{L_2} = 0 \qquad \text{or} \qquad -R_{1x} \times 40 + R_{1y} \times 40 = 0$$

These equations give the following:

$$R_{1x} = 10.71 \text{ kips} \qquad R_{3x} = -10.71 \text{ kips}$$
$$R_{1y} = 10.71 \text{ kips} \qquad R_{3y} = 14.29 \text{ kips}$$

To use the alternate solution procedure, we divide the structure into two parts, 12 and 23. The interactive forces at point 2 are R_{2x} and R_{2y}.

Applying the three equilibrium equations to the forces on parts 12, and 23, for part 2 we get

$$R_{1x} + R_{2x} = 0$$
$$R_{1y} - R_{2y} = 0$$
$$-R_{1x} \times 40 + R_{1y} \times 40 = 0$$

and for part 23

$$-R_{2x} + R_{3x} = 0$$
$$R_{2y} - 25 + R_{3y} = 0$$
$$-R_{2x} \times 60 + R_{2y} \times 80 - 25 \times 60 = 0$$

Solution of the above equations gives $R_{2x} = -10.71$ kips and $R_{2y} = 10.71$ kips, leading to the same reactions at supports 1 and 3 as before.

Example 2.10. Find the reactions of the three span beam with two hinges in the middle span as shown in Fig. 2.12.

Solution. Since no horizontal loads are acting on the beam, the horizontal reaction at support 1 is zero. Setting the sum of the vertical forces and the sum of the moments with respect to the right-hand end of the beam equal to zero, we find

$$R_1 - 10 + R_2 - 25 - 2 \times 10 + R_3 - 20 + R_4 = 0$$

$$R_1 \times 60 - 10 \times 50 + R_2 \times 40 - 25 \times 35 - 2 \times 10 \times 30 + R_3 \times 20 - 20 \times 10 = 0$$

The two equations of condition are found by setting the sum of the moments of the forces acting on the portion of the structure to the left of each

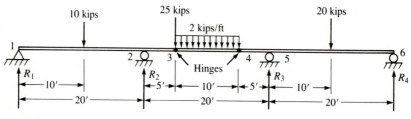

FIGURE 2.12

hinge, taken with respect to the hinge, equal to zero. This gives

$$R_1 \times 25 - 10 \times 15 + R_2 \times 5 = 0$$

and

$$R_1 \times 35 - 10 \times 25 + R_2 \times 15 - 25 \times 10 - 2 \times 10 \times 5 = 0$$

The solution of these four equations gives the four unknown reactions as follows,

$$R_1 = -3.75 \text{ kips} \qquad R_2 = 48.75 \text{ kips}$$

$$R_3 = 22.5 \text{ kips} \qquad R_4 = 7.5 \text{ kips}$$

In writing the second equation of condition, we calculated the sum of moments of the part of the structure from the left end to the second hinge. However, from the first equation of condition we knew that the sum of moments of the forces on the left of the first hinge with respect to the latter hinge is zero. Thus, we can calculate the shear force at the first hinge (sum of the forces on the left of that hinge) and add the moment of this shear force to those of the other forces between the two hinges.

The reactions for this structure can also be found by separating the structure into three parts, 13, 34 and 46 and by applying the equilibrium equations to each part to find the reactions and the interaction forces between the different parts.

2.4 STATICALLY DETERMINATE AND INDETERMINATE STRUCTURES

In order to establish whether a structure is unstable, statically determinate, or indeterminate we need to compare the number of unknown reactions with the number of equations of statics plus the equations of condition.

Let r = number of unknown reactions

n = number of equations of statics
($= 3$ for two-dimensional structures)

c = number of equations of condition

If

$r < n + c$ unstable structure

$r = n + c$ statically determinate structure

$r > n + c$ statically indeterminate structure

The number of unknown reactions in excess of equations of statics plus equations of condition, $r - (n + c)$, is the degree of indeterminacy. This is also equal to the number of reactions that must be removed in order to obtain a statically determinate structure. Removal of an additional reaction would then lead to a mechanism, i.e. an unstable structure.

It should be noted that the above conditions are necessary but not sufficient. Thus, in some cases, even though the above relations indicate a stable structure, the structure may still be unstable. lack of resistance against a force component, three hinges in one span, and converging reactions provide some such examples, as shown in Fig. L2.4 of Laboratory Experiments at the end of this chapter.

Statically indeterminate structures are generally more economical and the redundancy makes sudden failures less likely. However, when excessive settlements are possible, statically determinate structures provide a safer alternative, because settlements that are not considered in the design may induce very large stresses beyond the capacity of the structure.

The above conditions described in terms of the support reactions enable us to establish whether the structure is externally indeterminate. In Chapter 3 we discuss another type of indeterminacy for trusses having more bar forces than can be determined by the equations of statics. The latter condition is called *internal indeterminacy,* since it involves internal forces in the structure in contrast to external determinacy, which is due to redundant support reactions.

Example 2.11. Determine whether the structures shown in Figure 2.13 are statically determinate, indeterminate, or unstable. For the indeterminate structures establish the degree of indeterminacy.

Solution. The beams of Figs. 2.13a and 2.13b are unstable since they have fewer reactions than three (the number of equations of statics). The beams of Figs. 2.13c and 2.13d are statically determinate: each one has three reactions that can be found from the equations of statics. Figure 2.13e, with four reactions, is statically indeterminate to the first degree, while the beam of Fig. 2.13f is statically indeterminate to the third degree because it has six reactions. If there are no horizontal forces acting on this beam, then both horizontal reactions can be assumed to be zero. This would lead to two equations of statics and four unknown reactions and thus, indeterminacy to the second degree. The continuous beam of Fig. 2.13g is indeterminate to the second degree, since there are eight reactions versus three equations of statics plus three equations of condition for the three hinges. The continuous beam of Fig. 2.13h has four reactions and according to the necessary conditions for determinacy is indeterminate to the first degree. However, since it is supported by four rollers it cannot provide resistance against the horizontal component of the load. It is therefore unstable. Finally, the frame of Fig. 2.13i, with nine reactions and three equations of statics, is statically indeterminate to the sixth degree.

2.5   LOAD TRANSFER

The loads applied to a structure are transferred from some members to others before they are transmitted to the ground. Since structures are often designed by considering the individual components, we must be able to calculate the load carried by each member, and identify the manner in which the members transfer their loads to others. This usually amounts to finding the reactions of the members on the supporting members. Some common cases are discussed below.

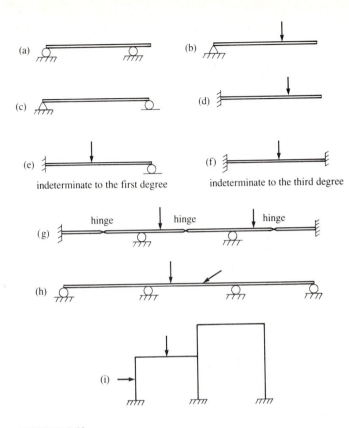

FIGURE 2.13
(a) $r = 2 < 3$, unstable. (b) $r = 2 < 3$, unstable. (c) $r = 3$, determinate. (d) $r = 3$, determinate. (e) $r = 4 > 3$, indeterminate to the first degree. (f) $r = 6 > 3$, indeterminate to the third degree. (g) $r = 8 > (3 + 3)$, indeterminate to the second degree. (h) $r = 4 > 3$, but unstable against horizontal forces. (i) $r = 9 > 3$, indeterminate to the sixth degree.

Example 2.12. Draw the free body diagrams for the beams and girders of the structure shown in Figs. 2.14a, b, c.

Solution. The girders provide a sufficient degree of restraint for the beams so that under horizontal forces the beams are not free to displace as rigid bodies. On the other hand, this restraint cannot prevent expansion due to large forces such as those due to temperature variations. This is similar to having a hinge at one end of the beams and a roller at the other. Thus, the free body diagram of the beams are drawn with one end hinged and the other on a roller, that is, simply supported. The loading on the beams consists of the concentrated load P plus the weight of the beam (Fig. 2.14d).

With a similar argument about axial restraint we can use a simply supported beam to model the girders if they are supported by walls. However, here the girders are built monolithically with the columns. Thus, a girder with its supporting columns must be analyzed as a frame. The girder loads consist of the

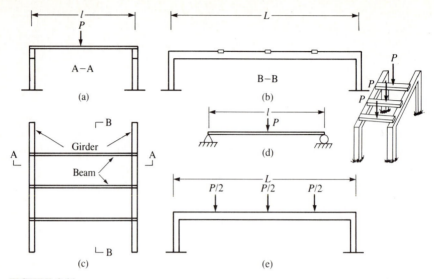

FIGURE 2.14

reactions of the beams on the girders (three loads of $P/2$), plus the weight of the girder itself (Fig. 2.14e).

In some cases, distributed loads are acting over a solid object such as a bridge pier or abutment. In such cases, the distribution through the thickness can be assumed with a 45° angle. A similar distribution is used for finding the loads on lintel beams. For distribution of soil pressure through the depth, a smaller angle is usually assumed.

Example 2.13. For the truss structure of Fig. 2.15, the weight of the roofing is 16 psf and the roof snow load is 30 psf. Find the load acting at joint 4 of the middle truss due to each of these loading conditions. The spacing of trusses is 20 ft.

Solution. For the weight of the roofing the length of the roof along the slope must be used. The slope of the roof is α with,

$$\cos \alpha = \frac{20}{\sqrt{20^2 + 5^2}} = 0.97$$

The tributary area for joint 4 is $20 \times 10/0.97 = 206.19 \text{ ft}^2$, and the joint load is $P = 16 \times 206.19 = 3\,299 \text{ lb} = 3.3 \text{ kips}$.

The tributary area for snow is the projection shown in the figure, that is, $20 \times 10 = 200 \text{ ft}^2$, and the joint load due to snow is

$$P = 30 \times 200 = 6\,000 \text{ lb} = 6 \text{ kips}$$

Example 2.14. Find the load on each member of the floor system in Fig. 2.16. The live load on the slab is 100 psf and the weight of the slab itself is 50 psf. Find the total loads acting on the beams and the girders.

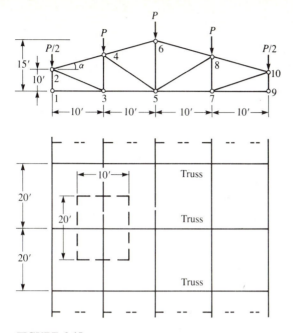

FIGURE 2.15

Solution. The slab transmits its load to the beams. The beams in turn transfer their loads to the girders and from the girders the loads are transmitted to the ground through the columns.

The total load on the slab due to the live load and the weight of the slab itself is 150 psf. The load per unit length of each beam may be taken as that acting over the width of the slab (mid-spacing of the beams), 10 ft. Thus, the load on the beam would be a uniform load $10 \times 150 = 1\,500$ lb $= 1.5$ kips plus the weight of the beam itself, as shown in Fig. 2.16d. The load of the beams is transferred to the girders through their reaction on the girders. This reaction is $1.5 \times 18 = 27$ kips. Thus, the loads on the girders would consist of the concentrated loads as shown in Fig. 2.16c.

The above procedure would be exact if the slab were supported only by the beams and not directly by the girders. Otherwise it would be an approximation. When the slab is supported by both the beams and the girders, as in the present case, a more exact procedure would be to draw 45° lines from the center of the columns to the mid-spacing of the beams, as shown on the left of the plan view. The tributary areas thus divided would transfer their loads to the beams or to the girders. The girders in addition to the slab load would carry the reactions of the beams. For the beams, the intensity of distributed loads thus vary linearly from zero at the girder to a maximum value of $10 \times 150 = 1\,500$ lb, and the distribution is as shown in Fig. 2.16f. The weight of the girders themselves would provide an additional uniformly distributed load. The girders would be having the triangular loads, varying from zero at the location of the beams to a maximum value of $10 \times 150 = 1\,500$ lb $= 1.5$ kips, plus the reactions of the beams, which are, from Fig. 2.16f equal 19.5 kips and the uniform weight of the girder itself (Fig. 2.16e).

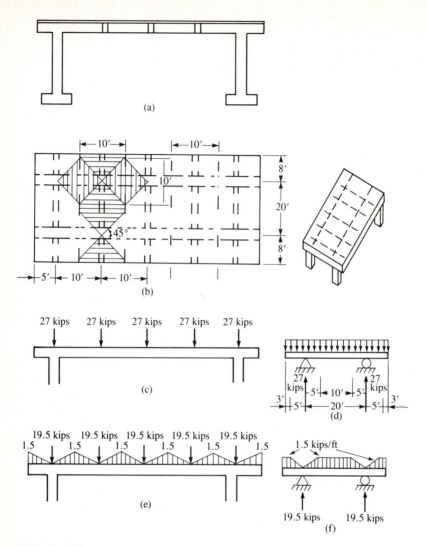

FIGURE 2.16

LABORATORY EXPERIMENTS

Proportionality of the Beam Reactions under a Concentrated Load

Place the ends of a small beam on two scales (Fig. L2.1). Apply a load at a distance l_1 from the left end and observe that the reactions are proportional to the distance of the load from the other end, that is, $R_1/R_2 = l_2/l_1$, or $R_1 = Pl_2/l$ and $R_2 = Pl_1/l$.

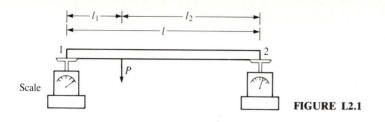

FIGURE L2.1

Superposition

Using the setup of Fig. L2.1, verify the principle of superposition by showing that the reaction of a support under two loads is equal to the sum of the reactions for both loads.

Deflection Effects on the Reactions

Using two beams with the same span, one flexible and the other rigid, and the setup of Fig. L2.1, show that the small deflections (less than 3–4% of the span length) do not alter the values of reactions.

Equations of Condition

Build models of the structures in Fig. 2.10 and show that if the net force on one side of a roller and net moment at a hinge from the forces on one side of the hinge are not zero then the structures are not stable. For example, in the structure of Fig. L2.2, consisting of two bars, hinged together at point 2, if end 1 is fixed then by moving end 3 we can observe that, because of the hinge at 2, the left part of the structure does not provide any resistance against moments produced at 2 by the forces on bar 23. Therefore, for equilibrium the net moment at 2 caused by the forces on part 23 must be zero.

Stability and Determinacy

For the structure of Fig. L2.3, two bars hinged together, if end 1 is hinged and end 3 is a roller, verify that the structure is not stable and the roller moves a large distance under any load, to cause the collapse of the structure. Show that by preventing the displacement of the roller, that is, having a hinge instead of a roller, we would have a stable structure capable of carrying loads. Show why this is the case. See Section 2.4.

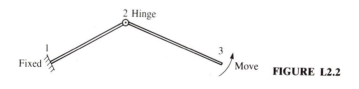

FIGURE L2.2

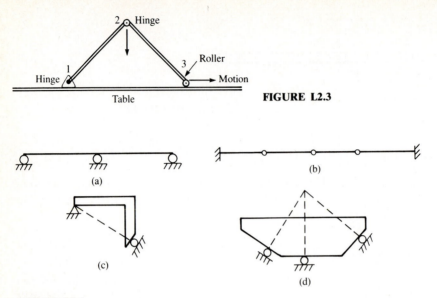

FIGURE L2.3

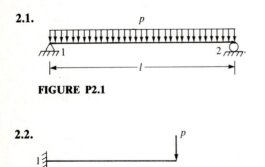

(a)

(b)

(c)

(d)

FIGURE L2.4
(a) No resistance against horizontal force. (b) More than two hinges in one span. (c) and (d) Converging reactions.

Unstable Structures Satisfying the Stability Conditions

Verify that the structures of Fig. L2.4 satisfy the static stability conditions, Section 2.4. Using models show that these structures are unstable.

PROBLEMS

Find the reactions of the following structures.

2.1.

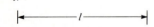

p

FIGURE P2.1

2.2.

p

FIGURE P2.2

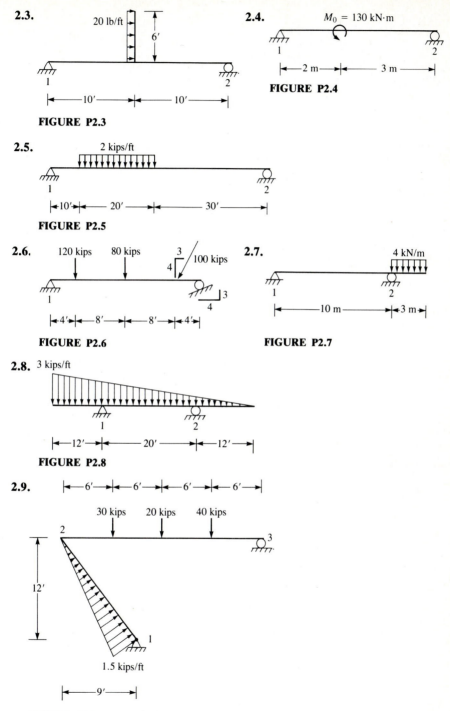

2.3.

20 lb/ft

6'

1

2

|←——— 10' ———→|←——— 10' ———→|

FIGURE P2.3

2.4.

$M_0 = 130$ kN·m

1

2

|←—— 2 m ——→|←——— 3 m ———→|

FIGURE P2.4

2.5.

2 kips/ft

1

2

|←10'→|←——— 20' ———→|←——— 30' ———→|

FIGURE P2.5

2.6.

120 kips   80 kips   100 kips

3

4

1

3

4

|←4'→|←— 8' —→|←— 8' —→|←4'→|

FIGURE P2.6

2.7.

4 kN/m

1

2

|←——— 10 m ———→|←3 m→|

FIGURE P2.7

2.8. 3 kips/ft

1

2

|←—12'—→|←——— 20' ———→|←—12'—→|

FIGURE P2.8

2.9.

|←— 6' —→|←— 6' —→|←— 6' —→|←— 6' —→|

30 kips   20 kips   40 kips

2

3

12'

1

1.5 kips/ft

|←——— 9' ———→|

FIGURE P2.9

2.10.

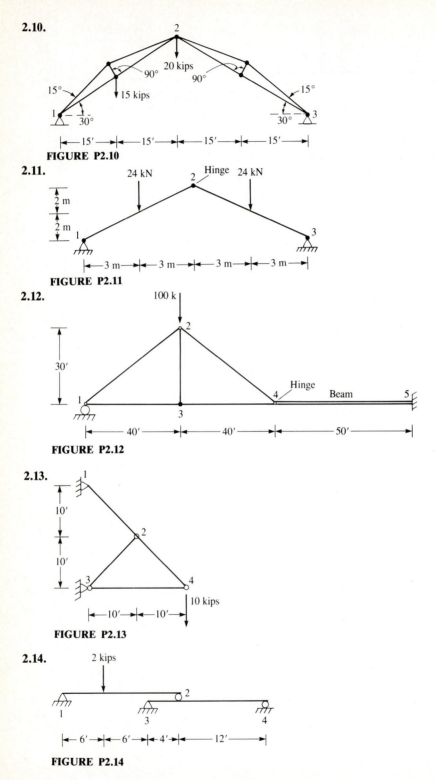

FIGURE P2.10

2.11.

FIGURE P2.11

2.12.

FIGURE P2.12

2.13.

FIGURE P2.13

2.14.

FIGURE P2.14

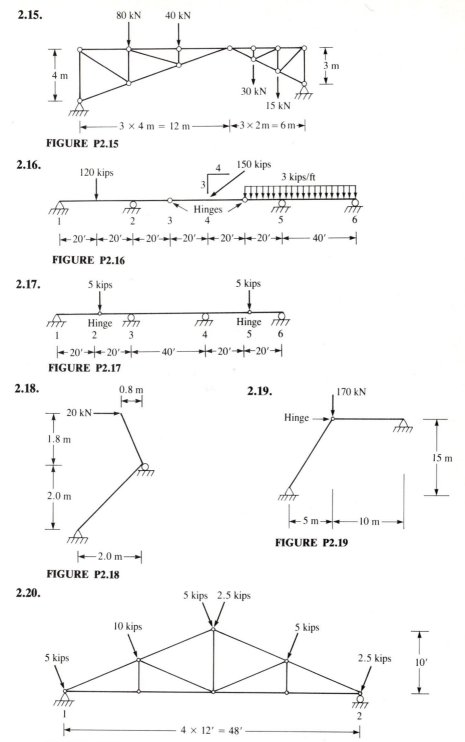

2.15.

80 kN   40 kN

4 m

3 m

30 kN

15 kN

|←——— 3 × 4 m = 12 m ———→|←3 × 2 m = 6 m→|

FIGURE P2.15

2.16.

120 kips

4

150 kips

3

3 kips/ft

Hinges

1   2   3   4   5   6

|←20'→|←20'→|←20'→|←20'→|←20'→|←20'→|←——— 40' ———→|

FIGURE P2.16

2.17.

5 kips   5 kips

Hinge   Hinge

1   2   3   4   5   6

|←20'→|←20'→|←——— 40' ———→|←20'→|←20'→|

FIGURE P2.17

2.18.

0.8 m

20 kN

1.8 m

2.0 m

|←—— 2.0 m ——→|

FIGURE P2.18

2.19.

170 kN

Hinge

15 m

|←5 m→|←—— 10 m ——→|

FIGURE P2.19

2.20.

5 kips   2.5 kips

10 kips

5 kips

5 kips

2.5 kips

10'

1   2

|←——————— 4 × 12' = 48' ———————→|

FIGURE P2.20

2.21.

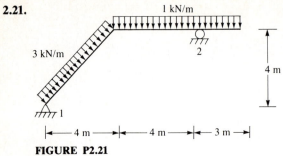

FIGURE P2.21

2.22.

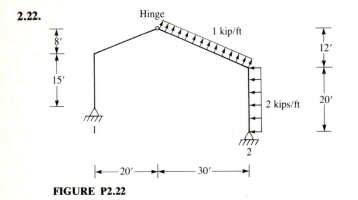

FIGURE P2.22

2.23.

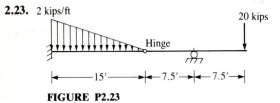

FIGURE P2.23

TRUSSES

Trusses are structures composed of straight bars whose members are in tension or compression only. They may be used to span distances from 30 ft to 1 000 ft. Trusses are usually made of steel or timber, and rarely of reinforced concrete because of its large dead load and low strength in tension. The bars of steel trusses may be single, double angles or channels, tubes, or even I-sections. Timber truss members have rectangular cross-sections.

Figure 3.1 shows some typical trusses used for steel and timber buildings and bridges. Since steel members are usually more slender than timber members, truss configurations with shorter compressive members are more suitable for steel structures. In this chapter we discuss two-dimensional (plane) trusses. Three-dimensional (space) trusses are covered in Chapter 18.

3.1 TRUSS ANALYSIS

Analysis of trusses involves the determination of the bar forces and possibly deflections. In the following sections we discuss methods of joints and sections. The method of joints is suitable for statically determinate trusses when all the bar forces are required. If only a few bar forces are needed, then the method of sections may be preferable. Application of the methods of joints and sections to statically indeterminate trusses involves the calculation of joint deflections. The procedures for finding deflections of trusses and solving indeterminate trusses are described in Section 6.3.

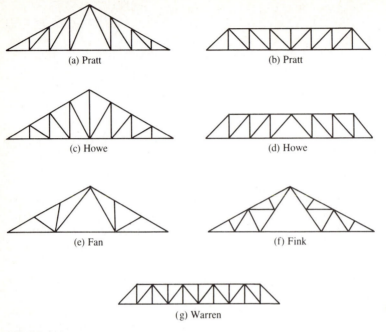

FIGURE 3.1
Truss types.

The stiffness method applies to statically determinate and indeterminate trusses. Furthermore, in this method the joint deflections are found first, from which the bar forces are calculated. The stiffness method for trusses is discussed in Chapter 12. Before analyzing a truss we must ensure that it is stable. Furthermore, if the method of joints or sections is used, we must verify that the truss is statically determinate.

Equations of static equilibrium (Eqs. (2.1)) can be used for finding the truss bar forces. For plane trusses these are

$$\sum F_x = 0 \qquad \sum F_y = 0 \qquad \sum M = 0$$

In the method of joints these equations are applied to the truss joints, while in the method of sections they are applied to a larger portion of the truss.

Truss Assumptions

The methods of truss analysis are based on the following assumptions.

1. The bars are pin-connected (i.e., they are hinged together at their ends).
2. The loads are applied at the joints of the truss only.

Although the joints of steel trusses are usually welded or bolted and are thus rigid or semi-rigid, and small loads are sometimes applied to the members, rather than the joints, the above assumptions generally produce sufficiently accurate results. This is because truss members are usually slender, so that at a short distance from the joints the forces are primarily axial without producing significant bending moments. For trusses with short members and rigid connections the above assumptions may not be accurate. The latter trusses must be analyzed as frames.

3.2 DETERMINACY AND INDETERMINACY

Trusses are statically determinate when all the bar forces can be found by the equations of statics alone. When there are redundant forces that cannot be found by the equations of statics, then the truss is statically indeterminate. Since a truss may also be statically indeterminate because the number of reactions is larger than the number of equations of statics plus equations of condition, two types of indeterminacy are possible. If there are redundant bar forces, then the truss is called *internally indeterminate*. If, on the other hand, there are redundant support reactions, then the truss is externally indeterminate. A truss may be both *internally* and *externally indeterminate*. Indeterminate trusses are more difficult to analyze with classical methods. The best solution procedure for such trusses would be the stiffness method. Furthermore, if there are fewer than a minimum number of bars, then the truss is unstable. In the following section we establish the requirements for stability and determinacy.

Truss Stability

Consider the four bars of Fig. 3.2a, which are pin-connected at joints 1, 2, 3, and 4. This configuration will collapse under a small load, as shown, and therefore is not stable. However, addition of bar 23 will render this system stable. To form such a stable configuration with joints 1, 2, 3, and 4, we can proceed as follows.

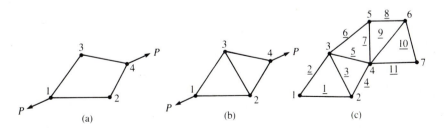

FIGURE 3.2

Starting with three of the joints, say 1, 2, and 3, connected together by three bars, we obtain a stable configuration. We then add the fourth joint using two additional bars. The resulting truss is also stable. This process can be continued by adding triangles with two bars and one joint. Trusses of a variety of configurations and with as many joints as necessary may be formed in this manner. Thus, for a truss with j joints the number of bars for a stable shape would be equal to the three bars of the original triangle, with three joints, plus two additional bars for each of the remaining $(j-3)$ joints. That is, the total number of bars must be, $b = 3 + 2(j-3)$, or

$$b = 2j - 3$$

For example, for the configuration of Fig. 3.2c, starting with a triangle with three bars 1, 2, 3, and three joints 1, 2, 3, we have constructed a truss with seven joints. That is, we have used four additional joints, each involving two additional bars. Therefore, the total number of bars is 11, which is equal to the 3 initial bars plus $2(7-3) = 8$.

A more general procedure for verifying the stability of truss configurations can be developed from the definition of internal determinacy. A common method for determining the bar forces of statically determinate trusses is the method of joints. This method is discussed in Section 3.3. In the joint method, each joint and its connecting bars is separated as a free body and its equilibrium is considered. Since all the bar forces converge at the joint, no moment is present. Thus, two equations of statics are available from which two unknown bar forces can be found. If the number of truss bars is b, then there will be b unknown bar forces. In addition, the support reactions must also be determined from the equilibrium equations. If the number of reaction components is r, then the total number of unknowns will be $b + r$, which must be calculated from $2j$ equations for j joints of the truss. Thus, we must have $2j = b + r$, or

$$\boxed{b = 2j - r} \tag{3.1}$$

This expression is the same as the previous one for a truss with three reaction components. Using Eq. (3.1), the following conditions can be distinguished:

$b < 2j - r$ The truss is unstable
$b = 2j - r$ The truss is statically determinate internally
$b > 2j - r$ The truss is statically indeterminate internally

It should be noted that Eq. (3.1) is necessary but not sufficient for internal determinacy. That is, in some cases even though this equation is satisfied the truss may still be unstable. Such an example is shown in Fig. 3.3a. This truss has nine members, six joints, and three reaction components, and therefore satisfies Eq. (3.1). However, it is easy to see that the right half of the truss, the rectangular shape, is unstable and can undergo large deformations

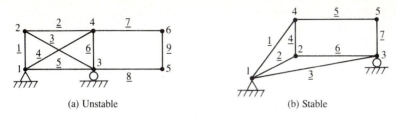

(a) Unstable   (b) Stable

FIGURE 3.3

with joints 5 and 6 displacing in the vertical direction. From Figs. 3.2a and 3.3a we may conclude that, when a truss has quadrilateral or rectangular shapes in its configuration, then we must verify that it is not unstable. We must emphasize, however, that the mere presence of quadrilateral or rectangular shapes in the configuration does not necessarily imply instability, since trusses with such configurations can be stable as is the case of Fig. 3.3b.

To verify stability of a truss that satisfies the above stability equation but arouses suspicion, we can proceed with the process we used in starting the discussion of truss stability. In that process we constructed stable trusses by adding two bars with an additional joint to an already stable configuration. In this process every stable part of the truss can be considered as a bar. Thus, we can follow the construction of the truss from the start or from a stable configuration. If the construction process at each stage continues with the addition of two bars (or two stable portions) and a joint to a previous, stable configuration, then the truss is stable. For example, in Fig. 3.3b we can start from triangle 123 then add the bars 14 and 24 with joint 4. This gives another stable configuration. To this we add bars 35 and 45 with joint 5. Therefore, the truss is stable. Applying the same process to Fig. 3.3a we observe that the left half of the truss (with six bars and four joints) is stable. However, to this stable configuration we are adding three bars (35, 56 and 64) and two joints (5 and 6). Thus, the truss is not stable. Further verification of truss stability may be done by building a small-scale model of the truss.

3.3 METHOD OF JOINTS

The method of joints is based on the premise that for a structure in equilibrium every portion thereof must be in equilibrium. It consists of considering each joint with the bars connected to it as a free body, and establishing its equilibrium. The truss members are assumed to be hinged together at the joints and all the bars connecting at a joint pass through the joint. Therefore the bar forces do not produce any moments at the joint, and thus the moment equilibrium equation is identically satisfied. Consequently, only two equations of statics, $\sum F_x = 0$ and $\sum F_y = 0$, are available for finding the unknown bar forces. No more than two unknown bar forces can therefore be present at the joint under consideration.

In the solution process we start at a truss joint with only two unknown bar forces (usually a support joint) and proceed to other joints where all the bar forces, except two, have previously been determined. The joints are isolated one by one and the bars connecting to it are replaced by the forces that they exert on the joint. The equilibrium equations are then applied to the joint and the process is continued until all the required bar forces are found. Notice that in this method determination of some bar forces may require computation of other bar forces.

Example 3.1. The two trusses of Fig. 3.4, are triangular brackets supporting the marquee of a hotel. The snow load, over the horizontal projection of the roof is 100 psf and is carried by the brackets. Each bracket is a two-bar truss.

Solution. As mentioned before, one of the assumtions of truss analysis is that the loads are applied at the joints only. Before starting the analysis we must determine the joint loads.

The tributary area for each truss is $10 \times 10 = 100 \, \text{ft}^2$. Hence the load on each truss is $100 \times 100 = 10\,000 \, \text{lb} = 10$ kips. This 10-kip load must be distributed between the two joints of the truss. Thus, a 5-kip load is applied at joint 1 and another 5-kip load is taken by joint 2 at the wall. The truss to be analyzed along with the applied loads is shown in Fig. 3.4c.

The truss of Fig. 3.4c has 2 bars, 3 joints and 4 reactions. Thus, Eq. (3.1) gives $2 = 2 \times 3 - 4 = 2$, indicating a stable truss. To analyze the truss using the method of joints, we consider the equilibrium of joint 1. We must have $\sum F_x = 0$ and $\sum F_y = 0$ at this joint.

In applying the equilibrium equations we assume (Fig. 3.4d) that the forces are positive (pulling on the joint), that is, that the bars are in tension. A negative value for a bar force would indicate compression in the bar. The first equilibrium

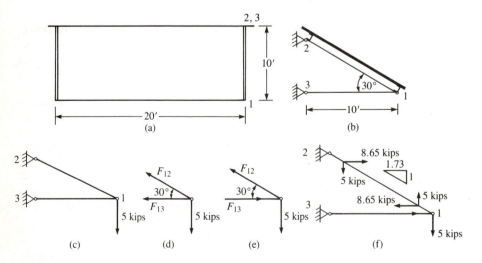

FIGURE 3.4

equation yields

$$F_{12} \times \cos 30° + F_{13} = 0$$

and the second equation results in

$$F_{12} \times \sin 30° - 5 = 0$$

From the last equation,

$$F_{12} = \frac{5}{\sin 30°} = \frac{5}{0.5} = 10 \text{ kips}$$

which upon substitution in the first equation gives

$$F_{13} = -F_{12} \cos 30° = -10 \times 0.87 = -8.7 \text{ kips}$$

The signs of F_{12} and F_{13} indicate that F_{12} is a tensile force as assumed, whereas F_{13} is a compressive force. The actual forces acting on joint 1 are shown in Fig. 3.4e. To demonstrate that positive forces are tensile, as assumed, we observe from Fig. 3.4e that the force F_{12} pulls on joint 1. In turn, joint 1 exerts a tensile force on bar 12. Hence, bar 12 is in tension. A similar reasoning would indicate that bar 13 is in compression.

An efficient way to carry out the calculations is to establish the equilibrium of the joints on the sketch of the truss, using the force components. For the present example this is done in Fig. 3.4f. The sketch of the truss with its joint loads is first drawn and the slopes of the bars are indicated on the sketch. For bar 12 the slope is 1/1.73. Considering the equilibrium of joint 1, we note that a vertical force of 5 kips is acting downward. The only other vertical force is the vertical component of F_{12}. Hence, this component must be 5 kips and must be acting upward. We indicate the magnitude and direction of this force on the sketch near joint 1. The horizontal component of F_{12} must thus act to the left. The magnitude of this component is in the same proportion to the vertical component as the horizontal and vertical ratios in the slope of the bar, that is, 1.73/1. This gives the horizontal component as $1.73 \times 5 = 8.65$ kips.

The horizontal forces at joint 1 consist of F_{13} and the horizontal component of F_{12}. Therefore, these two forces must balance each other. Thus, F_{13} must have the same magnitude as the horizontal component of F_{12}, that is, 8.65 kips, and must be acting in the opposite direction—that is, to the right. The joint force components exerted by the second end of a bar are equal to the components of the first end but act in the opposite direction. Forces in the inclined bars can be found from the components or by the proportions of the slopes. The direction of the bar forces can be found without calculating the force magnitudes. Once the directions of the joint forces are indicated, we can determine whether the bars are in tension or compression. The bar forces are the opposite of the joint forces denoted by the arrows. Thus, when the arrows at the ends of a bar indicate a pull the bar is in compression, and when they indicate a compressed bar the bar is in tension. This is because the bar forces are the negative of the joint forces. Thus, in Fig. 3.3f, we see that bar 12 is in tension and bar 13 is in compression.

Example 3.2. Find the bar forces for the truss of Fig. 3.5a, under the applied loads.

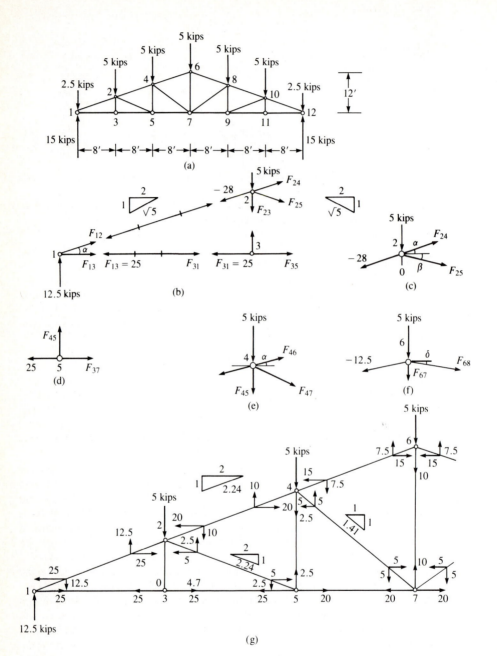

FIGURE 3.5

Solution. To check the determinacy of the truss, using Eq. (3.1) we have,

$$b = 21 = 2n - r = 2 \times 12 - 3 = 21$$

Therefore, the truss is statically determinate. To find the reactions, the first equilibrium equation, $\sum F_x = 0$, gives $R_{1x} = 0$. By symmetry or by setting the sum of the moments at the two supports equal to zero, we find

$$R_{1y} = R_{12y} = 15 \text{ kips}$$

Again using the joint method, we find the bar forces by satisfying the equilibrium of the joints. For the present example this is done in two different ways; first by writing the equilibrium equations for each joint, and then by satisfying the equilibrium of the joint forces directly on the sketch of the truss. The bars are denoted by their end joints (Fig. 3.5a).

Starting from a joint with only two unknown forces, joint 1 Fig. 3.56, we set the sum of the vertical forces equal to zero:

$$12.5 + F_{12} \sin \alpha = 0$$

which gives

$$F_{12} = -\frac{12.5}{\sin \alpha}$$

From Fig. 3.5a,

$$\sin \alpha = \frac{12}{\sqrt{12^2 + 24^2}} = 0.45 \qquad \cos \alpha = \frac{24}{\sqrt{12^2 + 24^2}} = 0.89$$

Thus,

$$F_{12} = -\frac{12.5}{0.45} = -28 \text{ kips}$$

The minus sign indicates that the direction of F_{12} is opposite to the assumed direction. Setting the sum of the horizontal forces equal to zero, we get

$$F_{12} \cos \alpha + F_{13} = 0 \qquad \text{or} \qquad -28 \times 0.89 + F_{13} = 0$$

which gives $F_{13} = 25$ kips.

Therefore, F_{13} acts in the assumed direction. Bar 13 applies a tension force F_{13} at the support. By the principle of action–reaction, the force in bar 13 (Fig. 3.5b) is acting in the opposite direction of the joint force F_{13}. That is, the support in turn applies a tension force to bar 13. In other words, bar 13 is in tension. F_{12}, on the other hand, exerts a compressive force at joint 1 and the joint applies a compressive force to bar 12. Therefore, bar 12 is in compression.

It is more convenient to calculate and denote the slopes of the members on the sketch of the truss and use them in calculating the force components. This provides a visual way of calculating the force components and reduces the chances of error. For example, the slope of bar 12 is 12/24 or 1/2. Thus, in the triangle denoting the slope the hypotenuse is $\sqrt{5}$. The equilibrium of the vertical forces at joint 1 can be written as

$$12.5 + \frac{1}{\sqrt{5}} F_{12} = 0 \qquad \text{which gives} \qquad F_{12} = -12.5\sqrt{5} = -28$$

We next consider another joint with two unknown forces. Since bar forces F_{12} and F_{13} are already determined, joint 3 has only two unknown forces which can be found from the two equilibrium equations. Figure 3.5b shows the force exerted at joint 1 by bar 13. By the principle of action–reaction the force at the left end of bar 13 is equal and opposite to joint force F_{13}. Equilibrium of bar 13 requires that the force at the right-hand end of this bar be also F_{13}, that is, the forces at the two ends of a bar must be equal and opposite, $F_{31} = F_{13}$. Again by the principle of action–reaction, the horizontal force applied on joint 3 by bar 13 must be the opposite of F_{31} acting at the right-hand end of that bar. Thus, the horizontal forces exerted on the joints at the two ends of a bar are equal and opposite. Once the joint force at one end of a bar is found, the opposite of that force must be applied at the joint on the other end of the bar. Therefore, we apply the opposite of F_{13}, 25 kips, at joint 3. The equilibrium equations with forces shown in Fig. 3.5b will yield the unknown values. From $\Sigma F_y = 0$ we find $F_{23} = 0$, and from $\Sigma F_x = 0$, $F_{35} = 25$ kips. Joint 2 now has only two unknown forces. Upon applying the opposite of F_{12} at this joint (F_{23} is zero), $\Sigma F_x = 0$ and $\Sigma F_y = 0$ give (Fig. 3.5c)

$$28\frac{2}{\sqrt{5}} + F_{24}\frac{2}{\sqrt{5}} + F_{25}\frac{2}{\sqrt{5}} = 0$$

and

$$28\frac{1}{\sqrt{5}} - 5 + F_{24}\frac{1}{\sqrt{5}} - F_{25}\frac{1}{\sqrt{5}} = 0$$

We thus get

$$28 + F_{24} + F_{25} = 0 \qquad \text{or} \qquad F_{24} + F_{25} = -28$$

and

$$28 - 5\sqrt{5} + F_{24} - F_{25} = 0 \qquad \text{or} \qquad F_{24} - F_{25} = -16.8$$

The solution of these two simultaneous equations gives

$$F_{24} = -22.4 \text{ kips} \qquad \text{and} \qquad F_{25} = -5.6 \text{ kips}$$

Applying the equilibrium equations to joints 5, 4 and 6, Figs. 3.5d, e, f, respectively, will result in all the bar forces for the left half of the truss. Since the truss and its applied loads are symmetric, the bar forces on the right half of the truss can be found by symmetry. Thus, $F_{12} = F_{10,12}$, $F_{13} = F_{11,12}$, etc. The forces in the truss bars are given below. A negative value indicates compression in the bar.

Bar	Force (kips)	Bar	Force (kips)
12	−28.00	45	2.50
13	25.00	46	−16.80
23	0	47	−7.00
24	−22.40	57	20.00
25	−5.60	67	10.00
35	25.00		

We now repeat the above calculations using the alternate presentation of the joint method by balancing the force components on the sketch of the truss. We first determine the slope of all the bars and indicate them on the sketch (Fig.

3.5g). Since joint 1 has only two unknown force components, we start the process there. At this joint there is a vertical force of 12.5 kips acting upward. The only other vertical force is the vertical component of F_{12}. The latter force must balance the 12.5-kip load. Thus, the magnitude of this vertical component must be 12.5 kips and it must be acting downward (opposite to the 12.5-kip force). We indicate the magnitude and direction of this force near joint 1 of the truss. The ratio of the horizontal component of F_{12} to its vertical component is the same as the horizontal to vertical values indicated for the slope of the bar, that is, 2/1. Therefore, the horizontal component of F_{12} is $(2/1)12.5 = 25$ kips. Note that once the direction of a force component is determined, the directions of the other component and the resultant are also known. The two components point away from each other. The magnitude of the bar force can be found either from the components or from a component and the slope. For bar 12 this is $(2.24/1)12.5 = 28$ kips. Inspection of the horizontal force components at joint 1 reveals that the force in bar 13 must be balancing the horizontal component of F_{12}. Therefore, F_{13} must be 25 kips acting to the right. Before proceeding with another joint we apply the opposite of the force components found to joints on the other ends of bars 12 and 13 (joints 2 and 3).

Moving on to joint 3 we observe that only bar (23) has a force in the vertical direction. Equilibrium of the vertical forces at this joint requires that this bar force be zero. The horizontal force F_{35} must thus be 25 kips to the right in order to balance the force applied to the joint by bar 13.

The next joint to be considered is 2. At this joint the unknown forces are F_{24} and F_{25}. Since each of these forces has a horizontal and vertical component, they can not be found directly on the sketch. These forces must be found by solving a system of two simultaneous equations. Since the directions of these forces are not readily known, we assume that they are positive (pulling on the joint). Thus, the horizontal forces at joint 2 consist of a 25-kip load acting to the right and the horizontal components of F_{24} and F_{25} also acting to the right. Each of these components will be expressed in terms of the unknown forces. For example, the horizontal and vertical components of F_{24} are, in view of the proportions of the slope of the bar, $(2/2.24)F_{24}$ and $(1/2.24)F_{24}$. Therefore, equilibrium of the horizontal and vertical forces at joint 2 gives

$$25 + \frac{2}{2.24} F_{24} + \frac{2}{2.24} F_{25} = 0 \qquad \text{or} \qquad F_{24} + F_{25} = -28$$

and

$$12.5 - 5 + \frac{1}{2.24} F_{24} - \frac{1}{2.24} F_{25} = 0 \qquad \text{or} \qquad F_{24} - F_{25} = -16.8$$

Note that with the unknown bar forces assumed as tension, the vertical component of F_{24} acts upward, while the vertical component of F_{25} acts downward. The solution of the above equations gives

$$F_{24} = -22.4 \text{ kips} \qquad \text{and} \qquad F_{25} = -5.6 \text{ kips}$$

Thus, the unknown forces are opposite to the assumed directions. The components of these forces are found by proportions of the slopes of the bars and indicated on the sketch of the truss. Proceeding in the same manner with other joints, we obtain the values shown in Fig. 3.5g.

3.4   METHOD OF SECTIONS

In this method a portion of the truss is isolated from the rest as a free body by passing a hypothetical section through the truss. The effect of the other part of the truss is considered by adding the forces that it applies on the first part. Since in the method of sections generally, not all the forces acting on the free body pass through the same point, three equilibrium equations, $\sum F_x = 0$, $\sum F_y = 0$, and $\sum M = 0$ are available. Therefore, three unknown bar forces can be determined. Note that moment equilibrium equations can be used instead of the two force equations. The points at which moments are taken are usually joints that some of the unknown forces pass through. These points may also be outside the free body diagram. As many sections can be taken as needed to find all the required bar forces. The slopes of the members can be used in calculating the force components.

The method of sections is particularly useful when only a small number of bar forces are needed. For example, when the same section is used for a number of bars, and it is possible to identify the location of the bar with the maximum force, then only that bar force must be calculated. The location of the maximum forces in the chords and diagonals of the truss can be established by comparison with beams. The compression and tension chord members act like the flanges of an I-beam carrying the bending moment caused by the loads. The diagonal members, on the other hand, act as the web of an I-beam, carrying the shear force. For single-span trusses under uniform gravity loads, the maximum forces in the chords are usually near the mid-span, while the maximum shear occurs near the supports. It should be noted that compressive forces, even though smaller than tensile forces, may be more critical because of the possibility of buckling.

Example 3.3. For the truss of Fig. 3.6a find the bar forces F_{46}, F_{56}, F_{57}.

Solution. In order to find the forces in these bars, an imaginary section is passed through the structure near the mid-span. This section, A–A, is chosen because it crosses the three bars in question. We note that two of the forces pass through point 6 and two through point 5. Thus, by setting the sum of the moments at 5 and 6 equal to zero, the forces F_{46} and F_{57} can be found directly. The right-hand side of the truss is hypothetically removed, and its effect is added through forces F_{46}, F_{56}, and F_{57}. Applying the three equilibrium equations to the free body of Fig. 3.6b will result in the three unknown forces. F_{46} can be found by setting the sum of moments at joint 5 equal to zero.

$$7.5 \times 20 - 5 \times 10 + F_{46} \times 8 = 0$$

or

$$F_{46} = -12.5 \text{ kips}$$

The moment equation, $\sum M_6 = 0$, gives F_{57},

$$7.5 \times 25 - 5 \times 15 - 5 \times 5 - F_{57} \times 8 = 0$$

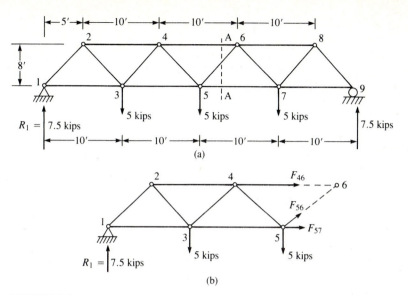

FIGURE 3.6

or

$$F_{57} = 10.9 \text{ kips}$$

The slope of the inclined members is 8/5 with a hypotenuse equal 9.43. Finally, $\Sigma F_x = 0$ results in

$$F_{56} \times \frac{5}{8.95} + F_{46} + F_{57} = 0$$

or

$$F_{56} \times 0.53 - 12.5 + 10.9 = 0$$

which gives $F_{56} = 3$ kips.

Example 3.4. For the truss of Fig. 3.7a, find the bar forces F_{24}, F_{25}, and F_{35}.

Solution. Again effecting the section A–A that crosses the three bars, the equilibrium equations can be applied to the free body diagram obtained in Fig. 3.7b. If we continue the lines of F_{24} and F_{35} to the left, they meet at point 0. Two of the unknown forces pass through point 0. Summing the moments of the forces acting on the free body at point 0 would result in F_{25}. The moment of F_{25} at 0 is calculated by using its components. Thus,

$$\frac{1}{\sqrt{2}} F_{25} \times 16 + \frac{1}{\sqrt{2}} F_{25} \times 4 + 200 \times 16 - 300 \times 12 = 0$$

or

$$F_{25} = 28.2 \text{ kN}.$$

Setting the sum of the vertical forces equal to zero gives

$$\frac{1}{\sqrt{17}} F_{24} - \frac{1}{\sqrt{2}} F_{25} - 200 + 300 = 0 \qquad \text{or} \qquad F_{24} = \frac{\sqrt{17}}{\sqrt{2}} F_{25} - \sqrt{17} \times 100$$

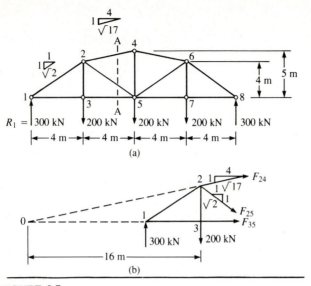

FIGURE 3.7

which results in $F_{24} = -330$ kN. Finally, setting the sum of the horizontal forces equal to zero results in

$$\frac{4}{\sqrt{17}} F_{24} + \frac{1}{\sqrt{2}} F_{25} + F_{35} = 0 \quad \text{or} \quad F_{35} = -\frac{4}{\sqrt{17}} F_{24} - \frac{1}{\sqrt{2}} F_{25}$$

or

$$F_{35} = \frac{4}{\sqrt{17}} \times 330 - \frac{1}{\sqrt{2}} 28.2 = 300 \text{ kN}$$

LABORATORY EXPERIMENTS

Stability

Unstable configurations not satisfying stability equations. Using strips of wood, build a model as in Fig. 3.2a and show that the configuration is not stable. Show that Eq. (3.1) indicates the instability. Then add an additional bar (Fig. 3.2b) and show that the resulting truss is stable. Again check Eq. (3.1).

Using additional strips, build a larger truss by adding sets of two bars and one joint, similar to Fig. 3.2c, and show that the resulting truss is stable. Check Eq. (3.1).

Now build a model similar to that of Fig. P3.7 (see later) and show that it is stable. Verify this also by Eq. (3.1).

Unstable configurations satisfying stability equation. Using strips of wood build a model similar to Fig. 3.3a and show that for this configuration as well as that of Fig. P3.14 the truss is unstable, even though Eq. (3.1) indicates otherwise.

Build models as in Figs. P3.1 and 3.12 and show that they are not stable. Check Eq. (3.1).

Bar Forces

A steel truss made of strips of approximately 1 ft length, 1 in width and $\frac{1}{16}$ in thickness with some strain gages on the members (Fig. L3.1) can be used for verifying the bar forces. Somewhat longer and thicker wooden strips as well as other shapes (Fig. L3.2) may also be used for this purpose. Since trusses carry large forces, smaller-sized members would be easier to use as long as they do not buckle under compression.

After recording the truss configuration, analyze the truss by using an assumed load at joint to be loaded during the experiment.

Set up the support condition, apply loads in increments, and measure and record the strain in the gage(s). The displacement may also be measured at one or more joints. From the measured strains and geometry, calculate the bar force(s) in the member(s) with the gage(s).

Verify that the bar forces vary linearly with load. That is, a doubled magnitude of load induces double the force.

Using the results of the analysis, scale them to find the bar forces for loads similar to the ones experimentally applied. Compare the measured and calculated values.

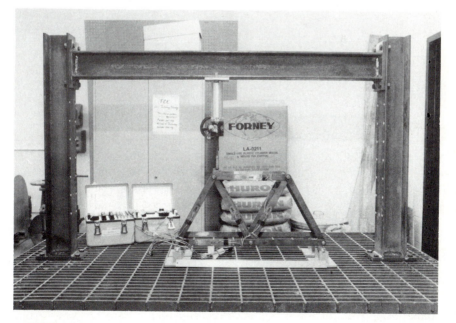

FIGURE L3.1

FIGURE L3.2

PROBLEMS

Classify the trusses of Figs. P3.1 to P3.15 for internal and external determinacy.

3.1.

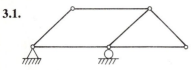

FIGURE P3.1

3.2.

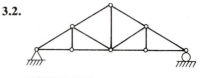

FIGURE P3.2

3.3.

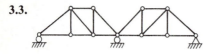

FIGURE P3.3

3.4.

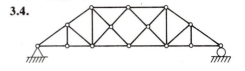

FIGURE P3.4

3.5.

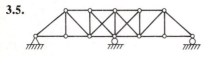

FIGURE P3.5

3.6.

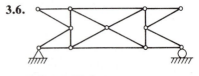

FIGURE P3.6

3.7.

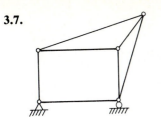

FIGURE P3.7

3.8.

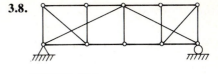

FIGURE P3.8

3.9.

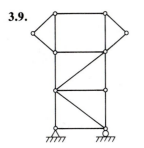

FIGURE P3.9

3.10.

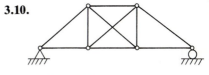

FIGURE P3.10

3.11.

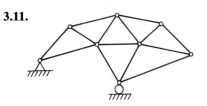

FIGURE P3.11

3.12.

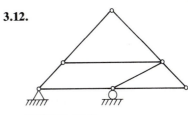

FIGURE P3.12

3.13.

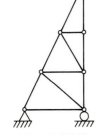

FIGURE P3.13

3.14.

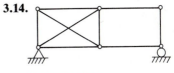

FIGURE P3.14

3.15

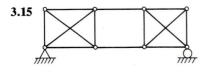

FIGURE P3.15

Find the bar forces for trusses of Figs. P3.16 to P3.27.

3.16.

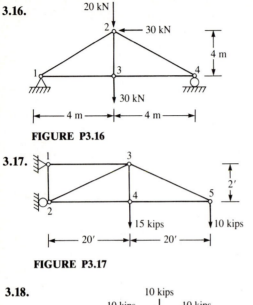

FIGURE P3.16

3.17.

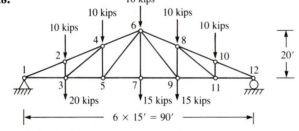

FIGURE P3.17

3.18.

FIGURE P3.18

3.19. Wind pressure = 60 psf

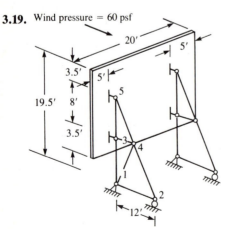

FIGURE P3.19

3.20.

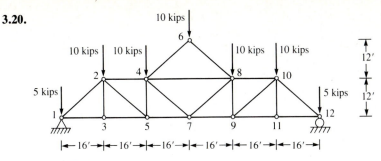

FIGURE P3.20

Hint: Consider the joints in this order: 1, 3, 2, 5, 6, 4, ..., and use symmetry.

3.21. Joints 1 to 8:

Hint: Consider joints 1, 2, 3. Then using the method of sections, with section A–A and taking moment at joint 8, find the force in bar 5–12.

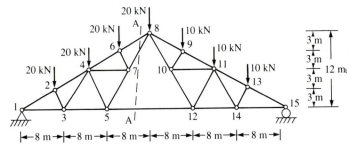

FIGURE P3.21

3.22.

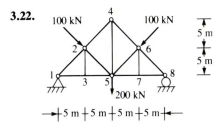

FIGURE P3.22

3.23.

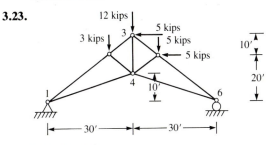

FIGURE P3.23

3.24.

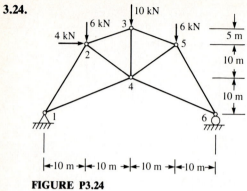

FIGURE P3.24

3.25.

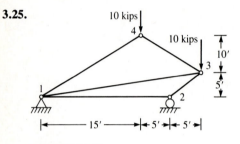

FIGURE P3.25

3.26.

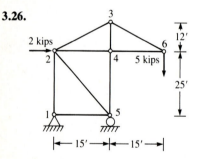

FIGURE P3.26

3.27.

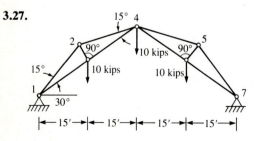

FIGURE P3.27

For the trusses of Figs. P3.28 to P3.40 find the bar forces indicated above each figure, using the method of sections.

3.28. Bars 12, 13, 46, 56, 57.

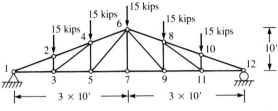

FIGURE P3.28

3.29. Bars 23, 78, 17.

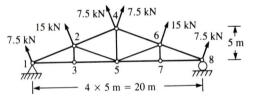

FIGURE P3.29

3.30. All bars.

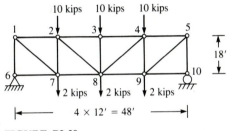

FIGURE P3.30

3.31. Bars 13, 14, 24.

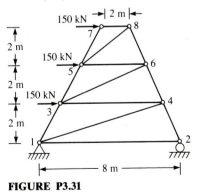

FIGURE P3.31

3.32. Bars 67, 6–10, 7–10

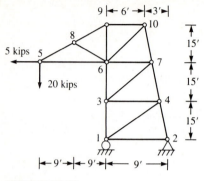

FIGURE P3.32

3.33. Bars 45, 57, 58, 68.

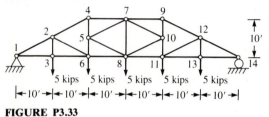

FIGURE P3.33

3.34. Bars 24, 34, 35

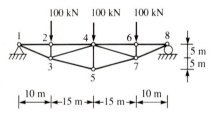

FIGURE P3.34

3.35. Bars 24, 14, 16.

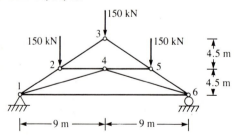

FIGURE P3.35

3.36. Bars 23, 24.

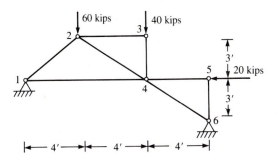

FIGURE P3.36

3.37. All bars.

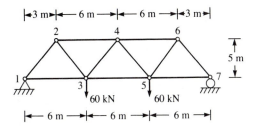

FIGURE P3.37

3.38. All bars.

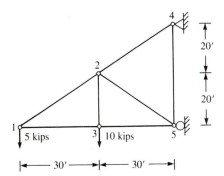

FIGURE P3.38

3.39. Bars 13, 14, 58.

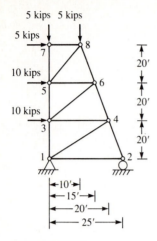

FIGURE P3.39

3.40. Bars 13, 23, 24.

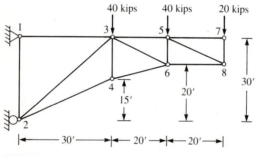

FIGURE P3.40

CHAPTER
4

SHEAR AND MOMENT IN BEAMS AND FRAMES

Beams are slender structural components, with a length several times larger than the cross-sectional dimensions. They carry their applied loads by developing axial and shear stresses (forces per unit area). The shape of the cross-section depends on the materials used. The cross-sections of concrete and timber beams are usually rectangular. The most common steel beams are wide-flange I-shaped beams.

Frames are structures composed of beam members with arbitrary orientations. The members are usually joined together by rigid connections. At a rigid connection the rotation of the connecting members is the same. Unlike truss members that undergo normal stresses, beams develop both normal and shear stresses. Normal stresses act parallel to the axis of the member, normal (perpendicular) to the cross-section, while shear stresses are perpendicular to the axis and in the plane of the cross-section. For design and analysis purposes, the stresses are represented by their resultant normal and shear forces and bending moments. Furthermore, the stresses caused by the applied axial loads in elastic beams are independent of bending. Thus, axial loads can be treated separately and their stresses can be superimposed on the bending stresses. Usually the stresses in beams and frames due to normal forces are negligible compared to bending stresses. On the other hand, if large compressive normal forces are acting on the members, they may cause buckling of the members or the system. This topic is discussed in Chapter 16 under beam–columns. Most beams and frames deflect primarily in one plane and are thus two-dimensional structures. If significant bending occurs in more than one plane, then we have a three-dimensional (space) beam or frame. Three-dimensional beams and frames are discussed in Chapter 18.

In this chapter we develop the basic relationships for beams and present the statics of determinate beams and frames. As discussed in Chapter 2, if the number of unknown reactions is larger than that of equations of statics plus equations of condition, then the beam or frame is statically indeterminate. Statically indeterminate beams and frames require consideration of deflections. Chapters 7, 9, 10, 11 and 12 present the methods of analysis for statically indeterminate beams and frames.

The first step in the analysis of a beam or frame is usually to determine the support reactions. The next step is the development of shear and moment diagrams. If axial loads are present, normal force diagrams may also be developed. These diagrams are used for design purposes for determining the section sizes. In addition design may involve stipulation of deflection limits.

4.1 DESIGN SIGN CONVENTION

The sign convention discussed here is appropriate for simple structures and is especially suitable for design purposes. This convention is also known as the strength of materials sign convention. In the course of design it is necessary to identify the side of a beam whose fibers are in compression and the side with fibers in tension. For example, this would allow the designer to identify the tension side of a reinforced concrete beam, where the reinforcing bars must be placed. For complex structures or for general methods of analysis the design sign convention is not easy to use. Therefore, in Chapter 10 the analysis sign convention is introduced and will be used for general methods of analysis such as the moment distribution and stiffness methods. Conversion from one convention to another is a trivial task. It should be noted that sign conventions are somewhat arbitrary and chosen at the designer's convenience. They can be changed from problem to problem.

In the design sign convention (Fig. 4.1) the x coordinate is chosen along the axis of the beam, usually from left to right. Positive loads are assumed downward, that is, in the direction of gravity. Reactions are positive upward, opposite to the gravity direction. Positive horizontal reactions are assumed acting to the right. Positive normal forces will induce tension in the member. Shear forces are positive when they cause deformations such that the left side of a beam segment deflects upward relative to its right end. Forces opposite to the above would be negative. Moments are positive when they cause compression in the top fibers of the beam, and are negative when they cause tension in the top fibers.

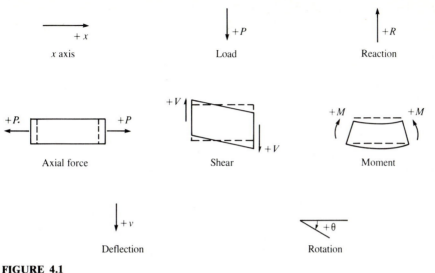

FIGURE 4.1
Positive values in the design sign convention.

In addition, downward deflections and clockwise rotations are positive. Fig. 4.1 shows the positive forces on a segment of a beam. Positive internal forces at a section are shown in Fig. 4.2b.

4.2 INTERNAL FORCES

The possible internal forces at a section of a beam are a normal force, a shear force, and a bending moment. When a structure is in equilibrium any portion thereof must also be in equilibrium under the external and internal forces. Thus, we can separate any portion of a structure with its external and internal forces as a free body. A free body of a beam is obtained by introducing one or more hypothetical cuts perpendicular to the axis of the beam. The internal forces are indicated on the free body diagram in the positive senses. The internal forces are then found by considering the equilibrium of one or more free bodies. A negative value of an internal force would indicate that the direction of the force is opposite to the assumed direction. If the equilibrium of a free body involves the support reactions, determination of support reactions should precede the analysis of the free body.

Consider, for example, the beam of Fig. 4.2a, with its applied loads and reactions. The internal forces at section 4, located at a distance x from the left end, are shown in Fig. 4.2b. These forces are indicated in the positive direction, so that the axial, shear, and bending deformations they induce in segments 1–4 and 4–2 are positive according to the sign convention of Fig. 4.1.

Since the external forces and reactions are in equilibrium, the internal forces on the left and right of a section, such as section 4, must balance each

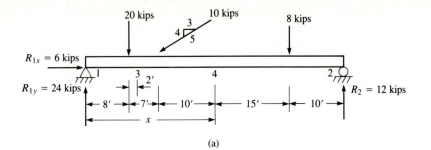

(a)

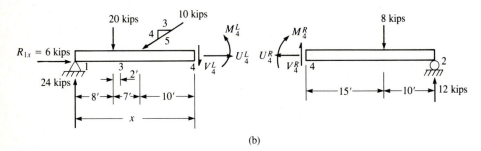

(b)

FIGURE 4.2

other in order for the equilibrium to remain enforced. This is known as the principle of action–reaction. Thus, U_4^L and U_4^R must be equal and opposite. Similarly, V_4^L must be equal and opposite to V_4^R and M_4^L equal and opposite to M_4^R. Therefore, the superscripts L and R may be dropped and the internal forces on both sides of a section may be denoted by the same symbols, e.g. U_4, V_4, and M_4, and shown in opposite directions.

To find the values of the internal forces, the left-hand or right-hand free body diagram in Fig. 4.2b may be used. The internal forces are found by considering the equilibrium of the free body, yielding the normal force, shear, and moment at section 4. Normal and shear forces as well as moments at other sections can be found in a similar manner. By considering the equilibrium of the free bodies in Fig. 4.2b, we can draw the following conclusions about the internal forces.

The normal force at a section is the sum of the axial forces acting on the left or right of that section. Shear at a section is the sum of the lateral forces acting on the left or right of the section. Moment at a section is the sum of the moments of all the forces on the left or right of the section taken at the section. The signs of the normal force, shear and moment at a section are determined from the deformation that they cause using the conventions of Fig. 4.1. Thus, a normal force causing tension at a section produces a positive internal normal force. A lateral force displacing the left end of a segment upward induces positive shear and a moment placing the top of a segment in compression is positive.

Example 4.1. Find the normal force, shear and moment at sections 3 and 4, 10 ft and 25 ft from the left end of the beam in Fig. 4.3a. Calculate the values from the left and right free bodies and verify that they give the same results.

Solution. To find the internal forces at section 3, a hypothetical cut is made at this section and the internal forces are indicated in the positive directions. The left-hand and right-hand segments of the beam are shown in Figs. 4.3b and 4.3c. Considering the left-hand segment of the beam (Fig. 4.3b) the equilibrium of the horizontal and vertical forces and that of moments give

$$\sum F_x = 0 \qquad\qquad 6 + U_3 = 0 \qquad U_3 = -6 \text{ kips}$$

$$\sum F_y = 0 \qquad\qquad 24 - 20 - V_3 = 0 \qquad V_3 = 4 \text{ kips}$$

$$\sum M_3 = 0 \qquad 24 \times 10 - 20 \times 2 - M_3 = 0 \qquad M_3 = 200 \text{ kip} \cdot \text{ft}$$

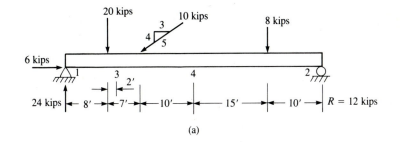

(a)

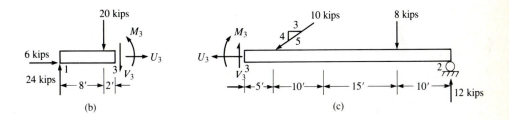

(b)   (c)

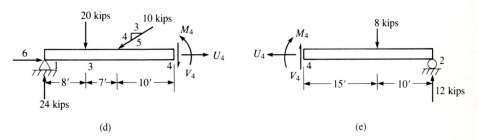

(d)   (e)

FIGURE 4.3

Calculating the forces at section 3 using the right segment (Fig. 4.3c) we have

$$\sum F_x = 0 \qquad U_3 + (3/5)10 = 0 \rightarrow U_3 = -6 \text{ kips}$$

$$\sum F_y = 0 \qquad V_3 - (4/5)10 - 8 + 12 = 0 \rightarrow V_3 = 4 \text{ kips}$$

$$\sum M_2 = 0 \qquad M_3 + V_3 \times 40 - (4/5)10 \times 35 - 8 \times 10 = 0$$

or

$$M_3 = -4 \times 40 + (4/5)10 \times 35 + 8 \times 10 = 200 \text{ kip} \cdot \text{ft}$$

To find the normal force, shear, and moment at section 4, we first consider the segment on the left of this section (Fig. 4.3d):

$$\sum F_x = 0 \qquad 6 - (3/5)10 + U_4 = 0 \rightarrow U_4 = 0 \text{ kips}$$

$$\sum F_y = 0 \qquad 24 - 20 - (4/5)10 - V_4 = 0 \rightarrow V_4 = -4 \text{ kips}$$

$$\sum M_4 = 0 \qquad 24 \times 25 - 20 \times 17 - (4/5)10 \times 10 - M_4 = 0, \ M_4 = 180 \text{ kip} \cdot \text{ft}$$

Calculating the forces from the right-hand segment (Fig. 4.3e), we have

$$\sum F_x = 0 \qquad\qquad\qquad U_4 = 0$$

$$\sum F_y = 0 \qquad\qquad V_4 - 8 + 12 = 0 \rightarrow V_4 = -4 \text{ kips}$$

$$\sum M_4 = 0 \qquad M_4 + V_4 \times 25 - 8 \times 10 = 0 \rightarrow M = 180 \text{ kip} \cdot \text{ft}$$

Thus, as expected, the normal force, shear, and moment calculated from the segments of the beam on the right and left of a section are the same.

It would be easy to check that the normal force, shear, and moment are the sum of the forces and their moments on the left (or right) of the section. For example, at section 4 the sum of axial forces on the left of the section is, $-6 + (3/5)10 = 0$; the sum of shear forces on the left of the section is, $24 - 20 - (4/5)10 = -4$; and sum of the moments of the forces on the left of the section taken at the section is, $24 \times 25 - 20 \times 17 - (4/5)10 \times 10 = 180$. These values are the same as before. Similarly, we can find these forces by summing up the values of forces on the right of the section. Since there are no axial forces on the right of section 4, the normal force at that section is zero. The shear force is $-12 + 8 = -4$, and moment is $12 \times 25 - 8 \times 15 = 180$. The signs of the normal force, shear, and moment are determined according to the sign convention of Fig. 4.1.

4.3 RELATION BETWEEN LOADS AND INTERNAL FORCES

To find the relation between the loads and internal forces and moments, we consider the free body of a beam segment with an infinitesimal length dx (Figs. 4.4b and 4.5b). As mentioned before, normal forces are independent of shear and moment. Here we discuss normal forces first.

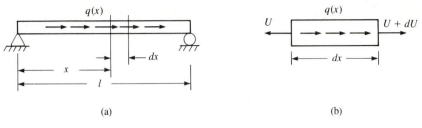

(a)   (b)

FIGURE 4.4

Consider the beam of Fig. 4.4a under a distributed axial load $q(x)$. An infinitesimal element of this beam is shown in Fig. 4.4b, with the internal forces at its two ends in the positive directions (producing tension).

Since the load on the infinitesimal length is small, the two end forces would differ only slightly. Thus, if the normal force on the left-hand section is U, on the right it would be $U + dU$. Equilibrium of the axial forces acting on the segment gives

$$-U + q(x)\, dx + U + dU = 0$$

or

$$\boxed{q(x) = -\frac{dU}{dx}}$$

(4.1)

That is, the derivative of the normal force is equal to the axial load; $q(x)$ may vary from section to section. However, in deriving the above equation we have assumed that the distributed axial load is constant over the length of the infinitesimal element. Integration of Eq. (4.1) gives,

$$U_2 - U_1 = -\int_{x_1}^{x_2} q(x)\, dx$$

(4.2)

Equation (4.2) indicates that the change of the normal force from point 1 to point 2 is equal to the negative of the sum of the axial forces between these two points, or the negative of the area under the axial load diagram between points 1 and 2. It should be mentioned that distributed axial loads are not common in structures. However, concentrated axial loads may be present. In truss members, such forces appear without lateral loads and moments, as we have seen in Chapter 3.

To find the relationship between the lateral loads, shear, and moment, we proceed in a similar manner. We separate a segment dx, located at x, from the left end of the beam (Fig. 4.5). The load $p(x)$ acting on the beam may vary throughout the span. However, over the infinitesimal segment dx the load can be assumed to remain constant. Shear and moment on the left of the segment are V and M. Because of the applied load, the shear and moment on the right

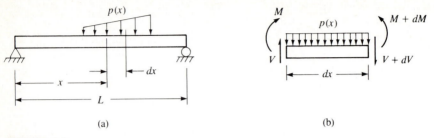

FIGURE 4.5
Free body diagram of a beam.

of the segment are slightly different. In Fig. 4.5b the load, shear forces, and moments on the segment are shown as positive.

Setting the sum of the vertical forces and the sum of moments with respect to the right-hand end of the segment equal to zero, we get

$$V - p(x)\,dx - (V + dV) = 0$$

and

$$M + V\,dx - p(x)\,(dx)^2/2 - (M + dM) = 0$$

Since dx is infinitesimal, the higher-order term, $p(x)\,(dx)^2/2$, is negligible and thus we find the following two relations:

$$p(x) = -\frac{dV}{dx} \qquad (4.3)$$

and

$$V = \frac{dM}{dx} \qquad (4.4)$$

Thus, the load is the negative of derivative of the shear, and shear is the derivative of the bending moment. Integration of Eqs. (4.3) and (4.4) from x_1 to x_2 gives

$$V_2 - V_1 = -\int_{x_1}^{x_2} p(x)\,dx \qquad (4.5)$$

and

$$M_2 - M_1 = \int_{x_1}^{x_2} V\,dx \qquad (4.6)$$

Equation (4.5) shows that the change in the shear force between points 1 and 2 is equal to the negative of the sum of the lateral forces between 1 and 2, or the negative of the area under the load diagram between these points.

Similarly, Eq. (4.6) indicates that the change in moment between points 1 and 2 is equal to the area under the shear diagram between these two points. Since shear and moment at the left-hand end of the beam are equal to the reactions of the left-hand support, if the integration is carried out from the left-hand end to an arbitrary section x of the beam, Eqs. (4.5) and (4.6) become

$$V = R_1 - \int_0^x p(x)\, dx \qquad (4.7)$$

and

$$M = M_1 + \int_0^x V\, dx \qquad (4.8)$$

From Eq. (4.7) we can see that shear at a section x is equal to the left reaction minus the area under the load diagram on the left of the section. Also, from Eq. (4.8) we see that the moment at a section x is equal to the moment at the left-hand support plus the area under the shear diagram to the left of the section. In calculating the moment, any concentrated moment on the left of the section must also be included. As mentioned earlier, shear and moment can also be calculated from the forces on the right of a section.

4.4 SHEAR AND MOMENT DIAGRAMS

Normal force, shear, and moment diagrams show the variation of the latter functions throughout a beam or frame. These diagrams are useful for design because they give the values at different sections of the beam or frame as well as the maximum values. In beams, and often in frames, axial loads are insignificant and therefore normal force diagrams are not as common as shear and moment diagrams.

Normal force, shear, and moment diagrams can be obtained either from their equations, or by calculating their values at key points. Equations of normal force, shear, and moment are obtained by calculating their values at an arbitrary section x along the member. Since frame members have different orientations, the x axis for each member will be different. For each member the axis would be the axis of the member itself, with a different origin, say the left-hand end of the member. The normal force and shear equations are obtained by summing all the normal or shear forces on the left of section x, including the support reactions. Similarly, the moment equation is found by calculating the moments of all the forces at the left of x, calculated at section x.

Normal force, shear and moment diagrams for frames are found in a similar manner, that is, by effecting a hypothetical cut at a section, indicating the internal forces (normal force, shear, moment), and finding the values of these internal forces by considering the equilibrium of the portion of the structure on one side of the section. This would indicate that normal force at a section is the sum of axial forces on the left or right of the section. Similarly, shear and moment at a section are the sum of the lateral forces and their moments on one side of the section. In calculating shear and moment in a frame member, the contribution of the forces acting on other members and the moment due to continuity of the joints must be considered. For example, normal and shear forces at the end of a vertical member transfer as shear and normal forces, respectively, to the connecting horizontal member.

Normal force and shear diagrams are constructed by calculating the normal and shear forces at different sections. These values are plotted normal to the axis of the members, with positive values plotted above the axis. Since for frames the axes of the members change directions, positive values may be plotted on either side of the members. However, in this case the signs must be indicated on the diagram. Usually positive moments are plotted on the tension side of the members. The following observations regarding normal-force, shear, and moment diagrams may be helpful in obtaining the correct plots of these actions.

For concentrated axial loads the normal-force diagram will consist of a number of steps. If axial forces act on the ends of members only, the normal-force diagram will be constant for each member.

Using Eqs. (4.3) and (4.4), which were derived with the sign convention adopted, or their counterparts Eqs. (4.5) and (4.6), we can make the following observations about shear and moment. These can be verified in the example problems. Reference to an example is made in each statement below.

If the load is zero over a segment, then shear is a constant there and moment varies linearly (left span in Fig. 4.9).

Under a uniform load shear varies linearly and moment as a parabola (right span of Fig. 4.9).

The change in shear is equal to the area under the load. Shear decreases for positive load and increases for negative load (right span of Fig. 4.9).

The change in moment is equal to the area under the shear diagram. Moment increases for positive shear and decreases for negative shear (left and right spans in Fig. 4.9).

Since derivative of moment is shear, moment is a maximum where shear is zero (Fig. 4.6c, d).

At a concentrated load, there is a jump in the shear diagram, and an abrupt change of slope in the moment diagram (Fig. 4.9).

At a concentrated moment, there is a jump in the moment diagram (Fig. 4.9).

On both sides of a frame connection, the moment is the same (Fig. 4.12f), unless there is a concentrated moment acting there.

The deflected shape of a segment indicates the sign of the moment. When the top of the beam is in compression the moment is positive, and when the top of the beam is in tension the moment is negative (Fig. 4.11).

Since bending is caused by moment, if the moment is zero over a segment of the beam then that segment remains straight (segment 56 in Fig. 4.11).

A sketch of the deflected shape of the beam or frame can be helpful in checking the moment diagram. Conversely, a moment diagram would be helpful in determining the deflected shape of a beam or frame.

Example 4.2. Draw the shear and moment diagrams for the simply supported beam of Fig. 4.6a under a uniformly distributed load over the span, and find the maximum values of shear and moment.

Solution. From the equilibrium of the free body diagram in Fig. 4.6b we have

$$\frac{pl}{2} - px - V = 0 \qquad \frac{pl}{2}x - \frac{px^2}{2} - M = 0$$

which yields the shear and moment equations

$$V = \frac{pl}{2} - px \qquad \text{and} \qquad M = \frac{plx}{2} - \frac{px^2}{2}$$

The plot of these equations are shown in Figs. 4.6c and 4.6d.

As we can see from Fig. 4.6c, the maximum shear occurs at the supports and is $pl/2$. To find the maximum moment, we set the derivative of the

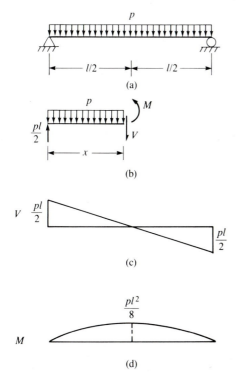

(a)

(b)

(c)

(d)   **FIGURE 4.6**

moment—that is, shear—equal to zero. The resulting equation gives x, the location of the maximum moment. Substitution of this value in the moment equation results in the maximum moment.

For this problem, $V = \frac{1}{2}pl - px = 0$, gives $x = l/2$. Thus, the maximum moment occurs at the mid-span. Its value is obtained by substituting $x = l/2$ in the moment equation, resulting in $M_{max} = pl^2/8$.

Example 4.3. Find the shear and moment equations for the simply supported beam of Fig. 4.7 with a concentrated load P acting at a from the left-hand end. Calculate the maximum values of shear and moment.

Solution. Different shear and moment equations must be written for the beam segments on the left and right of the load.

For $x \leq a$,

$$V = \frac{P(l - a)}{l} = P\left(1 - \frac{a}{l}\right)$$

$$M = \frac{P(l - a)}{l}x = P\left(1 - \frac{a}{l}\right)x$$

and for $x \geq a$,

$$V = P\left(1 - \frac{a}{l}\right) - P = -P\frac{a}{l}$$

$$M = P\left(1 - \frac{a}{l}\right)x - P(x - a) = Pa\left(1 - \frac{x}{l}\right)$$

Plots of the above equations are shear and moment diagrams shown in Figs. 4.7b and 4.7c. Since the above equations are not continuous functions, we cannot find their maximum values by setting their derivatives equal to zero. However, maximum values can be found from the diagrams of Figs. 4.7b and 4.7c. From Fig. 4.7b the maximum shear is $V_{max} = (1 - a/l)P$ for $a < l/2$ and $V_{max} = (a/l)P$ for $a \geq l/2$. If the ratio a/l were given, the larger of these two values would be the

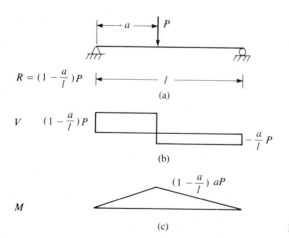

$$R = (1 - \frac{a}{l})P$$

(a)

$$V \quad (1 - \frac{a}{l})P$$

$$-\frac{a}{l}P$$

(b)

$$(1 - \frac{a}{l})aP$$

M

(c)

FIGURE 4.7

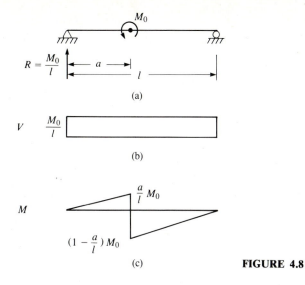

FIGURE 4.8

maximum shear. The maximum moment is $M_{max} = (1 - a/l)aP$ as shown in Fig. 4.7c. For the load acting at the mid-span of the beam, $M_{max} = Pl/4$ at $x = l/2$.

Example 4.4. Draw the shear and moment diagrams for the beam of Fig. 4.8a under a concentrated moment M_0 applied at distance a from the left-hand end of the beam.

Solution. For this problem, the shear force is $V = M_0/l$ throughout the beam. However, for the moment we must divide the beam into two parts.
    For $x \leq a$,

$$M = \frac{M_0}{l}x$$

and for $x \geq a$,

$$M = \frac{M_0}{l}x - M_0$$

The shear and moment diagrams for this beam are shown in Figs. 4.8b and 4.8c.

Example 4.5. Draw the shear and moment diagrams for the beam of Fig. 4.9a with the loads shown. An internal hinge connects the two parts of the beam at a distance 20 ft from the left-hand end.

Solution. The reactions are found from the three equations of statics plus the equation of condition for the hinge. The three equations of statics are

$$\sum F_x = 0 \quad \text{which gives} \quad R_{1x} = 0$$

$$\sum F_y = 0 \quad R_{1y} - 8 + R_{2y} - 1 \times 10 = 0 \quad \text{or} \quad R_{1y} + R_{2y} = 18$$

$$\sum M = 0$$

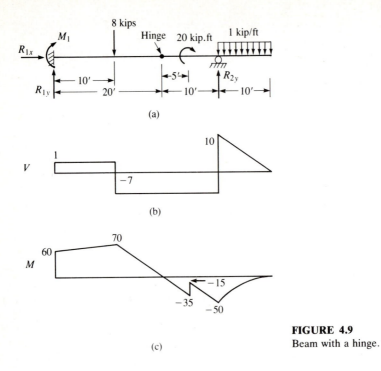

FIGURE 4.9
Beam with a hinge.

The sum of the moments of all the forces with respect to the left end of the beam gives

$$M_1 + 8 \times 10 + 20 - R_{2y} \times 30 + 1 \times 10 \times 35 = 0$$

The equation of condition is found by setting the sum of the moments of all the forces on the left of the hinge with respect to the hinge equal to zero. This gives

$$M_1 + 20 \times R_{1y} - 8 \times 10 = 0 \rightarrow M_1 + 20R_{1y} = 80$$

The solution of the above equations yields the following reactions:

$$R_{1y} = 1 \text{ kip} \qquad R_{2y} = 17 \text{ kips} \qquad M_1 = 60 \text{ kip} \cdot \text{ft}$$

The shear and moment diagrams are shown in Figs. 4.9b and 4.9c.

Example 4.6. Find the shear and moment diagrams for the beam of Fig. 4.10 under a linearly varying load and calculate the location and magnitude of the maximum moment.

Solution. The left reaction is found by setting the moment at the right-hand support equal to zero:

$$R_1 \times 30 - 1/2 \times 15 \times 30 \times 1/3 \times 30 = 0 \qquad \text{or} \qquad R_1 = 75 \text{ kN}$$

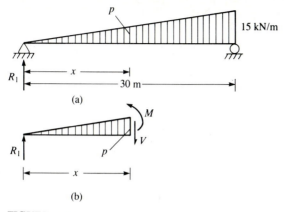

(a)

(b)

FIGURE 4.10

Referring to Fig. 4.10

$$p = \frac{x}{30} \times 15 = \frac{x}{2}$$

The shear force at x is

$$V = 75 - \frac{1}{2} \times \frac{x}{2} \times x = 75 - \frac{x^2}{4}$$

and the moment is

$$M = 75x - \frac{1}{2} \times \frac{x}{2} \times x \times \tfrac{1}{3} \times x = 75 - \frac{x^3}{12}$$

To find the maximum moment we set the shear equal to zero:

$$75 - \frac{x^2}{4} = 0$$

and find $x = 17.32$ m. Therefore,

$$M_{max} = 75(17.32) - \frac{(17.32)^3}{12} = 886 \text{ kN} \cdot \text{m}.$$

Example 4.7. Sketch the deflected shape of the beam of Fig. 4.11a and draw the qualitative moment diagram.

Solution. The deflected shape of segment 12 is similar to that of a cantilever beam. Since the left-hand end of the beam is fixed, the rotation at this point is zero. However, a significant change of slope takes place at the hinge. The loads at points 3 and 5 cause downward deflections. Because segment 12 provides a support at the hinge for the beam on the right of the hinge, a change of curvature occurs in segment 24. The exact location of this inflection point can only be determined by drawing the moment diagram for given values of loads and dimensions. No bending moment exists on segment 56 and therefore this segment rotates but remains straight.

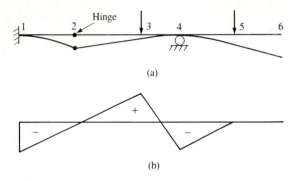

(a)

(b)

FIGURE 4.11

To sketch the moment diagram (Fig. 4.11b) we observe that there are only concentrated loads on the beam. Therefore, the moment diagram consists of straight lines. The top of the beam is in tension throughout segment 12 and thus the moment remains negative over that segment. It varies from the fixed-end moment at the left-hand end to zero at the hinge. Since there are no loads near the hinge, the moment diagram continues with the same slope to the right of the hinge. On the right of point 3 the slope changes because of the load at 3. Between points 3 and 4 (support) the moment diagram is a straight line, going from a positive value at 3 to a negative value at 4. The change of sign of the moment indicates an inflection point between 3 and 4. From point 4 to the next load at 5, the moment stays negative, but its magnitude decreases until it becomes zeros at 5. Beyond the last load, point 5, the moment is zero and the beam stays straight.

Example 4.8. Draw the normal force, shear, and moment diagrams for the frame of Fig. 4.12a.

Solution. The three reactions at the left support are shown in Fig. 4.12a. Since the frame is statically determinate, these reactions can be found by the equations of statics.

$$\sum F_x = 0 \qquad R_{1x} + 2 = 0 \qquad R_{1x} = -2 \text{ kips}$$

$$\sum F_y = 0 \qquad R_{1y} - 9 = 0 \qquad R_{1y} = 9 \text{ kips}$$

$$\sum M = 0 \text{ at point } 1 \rightarrow M_1 + 9 \times 10 + 2 \times 15 = 0 \qquad M_1 = -120 \text{ kip·ft}$$

To draw the normal force, shear, and moment diagrams for member 12 we choose the x axis from 1 to 2. The normal-force, shear, and moment equations for member 12 are found from equilibrium of the free body diagram in Fig. 4.12b. Thus,

$$9 + U = 0 \qquad \text{or} \qquad U = -9 \text{ kips}$$

$$2 - V = 0 \qquad \text{or} \qquad V = 2 \text{ kips}$$

$$-120 + 2x - M = 0 \qquad M = -120 + 2x$$

Thus, the normal and shear forces are constant in member 12 and the

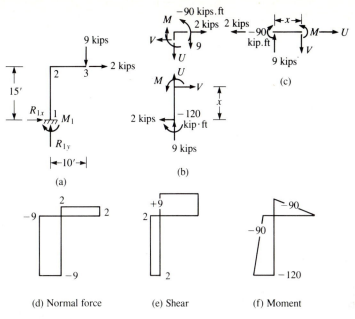

(a)

(b)

(c)

(d) Normal force   (e) Shear   (f) Moment

FIGURE 4.12

moment varies linearly. The moment diagram for this member can be determined from the values at two points. At point 1, $x = 0$ the moment is -120 kip·ft and at point 2, $x = 15$ ft, $M = -90$ kip · ft.

The axis for member 23 is taken from 2 to 3. Since the sum of axial forces on the right of any section in member 23 is 2, the normal force throughout that member is 2. To determine the shear forces in member 23, we note that the reaction 9 kip acting at point 1 transfers to 2, contributing to the shear force in member 23. Thus, the equation for shear in member 23 is, Fig. 4.12c,

$$9 - V = 0 \qquad \text{or} \qquad V = 9 \text{ kips}$$

Similarly, because of the continuity of the vertical and horizontal members at point 2, the moment at 2 in member 23 is the same as the moment at 2 in member 12, that is -90 kip · ft. The moment equation for member 23 is

$$-90 + 9x - M = 0 \qquad \text{or} \qquad M = -90 + 9x$$

With values -90 at $x = 0$ and 0 at $x = 10$ ft.

Note that, except for co-linear members, the normal and shear forces at the two sides of a connection are not the same, even in the absence of concentrated loads at the connection.

Example 4.9. Draw the normal force, shear, and moment diagrams for the frame of Fig. 4.13a.

Solution. There are four unknown reactions as shown in Fig. 4.13a. The three equations of statics plus one equation of condition, for the hinge, are sufficient for

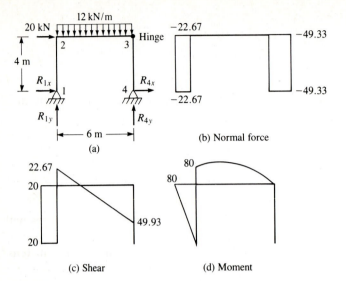

(a)

(b) Normal force

(c) Shear

(d) Moment

FIGURE 4.13

finding the reactions. The equation of condition is obtained by setting the sum of moments of the forces on part 123 or 34 at point 3 equal to zero. Thus,

$$\sum F_x = 0 \qquad R_{1x} + 20 + R_{4x} = 0$$

$$\sum F_y = 0 \qquad R_{1y} + R_{4y} - 12 \times 6 = 0$$

$$\sum M = 0, \text{ at 4 for the whole structure} \rightarrow R_{1y} \times 6 + 20 \times 4 - 12 \times \frac{6^2}{2} = 0$$

$$\sum M = 0, \text{ at 3 for part 34 (equation of condition)}, \qquad R_{4x} \times 4 = 0$$

gives

$$R_{4x} = 0$$

with

$$R_{1y} = 22.67 \text{ kN}$$
$$R_{4y} = 72 - R_{1y} = 72 - 22.67 = 49.33 \text{ kN}$$
$$R_{1x} = -20 - R_{4x} = -20 \text{ kN}$$

For member 12:

Normal force	$U = -22.67$ kN
Shear	$V = 20$ kN
Moment	$M = 20x$

For member 23:

$$U = 0$$
$$V = 22.67 - 12x$$
$$M = 80 + 22.67x - 6x^2$$

For member 34:

$$U = -49.33 \text{ kN}$$
$$V = 0$$
$$M = 0$$

The normal-force, shear, and moment diagrams are shown in Fig. 4.13b,c,d.

LABORATORY EXPERIMENTS

Simple Load Cell

A simple load cell can be used for measuring the internal axial forces with a strain indicator. It consists of a spring steel with a strain gage mounted on it and two holes to place the cell in line. This load cell can be used in tension areas and the load can be applied by a jack or turnbuckle.

This cell can be calibrated by putting it in line in a tension test and reading the strain indicator. Upon finding the best line fit for the load–voltage (strain indicator output) curve, the gage factor can be set equal to the slope of this line to allow the reading of the loads directly. The loads may be multiplied by a factor of 10 for more accuracy. In the latter case the strain indicator readings show the load times 10.

Internal Forces with a Load Cell

Using the above load cell, perform a deformation controlled tension test. That is, measure the force developed in a wire due to an imposed deformation.

Attach the load cell to one end of a piece of wire. Fasten one end of the wire to the testing table and the other one to the end of a loading jack (or turnbuckle) that is extended out. Connect the load cell to a strain indicator and set the gage factor equal to that of the load cell. Set the indicator to zero. After measuring the wire diameter (at several places), load the wire by tightening the jack (or the turnbuckle). Preload once to take the kinks out. Unload and set the indicator to zero again, then measure the deformation of a 1-ft wire segment using a ruler. Turn the jack (or the turnbuckle) slowly and record the data approximately every tenth turn. Do the reading as soon as you finish turning the jack to avoid inaccuracies due to creep deformations. Test the wire to failure.

Plot the stress–strain curve, determine the modulus of elasticity and the ultimate strength. Compare these values with the ones from the uniaxial test in Laboratory Experiments in Chapter 1.

Over what range of stress is the curve linear and what would you use for the allowable design stress of the wire if it were used as a hanger in a building?

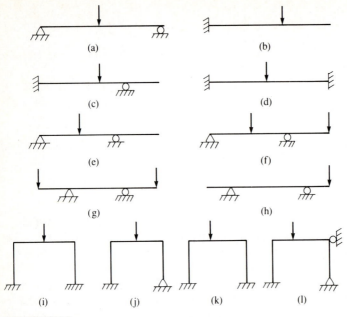

FIGURE L4.1

Deflected Shapes of Beams and Frames

Using beams made of spring steel, plastic, or wood strips, with spans of 1 to 2 ft (Fig. L4.1), apply loads as shown. Observe and sketch the deflected shapes. Indicate the sign of the moments on the diagrams and mark the approximate locations of the inflection points where moments are zero.

Draw the deflected shapes of the beams and frames of Fig. L4.2. Note that (principle of action–reaction) for the beams with hinges the shear forces at the two sides of the hinges are equal and opposite. Also note that the ends of the beams should be restrained from moving up at the supports. The author has designed a device, Fig. L10.1, that can be used for a variety of beams and frames and for demonstration of the moment distribution and stiffness methods.

Bending and Shear Failure of Concrete Beams

Bending and shear are two modes of failure. Another failure mode for concrete beams is loss of bond between the reinforcing bars and concrete (development length). Bending failure in concrete is more predictable than shear. Furthermore, bending failure initiated from the tensile steel bars is ductile, with slow progression and large deformations, while bending failure initiated from compressive concrete is brittle. In addition, bending cracks start in regions of high bending and progress perpendicularly to the axis of the beam (Fig. L4.3), whereas shear cracks start in regions of high shear usually near the supports and progress at an angle close to 45°.

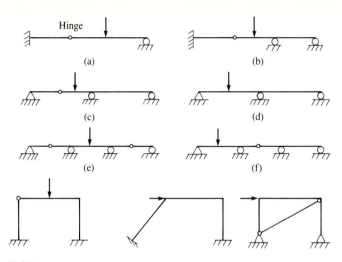

FIGURE L4.2

Fast cement (type 3) or Ultracal (a gypsum product with properties similar to those of concrete) may be used for making the concrete in order to reduce the hardening time. The specimens can thus be ready for testing after a day, as compared with 2–3 weeks for normal (type 1) cement. Beam specimens $1\frac{1}{2}$ in $\times \frac{3}{4}$ in $\times 12$ in may be poured in plexiglass or wood forms. Hanger wire or

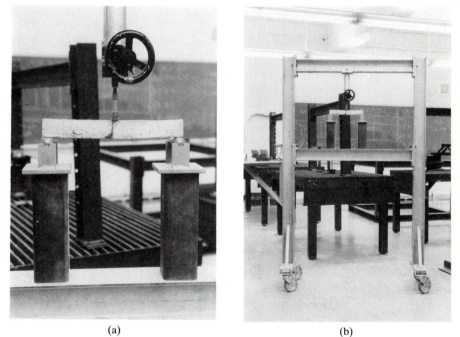

(a)   (b)

FIGURE L4.3

softer wires may be used for reinforcement. The wire surface may be roughened or rusted to avoid bond failure.

Test beams with different amounts of steel may be used to observe failure of over-reinforced (failure initiating in the compression side) and under-reinforced (failure initiating in the tension side), as well as plain concrete specimens. To produce shear failure, short spans loaded near the support may be used. Shear failure is sudden and occurs with noise.

PROBLEMS

Draw shear and moment diagrams for the following structures.

4.1.

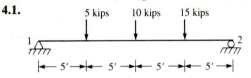

FIGURE P4.1

4.2.

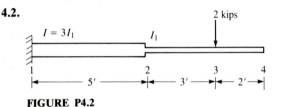

FIGURE P4.2

4.3.

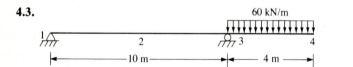

FIGURE P4.3

4.4.

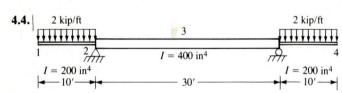

FIGURE P4.4

4.5.

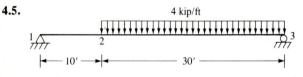

FIGURE P4.5

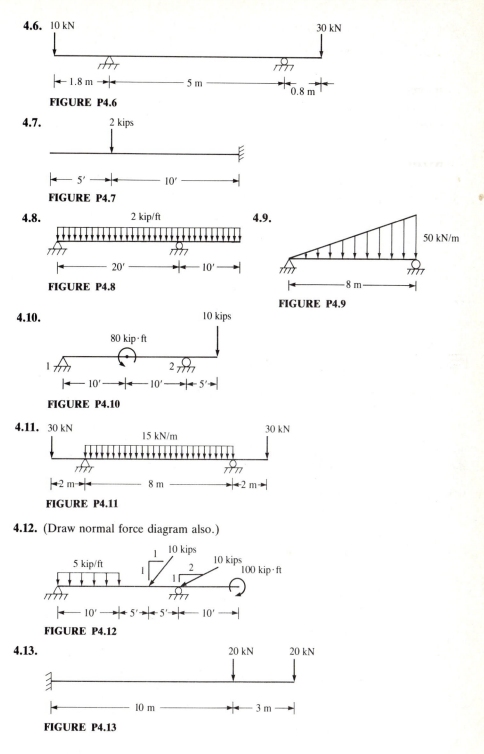

4.6.

FIGURE P4.6

4.7.

FIGURE P4.7

4.8.

FIGURE P4.8

4.9.

FIGURE P4.9

4.10.

FIGURE P4.10

4.11.

FIGURE P4.11

4.12. (Draw normal force diagram also.)

FIGURE P4.12

4.13.

FIGURE P4.13

4.14.

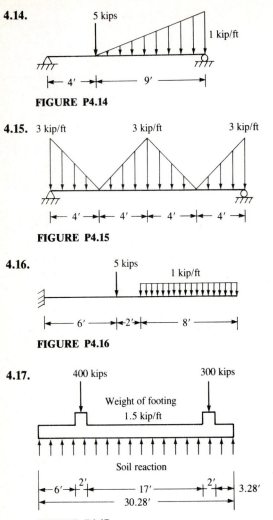

5 kips

1 kip/ft

|← 4' →|← 9' →|

FIGURE P4.14

4.15. 3 kip/ft   3 kip/ft   3 kip/ft

|← 4' →|← 4' →|← 4' →|← 4' →|

FIGURE P4.15

4.16.

5 kips

1 kip/ft

|← 6' →|←2'→|← 8' →|

FIGURE P4.16

4.17.

400 kips   300 kips

Weight of footing
1.5 kip/ft

Soil reaction

|←6'→|←2'→|← 17' →|←2'→|← 3.28'
|← 30.28' →|

FIGURE P4.17

Hint: Find the uniform reaction due to the concentrated loads and apply as upward load. The concentrated loads act as reactions to this uniform upward load. The weight of the footing counteracts its uniform reaction, and could be therefore left out.

4.18.

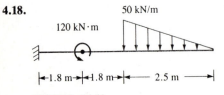

50 kN/m

120 kN·m

|←1.8 m→|←1.8 m→|← 2.5 m →|

FIGURE P4.18

4.19. (Also draw the normal force diagram.)

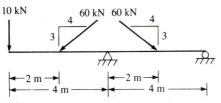

FIGURE P4.19

4.20.

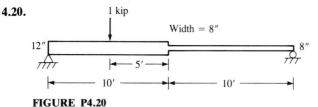

FIGURE P4.20

4.21.

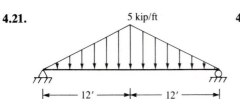

FIGURE P4.21

4.22.

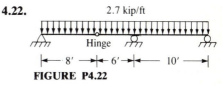

FIGURE P4.22

4.23.

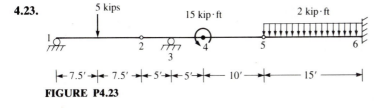

FIGURE P4.23

4.24.

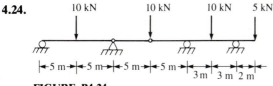

FIGURE P4.24

4.25.

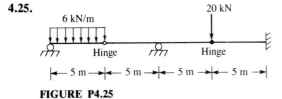

FIGURE P4.25

4.26. The values of shear at the left and right of points 2 are $V_2^- = -6.62$ kips and $V_2^+ = 6.94$ kips Find the reaction R_2.

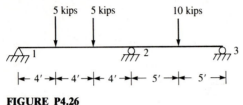

FIGURE P4.26

4.27. The moment at support 2 is $M_2 = -38.29$ kN $\cdot$ m.

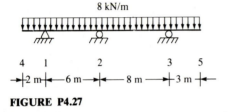

FIGURE P4.27

For the following beams find all the reactions and draw the shear and moment diagrams. In addition to the equations of statics use the given information.

4.28. Left reaction $R_1 = 15$ kips.

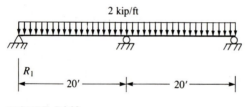

FIGURE P4.28

4.29. Reaction $R_2 = 13.75$ kN.

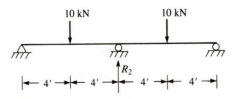

FIGURE P4.29

4.30. The fixed-end moment at 4, $M_4 = 32.2$ kip $\cdot$ ft; the reaction at 1, $R_1 = 7.6$ kips; and the reaction at 2, $R_2 = 24.2$ kips.

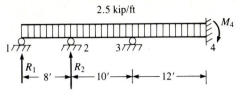

FIGURE P4.30

4.31. The fixed-end moments are $M_1 = -50$ kN · m and $M_2 = 50$ kN · m.

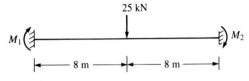

FIGURE P4.31

Find the normal force, shear and moment diagrams for the following frames.

4.32.

4.33.

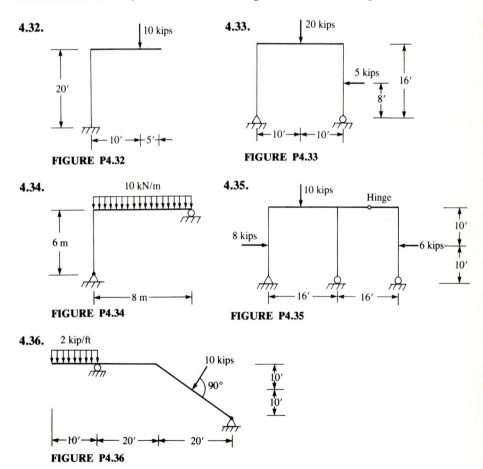

FIGURE P4.32

FIGURE P4.33

4.34.

4.35.

FIGURE P4.34

FIGURE P4.35

4.36.

FIGURE P4.36

CHAPTER

5

DEFLECTION

In this chapter we discuss the deformation of members under axial forces, followed by detailed presentation of bending deflections. The integration, moment–area, and conjugate beam methods are also presented. Energy methods and their application to deflection problems are left to Chapter 6 as well as shear deformation and deflection of trusses. The singularity method is covered in Chapter 9. It should be noted that if a single method is to be studied for the beam deflection then the virtual work method in conjunction with visual integration, Section 6.3, would be preferable since it applies easily to all deflection problems of beams and frames.

5.1 AXIAL DEFORMATION

Consider the bar of Fig. 5.1a under the axial forces shown. The normal force U at a section x can easily be calculated (Section 4.2). If at the left-hand end $U = U_1$, and the distributed axial load q is applied over the member, then the normal force is

$$U = -U_1 - \int_0^x q \, dx \tag{5.1}$$

Now consider the free body of Fig. 5.1b, a segment between x and

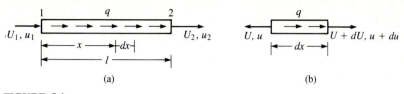

(a) (b)

FIGURE 5.1

$x + dx$. According to the design sign convention (Section 4.1), positive normal forces cause tension in the member. The force and displacement at the left of the infinitesimal segment are U and u and at the right of the segment $U + dU$ and $u + du$ (Fig. 5.1b).

The change in length of the segment is $u + du - u = du$, and its strain is

$$\varepsilon = \frac{du}{dx} \tag{5.2}$$

Assuming that the normal force over the length dx, $q\,dx$ is small compared to U, the stress–strain relation (Eq. (1.3)), $\sigma = E\varepsilon$, becomes

$$\frac{U}{A} = E\frac{du}{dx} \quad \text{or} \quad u' = \frac{U}{EA} \tag{5.3a, b}$$

$$du = \frac{U}{EA}dx \tag{5.3c}$$

where A is the area of the cross-section. Upon integration we can find the displacement at an arbitrary point:

$$u = \int_0^x \frac{U}{EA}dx \tag{5.4a}$$

In the absence of distributed axial loads ($q = 0$), $U = U_2 = -U_1$, and when the normal force and the cross-section are constant we get

$$u = \frac{U_2 x}{EA} + C_1$$

C_1 is calculated from the left boundary condition, i.e. at $x = 0$, $u = u_1$. Then the displacement at section x is

$$u = \frac{U_2 x}{EA} + u_1 \tag{5.4b}$$

At the right-hand end of a bar with length l, constant cross-section, and constant normal force U_2, the displacement u_2 is

$$u_2 = \frac{U_2 l}{EA} + u_1 \tag{5.4c}$$

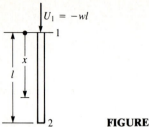

FIGURE 5.2

Example 5.1. Find the total elongation u_2 of a steel bar with length $l = 20$ ft and cross-sectional area $A = 2\,\text{in}^2$ (Fig. 5.2). The bar is hanging vertically under its own weight. The weight density of steel is $\gamma = 490\,\text{lb/ft}^3$ and its modulus of elasticity, $E = 29\,000$ ksi.

Solution. If the weight of the bar per unit length is w, at section x the tension force in the bar is equal to the weight of the segment below that section. This can be found by Eq. (5.1), as

$$U = wl - \int_0^x w\, dx = w(l - x)$$

By Eq. (5.4a),

$$u_2 = \int_0^l \frac{w(l - x)}{EA}\, dx = \frac{w}{EA} \int_0^l (l - x)\, dx$$

or

$$u_2 = \frac{wl^2}{2EA}$$

Since $Q = wl$ is the total weight of the bar, then

$$u_2 = \frac{Ql}{2EA}$$

For a bar with the same properties under an end load equal to the total weight Q, Eq. (5.4c) with $u_1 = 0$ and $U_2 = Q$ gives a displacement which is twice as large as that for the beam under its own weight.

For the given numerical values:

$$Q = wl = A\gamma l = \frac{2}{144} \times 490 \times 20 = 136.1\,\text{lb}$$

The end deflection is

$$u_2 = \frac{136.1 \times 20 \times 12}{2 \times 29 \times 10^6 \times 2} = 0.000014\,\text{in}$$

5.2 BENDING DEFORMATION

Under bending moment a beam deforms such that the end sections of a segment dx form an angle $d\theta$. As mentioned in Chapter 4, in the design sign convention used here positive moments cause compression in the upper fibres of the beam. Thus, under a positive moment the beam bends, putting the upper fibers in compression and the lower fibers in tension. Therefore, there is a plane where the fibers are under no stress. This is called the *neutral plane*. The intersection between the neutral plane and the longitudinal plane of bending is called the *neutral axis*. For symmetric sections the neutral axis is the same as the centroidal axis, which is called the *axis of the beam* or the *elastic curve*. The x axis is taken along the beam axis before deformation. The positive direction of the x coordinate is assumed from left to right and positive deflections are downward. The slope (or rotation) of the tangent to the beam axis is $\theta = dv/dx$ and is thus positive when clockwise. Under arbitrary loads the deformed shape of the beam axis is not circular. However, over an infinitesimal length the shape can be assumed circular. The radius of this circle is called the *radius of curvature*. If the radius of curvature of the beam axis at x is R, then for small deformations the curvature $1/R$ is small. The length of a fiber that was dx before deformation becomes ds after deformation takes place. For small deformations the slope θ is small and the arc ds can be assumed to have the same length as dx. Thus, referring to Fig. 5.3b, we can write $dx = ds = -R\,d\theta$. The negative sign is needed because for positive values of dx, R, θ_1 and θ_2 as shown in Fig. 5.3b, $d\theta = \theta_2 - \theta_1$ is negative. Thus,

$$\frac{1}{R} = -\frac{d\theta}{dx} \tag{5.5}$$

A fiber at distance y from the neutral axis (Fig. 5.3c), with length dx undergoes a strain to become dS, causing a strain

$$\varepsilon = \frac{dS - dx}{dx} \tag{5.6}$$

Since $dS = -(R + y)\,d\theta = dx - y\,d\theta$, substitution in Eq. (5.6) yields

$$\varepsilon = -y\frac{d\theta}{dx} \tag{5.7}$$

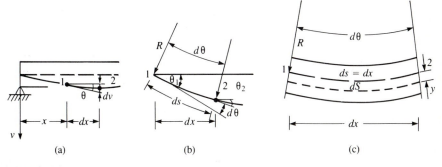

(a)   (b)   (c)

FIGURE 5.3

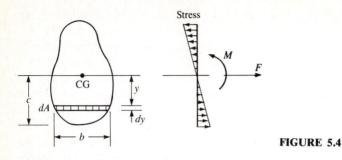

FIGURE 5.4

Because the fiber is under axial stress,

$$\sigma = \varepsilon E \quad \text{or} \quad \sigma = -Ey\frac{d\theta}{dx} \tag{5.8}$$

Equilibrium of the forces acting on the beam section (Fig. 5.4) gives

$$F = \int_A \sigma \, dA = 0 \tag{5.9}$$

and

$$M = \int_A \sigma y \, dA \tag{5.10}$$

For sections that are symmetric about the vertical axis, $dA = b \, dy$; b may vary with y, that is, $b = b(y)$. The first equilibrium equation gives the location of the neutral axis and the second one results in

$$M = \int \sigma by \, dy = -E\frac{d\theta}{dx} \int by^2 \, dy = -EI\frac{d\theta}{dx} \tag{5.11}$$

where

$$I = \int by^2 \, dy \tag{5.12}$$

is by definition the moment of inertia of the section.

According to our definition, and considering the axis of the beam (Fig. 5.5):

$$\boxed{\theta = \frac{dv}{dx} = v'} \tag{5.13}$$

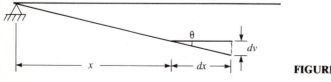

FIGURE 5.5

The derivative of Eq. (5.13) is

$$\frac{d\theta}{dx} = \frac{d^2v}{dx^2} = v'' \tag{5.14}$$

which in view of Eq. (5.5) gives

$$\frac{1}{R} = -v'' \tag{5.15}$$

The curvature is thus found to be the negative of change of slope or the negative of the second derivative of the deflection. If deformations are not small, then a more accurate form of curvature given in calculus texts must be used. Substitution from Eq. (5.14) into (5.11) gives $M = -EIv''$, or

$$\boxed{EIv'' = -M} \tag{5.16}$$

Now using Eqs. (4.3) and (4.4) in (5.16) we get

$$(EIv'')'' = p(x) \tag{5.17}$$

which for a constant cross-section becomes

$$EIv^{iv} = p(x) \tag{5.18}$$

Equation (5.16) (or its alternates (5.17), (5.18)) is used to find deflections and slopes of beams.

Bending Stress

If we find $d\theta/dx$ from Eq. (5.11) and substitute it into Eq. (5.8), we get

$$\sigma = \frac{My}{I} \tag{5.19a}$$

where M is the bending moment at the section, y is the distance of the point where stress is being calculated from the neutral axis, and I is the moment of inertia. The maximum stress for the extreme fiber of the section at a distance c from the neutral axis is

$$\sigma = \frac{Mc}{I} \tag{5.19b}$$

Tensile stresses are assumed positive.

Shear Stress

In the previous section we calculated the normal stress σ at a section due to bending moments. Since the bending moment varies from section to section, so does the stress. Considering an infinitesimal length of the beam, because of the

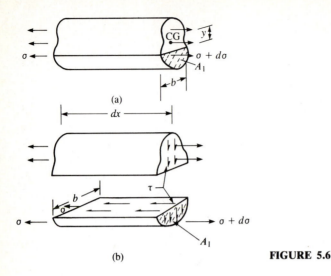

(a)

(b)

FIGURE 5.6

variation in the normal stress (Fig. 5.6a) equilibrium of a horizontal slice of this element requires that horizontal shear stresses develop on the horizontal planes of the beam.

Equilibrium of an infinitesimal cube from the element indicates that the same shear stress is also present on vertical sections (Fig. 5.6b). Equilibrium of the slice below distance y from the neutral axis, with width b and area of cross-section A_1 (the shaded area), gives

$$-\tau b\, dx + \int_{A_1} (\sigma + d\sigma)\, dA - \int_{A_1} \sigma\, dA = 0$$

$$-\tau b\, dx + \int_{A_1} d\sigma\, dA = 0$$

Since the change of stress at points with the same y is due to change of M, by Eq. (5.19a), $d\sigma = (y/I)\, dM$, therefore,

$$-\tau b\, dx + \int_{A_1} \left(\frac{y}{I}\right) dM\, dA = -\tau b\, dx + \frac{dM}{I} \int_{A_1} y\, dA = 0$$

which gives

$$\tau = \frac{dM}{dx} \frac{1}{Ib} \int_{A_1} y\, dA$$

Letting

$$Q = \int_{A_1} y\, dA \tag{5.20}$$

which is the first moment with respect to the neutral axis of the area of the section below y, where shear stress is being calculated. By Eq. (4.4),

$dM/dx = V$, and therefore the shear stress at a point of a section is

$$\tau = \frac{VQ}{Ib} \qquad (5.21)$$

In this equation, V is the shear force at the section, b is the width of the beam at the point where shear stress is being calculated, and I is the moment of inertia of the cross-section. Since the radius of gyration of a section is $r = \sqrt{I/A}$, where A is the area of the cross-section, we can write Eq. (5.21) as

$$\tau = \alpha \frac{V}{A} \qquad (5.22)$$

with

$$\alpha = \frac{Q}{r^2 b} \qquad (5.23)$$

where r is constant for a given section. However, Q varies with the point at which shear is being calculated and b may also vary with the location of the point.

For a rectangular beam, with the depth of the section equal to h, at a distance y from the neutral axis (Fig. 5.7),

$$Q = b\left(\frac{h}{2} - y\right) \times \frac{1}{2}\left(\frac{h}{2} + y\right) = \frac{b}{8}(h^2 - 4y^2)$$

$$I = \frac{bh^3}{12} \qquad A = bh \qquad r^2 = \frac{I}{A} = \frac{h^2}{12}$$

which gives (Eq. (5.23))

$$\alpha = \frac{3}{2}\left[1 - \left(\frac{2y}{h}\right)^2\right] \qquad (5.24)$$

and by Eq. (5.22),

$$\tau = \frac{3}{2}\frac{V}{bh}\left[1 - \left(\frac{2y}{h}\right)^2\right] \qquad (5.25)$$

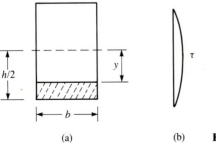

(a) (b) **FIGURE 5.7**

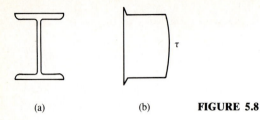

(a)   (b)   **FIGURE 5.8**

Thus, τ varies as a parabola (Fig. 5.7b) with a maximum value at the neutral axis and zero at the top and bottom fibers of the beam. Note that the maximum shear stress in this case is 50% larger than the average shear stress, found by dividing the shear force by the area of the cross-section.

For I and W sections the shear stress varies as shown in Fig. 5.8b. Thus, the shear carried by the flanges is negligible, and the variation of the shear stress in the web is small. Thus, for flanged sections it is assumed that all the shear is taken by the web. In the latter case it is common to take the shear area $A = b_w h$, with α equal to 1 and with b_w and h the width and depth of the web.

In the ultimate strength design of concrete beams, even for rectangular sections, uniform shear is assumed on the sections. This is because a section will not fail unless all its points have reached the ultimate stress.

5.3 BENDING DEFLECTION BY INTEGRATION

Once an equation is obtained for M, the differential equation (5.16) can be integrated to obtain expressions for slope and deflection. The constants of integration are found by applying the boundary conditions.

For a cantilever beam, the constants of integration will be zero, because the deflection and rotation are zero at $x = 0$.

> **Example 5.2.** Find the expressions for the slope and deflection of a cantilever beam (Fig. 5.9) under a uniform load. Calculate the value of the maximum deflection.
>
> **Solution.** The moment equation is
>
> $$M = M_0 + R_0 x - \frac{px^2}{2}$$

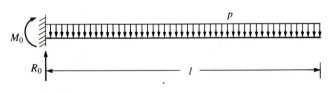

FIGURE 5.9

where the end reactions are $M_0 = -pl^2/2$ and $R_0 = pl$. The differential equation (5.16) becomes

$$EIv'' = -M_0 - R_0 x + \frac{px^2}{2}$$

Upon integration, expressions for the slope and deflection are found as

$$EIv' = -M_0 x - R_0 \frac{x^2}{2} + \frac{px^3}{6}$$

and

$$EIv = -\frac{M_0 x^2}{2} - \frac{R_0 x^3}{6} + \frac{px^4}{24}$$

Here the constants of integration are zero sine $v(0) = v'(0) = 0$.
Substitution for M_0 and R_0 results in

$$EIv' = \frac{pl^2}{2} x - \frac{pl}{2} x^2 + \frac{px^3}{6}$$

and

$$EIv = \frac{pl^2}{4} x^2 - \frac{pl}{6} x^3 + \frac{px^4}{24}$$

The maximum deflection occurs at the free end, $x = l$, and is

$$v_{max} = \frac{pl^4}{8EI}$$

Example 5.3. Find the expressions for the slope and deflection of the simply supported beam of Fig. 5.10 under a linearly varying load. Determine the maximum deflection. The span length is 30 ft and the maximum value of the load is 6 kip/ft.

Solution. The left-hand reaction is found by setting the moment of forces with respect to the right-hand end of the beam equal to zero:

$$R_1 \times 30 - \tfrac{1}{2} \times 6 \times 30 \times \frac{30}{3} = 0$$

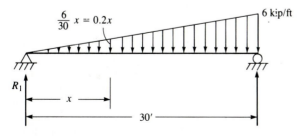

FIGURE 5.10

which gives $R_1 = 30$ kips. Then,

$$M = 30x - \tfrac{1}{2} \times 0.2x \times x \times \frac{x}{3} = 30x - \frac{x^3}{30}$$

and

$$EIv'' = -M = -30x + \frac{x^3}{30}$$

Upon integration,

$$EIv' = -15x^2 + \frac{x^4}{120} + C_1$$

and

$$EIv = -5x^3 + \frac{x^5}{600} + C_1 x + C_2$$

At $x = 0$ and $x = 30$ ft, $v = 0$, which gives

$$C_2 = 0$$

and

$$-5(30)^3 + \frac{(30)^5}{600} + 30C_1 = 0 \qquad \text{or} \qquad C_1 = 3\,150$$

Therefore, the expressions for the slope and deflection become

$$EIv' = -15x^2 + \frac{x^4}{120} + 3\,150$$

and

$$EIv = -5x^3 + \frac{x^5}{600} + 3\,150x$$

To find the maximum deflection, we must set the derivative of the deflection v (i.e. the slope) equal to zero. Thus,

$$-15x^2 + \frac{x^4}{120} + 3\,150 = 0 \qquad \text{or} \qquad x^4 - 1\,800x^2 + 378\,000 = 0$$

which gives

$$x^2 = \tfrac{1}{2}(1\,800 \pm \sqrt{1\,800^2 - 4 \times 378\,000}) = \tfrac{1}{2}(1\,800 \pm 1\,314.53)$$

or $x = 15.58$ and 39.46 ft.

The second value of x is larger than the span and therefore irrelevant. Thus, the maximum deflection occurs at $x = 15.58$ ft. Substitution of this value in the deflection equation gives

$$v_{max} = \frac{31\,697.8 \text{ kip} \cdot \text{ft}^3}{EI}$$

Example 5.4. Find the equations for the slope and deflection of a simply supported beam with a concentrated load P acting at a distance a from the left-hand end of the beam (Fig. 5.11). Calculate the values of the rotation at the left support and the deflection at the mid-span when the load is applied at the mid-span.

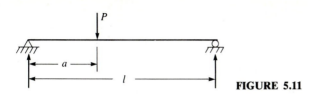

FIGURE 5.11

Solution. In the case of one or more concentrated loads, or partially distributed loads, as discussed under shear and moment in Section 4.4, there are more than one equations for moment. More specifically, there is one equation for each segment of the beam between two concentrated loads, and between a load and a reaction. The differential equation must be integrated separately for each segment of the beam between two forces. Because of the continuity of rotation and deflection, the rotations and deflections at the end of the segments must be the same whether they are calculated from the equation for the left or the right side of the load. The latter conditions plus the end boundary conditions provide a sufficient number of equations for determining all the unknown constants.

For a simply supported beam with a concentrated load, the equations for moment were found in Example 4.3. The differential equation (5.16), with $b = l - a$, becomes

$$EIv'' = -\frac{Pb}{l}x \qquad \text{for} \qquad x \le a$$

$$EIv'' = -\frac{Pb}{l}x + P(x - a) \qquad \text{for} \qquad x \ge a$$

Integration of these two equations will yield the slope and deflection of the beam. For $x \le a$,

$$EIv' = -\frac{Pb}{2l}x^2 + C_1$$

$$EIv = -\frac{Pb}{6l}x^3 + C_1x + C_2$$

and for $x \ge a$,

$$EIv' = -\frac{Pb}{2l}x^2 + \frac{Px^2}{2} - Pax + C_3$$

$$EIv = -\frac{Pb}{6l}x^3 + \frac{Px^3}{6} - \frac{Pax^2}{2} + C_3x + C_4$$

The constants of integration C_1 to C_4 are found by specifying that the deflections are zero at $x = 0$ and $x = l$. This results in

$$C_2 = 0 \qquad \text{and} \qquad -\frac{Pb}{6}l^2 + \frac{Pl^3}{6} - \frac{Pal^2}{2} + C_3l + C_4 = 0$$

Setting the values of deflection and rotation at $x = a$, calculated from the expressions for the left and right segments, equal gives

$$C_1 = -\frac{Pa^2}{2} + C_3$$

and

$$C_1 a = -\frac{Pa^3}{3} + C_3 a + C_4$$

These four equations give the values of the four unknown constants which upon substitution for b and simplification are

$$C_1 = \frac{Pb}{6l}(l^2 - b^2), \qquad C_2 = 0, \qquad C_3 = \frac{Pa}{6l}\left(l^2 - b^2 + \frac{3a^2 l}{b}\right) \qquad C_4 = -\frac{Pa^3}{3}$$

Substitution of these values into the expressions for slope and deflection yields for $x \leq a$,

$$EIv' = -\frac{Pb}{6l}[3x^2 - (l^2 - b^2)]$$

$$EIv = -\frac{Pb}{6l}[x^3 - (l^2 - b^2)x]$$

and for $x \geq a$,

$$EIv' = -\frac{Pb}{6l}\left[3x^2 - 3\frac{l}{b}(x - a) - (l^2 - b^2)\right]$$

$$EIv = -\frac{Pb}{6l}\left[x^3 - \frac{l}{b}(x - a)^3 - (l^2 - b^2)x\right]$$

From this example we can see that, for multiple loads, direct integration becomes tedious. This problem can be alleviated by using the principle of superposition and obtaining the solution as sum of the solutions for individual loads. The moment–area method discussed in the next section is a semi-graphical method, and avoids this problem. The singularity method, discussed in Chapter 9, allows us to write a single equation for the moment, irrespective of the number of loads, and carry out the integration of the beam differential equation in an efficient manner. The virtual work method, Section 6.3, also provides an efficient way to solve problems such as the one considered here, especially when it is used with visual integration, described under the same heading in Section 6.3.

5.4 MOMENT–AREA METHOD

The moment–area method is particularly suitable for cases where deflection or rotations are needed at a few points in the beam (which is usually the case in practice).

From Eq. 5.11 we can write

$$d\theta = -\frac{M}{EI}dx \tag{5.26}$$

Integrating both sides from point 1 to point 2, that is, from x_1 to x_2, we get

$$\int_{x_1}^{x_2} d\theta = \theta\big|_{x_1}^{x_2} = \int_{x_1}^{x_2} -\frac{M}{EI}dx$$

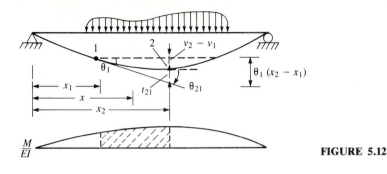

FIGURE 5.12

or

$$\theta_2 - \theta_1 = \theta_{21} = -\int_{x_1}^{x_2} \frac{M}{EI}\,dx \tag{5.27}$$

where θ_1 and θ_2 are the slopes of the elastic curve at points 1 and 2 (Fig. 5.12). θ_{21} denotes the angle between the tangents to the elastic curve at those points, or the change of slope from point 1 to 2. The integral on the right of Eq. (5.27) is the area under the moment diagram between those two points divided by EI. Note that by the area under the moment diagram we mean the area between the moment diagram and the axis on which the moment diagram is plotted. For negative moments this is actually the area above the moment diagram. The above equation is the statement of the first moment–area theorem and reads as follows.

First moment–area theorem

The change of slope from point 1 to point 2 on a beam is equal to the negative of the area under the M/EI diagram between those two points.

Now let us calculate the moment of the area under the M/EI diagram, between points 1 and 2, with respect to point 2. The right-hand side of Eq. (5.26) is the area of a differential element under the M/EI diagram at x. The product of this area by $(x_2 - x)$ and integration from x_1 to x_2 results in the moment of the area under the M/EI diagram between points 1 and 2 about point 2. Thus, multiplying both sides of Eq. (5.26) by $(x_2 - x)$ and integrating from x_1 to x_2, we get

$$\int_{x_1}^{x_2} (x_2 - x)\,d\theta = -\int_{x_1}^{x_2} (x_2 - x)\frac{M}{EI}\,dx$$

The left-hand side is

$$\int_{x_1}^{x_2} (x_2 - x)\,d\theta = x_2 \int_{x_1}^{x_2} d\theta - \int_{x_1}^{x_2} x\,d\theta$$

The first term on the right of the last equation gives $x_2(\theta_2 - \theta_1)$. The last integral can be integrated by parts to give

$$-\int_{x_1}^{x_2} x \, d\theta = -x\theta\big|_{x_1}^{x_2} + \int_{x_1}^{x_2} \theta \, dx = -(x_2\theta_2 - x_1\theta_1) + \int_{x_1}^{x_2} \theta \, dx$$

Since, according to Eq. (5.13), $\theta = dv/dx$, the last integral on the right of the last equation becomes

$$\int_{x_1}^{x_2} \theta \, dx = \int_{x_1}^{x_2} \frac{dv}{dx} \, dx = \int_{x_1}^{x_2} dv = v\big|_{x_1}^{x_2} = (v_2 - v_1)$$

Thus

$$\int_{x_1}^{x_2} (x_2 - x) \, d\theta = x_2(\theta_2 - \theta_1) - (x_2\theta_2 - x_1\theta_1) + (v_2 - v_1)$$

$$= -\theta_1(x_2 - x_1) + (v_2 - v_1)$$

Referring to Fig. 5.12, it is seen that $\theta_1(x_2 - x_1) - (v_2 - v_1)$ is t_{21}, the distance between point 2 and tangent of point 1. Thus, we get

$$t_{21} = \int_{x_1}^{x_2} (x_2 - x) \frac{M}{EI} \, dx \qquad (5.28)$$

Note that t_{21} is the distance along the normal to the axis of the beam before deflection takes place. Equation (5.28) reads as follows.

Second moment–area theorem

The tangent distance t_{21} between a point, 2, on the beam and the tangent of another point, 1, is equal to the moment of the M/EI diagram between points 1 and 2, with respect to point 2.

In calculating the deflection of beams with uniform or linearly varying loads, it is necessary to compute the areas under the moment diagrams, which are second- or third-degree parabolas. Table 5.1 gives the expressions for areas and centroidal distances of parabolas of different orders. More complex moment diagrams can be handled by dividing the areas into simpler shapes (see Example 5.10). The correct sign of deflection can be obtained if t_{21} is considered positive when point 2 is above the tangent of point 1, and negative when it is below the tangent. However, in general the calculations can be done ignoring the sign of t_{21} and referring to the deflected shape of the beam.

It should be noted that the moment–area theorems are not applicable to a beam segment containing a hinge. In the latter case the theorems are applied to the segments on the two sides of the hinge separately.

TABLE 5.1

Areas of parabolas and distances to centroids

	Area	Location of centroid $\bar{x}$
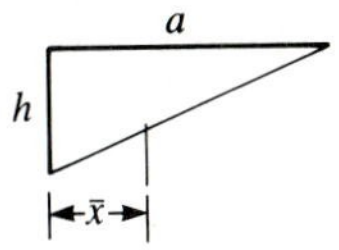	$\dfrac{ah}{2}$	$\dfrac{a}{3}$
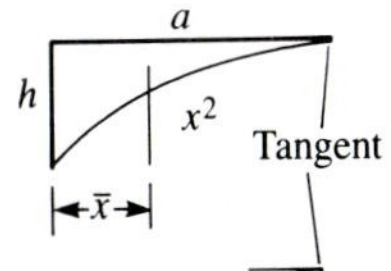 (Tangent, x^2)	$\dfrac{ah}{3}$	$\dfrac{a}{4}$
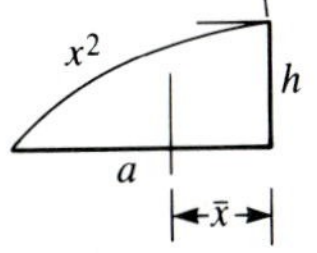 (x^2)	$\dfrac{2ah}{3}$	$\dfrac{3a}{8}$
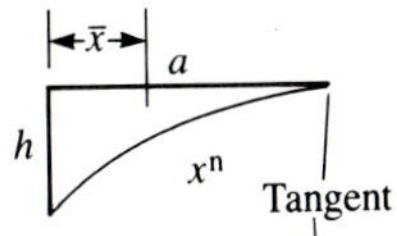 (x^n, Tangent)	$\dfrac{ah}{n+1}$	$\dfrac{a}{n+2}$
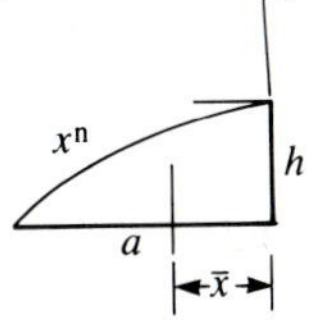 (x^n)	$\dfrac{nah}{n+1}$	$\dfrac{(n+1)a}{2(n+2)}$

Moment diagram for a uniform load p

$h = pl^2/8$ $h = pl^2/8$ or

	Area	Location of centroid
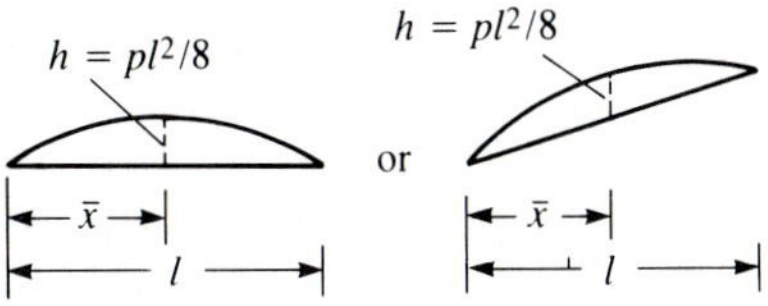	$\dfrac{pl^3}{12}$	$\dfrac{l}{2}$

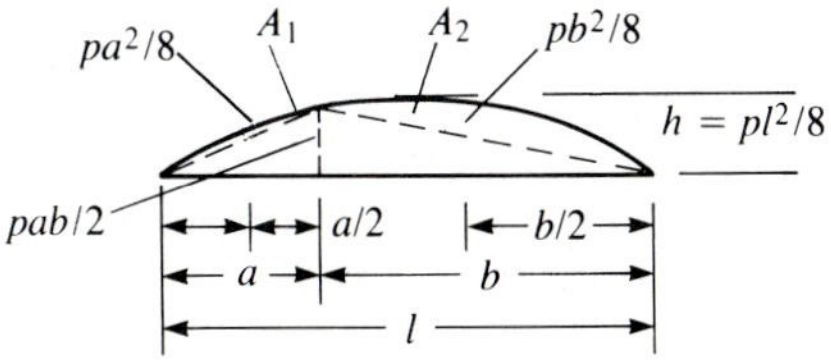

$pa^2/8$ A_1 A_2 $pb^2/8$ $h = pl^2/8$ $pab/2$ $a/2$ $b/2$ a b l

	Area	Location of centroid
	$A_1 = \dfrac{pa^3}{12}$	$\dfrac{a}{2}$
	$A_2 = \dfrac{pb^3}{12}$	$\dfrac{b}{2}$

A_1 A_2 $pb^2/8$ $pa^2/8$ $h = pl^2/8$ A_3 $\bar{x} = \dfrac{3l}{16}$

	Area	Location of centroid
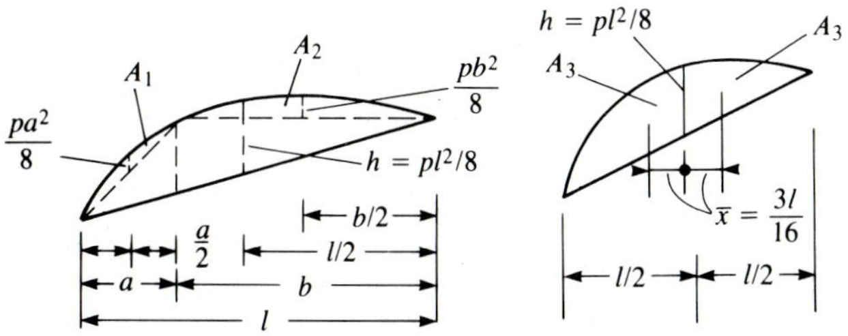	$A_3 = \dfrac{pl^3}{24}$	$\dfrac{3l}{16}$

125

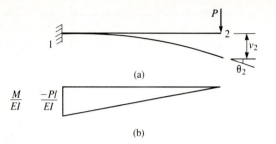

(a)

$\dfrac{M}{EI}$ $\dfrac{-Pl}{EI}$

(b)

FIGURE 5.13

Example 5.5. Find the slope and deflection of the cantilever beam with span l and load P applied at its free end (Fig. 5.13a).

Solution. The M/EI diagram is shown in Fig. 5.13b. At the fixed end the slope is zero and thus the tangent to the beam is horizontal. Thus, the angle θ_2 at 2, is equal to the change of angle between points 2 and 1, θ_{21}. Since $\theta_1 = 0$, $\theta_2 = \theta_1 + \theta_{21} = \theta_{21}$.

From the first moment–area theorem,

$$\theta_{21} = -\left(-\frac{Pl}{EI} \times \frac{l}{2}\right) = \frac{Pl^2}{2EI}$$

and by the second moment–area theorem (Eq. 5.28) with v_1 and θ_1 equal to zero,

$$v_2 = -t_{21} = -\left(-\frac{Pl}{EI} \times \frac{l}{2}\right) \times \tfrac{2}{3}l = \frac{Pl^3}{3EI}$$

Example 5.6. For the simply supported beam of Fig. 5.14a with a uniform load the M/EI diagram is shown in Fig. 5.14b. Find the rotation θ_1 of the left-hand end and the deflection v_3 of the mid-span.

Solution. The slope at the mid-span, point 3, is equal to zero and, by Eq. (5.27), $\theta_{31} = 0 - \theta_1$. Thus, the left end slope is equal to the angle between the tangents at the left end and the mid-span, and is thus equal to the left half of the area under

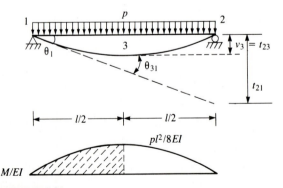

FIGURE 5.14

the M/EI diagram. Alternatively, the tangent distance, t_{21}, can be found, which yields the rotation upon division by the span length. The mid-span deflection is equal to t_{23}, the tangent distance between the right end and the mid-span. Thus, referring to Table 5.1,

$$\theta_1 = \theta_{31} = \frac{pl^2}{8EI} \times \frac{2}{3}\frac{l}{2} = \frac{pl^3}{24EI}$$

The mid-span deflection v_3, is equal to the moment of the area under the right half of the M/EI diagram with respect to the right end, that is,

$$v_3 = v_{max} = t_{23} = \frac{pl^2}{8EI} \times \frac{2}{3}\frac{l}{2} \times \frac{5}{8}\frac{l}{2} = \frac{5pl^4}{384EI}$$

Note that the distance of the centroid of the right half of the M/EI diagram from the right end is calculated by using $\bar{x}$ from Table 5.1.

v_3 can also be found by $v_3 = \frac{1}{2}t_{21} - t_{31}$. It is interesting to note that v_{max} for this problem is 9.6 times smaller than v_{max} for a cantilever beam (Example 5.2) with the same span length and load.

Example 5.7. Find the deflection of the variable-section beam of Fig. 5.15 under a concentrated load P at its mid-span. The moment of inertia of the left three-quarters of the span is $2I$, while the value for the right one-quarter is I. Find the numerical value of the mid-span deflection for $l = 40$ ft, $I = 200$ in^4, $P = 2$ kips and $E = 29\,000$ ksi.

Solution. Owing to lack of symmetry, the tangent to the elastic curve at the mid-span is not horizontal. The mid-span deflection is therefore found as

$$v_3 = \frac{1}{2}t_{21} - t_{31}$$

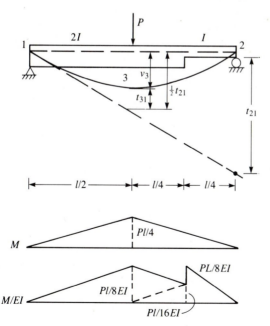

FIGURE 5.15

Referring to the M/EI diagram,

$$t_{21} = \frac{1}{EI}\left[\frac{1}{2} \times \frac{Pl}{8} \times \frac{l}{2}\left(\frac{l}{2} + \frac{1}{3}\frac{l}{2}\right) + \frac{1}{2}\frac{Pl}{8} \times \frac{l}{4}\left(\frac{l}{4} + \frac{2}{3}\frac{l}{4}\right)\right.$$

$$\left. + \frac{1}{2}\frac{Pl}{16} \times \frac{l}{4}\left(\frac{l}{4} + \frac{1}{3}\frac{l}{4}\right) + \frac{1}{2}\frac{Pl}{8} \times \frac{l}{4} \times \frac{2}{3}\frac{l}{4}\right]$$

or

$$t_{21} = \frac{25Pl^3}{768EI}$$

$$t_{31} = \frac{1}{EI}\left[\frac{1}{2} \times \frac{Pl}{8} \times \frac{l}{2} \times \frac{1}{3}\frac{l}{2}\right] = \frac{Pl^3}{192EI}$$

Thus

$$v_3 = \frac{1}{2}\frac{25Pl^3}{768EI} - \frac{Pl^3}{192EI} = \frac{17Pl^3}{1\,536EI}$$

The numerical value is

$$v_3 = \frac{17 \times 2 \times (40 \times 12)^3}{1\,536 \times 29\,000 \times 200} = 0.422 \text{ in}$$

Example 5.8. Calculate the maximum deflection for the 15-m span, simply supported beam of Fig. 5.16a under a concentrated load of 30 kN applied at 10 m from the left-hand end.

Solution. The maximum displacement occurs at the left of the load, say point 3, a distance x from the left-hand end. At this point the tangent is horizontal. Since $\theta_{31} = \theta_3 - \theta_1$, $\theta_3 = \theta_1 + \theta_{31} = 0$, we get $\theta_{31} = -\theta_1$.

Thus θ_{31}, the change of angle between the tangents of points 3 and 1, is equal $-\theta_1$. Referring to Fig. 5.16,

$$\theta_1 = \frac{t_{21}}{15}$$

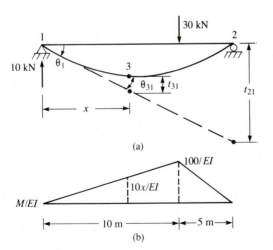

(a)

(b)

FIGURE 5.16

with

$$t_{21} = \frac{1}{2}\frac{100}{EI} \times 10\left(5 + \frac{10}{3}\right) + \frac{1}{2}\frac{100}{EI} \times 5 \times \frac{2}{3} \times 5 = \frac{5\,000\,\text{kN} \cdot \text{m}^3}{EI}$$

Thus,

$$\theta_1 = \frac{5\,000}{15EI} = \frac{333.3}{EI}$$

The angle between the tangents of points 1 and 3 (at x) is

$$\theta_{31} = -\frac{1}{2}\frac{10x}{EI}x = -\frac{5x^2}{EI}$$

setting $\theta_{31} = -\theta_1$, we get $-5x^2/EI = -333.3/EI$, which results in $x = 8.17\,\text{m}$. Referring to the M/EI diagram, at point 3,

$$\frac{M_3}{EI} = \frac{8.17}{10} \times \frac{100}{EI} = \frac{81.7}{EI}$$

Therefore, the maximum deflection is

$$v_3 = \frac{8.17}{15}t_{21} - t_{31}$$

With

$$t_{31} = \frac{1}{2}\frac{81.7}{EI} \times 8.17 \times \frac{1}{3} \times 8.17 = \frac{908.9}{EI}$$

we get

$$v_3 = \frac{8.17}{15} \times \frac{5\,000}{EI} - \frac{908.9}{EI} = \frac{1814.4\,\text{kN} \cdot \text{m}^3}{EI}$$

Example 5.9. For the beam of Fig. 5.17 calculate the deflection and rotation at points 3, the middle of the span, and 4, the end of the overhang. $I = 1\,200\,\text{in}^4$ and $E = 29\,000\,\text{ksi}$.

Solution. The reactions are $R_1 = 8.33\,\text{kips}$, $R_2 = 16.67\,\text{kips}$. Since $\theta_{31} = \theta_3 - \theta_1$, the slope at the mid-span can be found from

$$\theta_3 = \theta_1 + \theta_{31}$$

where (Fig. 5.17a)

$$\theta_1 = t_{21}/30.$$

To calculate t_{21}, referring to Fig. 5.17b, we can find the location of zero moment from similarity of the triangles:

$$\frac{37.43/EI}{x} = \frac{50/EI}{7.5 - x}$$

This gives $x = 3.21\,\text{ft}$. Then

$$t_{21} = \frac{1}{EI}[\tfrac{1}{2} \times 62.48 \times 7.5(22.5 + \tfrac{1}{3} \times 7.5) + \tfrac{1}{2} \times 62.48 \times 15(7.5 + \tfrac{2}{3} \times 15)$$

$$+ \tfrac{1}{2} \times 37.43 \times 15(7.5 + \tfrac{1}{3} \times 15) + \tfrac{1}{2} \times 37.43 \times 3.21(7.5 - \tfrac{1}{3} \times 3.21)$$

$$- \tfrac{1}{2} \times 50 \times 4.29 \times \tfrac{1}{3} \times 4.29]$$

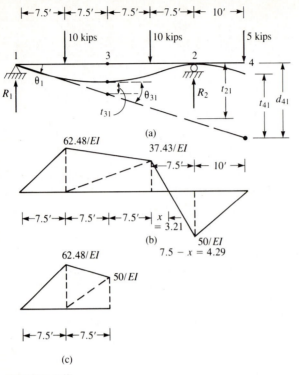

FIGURE 5.17

or

$$t_{21} = \frac{17\,798 \text{ kip} \cdot \text{ft}^3}{EI} = \frac{17\,798 \times 12^3}{29\,000 \times 1\,200} = 0.88 \text{ in}$$

Thus

$$\theta_1 = \frac{0.88}{30 \times 12} = 0.0024 \text{ rad}$$

Since (Fig. 5.17c), $M_3 = (62.48 + 37.43)/2 = 50$ kip · ft,

$$\theta_{31} = -\frac{1}{EI} [\tfrac{1}{2} \times 62.48 \times 7.5 + \tfrac{1}{2} \times 62.48 \times 7.5 + \tfrac{1}{2} \times 50 \times 7.5]$$

$$\theta_{31} = \frac{656 \text{ kip} \cdot \text{ft}^2}{EI} = \frac{656 \times 12^2}{29\,000 \times 1\,200} = 0.0027 \text{ rad}$$

and $\theta_3 = \theta_1 + \theta_{31} = 0.0024 - 0.0027 = -0.003$ rad.

The deflection v_3 at point 3 is $v_3 = \tfrac{1}{2}t_{21} - t_{31}$. Referring to Fig. 5.17c,

$$t_{31} = \frac{1}{EI} [\tfrac{1}{2} \times 62.48 \times 7.5(7.5 + \tfrac{1}{3} \times 7.5) + \tfrac{1}{2} \times 62.48 \times 7.5 \times \tfrac{2}{3} \times 7.5$$

$$+ \tfrac{1}{2} \times 50 \times 7.5 \times \tfrac{1}{3} \times 7.5]$$

or

$$t_{31} = \frac{3\,983 \text{ kip} \cdot \text{ft}^3}{EI} = \frac{3\,983 \times 12^3}{29\,000 \times 1\,200} = 0.198 \text{ in}$$

$$v_3 = \frac{0.88}{2} - 0.198 = 0.242 \text{ in}$$

Similarly, $\theta_4 = \theta_{41} + \theta_1$ with

$$\theta_{41} = -\frac{1}{EI} [\tfrac{1}{2} \times 62.48 \times 7.5 + \tfrac{1}{2} \times 62.48 \times 15 + \tfrac{1}{2} \times 37.43 \times 15$$

$$+ \tfrac{1}{2} \times 37.43 \times 3.21 - \tfrac{1}{2} \times 50 \times 4.29 - \tfrac{1}{2} \times 50 \times 10]$$

$$\theta_{41} = -\frac{687 \text{ kip} \cdot \text{ft}^2}{EI} = -\frac{687 \times 12^2}{29\,000 \times 1\,200} = -0.0028 \text{ rad}$$

Then $\theta_4 = 0.0024 - 0.0028 = -0.0004$ rad.

Also, $v_4 = d_{41} - t_{41} = (40/30)t_{21} - t_{41}$,

$$t_{41} = \frac{1}{EI} [\tfrac{1}{2} \times 62.48 \times 7.5(32.5 + \tfrac{1}{3} \times 7.5) + \tfrac{1}{2} \times 62.48 \times 15(17.5 + \tfrac{2}{3} \times 15)$$

$$+ \tfrac{1}{2} \times 37.43 \times 15(17.5 + \tfrac{1}{3} \times 15) + \tfrac{1}{2} \times 37.43 \times 3.21(14.29 + \tfrac{2}{3} \times 3.21)$$

$$- \tfrac{1}{2} \times 50 \times 4.29(10 + \tfrac{1}{3} \times 4.29) - \tfrac{1}{2} \times 50 \times 10 \times \tfrac{2}{3} \times 10]$$

$$t_{41} = \frac{25\,479 \text{ kip} \cdot \text{ft}^3}{EI} = \frac{25\,479 \times 12^3}{29\,000 \times 1\,200} = 1.27 \text{ in}$$

Therefore, $v_4 = \tfrac{4}{3} \times 0.88 - 1.27 = -0.097$ in.

Example 5.10. For the beam of Fig. 5.18 find the slope and deflection at the middle of the first span and at the end of the overhang.

Solution. The area of the M/EI diagram in the first span consists of a positive area and a negative area. To calculate the algebraic sum of these two areas, instead of determining the point of zero moment and finding each area separately, we can calculate the total area as the difference between the absolute values of the area of the parabola, whose base is the dashed line, and that of the triangle, whose hypotenuse is the dashed line. To show the validity of this assertion, let A_1 and A_2 be the absolute values of the positive and negative M/EI areas in the first span. The algebraic sum of these areas is thus $A_1 - A_2$. If the absolute value of the area under the parabola is A_p, that of the triangle is A_t, and that of the white area between the dashed line, the axis, and the parabola is A, then

$$A_p - A_t = (A_1 + A) - (A_2 + A) = A_1 - A_2$$

Similarly, the moment of the areas A_1 and A_2 with respect to any section can be found as the difference between the moment of the area under the parabola and the moment of the triangle with respect to that section. The latter can be easily verified for rectangular shapes.

Other arrangements for dividing up the areas are also possible as shown in

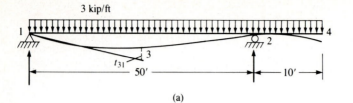

(a)

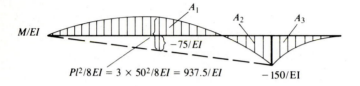

$$Pl^2/8EI = 3 \times 50^2/8EI = 937.5/EI$$

$-75/EI$

$-150/EI$

(b)

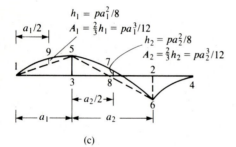

$$h_1 = pa_1^2/8$$
$$A_1 = \tfrac{2}{3}h_1 = pa_1^3/12$$
$$h_2 = pa_2^2/8$$
$$A_2 = \tfrac{2}{3}h_2 = pa_2^3/12$$

(c)

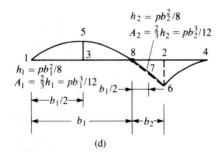

$$h_2 = pb_2^2/8$$
$$A_2 = \tfrac{2}{3}h_2 = pb_2^3/12$$
$$h_1 = pb_1^2/8$$
$$A_1 = \tfrac{2}{3}h_1 = pb_1^3/12$$

(d)

FIGURE 5.18

Figs. 5.18(c) and (d). In Fig. 5.18(c) the value of M/EI is needed at some point such as 5. The area over segment 23 is calculated as $A_{538} + A_{5\,765} - A_{826}$. In Fig. 5.18(d) the location of zero moment is needed which can be found by using the shear diagram or the moment equation. The area 82 678 above the parabola is found as the algebraic sum of the area of the triangle 826 and that of the parabola 8 768. That is $-A_{826} + A_{8\,768}$.

Using the first alternative, for the present problem,

$$t_{21} = [\tfrac{2}{3} \times 937.5 \times 50 \times (50/2) - \tfrac{1}{2} \times 150 \times 50 \times (50/3)]/EI = \frac{718\,750}{EI}$$

$$\theta_1 = \frac{t_{21}}{50} = \frac{14\,375}{EI}$$

$$\theta_{31} = -[\tfrac{2}{3} \times 937.5 \times 25 - \tfrac{1}{2} \times 75 \times 25]/EI = -\frac{14\,688}{EI}$$

$$\theta_3 = \theta_1 + \theta_{31} = -\frac{313}{EI}$$

$$t_{31} = [\tfrac{2}{3} \times 937.5 \times 25 \times \tfrac{3}{8} \times 25 - \tfrac{1}{2} \times 75 \times 25 \times (25/3)]/EI = \frac{138\,672}{EI}$$

$$v_3 = \frac{t_{21}}{2} - t_{31} = \frac{220\,703}{EI}$$

$$\theta_{41} = -[\tfrac{2}{3} \times 937.5 \times 50 - \tfrac{1}{2} \times 150 \times 50 - \tfrac{1}{3} \times 150 \times 10]/EI = -\frac{27\,000}{EI}$$

$$\theta_4 = \theta_1 + \theta_{41} = -\frac{12\,625}{EI}$$

$$t_{41} = \{\tfrac{2}{3} \times 937.5 \times 50 \times [(50/2) + 10] - \tfrac{1}{2} \times 150 \times 50 \times [(50/3) + 10]$$

$$- \tfrac{1}{3} \times 150 \times 10 \times \tfrac{3}{4} \times 10\}/EI = \frac{990\,000}{EI}$$

$$v_4 = \tfrac{6}{5} t_{21} - t_{41} = -\frac{127\,500}{EI}$$

Example 5.11. For the frame shown in Fig. 5.19, find the horizontal and vertical deflections, as well as the rotation of point 3.

Solution. The axial deformation is usually negligible in frames; therefore, we have two bending members, 12 and 23. Because of the continuity at the connection, point 2, at that point the rotation θ_2 at the ends of members 12 and 23 is the same. Referring to Fig. 5.19a, the rotation at end 3, θ_3, can be found as the sum of rotation of point 2 (θ_2, rotation of member 23 at 2) plus the rotation of point 3 relative to that at 2 (θ_{32}, rotation at the end of the cantilever 23). For small deflections, which was assumed in deriving the governing equation for beams, the horizontal deflection u_3 of point 3 is the same as the horizontal deflection u_2 of point 2. The vertical deflection of point 3, v_3, is equal to the vertical deflection d_{32} at 3 due to rotation of point 2 plus the vertical deflection v_{32} at end 3 of the cantilever 23.

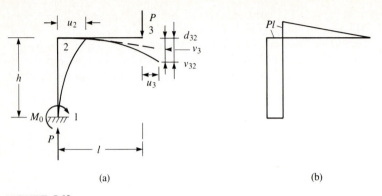

(a) (b)

FIGURE 5.19

The rotation at 2 is the change of slope with respect to the original axis 12. Neglecting the signs, by the first moment–area theorem applied to member 12 we get

$$\theta_2 = \frac{Pl \times h}{EI} = \frac{Plh}{EI}$$

and

$$\theta_{32} = \frac{Pl \times l}{2EI} = \frac{Pl^2}{2EI}$$

Thus,

$$\theta_3 = \theta_2 + \theta_{32} = \frac{Pl}{EI}\left(h + \frac{l}{2}\right)$$

The second moment–area theorem applied to member 12 gives,

$$u_2 = u_3 = \frac{Pl \times h \times h}{EI \times 2} = \frac{Plh^2}{2EI}$$

The relative deflection of point 3 from the tangent of point 2 is

$$v_{32} = \frac{Pl \times l}{2EI} \times \frac{2}{3}l = \frac{Pl^3}{3EI}$$

Therefore, the vertical deflection of point 3 is

$$v_3 = \theta_2 l + v_{32} = \frac{Pl^2}{EI}\left(h + \frac{l}{3}\right)$$

5.5 CONJUGATE BEAM METHOD

Consider a beam under an arbitrary loading and its M/EI diagram as shown in Fig. 5.20b. If in Eqs. (5.27) and (5.28) of the previous section we set the lower limits of the integrals equal to zero, that is, integrate the area under the M/EI diagram from the left end of the beam to an arbitrary point x_2, then Eq. (5.27)

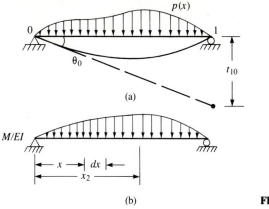

FIGURE 5.20

becomes,

$$\theta_2 - \theta_0 = -\int_0^{x_2} \frac{M}{EI}\,dx$$

or

$$\theta_2 = \theta_0 - \int_0^{x_2} \frac{M}{EI}\,dx \tag{5.29}$$

where θ_0 is the slope of the left end of the beam. This angle can be found as

$$\theta_0 = \frac{t_{10}}{l} = \frac{1}{l} \times \left(\begin{array}{l}\text{moment of the total area under the } M/EI \text{ diagram}\\ \text{with respect to point 1, the right-hand end}\end{array}\right)$$

Consider a conjugate beam, that is, a beam with the same length and in this case the same support conditions, but with a load diagram (elastic load) the same as the M/EI diagram (Fig. 5.20b). We see that θ_0 would be the left reaction of the latter beam, since this reaction is found by taking the moment of all the elastic loads with respect to the right-hand end of the beam and dividing it by the span of the beam. We further note that the right-hand side of Eq. (5.29) is the shear at an arbitrary point x_2 of the conjugate beam. This is true for any beam and thus the general statement can be made that:

> The shear force at a point of the conjugate beam is the slope of the actual beam at the same location.

Similarly, Eq. (5.28) and the expression just before it defining t_{21}, with $x_1 = 0$ and $v_1 = 0$, give

$$t_{20} = \theta_0 x_2 - v_2 = \int_0^{x_2} (x_2 - x) \frac{M}{EI}\,dx$$

or

$$v_2 = \theta_0 x_2 - \int_0^{x_2} (x_2 - x) \frac{M}{EI}\,dx \tag{5.30}$$

TABLE 5.2
Support conditions for conjugate beams

	Actual beam			Conugate beam	
Hinge		$\theta \neq 0$ $v = 0$			$V \neq 0$ $M = 0$
Roller		$\theta \neq 0$ $v = 0$			$V \neq 0$ $M = 0$
Fixed end		$\theta = 0$ $v = 0$			$V = 0$ $M = 0$
Free end		$\theta \neq 0$ $v \neq 0$			$V \neq 0$ $M \neq 0$
Interior hinge		$\theta \neq 0$ $v \neq 0$			$V \neq 0$ $M \neq 0$
Interior support		$\theta \neq 0$ $v = 0$			$V \neq 0$ $M = 0$

The right-hand side of the last equation is the left reaction of the conjugate beam times x_2 minus the moment of all the forces on the left of x_2 with respect to that section, that is, the moment in the conjugate beam. Thus,

> The moment at a point of the conjugate beam is the deflection of the actual beam at that point.

Since the shear and moment in the conjugate beam are the slope and deflection of the actual beam, for specified supports of the actual beam the supports shown in Table 5.2 must be used for conjugate beams.

Therefore, if an end of a beam is fixed (slope and deflection zero), the conjugate beam must have a free end at that point (shear and moment zero). Similarly, at an internal support the deflection of the beam is zero, but the rotation is not. A hinge in the conjugate beam would provide a zero moment in the latter and a nonzero shear.

Example 5.12. Find the left-end slope and the mid-span deflection of the simply supported beam of Fig. 5.21a. The span is 24 ft, with two 10-kip loads acting at 8 ft from the ends of the beam.

Solution. The conjugate beam for this problem is shown in Fig. 5.21b. The slope at the left end of the actual beam is the shear at that point in the conjugate beam, that is,

$$\theta_1 = R'_1 = \frac{1}{2} \times \frac{80}{EI} \times 8 + \frac{80}{EI} \times 4 = \frac{640 \text{ kip} \cdot \text{ft}^2}{EI}$$

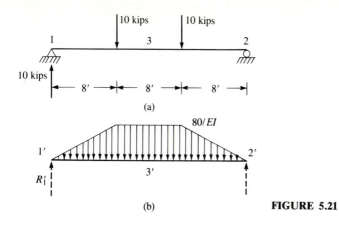

FIGURE 5.21

The mid-span deflection is equal to the moment at the mid-span of the conjugate beam, that is,

$$v_3 = M_3' = R_1' \times 12 - \frac{1}{2} \times \frac{80}{EI} \times 8\left(4 + \frac{8}{3}\right) - \frac{80}{EI} \times \frac{4^2}{2}$$

or

$$v_3 = \frac{4\,907 \text{ kip} \cdot \text{ft}^3}{EI}$$

Example 5.13. Find the deflection of the beam shown in Fig. 5.22a at $x = l/2$. The beam has a hinge at this point.

Solution. The conjugate beam for this problem is shown in Fig. 5.22b. The moment in this conjugate beam at $l/2$ is the deflection at that point in the actual beam. To find the left reaction of the conjugate beam we set the moment of the elastic loads (Fig. 5.22b) with respect to the right support of the conjugate beam equal to zero.

$$R_1' \times \frac{l}{2} - \frac{1}{2} \times \frac{Pl}{8} \times \frac{l}{2} \times \frac{l}{4} - \frac{1}{2} \times \frac{Pl}{4} \times \frac{l}{2} \times \frac{2}{3} \times \frac{l}{2} = 0$$

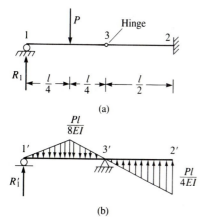

FIGURE 5.22

Thus, $R_1' = 11Pl^2/192EI$ which gives

$$v_3 = R_1' \times \frac{l}{2} - \frac{1}{2} \frac{Pl}{8EI} \times \frac{l}{2} \times \frac{l}{4} = \frac{Pl^3}{48EI}$$

Note that this problem can be solved in an easier way by observing that the load acting on the cantilever 32 at point 3 ia $P/2$, and that the portion 13 of the beam does not provide any resistance to deflection of point 3. Thus, the deflection of point 3 can be found by considering the cantilever 23 under a load $P/2$ at point 3, which would give the same result as obtained above.

LABORATORY EXPERIMENTS

Principle of Superposition

Select a structure (e.g., a steel strip with a strain gage installed at the mid-span used as a simply supported beam). Apply a load P_1 at point 1 and measure the strain in the gage and the deflection at a point 3 using a dial gage. Remove P_1 and apply another load P_2 at point 2 and measure the deflection in the gage and the deflection at point 3. Now apply P_1 and P_2 simultaneously and again measure the strain in the gage and the deflection at point 3. Verify that the sum of the strains measured for the individual loads is equal to the strain measured when both loads are applied. Do the same for the deflection.

Alternatively, moments can be calculated from strains and used to verify the principle of superposition.

Bending Strain–Curvature Relation

In linear elastic structures, strain is proportional to the load causing it. Furthermore, the structure returns to its initial position upon unloading. By Eq. (5.15) for small deformations the curvature at any point is

$$\frac{1}{R} = -v''$$

In view of Eq. (5.16),

$$\frac{1}{R} = \frac{M}{EI}$$

Consider an element of the beam under gravity loading with a length of one unit (Fig. L5.1). Upon deformation, the lengths of the bottom and top fibers become, $1 + \varepsilon_1$ and $1 + \varepsilon_2$, where ε_2 has a negative value. Thus,

$$1 + \varepsilon_1 = \left(R + \frac{h}{2}\right) d\theta \qquad R \, d\theta = 1 + \varepsilon_1 - \frac{h}{2} d\theta$$

and

$$1 + \varepsilon_2 = \left(R - \frac{h}{2}\right) d\theta \qquad \text{or} \qquad R \, d\theta = 1 + \varepsilon_2 + \frac{h}{2} d\theta$$

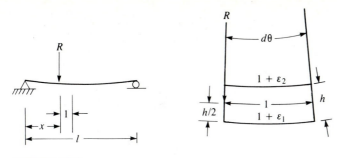

FIGURE L5.1

Then

$$1 + \varepsilon_1 - \frac{h}{2} d\theta = 1 + \varepsilon_2 + \frac{h}{2} d\theta$$

or

$$h \, d\theta = \varepsilon_1 - \varepsilon_2 \qquad \text{but from the above figure} \qquad d\theta = \frac{1}{R}$$

Therefore,

$$\frac{1}{R} = \frac{\varepsilon_1 - \varepsilon_2}{h}$$

Load a beam with strain gages on it and calculate the curvature.

Linear Elastic Beam Theory

Use a 1–2 ft long beam (Fig. L5.2) with one or more strain gages. Load the beam at a point, measure the strain in the gage and deflection at one or more points. Do this for a load at the mid-span and one at another point. Calculate the moment from the measured strain(s) and compare it with the moment found analytically. If only one gage is available at a section, the strain on the opposite side of the section may be assumed to be equal to the negative of the measured strain. Also compare the analytical values of deflections to the measured values.

Verify the linear relationship between the load and deflection at a point by measuring the deflection (with a dial gage) for increasing values of a load and checking that at each step the ratio load/deflection is a constant. Plot the load deflection curve. Continue to increase the load and find the point where the linear relation ceases to be valid.

Using a beam with sufficient depth (say 2 in with $\frac{1}{2}$ in width), with a few strain gages throughout the depth (on one side) and gages at the top and bottom, verify that the bending stress varies linearly through the depth of the section unless plastic (permanent) deformations occur.

FIGURE L5.2

FIGURE L5.3

Verification of the Frame Behavior

Using a one-story one-bay frame (Fig. L5.3), with strain gages at the two sides of the rigid connection. Apply a lateral load at the other connection and measure the strains. Also measure the lateral deflection at the connection. Show that the moments on both sides of the connection are the same, and compare the analytical and experimental moments and deflections.

NOTE. In running these tests, perform preliminary calculations to make sure that the material remains within the elastic range and that the deflections are not very large (more than 1% of the span), except when you want to verify the limit of applicability of the beam theory.

Make sure that the loading device is not in contact with the strain gage. If two gages are installed on the top and bottom at a section, rewire the two strain gages into a half-bridge configuration. Thus, the strains are added for those gages placed on the opposite sides of the bridge. Calibrate the gages by applying known loads and measuring the voltage. Set the gage factor so that the value read on the strain indicator is the moment when a load is applied at the gage. This can be used as a bending moment transducer.

PROBLEMS

Problems 5.1 to 5.5 are concerned with axial deformations.

5.1. Find the deflection at the top of a two-step square column (Fig. P5.1) under a load $P = 100$ kips. Assume there is no possibility of buckling. $E = 3\,100$ ksi.

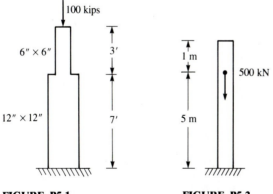

FIGURE P5.1   **FIGURE P5.2**

5.2. A circular bar with a constant cross-section ($A = 100$ cm^2) is under a load that acts 1 meter below the top (Fig. P5.2). Assuming that the bar will not buckle, find the vertical displacement of the top. $E = 80\,000$ MPa.

5.3. The steel bar of Fig. P5.3 is hanging from the ceiling. It has a square section with dimensions varying from 10.0 cm to 2.5 cm. Find the deflection at the bottom end

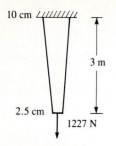

FIGURE P5.3

of this bar under its own weight and under an end force $P = 1\,277$ N (equal to its total weight).

5.4. Find the displacement of the ends of the vertical bars shown in Fig. P5.4. $E = 29\,000$ ksi.

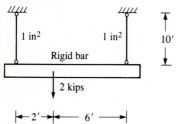

FIGURE P5.4

5.5. Find the displacement of end 3 of the rigid bar hinged at the wall and supported by a steel tie-bar at 3 (Fig. P5.5). The tie-bar has an area $A = 1\,\text{in}^2$. $E = 29\,000$ ksi.

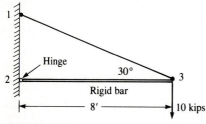

FIGURE P5.5

The following problems are concerned with bending deformations.

Do Problems 5.6 to 5.11 by the integration method.

5.6. Find the free end deflection and rotation in Fig. P5.6 in terms of P, l, E, and I.

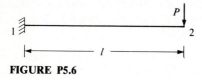

FIGURE P5.6

5.7. Determine the end rotation and maximum deflection of the simply supported beam in Fig. P5.7 with a span l under a distributed load p. Find the values for $p = 3$ kip/ft, $l = 25$ ft, $E = 29\,000$ ksi, and $I = 300$ in^4.

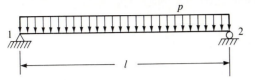

FIGURE P5.7

5.8. Determine the maximum deflection of the beam of Fig. P5.8 in terms of w_0, l, E, and I.

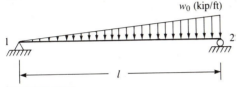

FIGURE P5.8

5.9. Find the free-end rotation and deflection at 6 m from the left-hand end of the beam in Fig. P5.9. $E = 200\,000$ MPa; $I = 5\,000$ cm^4.

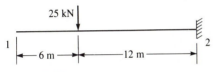

FIGURE P5.9

5.10. Find the free-end rotation and deflection of the cantilever beam in Fig. P5.10 under a concentrated moment M_0. Calculate the values for $M_0 = 300$ kip · ft, $l = 10$ ft and $I = 100$ in^4, $E = 29\,000$ ksi.

FIGURE P5.10

5.11. Find the rotations θ_1 and θ_2 at the left and right ends of the beam in Fig. P5.11 and determine the deflection at 2 in terms of EI.

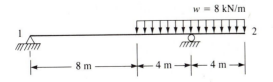

FIGURE P5.11

For Problems 5.12 to 5.20 use the moment–area method and calculate the values in terms of E and I.

5.12. Find the rotation and deflection of the free end in Fig. P5.12.

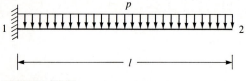

FIGURE P5.12

5.13. Find the left-end rotation and deflection at 10 ft from the left support in Fig. P5.13.

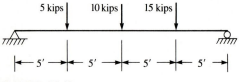

FIGURE P5.13

5.14. Find the rotation and deflection at the mid-span in Fig. P5.14.

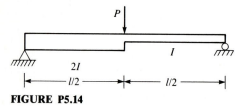

FIGURE P5.14

5.15. Find the deflection at the free end in Fig. P5.15.

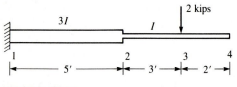

FIGURE P5.15

5.16. Find the deflections and rotations at points 2 (mid-span) and 4 in Fig. P5.16.

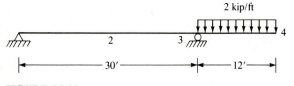

FIGURE P5.16

5.17. Find the deflections at the mid-span and at the end of the cantilever in Fig. P5.17.

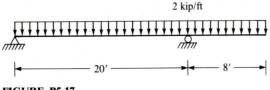

FIGURE P5.17

5.18. Find the deflections at the hinge and at the location of the load in Fig. P5.18.

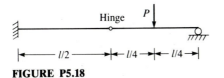

FIGURE P5.18

5.19. Find the deflections at points 1, 3, and 5 in Fig. P5.19.

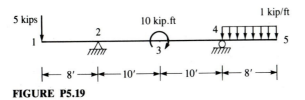

FIGURE P5.19

5.20. Find the deflections and rotations at points 3 (mid-span) and 5 in Fig. P5.20.

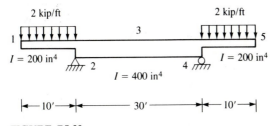

FIGURE P5.20

For Problems 5.21 to 5.28 find the solution of the beams by the conjugate beam method in terms of E and I.

5.21. Find the deflection and rotation at the mid-span in Fig. P5.21.

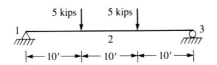

FIGURE P5.21

5.22. Find the deflections at points 2 and 4 in Fig. P5.22.

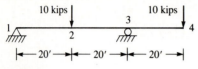

FIGURE P5.22

5.23. For Fig. P5.23, find the vertical deflection v_3 of point 3, (a) assuming that the steel cable is inextensible and (b) considering the extension of the cable. $I = 10\,000\ cm^2$, $E = 200\,000\ MPa$, and the area of the cable is $1\ cm^2$.

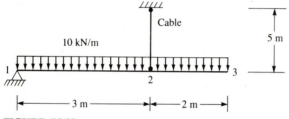

FIGURE P5.23

5.24. Find the deflection of point 2 in Fig. P5.24. $E = 29\,000$ ksi; $I = 200\ in^4$.

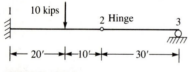

FIGURE P5.24

5.25. Find the vertical deflection of the hinge in Fig. P5.25. $I = 200\ in^4$; $E = 29\,000$ ksi.

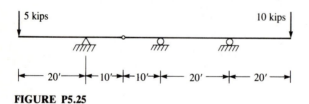

FIGURE P5.25

5.26. Find the rotation of the right end and the deflection at the concentrated load in Fig. P5.26.

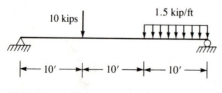

FIGURE P5.26

5.27. Find the deflection and rotation of point 3 in Fig. P5.27.

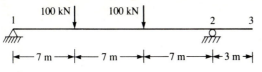

FIGURE P5.27

5.28. Find the deflection at the hinge in span 1 and at the middle of span 2 in Fig. P5.28.

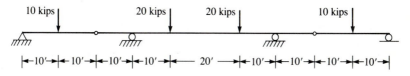

FIGURE P5.28

5.29. A cantilever beam of length $2l$ and flexural rigidity EI rests at its mid-length on a roller that is supported on a shorter cantilever also of flexural rigidity EI as shown in Fig. P5.29. If the cantilevers are initially unstressed, calculate the vertical deflection of the free end of the longer cantilever when a vertical load P is placed at that point. Also calculate the largest bending moments in each cantilever.

Hint: the two beams deflect by the same amount at the contact point. Denote the interactive force by R. Calculate the deflection of the contact point for each cantilever separately, in terms of R. Find R by setting the two deflections equal.

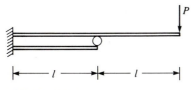

FIGURE P5.29

Find the deflection of the following frames in terms of EI. EI is the same for all members.

5.30. Calculate the vertical deflection of point 3 in Fig. P5.30.

5.31. Find the vertical and horizontal deflections of point 3 in Fig. P5.31.

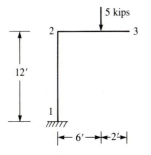

FIGURE P5.30

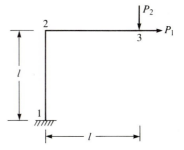

FIGURE P5.31

5.32. Determine the horizontal deflection of point 4 in Fig. P5.32.

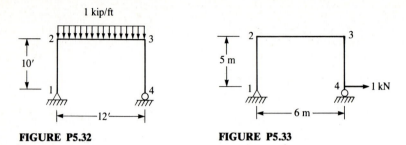

1 kip/ft

FIGURE P5.32

FIGURE P5.33

5.33. Calculate the horizontal deflection of point 4 in Fig. P5.33.

CHAPTER

6

ENERGY
METHODS

Energy methods are powerful techniques for both formulation and solution of structural problems. In this chapter, following a discussion of work and energy, and starting with the principle of conservation of energy, we discuss the application of energy methods. The energy methods include the methods of real work and virtual work, Castigliano's theorems, Maxwell-Betti's reciprocal theorem, and the Rayleigh–Ritz method for approximate solutions. In addition to the above theorems, the Muller–Breslau principle, discussed in Chapter 8, is based on energy principles.

From the methods discussed below, the method of virtual displacements and Castigliano's first theorem, which is directly related to it, apply to nonlinear elastic materials. On the other hand, the method of virtual forces and Castigliano's second theorem, which is derived from it, apply only to linear elastic materials. They can be extended to nonlinear elastic materials by using the complementary work and energy instead of elastic work and energy. The virtual work methods for deflection of beams and trusses are also based on the method of virtual forces and thus in their present form are only applicable to linear elastic materials.

6.1 WORK AND ENERGY

The work done by a force is the product of the force and the distance it moves. Under the applied loads, structures will deform and their fibers will develop strains and stresses. The product of the internal forces and their associated displacements is the internal work done by the system.

External Work

If a structure with linear elastic material is under a load P_i at a point i and an additional infinitesimal deformation dv_i is induced at i, by another load, then P_i remains constant (Fig. 6.1a) and the work done by P_i because of the displacement dv_i is $dW = P_i\,dv_i$. If the additional displacement is v_i then the work is the area under the force–displacement curve, that is,

$$W = \int_0^{v_i} P_i\,dv_i = P_i \int_0^{v_i} dv_i = P_i v_i \tag{6.1}$$

Thus, the work in this case is the product of the force and displacement. Note that the second load may be real or virtual (hypothetical). In the latter case the work is virtual and is the product of a real force and a virtual displacement.

On the other hand, if the deformation is induced by the load itself, then for a linear elastic material the displacement is proportional to the load as indicated by Fig. 6.1b and $v_i = P_i/K$, where K is a proportionality constant, depending on the geometry and material properties. The elastic work of P_i for a deflection v_i is thus the area under the force–displacement curve, or

$$W = \int_0^{v_i} P_i\,dv_i = \int_0^{v_i} Kv_i\,dv_i = K \int_0^{v_i} v_i\,dv_i = \tfrac{1}{2}Kv_i^2 = \tfrac{1}{2}P_i v_i \tag{6.2}$$

Hence, if the deflection in a linear elastic structure is caused by the load itself, rather than some other effect, then the work is equal to the product of the load and its displacement times $\tfrac{1}{2}$. For a nonlinear elastic material (Fig. 6.1c), we can also calculate the elastic work as the area under the force–displacement curve.

Thus, the area under the force–displacement diagram, W_e, is the elastic work. We can also calculate the area above the force–displacement diagram,

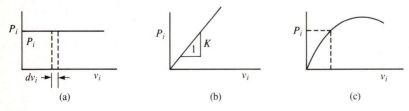

(a)   (b)   (c)

FIGURE 6.1

called the complementary work,

$$W^* = \int_0^{P_i} v_i \, dP_i$$

It can be seen from Fig. 6.1b, that for linear elastic materials this is the same as the elastic work. However, for nonlinear elastic materials the complementary and elastic works are different.

Internal Work

The internal forces developed in an elastic structure in response to the applied loads and their deformations have the capacity to do work and restore the structure to its original configuration once the loads are removed. This internal energy is stored in the system in the form of strain energy U_S, or the potential of the system for doing work. Since deformations are opposed by the internal forces, the strain energy and the external work have opposite signs. We consider the internal work as positive. Therefore, external work will be negative.

Consider an infinitesimal element of a structure under forces causing normal stress σ. The normal force acting on the section (Fig. 6.2), is $\sigma \, dy \, dz$, and the change of length is the product of the strain and the length of the element. Since the loads are increased from zero to their actual values, so are the stresses and strains. Thus, the internal work of this infinitesimal element when the full load is applied causing a strain ε is

$$\int_0^\varepsilon \sigma \, dy \, dz \, d\varepsilon \, dx = dx \, dy \, dz \int_0^\varepsilon \sigma \, d\varepsilon = d(\text{vol}) \int_0^\varepsilon \sigma \, d\varepsilon$$

For linear elastic materials, $\sigma = E\varepsilon$, and the internal work of the infinitesimal element is

$$E \, d(\text{vol}) \int_0^\varepsilon \varepsilon \, d\varepsilon = \tfrac{1}{2} E\varepsilon^2 \, d(\text{vol}) = \tfrac{1}{2}\sigma\varepsilon \, d(\text{vol})$$

The elastic energy (internal work) of a system under normal or axial stress is the integral of the energy for the infinitesimal element over the volume of the system,

$$U_a = \tfrac{1}{2}\int (\sigma \, dy \, dz)(\varepsilon \, dx) = \int \sigma\varepsilon \, dx \, dy \, dz = \int_{\text{vol}} \sigma\varepsilon \, d(\text{vol})$$

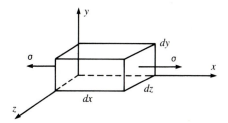

FIGURE 6.2

or

$$U_a = \tfrac{1}{2} \int_{\text{vol}} \sigma\varepsilon \, d(\text{vol}) \tag{6.3}$$

The elastic energy due to shear deformation can be found in a similar manner, by calculating the internal work for an infinitesimal element. The result would be similar to Eq. (6.3), except that normal stress and strain would be replaced by shear stress and strain. Therefore,

$$U_s = \tfrac{1}{2} \int_{\text{vol}} \tau\gamma \, d(\text{vol}) \tag{6.4}$$

For linear elastic structures, the stress–strain curves are linear, similar to Fig. 6.1b. The elastic energy of the system is related to the area under this curve, and the complementary energy is related to the area above the curve. For linear elastic materials, the elastic energy and complementary energy are the same.

Elastic Energy Expressions

For an elastic spring under axial deformation, the relation between the applied force P and displacement v is linear, that is, $P = Kv$, where K is the stiffness of the spring. The elastic energy is

$$U_a = \int_v P \, dv = \int_v Kv \, dv = \tfrac{1}{2}Kv^2 = \tfrac{1}{2}Pv = \frac{P^2}{2K} \tag{6.5}$$

which is equal to the area under the load–displacement curve.

For a member under axial load U inducing deformation u, the stress is $\sigma = U/A$, $\varepsilon = \sigma/E$ and $d(\text{vol}) = A \, dx$. The elastic energy (Eq. (6.3)), invoking Eq. (5.3b), becomes

$$U_a = \tfrac{1}{2} \int_0^l \frac{U^2}{EA} \, dx = \tfrac{1}{2} \int_0^l EA(u')^2 \, dx \tag{6.6}$$

For bending or flexural members, $\sigma = My/I$ and $\varepsilon = \sigma/E$, the elastic energy (Eq. (6.3)) becomes

$$U_b = \tfrac{1}{2} \int_{\text{vol}} \frac{1}{E} \left[\frac{My}{I}\right]^2 d(\text{vol}) = \tfrac{1}{2} \int_0^l \frac{1}{E} \left[\frac{M}{I}\right]^2 dx \int_A y^2 \, dA$$

Since by definition the last integral is I, the above expression for the elastic energy, using Eq. (5.16), becomes

$$U_b = \tfrac{1}{2} \int_0^l \frac{M^2}{EI} \, dx = \tfrac{1}{2} \int_0^l EI(v'')^2 \, dx \tag{6.7}$$

The elastic energy of a beam due to shear can be found by substituting for shear stress and strain in Eq. (6.4). The relation between the shear stress τ and strain γ is $\tau = G\gamma$, where G is the shear modulus. In terms of the modulus of elasticity E and Poissons ratio v, $G = E/2(1 + v)$; $v = 0.3$ for steel and about 0.2 for concrete. Thus, Eq. (6.4) becomes

$$U_s = \tfrac{1}{2} \int_{vol} \frac{\tau^2}{G} d(vol) = \tfrac{1}{2} \int_A \int_l \frac{\tau^2}{G} dA \, dx$$

From Chapter 5 (Eq. (5.22)), shear stress at any point in a section can be expressed as $\tau = \alpha V/A$, where α is a function of the location of the point in the section, given by Eq. 5.23. The last expression upon substitution for τ becomes,

$$U_s = \tfrac{1}{2} \int_A \int_l \alpha^2 \frac{V^2}{G} dA \, dx = \tfrac{1}{2} \int_l \frac{V^2}{G} dx \int_A \alpha^2 \, dA$$

We define the dimensionless constant

$$\kappa = \frac{1}{A} \int_A \alpha^2 \, dA \tag{6.8a}$$

called the *shape* (or *form*) *factor*, which can be calculated once α is determined by Eq. (5.23). A/κ is called the shear area. Therefore, the expression for the strain energy due to shear becomes

$$U_s = \tfrac{1}{2}\kappa \int_l \frac{V^2}{GA} dx \tag{6.8b}$$

As found in the following example, the shape factor for a rectangular section is 1.2. For flanged sections such as I-beams under shear stress, as mentioned in Chapter 5, shear is mainly carried by the web in a uniform manner. Thus, for the latter sections $\alpha = 1$, which results in $\kappa = 1$. For circular sections κ is 1.11.

As discussed in Chapter 17, for a member with length l, under end torques T, the relation between the torque and the angle of twist per unit length, ψ', is $T = GJ\psi'$, where $G = E/2(1 + v)$ is the shear modulus. v is Poisson's ratio with values given above, E is the modulus of elasticity with values given in Table 1.1, and J is the torsion constant. Table 17.1 gives the values of torsion constants for different sections. Since torsion causes shear stresses and strains, the elastic energy of torsion (Eq. (6.4)) becomes Eq. (17.6), that is,

$$U_t = \tfrac{1}{2} \int_l \frac{T^2}{GJ} dx = \tfrac{1}{2} \int_l GJ(\psi')^2 \, dx \tag{6.9}$$

If more than one deformation mode is present then the total strain energy is the sum of the energy for the different deformation modes. For example, if a member is under axial force and bending, then its elastic energy is the sum of the expressions in Eqs. (6.6) and (6.7).

Example 6.1. Find the shape factor in the shear energy expression for a rectangular beam with width b and depth h.

Solution. The expression for α for rectangular sections was found in Chapter 5, Eq. (5.24), as

$$\alpha = \frac{3}{2}\left[1 - \left(\frac{2y}{h}\right)^2\right]$$

By Eq. (6.8a),

$$\kappa = \frac{1}{bh}\int_A \alpha^2 \, dA = \frac{9}{4bh}\int_{-h/2}^{h/2}\left[1 - \frac{4y^2}{h^2}\right]^2 b \, dy$$

$$= \frac{9}{4h}\int_{-h/2}^{h/2}\left[1 + \frac{16y^4}{h^4} - \frac{8y^2}{h^2}\right]dy = \frac{9}{4h}\left[y + \frac{16y^5}{5h^4} - \frac{8y^3}{3h^2}\right]_{-h/2}^{h/2}$$

$$= \frac{9}{4h}\left[\frac{h}{2} + \frac{16h^5}{160h^4} - \frac{h^3}{3h^2} + \frac{h}{2} + \frac{16h^5}{160h^4} - \frac{h^3}{3h^2}\right]$$

$$= \frac{9}{2}\left[\frac{1}{2} + \frac{1}{10} - \frac{1}{3}\right] = \frac{9}{2}\left[\frac{15 + 3 - 10}{30}\right] = \frac{12}{10}$$

Thus, $\kappa = 1.2$ for rectangular sections.

6.2 METHOD OF REAL WORK

The principle of conservation of energy is a special case of the first law of thermodynamics for isolated systems. That is, systems with no absorption or dissipation of energy by heat or other effects. The principle of conservation of energy states that when a structure deforms there is no change in the total energy of the system. Structures supported by rigid supports can be considered as isolated systems. For structures on elastic supports, the energy of deformation of the support must be added.

When a structure deforms, the external loads acting on it perform *external work* W_e because their points of application move. Upon deformation, the structure develops internal forces that resist the action of the external forces. The work of the internal forces, through their respective deformations W_i is called *internal work*. Since internal forces oppose the deformations caused by the external loads, the internal work has the opposite sign of the external work. The change in the total energy of the system is the algebraic sum of the external and internal works. Since, by the principle of conservation of energy, the total energy of the system does not change, the magnitude of the external and internal works must be equal. Thus,

$$\boxed{W_e = W_i}$$

(6.10)

Equation (6.10) is called the principle of work, and states that for a system under deformation the external work is equal to the internal work.

It should be noted that in structural engineering the magnitude of the total energy of the system is immaterial. Solutions of problems are obtained by considering the *change* of energy using Eqs. (6.10). Thus, the energy of the undeformed system may be taken equal to zero.

Example 6.2. Find the deflection of the cantilever beam of Fig. 6.3, with span l under a load P at its free end. The moment of inertia of the beam is I and its modulus of elasticity is E.

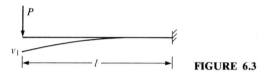

FIGURE 6.3

Solution. Using the principle of work, Eq. (6.10) can be used to find the end deflection of this beam. Since the deflection v_1 at the load is caused by the load P, the external work is $W_e = \frac{1}{2}Pv_1$. The internal work is the elastic energy of the beam due to bending. With $M = -Px$, Eq. (6.7) gives

$$U_b = \frac{1}{2}\int_0^l \frac{M^2}{EI}\,dx = \frac{P^2}{2EI}\int_0^l x^2\,dx = \frac{P^2l^3}{6EI}$$

$W_e = U_b$ gives

$$\frac{1}{2}Pv_1 = \frac{P^2l^3}{6EI} \quad \text{or} \quad v_1 = \frac{Pl^3}{3EI}$$

6.3 METHOD OF VIRTUAL WORK

A structure that is in equilibrium under its applied loads deforms and develops internal forces. If such a structure is subjected to an additional virtual displacement or virtual force, additional displacements and internal forces will result. The work of the real forces through the virtual displacements, or that of the virtual forces through the real displacements, is the virtual work done by the system. According to the principle of virtual work for such a system, the virtual works of the external and internal forces are equal. It is clear from the above discussion that we can induce virtual work by imposing virtual displacements or virtual forces. As we will see, virtual displacements must be compatible with the boundary conditions.

Although the virtual work principle can be proved in its general form, here we carry out a proof for the special case of a bar under axial forces by applying a virtual displacement. For the bar of Fig. 6.4a, which is in equilibrium under the end forces U_1 and U_2, equilibrium of forces requires that $U_1 = U_2 = U$, where U is the axial force at an arbitrary section x.

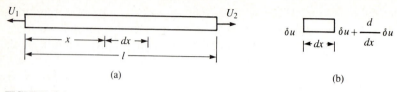

FIGURE 6.4

Under this load the axial displacement at section x is u. If we keep the forces constant and induce a small additional virtual displacement in the bar so that the displacement at x is δu, the displacements at the two ends of an infinitesimal element (Fig. 6.4b) are δu and $\delta u + d(\delta u)/dx$ causing a change of length in the infinitesimal element equal to $d(\delta u)/dx$. The virtual work of the infinitesimal element is $U\,d(\delta u)/dx$ and for the whole bar the internal virtual work is

$$\delta W_i = \int_0^l U \frac{d}{dx}(\delta u)\,dx = U\int_0^l \frac{d}{dx}(\delta u)\,dx = U\,\delta u\Big|_0^l = U(\delta u_2 - \delta u_1)$$

The virtual work of the external forces (external virtual work) is

$$\delta W_e = U_2\,\delta u_2 - U_1\,\delta u_1 = U(\delta u_2 - \delta u_1) = \delta W_i$$

Thus, we find that the external and internal virtual works are equal, which is the virtual work principle.

Since we used the boundary conditions in the above proof, the virtual displacements must be compatible with the boundary conditions. In other words, the virtual displacements at unyielding supports must be zero. Furthermore, the equilibrium equation was invoked in the above derivations. Thus, the virtual work equation implies the equilibrium of the structure. In fact, the virtual work equation can be viewed as a substitute for the equilibrium equations.

The internal virtual work can also be written in terms of stresses and strains rather than forces and displacements. To achieve this end we can substitute $U = \sigma A$, $\delta\varepsilon = d(\delta u)/dx$ and $d(\text{vol}) = A\,dx$ in the above equation for the internal virtual work, where σ is the normal stress, δu is the virtual displacement, $\delta\varepsilon$ the virtual strain, and A is the area of the cross-section. Therefore, the internal virtual work becomes

$$\delta W_i = \int_{\text{vol}} \sigma\,\delta\varepsilon\,d(\text{vol})$$

Referring to the elastic energy of a bar under an axial force (Eq. (6.3)) we can show that the right-hand side of the above equation is δU_a. That is, the internal energy is the elastic energy of the system. Therefore, the internal virtual work is equal to the variation of the elastic energy of the system.

Following a similar procedure the virtual work equation for any structure can be written as

$$\boxed{\delta U_S = \delta W_e}$$

(6.11)

Equation (6.11) states that, for a system in equilibrium the variation of the elastic energy is equal to the variation of the external work.

Consider a structure under n real loads P_i inducing stress σ. If this structure is subjected to virtual displacements, which cause displacements δv_i at and in the direction of the loads, and virtual strain $\delta \varepsilon$, then Eq. (6.11) becomes

$$\int_{\text{vol}} \sigma(\delta \varepsilon) \, d(\text{vol}) = \sum_{i=1}^{n} P_i \, \delta v_i$$

(6.12)

Similarly, if a structure that is in equilibrium under virtual forces δP_i causing stress $\delta \sigma$ is subjected to real loads P_i inducing real displacements v_i at the locations (and in the direction) of the virtual forces, Eq. (6.11) gives

$$\int_{\text{vol}} \varepsilon(\delta \sigma) \, d(\text{vol}) = \sum_{i=1}^{n} v_i \, \delta P_i$$

(6.13)

Note that ε in Eq. (6.13) is strain due to the real external loads.

The virtual work principle can be used for calculating deflection of structures. It is especially suitable for structures under combined stresses and discontinuous loads. In this chapter we discuss its application to deflection of trusses and beams.

Deflection of Trusses by Virtual Work

In buildings, and especially in bridges, excessive deflection will cause undesirable vibrations. Deflections are also used in finding the bar forces of statically indeterminate trusses. The virtual work method (virtual forces) is used here to derive the expression for calculating deflection of statically determinate trusses. The procedure can be developed using Eqs. (6.11) or (6.13). The former equation is used here.

Consider the truss of Fig. 6.5a under the applied loads shown, and suppose that we want to calculate the vertical deflection at joint 5. We use this truss for the purpose of presentation and proceed as follows to find a general formula for calculating truss deflections.

Assume that before the external loads are applied, a virtual (imaginary) force of one unit is acting in the vertical direction at joint 5 (Fig. 6.5b). Internal bar forces f_i are developed under this force that can be calculated by either the method of joints or method of sections, discussed in Chapter 3. After the external loads P_i are applied, additional bar forces F_i are developed. F_i can also be found by analyzing the truss of Fig. 6.5a using the method of joints or sections. Once the loads P_i are added, the truss bars with original

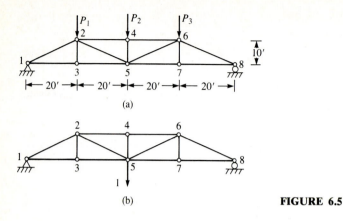

(a)

(b)

FIGURE 6.5

lengths l_i, under internal forces, f_i, undergo changes of length Δl_i, while the joint with the unit force undergoes an additional deflection v. Setting the external virtual work equal to the internal virtual work, Eq. (6.11) leads to an expression for calculating the truss deflections.

The external virtual work done by the unit force is $1 \times v$, while the internal virtual work done by the virtual bar forces f_i is $\sum_i f_i \, \Delta l_i$, where the summation over i indicates the total work for all the bars. Thus, the virtual work equation becomes

$$v = \sum_i f_i \, \Delta l_i$$

Δl_i are the changes of lengths of the truss bars due to bar forces F_i induced by the external loads P_i. For each bar with cross-sectional area A_i, length l_i, and internal force F_i, the change of length is, by Eq. (5.4),

$$\Delta l_i = \frac{F_i l_i}{EA_i}$$

Substitution of the last relation in the equation for v gives

$$v = \frac{1}{E} \sum_i \frac{F_i f_i l_i}{A_i} \qquad (6.14)$$

Equation (6.14) can be used to find the deflection of any joint of a truss in any direction. The steps in the solution are as follows.

1. Find the bar forces F_i under the applied loads.
2. Remove the loads, apply a unit load at the joint and in the direction that the deflection is sought, and find the bar forces f_i due to this unit load.
3. Using Eq. (6.14) calculate the deflection.

Note that if the deflection is needed at more than one joint or in more than one direction, steps 2 and 3 must be repeated for each deflection.

Example 6.3. Find the vertical deflection of joint 5 and horizontal deflection of joint 8 for the truss of Fig. 6.5, with $P_1 = P_2 = P_3 = 10$ kips, $E = 29\,000$ ksi. The areas of cross-sections of the bars are shown in Table 5.3, where the deflection calculations are performed.

Solution. The bar forces due to the applied loads, a vertical unit load at 5 and a horizontal unit load at 8 are found by the method of joints and recorded in the table.

TABLE 5.3
Data and calculations for Example 6.3

Bar#	l_i	A_i	F_i	f_{i5}	f_{i8}	$F_i f_{i5} l_i / A_i$	$F_i (f_{i8}) l_i / A_i$
12	268.33	6	−33.54	−1.12	0	1 679.96	0
13	240	6	30	1	1	1 200	1 200
23	120	4	0	0	0	0	0
24	240	6	−40	−2	0	3 200	0
25	268.33	6	11.8	1.12	0	591.04	0
35	240	6	30	1	1	1 200	1 200
45	120	4	−10	0	0	0	0
46	240	6	−40	−2	0	3 200	0
56	268.33	6	11.8	1.12	0	591.04	0
57	240	6	30	1	1	1 200	1 200
67	120	4	0	0	0	0	0
68	268.33	6	−33.54	−1.12	0	1 679.96	0
78	240	6	30	1	1	1 200	1 200
					$(1/E) \sum F_i f_i l_i / A_i = 0.54$ and		0.17

The columns of Table 5.3 headed f_{i5} and f_{i8} give the bar forces due to unit loads at joints 5 and 8. Each term of Eq. (6.14) is calculated in last columns of the table respectively. The values of deflection at 5 and 8, given by Eq. (6.14), are found in the last row of the table. Thus, the vertical deflection of joint 5 is 0.54 in and the horizontal displacement of joint 8 is 0.17 in to the right.

Internally Indeterminate Trusses

As we saw in Chapter 3, trusses that are externally or internally indeterminate cannot be analyzed by the equations of statics alone. Their analysis requires consideration of deflections. If, for example, a truss is externally indeterminate, because of redundant reactions, then it can be analyzed by the method of consistent deformations (Chapter 7) using the previous procedure for finding the deflections needed in the solution process.

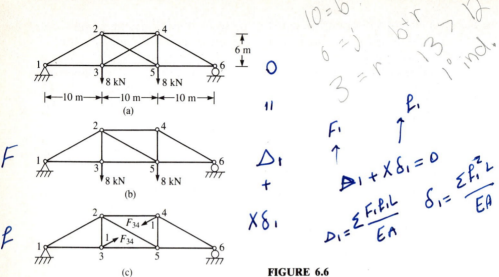

FIGURE 6.6

Analysis of internally indeterminate trusses also requires determination of deflections. Consider, for example, the truss of Fig. 6.6a, with one redundant bar force. If we remove the redundant bar jk (34 for the truss of Fig. 6.6c) we obtain a statically determinate truss called the primary truss (Fig. 6.6b).

The solution of the internally indeterminate truss can be found by superposition of the primary truss and one including the effect of the redundant bar force, F_{jk} (Fig. 6.6c). Under the applied loads a displacement Δ_{jk} would occur in the primary truss between joints j and k along the line of the redundant force F_{jk}. The effect of the bar jk is to keep this deflection from taking place. That is, the bar force F_{jk} must induce a displacement equal and opposite to Δ_{jk} so that the final displacement is zero. If the displacement along bar jk due to a unit force in bar jk is δ_{jk}, then for compatibility of deformations we must have $F_{jk}\delta_{jk} + \Delta_{jk} = 0$. This equation enables us to calculate the redundant force F_{jk}.

Therefore, to find the redundant force F_{jk} we first remove this bar and calculate the displacement Δ_{jk} along the direction jk. Then we calculate the deflection along jk due to unit forces acting at j and k along jk.

To find the deflections Δ_{jk} and δ_{jk} we can use the method of virtual forces, Eq. (6.11). Δ_{jk} is the displacement of the primary truss along jk. To calculate this displacement we find the bar forces F_i in the primary truss under the applied loads. We then subject joints j and k of the primary truss to unit virtual forces along the line jk. These unit forces are equivalent to a unit force in the bar jk. If they point to each other, then we are assuming a unit tension force in bar jk. The external virtual work is the product of the deflection along *jk* due to the applied loads, Δ_{jk}, and the unit virtual force. The internal virtual work is the product of the virtual bar forces, induced in the primary truss by the pair of unit forces, and the stretching of the bars due to the applied loads. Thus, if the bar forces due to the unit loads are f_i and the bar forces due to the external loads are F_i, the bar extensions under the applied loads are, by Eq. (5.4),

$$\frac{F_i l_i}{EA_i}$$

and the virtual work equation becomes

$$\sum_i \frac{f_i F_i l_i}{EA_i} = 1 \times \Delta_{jk}$$

or

$$\Delta_{jk} = \sum_i \frac{f_i F_i l_i}{EA_i} \tag{6.15a}$$

Similarly,

$$\delta_{jk} = \sum_i \frac{f_i^2 l_i}{EA_i} \qquad Djk + f_j k \, \delta_j k = 0 \tag{6.15b}$$

and from the above compatibility equation,

$$F_{jk} = -\Delta_{jk}/\delta_{jk}. \tag{6.15c}$$

Thus, to find a redundant bar force in an internally indeterminate truss we must follow the steps outlined below.

1. Remove the redundant bar *jk*; find the bar forces F_i of the resulting primary truss under the applied loads.
2. Remove the applied loads and apply a pair of unit loads at joints *j* and *k*, along the removed bar *jk* pointing to each other, and find the bar forces, f_i due to these unit loads.
3. Find the displacements Δ_{jk}, due to the applied loads, and δ_{jk}, due to the unit loads, by Eqs. 6.15a and 6.15b. The redundant bar force can then be found by Eq. 6.15c.

Example 6.4. Find the redundant bar force F_{34} for the truss of Fig. 6.6. Two equal vertical loads of 8 kN are applied at joints 3 and 5. The areas of cross-sections of the bars are shown in Table 5.4 and $E = 200\,000$ MPa $= 20\,000$ kN/cm².

Solution. The bar forces F_i in the primary truss (with bar 34 removed) under the external loads are found by the joint method. Similarly, the bar forces f_i in the primary truss, due to unit forces applied at 3 and 4 in directions 34 and 43, are calculated and shown in Table 5.4.

TABLE 5.4
Data and calculations for Example 6.4

Bar #	l_i	A_i	F_i	f_i	$f_iF_il_i/A_i$	$f_i^2l_i/A_i$
12	1 166.19	30	−15.55	0	0	0
13	1 000	30	13.33	0	0	0
23	600	20	8.00	−0.51	−122.40	7.8
24	1 000	30	−13.33	−0.86	382.12	24.65
25	1 166.19	30	0	1	0	38.87
35	1 000	30	13.33	−0.86	−382.12	24.65
45	600	20	8.00	−0.51	−122.40	7.8
46	1 166.19	30	−15.55	0	0	0
56	1 000	30	13.33	0	0	0
					$\Delta_{34} = -0.0122,$	$\delta_{34} = 0.0052$

The deflections along the line 34 due to the applied loads and due to the pair of unit loads are calculated in the last row of the table using equations (6.15a) and (6.15b). The redundant bar force (Eq. (6.15c)) is $F_{34} = 0.0122/0.0052 = 2.35$ kN. Once the redundant bar force is found, the other bar forces can be calculated by the method of joints or sections.

Deflection of Beams by Virtual Work

The virtual work method discussed here can be easily applied to deflection problems of beams and frames. In fact it may be preferable to the deflection methods discussed in Chapter 5 since it applies uniformly to all beams and frames as well as other structures. For finding the deflection at a fixed point of a beam or frame, the visual integration discussed in the following section would make the process simpler. The dead load deflections of beams are often compensated by an initial opposite deflection (camber). Consider the beam of Fig. 6.7 and assume that we want to calculate its deflection at point 3. We use equation of virtual forces, Eq. (6.13), for this purpose. If we let the system of virtual forces be a single unit vertical force at 3, $\delta P_i = 1$, then the virtual stress due to this unit force is

$$\delta\sigma = my/I$$

where m is the moment in the beam due to the unit virtual force. The strain

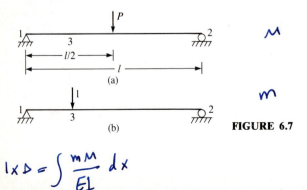

FIGURE 6.7

$$1 \times \Delta = \int \frac{mM}{EI} \, dx$$

due to the applied loads is

$$\varepsilon = \frac{\sigma}{E} = \frac{My}{EI}$$

M is the moment in the beam due to the applied loads. Equation (6.13) thus becomes

$$v_i = \int_{\text{vol}} \frac{My}{EI} \frac{my}{I} \, d(\text{vol}) = \int_0^l \frac{mM}{EI^2} \, dx \int_A y^2 \, dA$$

Since the last integral is by definition I, we have

$$\boxed{v_i = \int_0^l \frac{mM}{EI} \, dx} \tag{6.16}$$

The virtual work procedure just discussed is also called the *unit load method*. To calculate the deflection of a beam at a point along the axis of the beam, we must proceed with the following steps.

1. Find the expression for the moment—due to the applied loads throughout the beam.
2. Remove the applied loads and add a unit load at the point (and in the direction) where deflection is sought. Find the expression for moment m *due* to this unit load throughout the beam. To calculate the slope at a point, instead of deflection, m is found for a unit moment there.
3. Calculate the deflection v_i by Eq. (6.16). A positive value indicates that the deflection is in the same direction as that of the unit load. If m is due to a unit applied moment at point i then Eq. (6.16) yields the slope θ_i instead of deflection v_i.

Note that for calculating the maximum deflection, the deflection at an arbitrary distance ξ is found. In this case the integral in Eq. (6.16) is split into two parts; from zero to ξ and from ξ to l. The location of the maximum moment is found by setting the derivative of the deflection (in terms of ξ) equal to zero.

Example 6.5. Find the deflection of a simply supported beam (Fig. 6.7) at point 3, $l/4$ from the left-hand end. The beam is under a load P applied at its mid-span.

Solution. The equations for moments M and m due to the external load P and a unit vertical force at $l/4$ are as given below.

$$M = \frac{P}{2}x \qquad\qquad 0 \le x \le \frac{l}{2}$$

$$M = \frac{P}{2}x - P\left(x - \frac{l}{2}\right) = P\frac{l-x}{2} \qquad \frac{l}{2} \le x \le l$$

and

$$m = \tfrac{3}{4}x \qquad\qquad 0 \le x \le \frac{l}{4}$$

$$m = \tfrac{3}{4}x - \left(x - \frac{l}{4}\right) = \frac{l-x}{4} \qquad \frac{l}{4} \le x \le l$$

The deflection at point 3 is

$$v_3 = \int_0^l \frac{mM}{EI}\, dx$$

which gives

$$EIv_3 = \int_0^{l/4} \tfrac{3}{4}x \times \frac{P}{2}x\, dx + \int_{l/4}^{l/2} \frac{l-x}{4} P\frac{x}{2}\, dx + \int_{l/2}^{l} \frac{l-x}{4} P\frac{l-x}{2}\, dx$$

or

$$v_3 = \frac{11Pl^3}{768EI}$$

Example 6.6. Find the deflection at the end of the overhang of the beam in Fig. 6.8 under a uniformly distributed load.

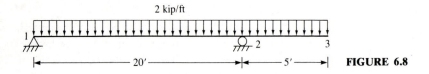

2 kip/ft

|← 20' →|← 5' →| **FIGURE 6.8**

Solution. The reactions under the external loads are

$$R_1 \times 20 - 2 \times 25(20 - 12.5) = 0 \qquad \text{or} \qquad R_1 = 18.75 \text{ kips}$$

$$R_2 = 2 \times 25 - 18.75 = 31.25 \text{ kips}$$

and under the unit load applied at the end of the beam they are

$$r_1 \times 20 + 1 \times 5 = 0 \qquad \text{or} \qquad r_1 = -0.25$$

$$r_2 = 1 - (-0.25) = 1.25 \text{ kips}$$

The moments due to the applied load and a unit load acting at the free end of the beam are as follows.

For $x \le 20$ ft:

$$M = 18.75x - 2x^2/2 = 18.75x - x^2$$

$$m = -0.25x$$

For $x \ge 20$ ft:

$$M = 18.75x - 2x^2/2 + 31.25(x - 20) = -625.0 + 50x - x^2$$

$$m = -0.25x + 1.25(x - 20) = -25 + x$$

By Eq. (6.15)

$$v_3 = \frac{1}{EI} \int_0^{20} (18.75x - x^2)(-0.25x)\, dx$$

$$+ \frac{1}{EI} \int_{20}^{25} (-625 + 50x - x^2)(-25 + x)\, dx$$

which gives

$$v_3 = -\frac{2343.75 \text{ kip} \cdot \text{ft}^3}{EI}$$

Visual integration of the virtual work equation for beams and frames. In Eq. (6.16), m is the moment due to a unit load applied at the location and in the direction of the desired deflection. This moment diagram is composed of a series of lines. In the following derivation we use a single line. However the results can be extended to the case of multiple line segments. When m is a single line over the length it can be described by $m = a + bx$, with a and b as constants. Then if M is also a single function over the length Eq. (6.16) becomes

$$v_i = \int_0^l (a + bx) \frac{M}{EI}\, dx = a \int_0^l \frac{M}{EI}\, dx + b \int_0^l \frac{M}{EI} x\, dx$$

In the above equation, the first integral is the area under the M/EI diagram. Let us denote this area by A. The second integral is the moment of the area A with respect to the left hand end. If the centroid of the M/EI diagram from the left hand end is at $\bar{x}$, then $v_i = aA + b\bar{x}A = A(a + b\bar{x})$. Since $a + b\bar{x} = \bar{m}$ is the value of m at $\bar{x}$, the location of the centroid of the M/EI diagram, the deflection $v_i = A\bar{m}$. Therefore, we can calculate the deflection as a product of the area of the M/EI diagram and the value of m (from the moment diagram for the unit load) at the location of the centroid of the M/EI diagram. Expressions for calculating areas and their centroids are given in Table 5.1. Some suggestions for dividing up complex moment areas were made in Example 5.10.

When M and/or m diagrams are discontinuous functions, the above process must be applied to each segment separately and the results added. That is, the deflection is $v_i = \sum A_j \bar{m}_j$. It should also be mentioned that the same sign convention must be used for plotting both the M and m moment diagrams.

Example 6.7. Find the vertical deflection v_0 at the mid-span of a simply supported beam (Fig. 6.9a) with a constant cross-section under a uniform load p.

Solution. The M/EI diagram is shown in Fig. 6.9b. Figure 6.9c gives the moment diagram m, for a unit vertical load acting at the mid-span (where we want to calculate the deflection). Since the m diagram is discontinuous, we apply the procedure to each half of the diagrams separately and add the results. The area of

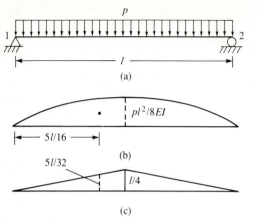

(a)

(b)

51/16

51/32

(c)

1/4

FIGURE 6.9

the M/EI diagram for the left half of the beam, from Table 5.1, is

$$A = \frac{2}{3}\frac{Pl^2}{8EI} \times \frac{l}{2} = \frac{Pl^3}{24EI}$$

and its centroid is at $\bar{x} = (5/8)(l/2) = 5l/16$ from the left end. The ordinate of m at this centroid is

$$\bar{m} = \left(\frac{5l}{16} \times \frac{l}{4}\right) / \left(\frac{l}{2}\right) = \frac{5l}{32}$$

By symmetry,

$$v_0 = 2A\bar{m} = 2 \times \frac{Pl^3}{24EI} \times \frac{5l}{32} = \frac{5Pl^4}{384EI}$$

Example 6.8. For the beam of Example 6.6 (Fig. 6.8) find the deflection and the slope at the end of the overhang.

Solution. The moment diagram for the load is given in Fig. 6.10a. Figures 6.10b and 6.10c show the moment diagrams due to a clockwise unit moment and a

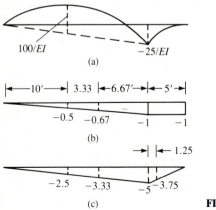

100/EI

−25/EI

(a)

|←10'→| 3.33 |←6.67'→|←5'→|

−0.5   −0.67   −1   −1

(b)

→| |←1.25

−2.5   −3.33   −5 −3.75

(c)

FIGURE 6.10

downward vertical load applied at the right-hand end of the beam, respectively. The area of the moment diagram for the left span is found as the difference between the area of the parabola and the triangle, see Example 5.10. Each one of these areas is multiplied by the value of m at the centroid of that area. Thus,

$$EI\theta_3 = \tfrac{2}{3} \times 100 \times 20(-0.5) - \tfrac{1}{2} \times 25 \times 20(-0.67) - \tfrac{1}{3} \times 25 \times 5(-1) = -457.50$$

$$EIv_3 = \tfrac{2}{3} \times 100 \times 20(-2.5) - \tfrac{1}{2} \times 25 \times 20(-3.33) - \tfrac{1}{3} \times 25 \times 5(-3.75) = -2\,344.58$$

Example 6.9. Find the vertical deflection at the free end of the frame of Fig. 6.11a.

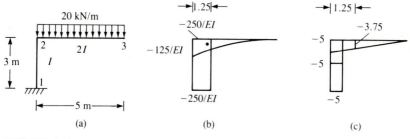

(a) (b) (c)

FIGURE 6.11

Solution. The M/EI diagram for the load is shown in Fig. 6.11b. The moment diagram due to a unit downward load is given in Fig. 6.11c. The vertical deflection at point 3 is found as the sum of the products of M/EI area for each segment and the value of m corresponding to the centroid of that M/EI segment.

$$v_3 = -\frac{250}{EI} \times 3 \times (-5) - \frac{1}{3} \times \frac{125}{EI} \times 5 \times (-3.75) = \frac{4531}{EI}$$

6.4 CASTIGLIANO'S THEOREMS

Castigliano's theorems are useful devices for calculating deflection of structures, especially those under combined stresses. They are also used in the solution of statically indeterminate structures. Castigliano's second theorem is suitable for determining the stiffness matrix of complex elements. In Chapters 15 and 17 it is used for deriving stiffness matrices for arches and bow-girders.

The virtual work equations, for virtual displacements and virtual forces, are given by Eqs (6.12) and (6.13), respectively. The left-hand sides of these equations are the variation of the elastic energy of the system, δU_S, as indicated by Eq. (6.11). In the virtual work expression in terms of virtual displacements (Eq. 6.12), the left-hand side, δU_S, can be expressed as a product of the internal real forces and virtual displacements or variation of displacements δv_i. This can be done by assuming U_S to be a function of the virtual displacements, that is, by writing Eq. (6.12) as

$$\delta U_S(v_i) - \sum_{i=1}^{n} P_i \, \delta v_i = 0 \tag{6.17}$$

Considering v_i as the basic variables of the problem, the variation of U_S is

$$\delta U_S = \sum_{i=1}^{n} \frac{\partial U_S}{\partial v_i} \delta v_i \qquad (6.18)$$

Thus, Eq. (6.17) becomes

$$\sum_{i=1}^{n} \left[\frac{\partial U_S}{\partial v_i} - P_i \right] \delta v_i = 0$$

δv_i is an arbitrary variation (or arbitrary virtual displacement), and is therefore nonzero except for the trivial solution. We thus obtain

$$\frac{\partial U_S}{\partial v_i} - P_i = 0$$

or

$$\boxed{P_i = \frac{\partial U_S}{\partial v_i}} \qquad (6.19)$$

Equation (6.19) is Castigliano's first theorem, stating that the partial derivative of the elastic energy of a structure with respect to the displacement of a point is equal to the force at that point in the direction of the displacement v_i.

Similarly, in the virtual work expression in terms of virtual forces (Eq. (6.13)), the elastic energy is a function of virtual forces. Thus, we have

$$\delta U_S(P_i) - \sum_{i=1}^{n} v_i \, \delta P_i = 0 \qquad (6.20)$$

P_i are the basic variables of the problem and the variation of U_S in terms of P_i is

$$\delta U_S = \sum_{i=1}^{n} \frac{\partial U_S}{\partial P_i} \delta P_i \qquad (6.21)$$

Therefore, Eq. (6.20) becomes

$$\sum_{i=1}^{n} \left[\frac{\partial U_S}{\partial P_i} - v_i \right] \delta P_i = 0$$

Again δP_i are arbitrary variations of forces P_i (or arbitrary virtual forces) and thus we have

$$\frac{\partial U_S}{\partial P_i} - v_i = 0$$

or

$$\boxed{v_i = \frac{\partial U_S}{\partial P_i}} \qquad (6.22)$$

This is Castigliano's second theorem, stating that the partial derivative of the strain energy with respect to a force P_i is equal to the deflection at the location and in the direction of the force. In addition to finding deflections, Eq. (6.22) can be used to find the redundant reactions of structures. This is done by expressing the energy in terms of the unknown redundant reactions. Equations for finding the redundant reactions are obtained by calculating the deflections at the location of the reactions and setting them equal to zero.

For beams with $U_S = U_b$ given by Eqs. (6.7), Castigliano's second theorem becomes

$$v_i = \int_0^l \frac{M}{EI} \frac{\partial M}{\partial P_i} dx \tag{6.23}$$

Example 6.10. The structure shown in Fig. 6.12 is in the horizontal plane. Find the vertical deflection at the free end due a vertical load P acting there. Both members have the same constant cross-section, with moment of inertia I and torsion constant $J = I/4$. The modulus of elasticity is E and the shear modulus $G = E/2$.

Solution. We note that beam 12 is under bending and torsion while beam 23 is under bending only. Castigliano's second theorem (Eq. (6.22)) is used to find the end deflection. The elastic energy of this structure includes the energies due to bending and torsion. That is, $U_S = U_b + U_t$. The expressions for the latter energies are given by Eqs. (6.7) and (6.9) in terms of the moment and torque. Thus, for the two bars the elastic energy is

$$U_S = \frac{1}{2} \int_0^l \frac{M_{12}^2}{EI} dx + \frac{1}{2} \int_0^l \frac{T_{12}^2}{GJ} dx + \frac{1}{2} \int_0^{l/2} \frac{M_{23}^2}{EI} dx$$

where the first two terms are the bending and torsion energies of member 12, and the last term is the bending energy of member 23. Here

$$M_{12} = -Pl + Px$$

$$T_{12} = Pl/2$$

$$M_{23} = P\left(\frac{l}{2} - x\right)$$

Thus,

$$U_S = \frac{1}{2} \int_0^l \frac{P^2(l-x)^2}{EI} dx + \frac{1}{2} \int_0^l \frac{P^2l^2/4}{EI/8} dx + \frac{1}{2} \int_0^{l/2} \frac{P^2(l/2-x)^2}{EI} dx$$

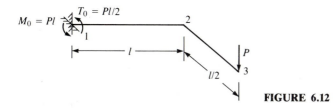

FIGURE 6.12

or

$$U_S = 19P^2 l^3/16EI$$

Equation (6.22) then gives

$$v_3 = \frac{\partial U_S}{\partial P} = \frac{19Pl^3}{8EI}$$

Example 6.11. Find the vertical mid-span deflection v of the structure in Fig. 6.13. The members are hinged at the ends and at the middle with a span length $2l$. A vertical load P is applied at the mid-span.

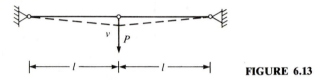

FIGURE 6.13

Solution. If we try to find the solution by ignoring the deflection, the vertical reactions are found to be zero, resulting in zero number forces. However, we know that, with appropriate member sizes, the structure will carry the load, developing internal forces. The problem is that the assumption of linear force–displacement relationship is inadequate in this case. However, since Castigliano's first theorem is applicable to nonlinear elastic problems as well, we can use it to find the solution of this problem.

After deflection v has taken place, the members stretch and the length of each member changes from l to $\sqrt{l^2 + v^2}$. In the absence of other loads the strain in each length l remains constant and equal to

$$\varepsilon = \frac{\Delta l}{l} = \frac{\sqrt{l^2 + v^2} - l}{l} = \sqrt{1 + (v/l)^2} - 1$$

For small deflections, with Taylor expansion, the above expression can be approximated by $\varepsilon = \frac{1}{2}(v/l)^2$.

The elastic energy of the two bars, by Eq. (6.6), with $u' = \varepsilon$, is

$$U_a = \int_0^l EA\varepsilon^2 \, dx = \frac{1}{4}EAl\left(\frac{v}{l}\right)^4$$

By Castigliano's first theorem, $\partial U_a/\partial v = P$, which gives $EAv^3/l^3 = P$. Therefore, the vertical displacement at the mid-span is $v = l(P/EA)^{1/3}$. As we see, v is not linearly related to P and we have a nonlinear relation.

Shear Deflection in Beams

For beams under significant shear forces and small length to depth ratios, for example, deep beams, the adjacent sections undergo relative deformations under the shear force. The shear deformation can be calculated by an energy method. Here Castigliano's second theorem is used for this purpose.

The strain energy due to shear distortion in a beam is given by Eq.

(6.8b). For a section at x this is

$$U_s = \tfrac{1}{2}\kappa \int_0^x \frac{V^2}{GA}\, dx$$

where V is the shear force at the section, A is the area of the cross-section, G is the shear modulus, and κ is the shape factor, defined by Eq. (6.8a). The deflection due to shear is, by Castigliano's second theorem (Eq. (6.22)),

$$v = \frac{\partial U_s}{\partial V} = \frac{\kappa}{G} \int_0^x \frac{V}{A}\, dx$$

or

$$\boxed{v = \frac{\kappa}{G} \int_0^x \frac{V}{A}\, dx} \qquad\qquad (6.24)$$

Example 6.12. In order to design an economical section for a 40-ft span beam under a uniform load of 1 kip/ft, W8 × 35 sections are used and made into the configuration shown in Fig. 6.14a, by welding short pieces to the chords. Calculate the maximum bending and shear deflections of this beam.

Solution. Since for most of the length the cross-section consists of that of the W8 sections, the latter section is used in calculating the section properties of the beam. In view of Eq. (5.23), the shear at a section of this beam would have a distribution as shown in Fig. 6.14c. Thus, the area of the webs of the two W8 sections, the top and bottom chords, is used as the shear area for the beam. Furthermore, the shear stress is assumed constant over these areas.

For W8 × 35, the moment of inertia is 127 in^4, and the area of cross-section is 10.3 in^2 (from the AISC Manual of Steel Construction). Therefore, with $d = 8$ in and $t_w = 0.31$ in, the shear area is $A = 2 \times 8 \times 0.31 = 4.96$ in^2. From Example 5.3, the maximum bending deflection at the mid-span of the beam is

$$v_b = \frac{5Pl^4}{384EI}$$

$$I = 2[127 + (10.3)(12 + 4.0)^2] = 5\,567.2 \text{ in}^4$$

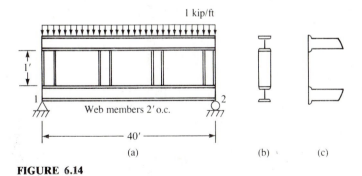

FIGURE 6.14

The bending deflection is

$$v_b = \frac{5(1)(40)^4(12)^3}{384(29\,000)(5\,567.2)} = 0.357 \text{ in}$$

The shear deflection (Eq. (6.24)) is

$$v_s = \frac{\kappa}{G} \int_0^x \frac{V}{A} dx$$

For W sections taking $\kappa = 1.0$, with shear force $V = 20 - x$, we get the shear deflection

$$v_s = \frac{\kappa}{G} \int_0^x \frac{V}{A} dx = \frac{1}{G} \int_0^x \frac{20 - x}{4.96} dx = \frac{1}{G} \left(\frac{20x}{4.96} - \frac{x^2}{9.92} \right)$$

The location of the maximum shear deflection is found from $(dv_s/dx) = 0$, as

$$\frac{20}{4.96} - \frac{x}{4.96} \qquad \text{or} \qquad x = 20 \text{ ft}$$

Then the maximum shear deflection is

$$v_s = \frac{1}{11\,200} \left(\frac{20(20)(12)}{4.96} - \frac{(20 \times 12)^2}{9.92} \right) = 0.432 \text{ in}$$

Note that the common assumption that shear deflection is negligible does not apply to this beam. In fact the shear deformation is larger than the bending deflection. The beam of this example can provide an economical design. However, precaution must be taken so that buckling of the chord and lateral buckling of the beam is not a problem.

6.5 MAXWELL–BETTI'S RECIPROCAL THEOREM

This theorem was initially presented by Maxwell and was generalized by Betti. One of the applications of the theorem is to show that stiffness matrices for structural systems are symmetric. Betti's theorem reads as follows.

If v_{12} is the deflection produced at point 1 by a load P_2 acting at point 2, and v_{21} is the deflection produced at 2 in the direction of P_2 by a load P_1 acting at point 1 in the direction of v_{12}, then

$$P_1 v_{12} = P_2 v_{21} \qquad (6.25)$$

If P_1 and P_2 are equal one, then we obtain Maxwell's theorem, which states that the deflection v_{21} at 2 due to a unit load at 1, is equal to the deflection v_{12} at 1 due to the unit load applied at 2.

Maxwell–Betti's reciprocal theorem can be proved by the virtual work method. Here we present a proof based on the fact that for an elastic system the total work is independent of the order of application of the loads.

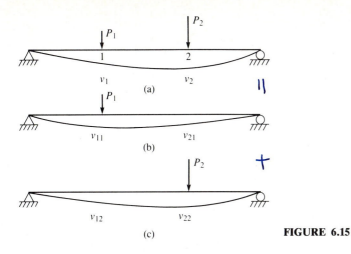

FIGURE 6.15

Consider the beam of Fig. 6.15a under two loads P_1 and P_2 acting at 1 and 2, respectively. To prove Maxwell–Betti's theorem we obtain the final state of the beam by two different loading sequences. In the first sequence we apply P_1 first and then P_2. In the second sequence we apply P_2 first and then P_1. Since the final state of the beam is the same, the total work done in both cases must be the same.

In the first sequence, after the load P_1 is applied the deflection at point 1 is v_{11} and at point 2 is v_{21} (Fig. 6.15b). The external work done in this step is $W_{11} = \frac{1}{2}P_1 v_{11}$. Upon addition of load P_2 at point 2, additional deflections v_{12}, and v_{22} take place at points 1 and 2. The external work done due to the additional load is $W_{12} = \frac{1}{2}P_2 v_{22} + P_1 v_{12}$. The last term does not have the factor $\frac{1}{2}$ because the load P_1 was already there and thus rides through the deflection v_{12} (it does not start from zero). The total work done after both loads are applied is $W_1 = \frac{1}{2}P_1 v_{11} + \frac{1}{2}P_2 v_{22} + P_1 v_{12}$.

In the second sequence of loading, the load P_2 is applied first, producing deflections v_{21} at point 1 and v_{22} at point 2 (Fig. 6.12c). The work done is $W_{21} = \frac{1}{2}P_2 v_{22}$. After the load P_1 is applied it poduces additional deflections v_{11} at 1 and v_{21} at 2. The work done in this step of loading is $W_{21} = \frac{1}{2}P_1 v_{11} + P_2 v_{21}$. The total work in the second loading sequence is $W_2 = \frac{1}{2}P_2 v_{22} + \frac{1}{2}P_1 v_{11} + P_2 v_{21}$. Setting the two values of work W_1 and W_2 from the two sequences of loading equal, we get $P_1 v_{12} = P_2 v_{21}$, which proves the theorem.

Example 6.13. For the beam of Fig. 6.16a under a load $P_1 = 20$ kips at point 1, 5 ft from the left-hand end, calculate v_{21}, the deflection at point 2, the mid-span. Then for the same beam with a load $P_2 = 10$ kips at point 2 find the deflection v_{12} at point 1 (Fig. 6.16c). Verify Maxwell–Betti's theorem, $P_1 v_{12} = P_2 v_{21}$ and Maxwell's law $v_{12} = v_{21}$, for $P_1 = P_2 = 1$.

Solution. The deflections are calculated by the conjugate beam method of Section 5.5. The conjugate beams for the two loding conditions (Fig. 6.16a and

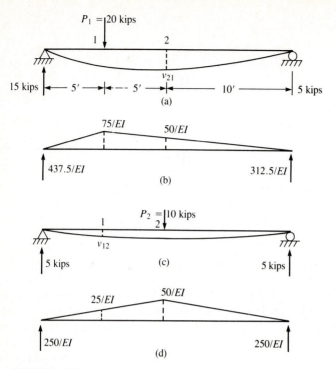

FIGURE 6.16

6.16c) are shown in Figs. 6.16b and 6.16d. For the first load the deflection at point 2 is equal to the moment in the conjugate beam (Fig. 6.16b) at 2. Calculating this moment from the right we get

$$v_{21} = M_2' = \frac{312.5}{EI} \times 10 - \frac{1}{2} \times \frac{50}{EI} \times 10 \times \frac{10}{3} = \frac{2\,291.66}{2EI}$$

and for the second load finding the moment in the conjugate beam at 1 (Fig. 6.16d) from the left, we have

$$v_{12} = M_1' = \frac{250}{EI} \times 5 - \frac{1}{2} \times \frac{25}{EI} \times 5 \times \frac{5}{3} = \frac{1\,145.833}{EI}$$

By Maxwell–Betti's theorem we must have

$$P_1 v_{12} = P_2 v_{21},$$

Hence,

$$20 \times \frac{1\,145.833}{EI} = 10 \times \frac{2\,291.67}{EI} = \frac{29\,916.7}{EI}$$

which verifies the theorem. To verify Maxwell's law we can use unit loads. If both loads are taken as 1 kip, then $v_{21} = \frac{1}{2} \times 229.167/EI = 114.5834/EI = v_{12}$.

6.6 PRINCIPLE OF MINIMUM POTENTIAL ENERGY

For a structure in equilibrium the virtual work equation was given by Eq. (6.11), which can be written as

$$\delta U_S - \delta W_e = 0$$

This equation can also be cast in the following form:

$$\delta(U_S - W_e) = 0 \tag{6.26a}$$

or

$$\delta\Pi = 0 \tag{6.26b}$$

where

$$\Pi = U_S - W_e \tag{6.27}$$

The term Π here is the algebraic sum of the elastic energy of the structure and the potential energy of the load, and is called the *total potential energy* of the system.

Note that in the equation of virtual work, from which the above equation for potential energy was derived, the loads are kept constant during the application of the virtual displacements or virtual forces. This implies that the potential energy of the loads is the product of the loads and the pertinent displacements. Thus, if the deflection is given by the function v, for a number of concentrated loads and a distributed load the potential energy of the loads is

$$W_e = \sum P_i v_i + \int_0^l p(x)\, v\, dx \tag{6.28}$$

On the other hand, from the variation δU_S of the elastic energy given by the left-hand side of Eqs. (6.12) and (6.13), we can find that

$$U_S = \tfrac{1}{2} \int_{\text{vol}} \sigma\varepsilon\, d(\text{vol})$$

That is, the elastic energy expression such as given by Eqs. (6.3) to (6.9). For example, from Eq. (6.12),

$$\delta U_S = \int \sigma\, \delta\varepsilon\, d(\text{vol}),$$

since for an elastic system $\sigma = E\varepsilon$,

$$\delta U_S = \int E\varepsilon\, \delta\varepsilon\, d(\text{vol}) = \tfrac{1}{2} \int E\, \delta(\varepsilon^2)\, d(\text{vol}) = \delta\left[\tfrac{1}{2} \int E\varepsilon^2\, d(\text{vol})\right]$$

$$= \delta\left[\tfrac{1}{2} \int \sigma\varepsilon\, d(\text{vol})\right]$$

The last derivation is based on calculus of variations, from which we can conclude that the form of the strain energy used in the total potential energy is as stated by the equation following Eq. (6.28) or Eq. (6.3).

For example, the total potential energy of a beam under a distributed load $p(x)$ can be written as

$$\Pi = \tfrac{1}{2} \int_0^l [EI(v'')^2 - 2p(x)v]\, dx \tag{6.29}$$

Two important principles pertaining to potential energy have significant applications in structural engineering. The principle of stationary potential energy (Eq. (6.26b)) states that, for an isolated system, defined in Section 6.2, which is generally the case of structures, the variation or change of total potential energy is zero. This means that the total potential energy of structures is stationary, that is, it has a constant value.

The second principle is that of minimum potential energy, which states that for an isolated system in equilibrium the total potential energy is a minimum. These principles have many applications in structural analysis.

6.7   RAYLEIGH–RITZ METHOD

The Rayleigh–Ritz method is used for finding approximate solutions of structures. In applying the principle of stationary potential energy (Eq. (6.26b)) we can assume a deflected shape in terms of a set of specified functions ϕ_j, each satisfying the geometric (displacement and rotation) boundary conditions, multiplied by an unknown parameter a_j; that is,

$$v = \sum_{j=1}^{J} a_j \phi_j \tag{6.30}$$

Here J is the number of terms used. The elastic energy and the work of the external forces become functions of a_j, $U_S = U_S(a_j)$ and $W_e = W_e(a_j)$. Considering a_j as the basic variables of the problem, the variations become

$$\delta U_S = \sum_{j=1}^{J} \frac{\partial U_S}{\partial a_j} \delta a_j \tag{6.31}$$

and

$$\delta W_e = \sum_{i=1}^{n} \sum_{j=1}^{J} P_i \frac{\partial v_i}{\partial a_j} \delta a_j \tag{6.32}$$

Equation (6.32) is for a set of concentrated external loads. For a uniformly distributed load p, the external virtual work is

$$\delta W_e = \int_0^l p\, \delta v\, dx = \int_0^l p \sum_{j=1}^{J} \frac{\partial v}{\partial a_j} \delta a_j\, dx \tag{6.33}$$

If both concentrated and uniform loads are present, then the external work would be the sum of Eqs. (6.32) and (6.33). For the latter case, substitution of

Eqs. (6.31), (6.32), and (6.33) into Eq. (6.26a), or its equivalent, Eq. (6.11), gives

$$\sum_{j=1}^{J} \frac{\partial U_S}{\partial a_j} \, \delta a_j - \sum_{i=1}^{n} P_i \sum_{j=1}^{J} \frac{\partial v_i}{\partial a_j} \, \delta a_j - \int_0^l p \sum_{j=1}^{J} \frac{\partial v}{\partial a_j} \, \delta a_j \, dx = 0$$

or upon collecting terms

$$\sum_{j=1}^{J} \left[\frac{\partial U_S}{\partial a_j} - \sum_{i=1}^{n} P_i \frac{\partial v_i}{\partial a_j} - \int_0^l p \frac{\partial v}{\partial a_j} \, dx \right] \delta a_j = 0$$

Since δa_j are arbitrary variations, the last equation gives

$$\frac{\partial U_S}{\partial a_j} - \sum_{i=1}^{n} P_i \frac{\partial v_i}{\partial a_j} - \int_0^l p \frac{\partial v}{\partial a_j} \, dx = 0 \qquad (6.34)$$

Equation (6.34) provides J relationships for finding the J unknown parameters a_j. For bending of beams, the elastic energy is (Eq. (6.7))

$$U_S = U_b = \tfrac{1}{2} \int_0^l EI v''^2 \, dx$$

and

$$\frac{\partial U_b}{\partial a_j} = \frac{\partial U_b}{\partial v''} \frac{\partial v''}{\partial a_j} = \int_0^l EI v'' \frac{\partial v''}{\partial a_j} \, dx$$

Then Eq. (6.34) becomes

$$\int_0^l EI v'' \frac{\partial v''}{\partial a_j} \, dx - \sum_{i=1}^{n} P_i \frac{\partial v_i}{\partial a_j} - \int_0^l p \frac{\partial v}{\partial a_j} \, dx = 0 \qquad j = 1 \text{ to } J \qquad (6.35)$$

P_i are concentrated forces applied at x_i and v_i are deflections at those points; p is a uniformly distributed load and v is the deflection at x.

Example 6.14. For the simply supported beam with a uniform load of Fig. 6.17 find the approximate values of the mid-span deflection and moment, and compare them with the exact values.

Solution. Assume $v = a_1 \sin(\pi x / l)$, which satisfies the geometric boundary conditions. Since there are no concentrated forces, Eq. (6.35) becomes

$$\int_0^l EI v'' \frac{\partial v''}{\partial a_1} \, dx - \int_0^l p \frac{\partial v}{\partial a_1} \, dx = 0$$

$$v'' = -a_1 \frac{\pi^2}{l^2} \sin \frac{\pi x}{l} \qquad \frac{\partial v''}{\partial a_1} = -\frac{\pi^2}{l^2} \sin \frac{\pi x}{l} \qquad \text{and} \qquad \frac{\partial v}{\partial a_1} = \sin \frac{\pi x}{l}$$

l, E, I **FIGURE 6.17**

Thus,

$$\int_0^l EIa_1 \frac{\pi^4}{l^4} \sin^2 \frac{\pi x}{l} \, dx - \int_0^l p \sin \frac{\pi x}{l} \, dx = 0$$

$$EIa_1 \frac{\pi^4}{l^4} \int_0^l \left[\tfrac{1}{2} - \tfrac{1}{2} \cos \frac{2\pi x}{l} \right] dx - p \int_0^l \sin \frac{\pi x}{l} \, dx = 0$$

$$EIa_1 \frac{\pi^4}{l^4} \left[\frac{x}{2} - \frac{l}{4\pi} \sin \frac{2\pi x}{l} \right]_0^l + \frac{pl}{\pi} \left[\cos \frac{\pi x}{l} \right]_0^l = 0$$

$$EIa_1 \frac{\pi^4}{l^4} \frac{l}{2} - \frac{2pl}{\pi} = 0$$

$$a_1 = \frac{4pl^4}{\pi^5 EI}$$

and

$$v = \frac{pl^4}{\pi^5 EI} \sin \frac{\pi x}{l}$$

At the mid span,

$$v_0 = \frac{4pl^4}{\pi^5 EI} = \frac{pl^4}{76.5EI}$$

The exact deflection, from Example 6.7, is

$$v_0 = \frac{5pl^4}{384EI} = \frac{pl^4}{76.8EI}$$

The difference is 0.4%. The equation for the moment is

$$M = -EIv'' = \frac{4pl^2}{\pi^3} \sin \frac{\pi x}{l}$$

resulting in a mid-span moment of $pl^2/7.75$. This differs by 3.2% from the exact solution, $pl^2/8$. As we can see, the approximate value of the moment is much less accurate than that of the deflection. Of course the accuracy can be improved by using more than one term in the assumed displacement function. The results improve substantially with each additional term. Generally, using two or three terms gives very accurate results.

LABORATORY EXPERIMENTS

Energy of a Spring

Hang a spring and use a weight to apply a load at the lower end of the spring. Apply the weight suddenly and observe the vibration that the sudden transfer of the potential energy of the weight causes. Then apply the weight W slowly and observe the deformation of the spring. Measure the final deflection v. Calculate the stiffness of the spring $K = W/v$.

Apply increasing values of loads in several steps. Measure the deflection at each step and plot the load–deflection curve. Observe that this is a linear curve. Calculate the area under this curve and show that it is equal to the elastic energy of the spring, $\frac{1}{2}Kv^2$.

Principle of Real work

Use a model of the structure in Fig. 6.12 made out of a steel strip with $\frac{1}{16}$ in or larger thickness, or a circular bar, so that lateral buckling is not a problem. Calculate the internal energy of this system under an applied load at the end. Show that it is equal to the value of the load times the end deflection/2.

Truss Deflection

Use the truss of the Bar Forces experiment in Chapter 3 or one allowing larger deflections. Apply a load at a joint, measure the deflection at that or another joint and compare the measured and calculated deflections.

Castigliano's Theorem

Express the energy of the system in Fig. 6.12 in terms of a parameter P, the load applied at the end of the structure. Find the end deflection by Castigliano's second theorem, Eq. (6.22). Now apply an end load, measure the end deflection and compare the measured and calculated values.

A Nonlinear Problem

Set up a system with three hinges as in Example 6.11. Observe that under load the displacement is not zero as the small deformation theory would indicate. Place a simple load cell, such as described in the Laboratory Experiments of Chapter 4, in line with one end of the structure and show that the internal force is not zero as the linear theory would predict.

Shear Deflection

Shear deflection becomes significant in beams with low span-to-depth ratios and those with small shear areas, such as beams with web cut-outs. Use a rather deep thin rectangular beam (not so thin to buckle laterally), or a beam with large cut-outs, with fixed ends. Apply a load at the mid-span and measure the deflection there. Compare the measured deflection with the calculated value of bending deflection only. Calculate the shear deflection (see the last heading in Section 6.4), and compare the measured value to the total of bending and shear deflections. Use a loading jack with a load cell or weights to load the beam, and measure the deflection with a dial gage.

Maxwell–Betti's Reciprocal Theorem

Use a bar with circular cross-section to construct the structure of Fig. L6.1. The fixed end may be produced by bending the bar a distance and clamping it to a table at two points on this distance. Apply a load P_1 at point 1 and measure the deflection v_{21} at point 2 (deflection at point 2 due to a load at point 1). Remove the load P_1 and apply a load P_2 at point 2 and measure the deflection v_{12} at point 1 (deflection at point 1 due to a load at 2). Show that Betti's law is valid, that is, $P_1 v_{12} = P_2 v_{21}$.

Use the same (or unit) load in both cases and verify Maxwell's theorem that is $v_{12} = v_{21}$.

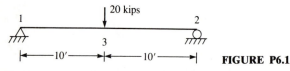

FIGURE L6.1

PROBLEMS

6.1. Find the deflection of the beams of Fig. P6.1 at point 3 by the method of real work.

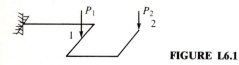

FIGURE P6.1

6.2. Repeat the problem of Example 6.10 using the method of real work.

6.3. Calculate the vertical deflection of joint 5 of the truss of Fig. P6.3. The cross-sectional area of all the bars is $2\ \text{in}^2$, and $E = 29\,000$ ksi.

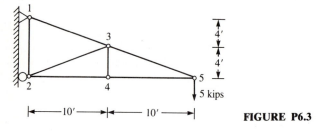

FIGURE P6.3

6.4. Find the vertical and horizontal deflections of joint 5 of Fig. P6.4. The

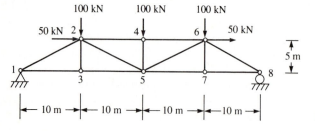

FIGURE P6.4

cross-sectional area of bars 23, 45 and 67 is 10 cm² and that of all the other bars 16 cm². $E = 200\,000$ MPa.

6.5. Find the bar forces for the truss of Fig. P6.5. The area of the cross-section of all the bars is 6 in² and $E = 29\,000$ ksi.

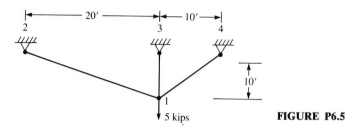

FIGURE P6.5

6.6. Calculate the bar forces for the truss of Fig. P6.6.

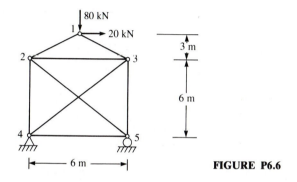

FIGURE P6.6

6.7. In the pin-jointed plane truss of Fig. P6.7 all the bars are of the same material with the same cross-section. A vertical load P is applied at joint 1 and an adjustable tie between joints 2 and 4 is then shortened until joint 1 returns to its original height. Find the final tension in the tie.

Hint: Calculate the deflection v_1 of 1 without the tie-bar. Then find the deflection δ_1 of 1 due to a unit force in the tie bar (unit forces at 2 and 4). Find the force F_{24} in the tie bar from $v_1 + F_{24}\delta_1 = 0$.

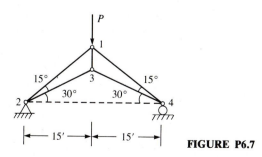

FIGURE P6.7

6.8. Find the deflection at point 3 of the beam in Fig. P6.8 in terms of EI by the virtual work method.

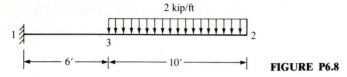

FIGURE P6.8

6.9. Calculate the deflection at point 3 of the beam of Fig. P6.9 in terms of EI by the virtual work method.

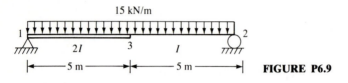

FIGURE P6.9

6.10. Using the virtual work method, find the maximum deflection of the beam in Fig. P6.10.

    Hint: Calculate the deflection at distance ξ and find the value of ξ giving the maximum deflection.

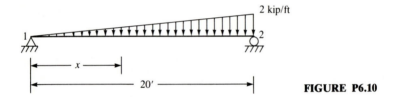

FIGURE P6.10

6.11. Repeat Example 6.2 using the virtual work method.

Do Problems 6.12 to 6.16 by the virtual work method with visual integration.

6.12. Calculate the deflection of the free end of the beam of Fig. P6.12. $E = 29\,000$ ksi.

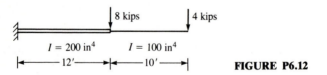

FIGURE P6.12

6.13. Find the deflection of the free end of the beam of Fig. P6.13 in terms of E and I.

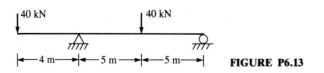

FIGURE P6.13

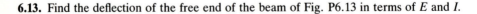

6.14. Determine the deflection at the free end of the beam of Fig. P6.14 in terms of p_0, l, E and I.

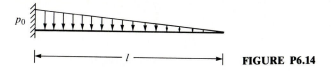

FIGURE P6.14

6.15. Find the horizontal deflection of the roller end in Fig. P6.15 (a) for the applied horizontal load of 10 kips at joint 2; (b) due to a horizontal load of 1 kip at the roller.

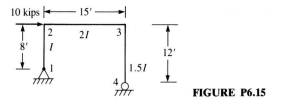

FIGURE P6.15

6.16. Calculate the free-end deflection of the beam of Fig. P6.16 in terms of constant E and I.

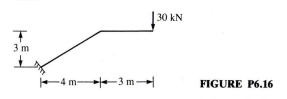

FIGURE P6.16

6.17. For the beam of Fig. P6.17, find the deflection at the free end of the overhang by Castigliano's theorem in terms of EI.

> *Hint*: Do the problem in terms of P and substitute for P in the expression found for the deflection.

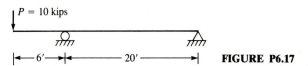

FIGURE P6.17

6.18. Using Castigliano's theorem find the deflection of the beam of Fig. P6.18 at point 3.

> *Hint*: Use P instead of the load at point 3 and substitute for P in the expression found for the deflection.

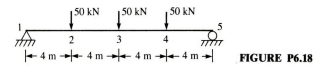

FIGURE P6.18

6.19. Find the rotation at the left end of the beam in Fig. P6.19 in terms of M_1, l, E, and I, using Castigliano's theorem.

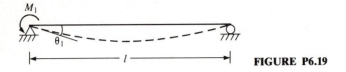

FIGURE P6.19

6.20. Find the bending and shear deflections at the free end of the deep beam (shear wall) in Fig. P6.20. The width of the beam is 0.5 ft, and $E = 3\,000$ ksi, $v = 0.2$.

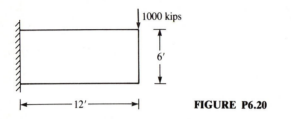

FIGURE P6.20

6.21. Assuming one trigonometric term (see Example 6.14) find the deflection of point 3 in Problem 6.1.

6.22. Verify that $v = a[1 - \cos(2\pi x/l)]$ can be used to approximate the deflected shape of a fixed-end beam under a uniform load by checking that it satisfies the geometric boundary conditions (zero slope and deflection at the ends), and gives nonzero deflection at the mid-span. Using this expression with a as the unknown parameter, find the value of a by the Rayleigh–Ritz method. Then calculate the deflection at the mid-span and find the values of moment and shear at the end as well as at the mid-span, (denoted as points 1, 2 and 3).

   Find the percentage error, compare with the exact solution, which is $v_3 = pl^4/384EI$, $M_1 = M_2 = -2M_3 = -pl^2/12$, and $V_1 = V_2 = pl/2$, $V_3 = 0$.

6.23. Using the approximate displacement function $v = a[1 - \cos(\pi x/2l)]$, where a is an unknown parameter, find the displacement at the free end of the cantilever beam of Problem 6.8.

CHAPTER 7

METHOD OF CONSISTENT DEFORMATIONS

Three Moment Equation
and Portal Method

7.1 METHOD OF CONSISTENT DEFORMATIONS

In discussing reactions and determinacy of beams in Chapter 2 we noted that the degree of indeterminacy or the number of redundant forces is equal to the number of unknown reactions in excess of the equations of statics (plus equations of condition). Once the redundant forces are determined, then the forces at any point in the structure can be calculated by statics and deflections can be obtained by one of the methods discussed in Chapters 5 or 6. The redundant forces can be chosen as reactions or moments. Additional equations for finding the redundant forces are obtained by considering the deformation of the beam. Consider, for example, the two-span beam of Fig. 7.1a. This beam has one redundant force. Let us use the middle reaction, R_1, as the redundant force. To find this reaction we proceed as follows. First assume that support 1 is removed. This would lead to a statically determinate structure called the *primary structure*, Fig. 7.1b, with a deflection v_{10} at point 1. The subscript zero indicates that the redundant force is set equal to zero, and the deflection of the primary structure is found under the applied loads.

If we apply the reaction R_1, without the external load, it will produce the deflection v_{11} (Fig. 7.1c). If the load and the reaction force are acting simultaneously (Fig. 7.1a) the deflection at support 1 must be zero. That is, we

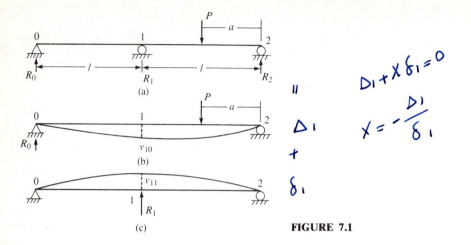

$$\Delta_1 + X \delta_1 = 0$$

$$\|$$

$$\Delta_1 \qquad X = -\frac{\Delta_1}{\delta_1}$$

$$+$$

$$\delta_1$$

FIGURE 7.1

must have

$$v_{10} + v_{11} = 0 \tag{7.1}$$

To find the deflection of point 1 due to the reaction, we can first calculate the deflection δ_{11} at 1 due to a unit load (acting downward) at 1. The deflection due to the reaction is then

$$v_{11} = -\delta_{11} R_1 \tag{7.2}$$

Equation (7.1) can then be written as

$$v_{10} - \delta_{11} R_1 = 0 \tag{7.3a}$$

The solution of Eq. (7.3a) gives

$$R_1 = \frac{v_{10}}{\delta_{11}}. \tag{7.3b}$$

If there are more than one redundant reactions, the primary structure is obtained by removing them all. The deflections at all the points of application of the redundant reactions in the primary structure, v_{10}, v_{20}, ..., are found. Then the loads are removed and unit downward loads are applied, one at a time, at the location of the redundant reactions. The deflections at the locations of the redundants, δ_{11}, δ_{21}, ..., δ_{n1} are calculated for a unit load at the location of the first redundant. Similarly δ_{12}, ..., δ_{nn} are found for unit loads at the locations of other redundants. Setting the deflections of all the supports equal to zero we obtain the following system of equations.

$$\delta_{11} R_1 + \delta_{12} R_2 + \cdots + \delta_{1n} R_n - v_{10} = 0$$
$$\cdots$$
$$\cdots \tag{7.4}$$
$$\delta_{n1} R_1 + \delta_{n2} R_2 + \cdots + \delta_{nn} R_n - v_{n0} = 0$$

If the deflection at a given support is not zero, but a specified value, for example due to settlement, then the right-hand side of the equation will not be zero, but equal to the specified deflection. Note that in Eq. (7.4), δ_{12} is the deflection at the location of the first redundant reaction due to a unit force at the second redundant reaction. The solution of Eq. (7.4) gives the values of the redundant forces $R_1, R_2, \ldots, R_n$.

The choice of redundant forces is somewhat arbitrary. The amount of work will be reduced by taking advantage of symmetry if it exists and by selecting the primary structure so that the effect of the loads is localized. For the beam of Fig. 7.1a the reactions R_0 or R_2, or moment M_1, at the middle support, may be used as the redundant instead of R_1. When support moments are used as redundants, then the primary structure consists of a number of simply supported spans under the applied loads and consistent deformations involve the continuity of the slopes. That is, the values of slope calculated at the left and right of the support i, θ_i' and θ_i'', are set equal. Thus, instead of Eq. (7.4), Eq. (7.5) below is used for satisfying the compatibility of slopes:

$$\theta_i' = \theta_i'' \tag{7.5}$$

where

$$\theta_i' = \theta_i'^0 + \theta_i'^{i-1} M_{i-1} + \theta_i'^i M_i \tag{7.6}$$

and

$$\theta_i'' = \theta_i''^0 + \theta_i''^i M_i + \theta_i''^{i+1} M_{i+1} \tag{7.7}$$

$\theta_i'^0$ is the slope at the left of support i for the primary structure (simply supported) under the applied loads. $\theta_i''^0$ is the slope at the right of support i for the primary beam. $\theta_i'^{i-1}$ and $\theta_i'^i$ are slopes at the right end of the left span (left of support i) due to unit moments acting at the left and right ends of that span. $\theta_i''^i$ and $\theta_i''^{i+1}$ are slopes at the left end of the right span (right of support i) due to unit moments at the left and right ends of that span.

Note that a combination of support reactions and moments can be used as redundants. Once the redundant forces or moments are found, all the other reactions as well as shear and moment at any point can be found by statics.

Steps in the Method of Consistent Deformations

The following steps are involved in the application of the method of consistent deformations.

1. Select the redundant reactions or moments.
2. Remove the redundant reactions to obtain a primary structure. Find $v_{10}, v_{20}, \ldots$, (or $\theta_1'^0, \theta_1''^0, \ldots, \theta_2'^0, \theta_2''^0, \ldots$), deflections (or rotations) of the primary structure at the redundant forces under the applied loads.
3. Remove the applied loads from the primary structure and apply a unit load (or moment) at and in the direction of the first redundant force. Find $\delta_{11}, \delta_{21}, \ldots$, (or θ_i', θ_i'') deflections (or rotations) at all the redundant forces (or moments).

4. Repeat step 3 for all the redundant forces to obtain $\delta_{12}, \delta_{22}, \ldots, \delta_{31}$, $\delta_{32}, \ldots, \delta_{1n}, \delta_{2n}, \ldots$, etc. (or $\theta_i'^1, \theta_i''^1$, etc.).

5. Set up the compatibility equations, Eq. (7.4) (or Eq. (7.5)), and solve for the redundant forces or moments.

6. Calculate the other reactions by statics and if necessary find shear, moment and deflection throughout the beam.

The deflections and slopes for some common loads are given in Table 7.1.

Example 7.1. Find the reaction R_2 of the two-span beam of Fig. 7.2a under a concentrated load of 5 kips acting at 10 ft from the right-hand end of the beam. The modulus of elasticity and moment of inertia of the beam are E and I.

Solution. Following the steps in the method of consistent deformations, we find the solution by superposition. We obtain the primary structure by removing the reaction R_2. The deflection of the primary structure at point 2 (Fig. 7.2b) is found by the moment-area method. The left reaction of the primary structure in Fig. 7.2b is

$$R_1^0 = 5 \times \frac{10}{40} = 1.25 \text{ kips}$$

The moment diagram for the primary structure is given in Fig. 7.2c, yielding the deflection v_{20} as

$$v_{20} = \tfrac{1}{2}t_{31} - t_{21}$$

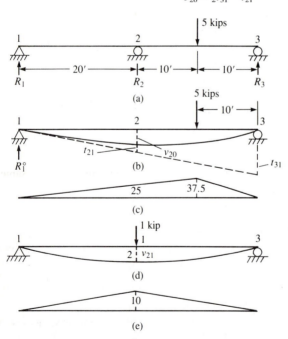

(a)

(b)

(c)

(d)

(e)

FIGURE 7.2

By the second moment–area theorem,

$$t_{31} = \frac{1}{2} \times \frac{37.5}{EI} \times 30 \left(\frac{30}{3} + 10 \right) + \frac{1}{2} \times \frac{37.5}{EI} \times 10 \times \frac{2}{3} \times 10 = \frac{12\,500}{EI}$$

and

$$t_{21} = \frac{1}{2} \times \frac{25}{EI} \times 20 \times \frac{1}{3} \times 20 = \frac{1\,667}{EI}$$

$$v_{20} = \frac{1}{2} \times \frac{12\,500}{EI} - \frac{1\,667}{EI} = \frac{4\,583}{EI}$$

To find the deflection at 2 due to reaction R_2, we apply a unit downward load to the primary structure at this point (Fig. 7.2d). The moment diagram for the unit load is shown in Fig. 7.2e. The deflection at point 2 due to this unit load is

$$\delta_{21} = t_{32} = \frac{1}{2} \left(\frac{10}{EI} \right) \times 20 \times \frac{2}{3} \times 20 = \frac{1\,333}{EI}$$

Thus, the deflection due to R_2 is

$$v_{21} = -\delta_{21} R_2 = -\frac{1\,333}{EI} R_2$$

The total deflection at support 2, which is the sum of the two displacements, must be equal to zero,

$$v_{20} + v_{21} = v_{20} - \delta_{21} R_2 = 0 \qquad \text{or} \qquad \frac{4\,583}{EI} - \frac{1\,333}{EI} R_2 = 0$$

which gives $R_2 = 3.44$ kips.

Example 7.2. Using the support moments as redundants in Fig. 7.3, find the values of the redundants.

Solution. Because of symmetry, the redundant moments are the same at supports 2 and 3. Thus, M_3 is equal to M_2. The redundant moments are shown in the positive direction (causing compression at the top of the beam). The redundants provide continuity of the slopes at the supports.

The primary structure is obtained by removing the redundant moments. This will result in three single-span beams under uniform loads. Since there is a single redundant moment, only one equation is needed to find its value. This equation can be provided by specifying the continuity of rotation at support 2, that is, the rotation θ_2' calculated at the left of support 2 and θ_2'' calculated at the right of 2 must be equal. Therefore, $\theta_2' = \theta_2''$.

The slopes θ_2' and θ_2'' can be calculated by the moment–area method. The moment diagrams for the external loads and unit value of the redundant moment are shown in Figs. 7.3c and 7.3d. The angles θ_2' and θ_2'' are equal to the sum of

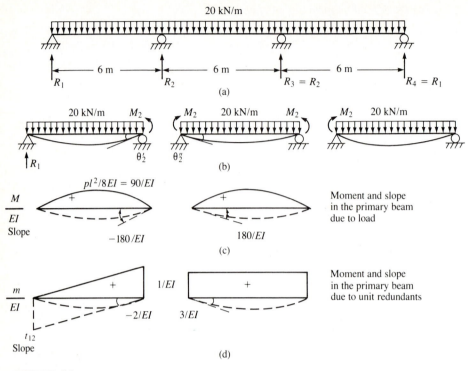

FIGURE 7.3

the rotations due to the external loads and due to the redundant moment. That is,

$$\theta_2' = \theta_2'^0 + \theta_2'^1 M_2 \quad \text{and} \quad \theta_2'' = \theta_2''^0 + \theta_2''^1 M_2 + \theta_2''^2 M_2$$

where superscript 0 indicates the rotation due to the external loads and superscripts 1 and 2 denote the rotations due to unit values of M_2 and M_3 ($= M_2$ here). The magnitudes of rotations are indicated below the moment diagrams (Figs. 7.3c and 7.3d): $\theta_2'^0$ due to the uniform load (Fig. 7.3c) is equal to the change of angle between the mid-span and the end of this span; $\theta_2'^1$ due to the unit redundant moment (Fig. 7.3d) is t_{12}/l. Where t_{12} is found by taking the moment of the M/EI diagram with respect to the left hand end. The value of θ_2' due to the redundant moment is thus $\theta_2'^1 M_2$. Thus,

$$\theta_2'^0 = -\left(\frac{2}{3} \times \frac{90}{EI} \times \frac{6}{2}\right) = -\frac{180}{EI}$$

$$\theta_2'^1 = -\left(\frac{1}{2} \times \frac{6}{EI} \times \frac{2}{3} \times \frac{6}{6}\right) = -\frac{2}{EI}$$

and

$$\theta_2''^0 = \left(\frac{2}{3} \times \frac{90}{EI} \times \frac{6}{2}\right) = \frac{180}{EI}$$

$$\theta_2''^1 + \theta_2''^2 = \left(\frac{1}{EI} \times \frac{6}{2}\right) = \frac{3}{EI} \qquad \text{due to unit moments at both ends}$$

$$\theta_2' = -\frac{180}{EI} - \frac{2}{EI}M_2 \qquad \text{and} \qquad \theta_2'' = \frac{180}{EI} + \frac{3}{EI}M_2$$

Continuity of slope, $\theta_2' = \theta_2''$, gives

$$-\frac{180}{RI} - \frac{2}{EI}M_2 = \frac{180}{EI} + \frac{3}{EI}M_2$$

which leads to $5M_2 = -360$, or $M_2 = -72 \text{ kN} \cdot \text{m}$

7.2 SHEAR AND MOMENT DIAGRAMS FOR CONTINUOUS BEAMS

Shear and moment diagrams were discussed in detail in Chapter 4. Here we mention some additional points pertaining to continuous beams and frames. After the redundant forces are calculated, other reactions can be found by the equations of statics. Shear and moment diagrams for continuous beams can be constructed in a similar manner to those of statically determinate beams. Thus, shear at a section can be found by summing the forces on the left of that section. Similarly, moment at a section can be determined by finding the sum of moments at that section of all the forces on the left of that section.

Usually the negative moments at the supports and the maximum positive moments in the spans are needed for design. The values of the moments at mid-spans are often close to the maximum value and may suffice for design purposes. Moment and shear diagrams with values at the supports and at the mid-spans can be readily obtained by superposition. This is done by first drawing the moment diagram due to the support moments. This diagram consists of straight lines connecting the support moments. Positive moments due to the external loads in each span are then plotted and superimposed on the support moment diagram. It is convenient to use the lines of the support moment diagram as the x axis. The validity of this process can be shown by noting that each span of the continuous beam is a simply supported beam under the loads plus the support moments acting at the two ends. The moment diagram for the span would be that for the simply supported beam with the loads plus the diagram for the end moments.

Example 7.3. Find the shear and moment diagrams for the beam of Example 7.2, Fig. 7.3a.

Solution. To find the left reaction R_1 consider the left span in Fig. 7.3b. Equilibrium of moments at the right hand end, gives

$$R_1 \times 6 - 20 \times \frac{6^2}{2} - M_2 = 0$$

The redundant moment was found in Example 7.2, $M_2 = -72 \text{ kN} \cdot \text{m}$. Thus,

$$R_1 = 60 + \frac{M_2}{6} = 60 - \frac{72}{6} = 48 \text{ kN}$$

R_2 can be found as the sum of the reactions at point 2 from the left and right spans (Fig. 7.3b) or by considering the equilibrium of the three-span beam (Fig. 7.3a). Equilibrium of vertical forces for the latter beam considering the symmetry, gives

$$2R_1 + 2R_2 - 20 \times 18 = 0$$
$$R_2 = 180 - R_1$$
$$= 180 - 48 = 132 \text{ kN}$$

In Fig. 7.3a, shear at the left of support 2 is found as sum of the forces on the left of this section. The reaction R_2 is equal to the jump in the shear diagram at 2, shear at the right of support 2 is found by adding R_2 to the shear at the left of 2. The moment diagram is obtained by first plotting the negative support moments. The moment diagram for the first span is the algebraic sum of the positive and negative moment diagrams, that is, Fig. 7.3c and Fig. 7.3d times M_2 $(-72 \text{ kN} \cdot \text{m})$.

The static moment diagram (Fig. 7.3c) for a uniform load is a parabola. The maximum static moment for the load is $pl^2/8 = 90 \text{ kN} \cdot \text{m}$ at the mid-span. The negative moment at the middle of the first span due to M_2 is $\frac{1}{2}(-72) = -36 \text{ kN} \cdot \text{m}$. Thus, the positive moment at the middle of the first span is $90 - 36 = 54 \text{ kN} \cdot \text{m}$. Similarly, the positive moment at the middle of the second span is $90 - 72 = 18 \text{ kN} \cdot \text{m}$. The final shear and moment diagrams are shown in Figs. 7.4a and 7.4b.

Note that the support reactions are equal to the algebraic sum of the shear on the two sides of the support, that is the jumps in the shear diagram.

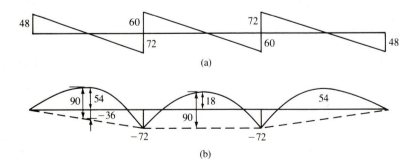

FIGURE 7.4

Analysis of Frames by the Method of Consistent Deformations

The method of consistent deformations can be used for simple frames with a small number of redundant forces. For more complex frames the slope deflection or moment distribution methods (Chapter 10) may be used. However, the stiffness method programmed on a computer is usually used for large frames. Application of the method of consistent deformations consists of establishing the compatibility of displacements and rotations at the supports according to the support conditions.

Example 7.4. Find the unknown reactions of the frame of Fig. 7.5.

Solution. There are four unknown reactions R_{1x}, R_{1y}, M_1, and R_3. Therefore, the frame is statically indeterminate to the first degree. To find an additional equation we remove the reaction R_3 to obtain a primary frame. The latter frame undergoes a vertical deflection v_3 at point 3 under the load P. The effect of the reaction R_3 is to restore this displacement to zero, that is, R_3 produces a vertical displacement at 3 such that the sum of the two displacements is zero. Therefore, we calculate the deflection of the primary structure at point 3 due to the applied load P (Fig. 7.5b), and due to reaction R_3 (Fig. 7.5c). The latter deflection would be equal $-\delta_{11}R_3$. Where δ_{11} is the vertical deflection at 3 due to a unit downward load there. Setting the sum of these two deflections equal to zero yields R_3.

To find the deflection of the frame in Fig. 7.5b, we note that v_3 is the sum of the end deflection of the cantilever 23 (with end 2 fixed) and the deflection at 3 due to the rotation of the axis of member 23. If the rotation of point 2 is θ_2, then the deflection at 3 due to this rotation is $\theta_2 l$. However, because of the rigid connection at 2, the rotation of end 2 of member 23 is the same as the rotation of end 2 of member 12. Using the moment–area method (Section 5.2), this rotation

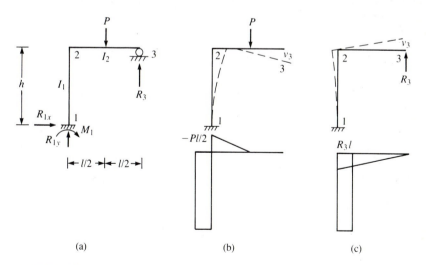

(a)   (b)   (c)

FIGURE 7.5

is

$$\theta_2 = -\frac{(-Pl/2) \times h}{EI_1} = \frac{Plh}{2EI_1}$$

Thus

$$v_{30} = \frac{Plh}{2EI_1}\,l + \frac{Pl}{2EI_2} \times \frac{1}{2} \times \frac{l}{2} \times \left(\frac{l}{2} + \frac{2}{3} \times \frac{l}{2}\right) = \frac{Pl^2h}{2EI_1} + \frac{5Pl^3}{48EI_2}$$

$$\delta_{31} = \frac{l}{EI_1} \times h \times l + \frac{l}{EI_2} \times \tfrac{1}{2}l \times \tfrac{2}{3}l = \frac{l^2h}{EI_1} + \frac{l^3}{3EI_2}$$

$$v_{30} - \delta_{31}R_3 = 0, \qquad R_3 = \frac{v_{30}}{\delta_{31}}$$

or

$$R_3 = \frac{(h/2I_1) + (5l/48I_2)}{(h/I_1) + (l/3I2)}\,P$$

Let

$$\beta = \frac{hI_2}{lI_1} \qquad \text{then} \qquad R_3 = \frac{5 + 24\beta}{1 + 3\beta}\frac{P}{16}$$

All the other reactions, as well as shear and moment diagrams can be determined by the equations of statics.

Note that for more complex frames it would be easier to use the numerical values of lengths and loads in the calculations.

7.3   THREE-MOMENT EQUATION

The three-moment equation provides a simple method of solution for continuous beams. It is derived by applying the consistent deformation method to a support of a continuous beam under arbitrary loads. This results in an equation in terms of the moments at three adjacent supports. The solution of any continuous beam can be found by repeated application of this equation until a sufficient number of equations is found for calculating the unknown moments at the supports.

Consider the continuous beam of Fig. 7.6 under arbitrary loads. The moments at supports $i-1$, i and $i+1$ are M_{i-1}, M_i, and M_{i+1}. The moment diagrams in the spans are the sum of moments due to the loads (on simply supported spans) plus that due to the end moments (Figs. 7.6b and 7.6c).

The end rotations for each span are sum of the rotations due to the external loads, plus the rotations due to the support moments. Thus, the rotation at the right-hand end of the left span is

$$\theta_i' = \theta_i'^0 + \theta_i'^{i-1}M_{i-1} + \theta_i'^i M_i$$

where $\theta_i'^0$ is the rotation of the right-hand end due to the external loads, $\theta_i'^{i-1}$ is the rotation there due to a unit value of M_{i-1} and $\theta_i'^i$ is the rotation at the right-hand end due to a unit value of M_i. Similarly, the rotation at the

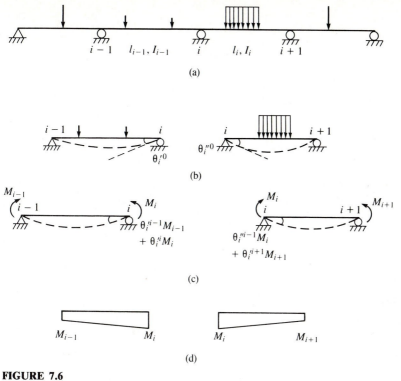

FIGURE 7.6

left-hand end of the right support is

$$\theta_i'' = \theta_i''^0 + \theta_i''^i M_i + \theta_i''^{i+1} M_{i+1}$$

Using the conjugate beam method we can calculate the rotations due to the unit moments. Therefore, the rotations at the right end of the left span due to unit moments at the left and right ends are (Figs. 7.6c and 7.6d), with clockwise rotations positive according to our sign convention,

$$\theta_i'^{i-1} = -\frac{l_{i-1}}{6EI_{i-1}} \quad \text{and} \quad \theta_i'^i = -\frac{l_{i-1}}{3EI_{i-1}}$$

and the rotations at the left end of the right span due to unit moments are

$$\theta_i'''^i = \frac{li}{3EI_i} \quad \text{and} \quad \theta_i'''^i = \frac{li}{6EI_i}$$

Because of the continuity of the continuous beam at support i, the rotations calculated at the left and right of i must be the same, $\theta_i' = \theta_i''$, or

$$\theta_i'^0 - \frac{l_{i-1}}{6EI_{i-1}} M_{i-1} - \frac{l_{i-1}}{3EI_{i-1}} M_i = \theta_i''^0 + \frac{l_i}{3EI_i} M_i + \frac{l_i}{6EI_i} M_{i+1}$$

TABLE 7.1
Deflections and end rotations

Case	θ_1^{r0}	θ_2^{r0}	Deflection (at mid-span) unless otherwise specified
Distributed load	$\dfrac{pl^3}{24EI}$	$-\dfrac{pl^3}{24EI}$	$\dfrac{5pl^4}{384EI}$
Concentrated load at mid-span	$\dfrac{Pl^2}{16EI}$	$-\dfrac{Pl^2}{16EI}$	$\dfrac{Pl^3}{48EI}$
Concentrated load at a	$\dfrac{Pa}{6EIl}(l-a)(2l-a)$	$-\dfrac{Pa}{6EIl}(l^2-a^2)$	$\dfrac{Pa^2b^2}{3EIl}$ (at $x=a$)
Linearly varying load	$\dfrac{7pl^4}{360EI}$	$-\dfrac{8pl^4}{360EI}$	$\dfrac{5pl^5}{768EI}$ $\dfrac{pl^5}{153.2EI}$ (max defl at $x=0.519l$)
Partially distributed load	$\dfrac{p}{24EIl}[2bl^2(b+2c)-(l-a)^4+c^4]$	$-\dfrac{p}{24EIl}\big[4bl^2(b+2c)$ $-(l-a)^3(3l+a)+c^3(4l-c)\big]$	$\dfrac{pb}{384EI}[8l^3-b^2(4l-b)]$ (for $c=a$)

which gives

$$\frac{l_{i-1}}{I_{i-1}} M_{i-1} + 2\left(\frac{l_{i-1}}{I_{i-1}} + \frac{l_i}{I_i}\right) M_i + \frac{l_i}{I_i} M_{i+1} = -6E(\theta_i''^0 - \theta_i'^0) \qquad (7.8)$$

If the moment of inertia is constant throughout the beam,

$$l_{i-1} M_{i-1} + 2(l_{i-1} + l_i) M_i + l_i M_{i+1} = -6EI(\theta_i''^0 - \theta_i'^0) \qquad (7.9)$$

For equal span beams with a constant cross-section,

$$M_{i-1} + 4M_i + M_{i+1} = -\frac{6EI}{l}(\theta_i''^0 - \theta_i'^0) \qquad (7.10)$$

$\theta_i''^0$ and $\theta_i'^0$ are the rotations at the right and left of support i calculated from simply supported beams under the external loads. The rotations for some common loads are given in Table 7.1. For other loads the rotations can be found by integration, moment–area, conjugate beam or virtual work method.

Note that if the left-hand support is fixed, then a span with zero values of span length, load, and M_{i-1} is assumed to the left of that support. If the right-hand support is fixed, then a span with zero values of length, load, and M_{i+1} is assumed on the right of that support. For fixed-end single-span beams, such zero spans are assumed on both sides of the span. On the other hand, overhangs are replaced by the moment that they produce at the support. If the beam and the loads are symmetric, then the moments at the symmetric supports will be the same, and the number of unknown moments can be reduced.

Example 7.5. Find the support moments for the beam of Fig. 7.7. EI is constant.

Solution. With subscripts 1, 2, 3, and 4 denoting the moments at the four supports, we apply the three-moment equation (Eq. (7.10) since I and l are constant) to supports 2 and 3. The end slopes for uniform and concentrated loads are given in Table 7.1. Thus,

$$M_1 + 4M_2 + M_3 = -\frac{6EI}{10}\left[\frac{8 \times 10^2}{16EI} + \frac{2 \times (10)^3}{24EI}\right]$$

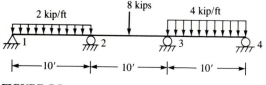

FIGURE 7.7

and

$$M_2 + 4M_3 + M_4 = -\frac{6EI}{10}\left[\frac{4 \times (10)^3}{24EI} + \frac{8 \times 10^2}{16EI}\right]$$

At supports 1 and 4 the moments are zero. Thus, $M_1 = M_4 = 0$ and the above equations yield

$$4M_2 + M_3 = -80$$

$$M_2 + 4M_3 = -130$$

or

$$M_2 = -12.67 \text{ kip} \cdot \text{ft}, \qquad M_3 = -29.33 \text{ kip} \cdot \text{ft}.$$

7.4 PORTAL METHOD

Approximate methods are often used for preliminary design of multistory buildings. Determination of the internal forces is done separately for the vertical and horizontal loads. Higher stresses in the beams are due to vertical loads, while the columns are stressed primarily by lateral loads. To find the beam stresses due to vertical loads it is assumed that all the beams are fixed at

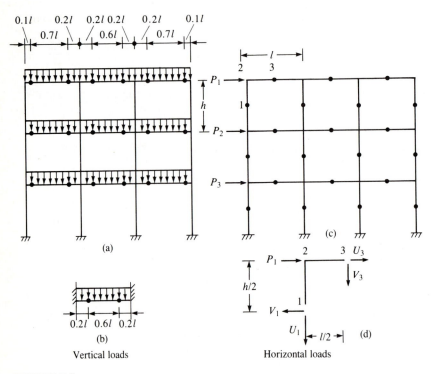

Vertical loads   Horizontal loads

FIGURE 7.8

the columns and except for the exterior beams they develop inflection points at $0.2l$ from the ends of the span (Fig. 7.8a). At the external sides of the exterior beams, the inflection point is assumed at $0.1l$. This renders the beams statically determinate and allows the determination of the shear and moment throughout the beam.

Thus, for the vertical loads the positive moment for the interior beams is (Fig. 7.8b)

$$M^+ = \frac{p(0.6l)^2}{8} = 0.045pl^2$$

The negative moment is

$$M^- = (-0.3pl)(0.2l) - \frac{p(0.2l)^2}{2} = -0.08pl^2$$

For analysis of the frame under lateral loads, either the portal or the cantilever method is used. Both methods assume inflection points or hinges at the mid-lengths of all the beams and columns. An additional condition is needed for finding the internal forces, Fig. 7.8c.

In the portal method the shear at the hinges of the interior columns is assumed to be twice that of the two exterior columns. The total shear force must balance the horizontal force acting above that level. In the cantilever method the additional condition is provided by assuming that the vertical forces in the columns vary in proportion to their distance from the centroid of all the columns, and that the moment these forces create must balance the moment of the horizontal force above that level. Starting from the top corner of the frame where a lateral load is applied, each segment is separated at its nearest hinge and the shear and normal forces at the hinge are found by equilibrium of the segment. The moments at the junction of the members can then be found. For example, consider segment 123 in (Fig. 7.8d). The internal forces are indicated such that they produce positive shear and normal forces. Equilibrium of the free body requires that

$$\sum F_x = 0 \qquad P_1 - V_1 + U_3 = 0$$

$$\sum M_3 = 0 \qquad \frac{V_1 h}{2} - \frac{U_1 l}{2} = 0$$

$$\sum F_y = 0 \qquad U_1 + V_3 = 0$$

Using the portal method, we assume that shear in the exterior columns is V_1 and in the interior columns $2V_1$. Equilibrium of the top half story of the frame with respect to horizontal forces gives

$$6V_1 - P_1 = 0 \qquad \text{or} \qquad V_1 = \frac{P_1}{6}$$

Substitution of V_1 in the above equilibrium equations yields

$$U_1 = \frac{h}{l}V_1 = \frac{hP_1}{(6l)}$$

$$V_3 = -U_1 = -\frac{hP_1}{(6l)}$$

$$U_3 = V_1 - P_1 = \frac{P_1}{6} - P_1 = -\frac{5P_1}{6}$$

We apply the opposite of the forces to the adjacent segments and proceed with finding the unknown forces in those segments.

LABORATORY EXPERIMENTS

Consistent Deformations

Use a steel strip to set up a two-span beam. Connect a wire to the beam at the location of the middle support, pass it over a pulley and attach a can to the other end of the wire. Apply a load in one of the spans. Fill the can gradually with sand or water until the deflection of the middle support becomes zero. Weigh the can and verify that it is equal to the calculated reaction of the middle support.

Portal Method

Use a two-story one-bay frame such as the one in Fig. L1.6. Apply lateral loads at the story levels. Observe that the inflection points (points of curvature change) occur at the mid-points of the columns and beams. Try an unsymmetric frame and observe that the latter statement is no longer true.

PROBLEMS

7.1. For the beam shown in Fig. P7.1, find the redundant force R_2 using the method of consistent deformations and draw the shear and moment diagrams.

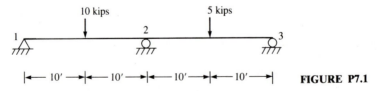

FIGURE P7.1

7.2. The supported cantilever beam of Fig. P7.2 is statically indeterminate to the first degree. Choose a redundant and using the method of consistent deformations calculate the value of this redundant force or moment.

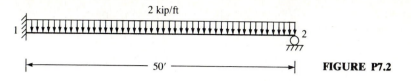

2 kip/ft

50'

FIGURE P7.2

7.3. Find the redundant force (i.e., the vertical reaction at point 2) for the beam in Fig. P7.3. E is constant.

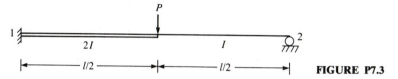

P

$2I$ I

$l/2$ $l/2$

FIGURE P7.3

7.4. For the beam shown in Fig. P7.4, find the redundant M_2 using the method of consistent deformations and draw the shear and moment diagrams.

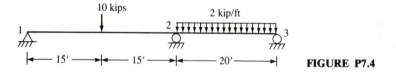

10 kips   2 kip/ft

15'   15'   20'

FIGURE P7.4

7.5. Find the unknown reactions in Fig. P7.5. Then draw the shear and moment diagrams.

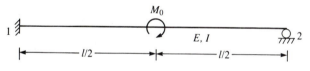

M_0

E, I

$l/2$ $l/2$

FIGURE P7.5

7.6. For the beam shown in Fig. P7.6, compute the moments at the fixed ends and draw the shear and moment diagrams.

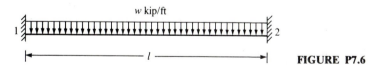

w kip/ft

l

FIGURE P7.6

7.7. For the beam in Fig. P7.7 find the redundant forces.

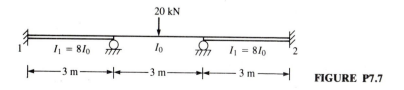

20 kN

$I_1 = 8I_0$ I_0 $I_1 = 8I_0$

3 m   3 m   3 m

FIGURE P7.7

7.8. For the beam of Fig. P7.8, (*a*) find the moment at point 2 using the three-moment equation; (*b*) find the reactions at points 2 and 3; (*c*) draw the moment diagram.

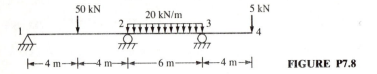

FIGURE P7.8

7.9. Using the reaction of point 3 (Fig. P7.9) as redundant, find this reaction with the method of consistent deformations. Draw the shear and moment diagrams. The moment of inertia of the beam is 200 in^4 and that of the column 100 in^4.

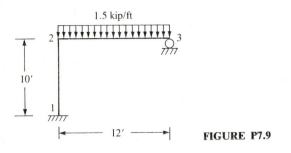

FIGURE P7.9

7.10. For the truss of Fig. P7.10 find the redundant reaction R_5 and calculate the bar forces. For all the bars $A = 5$ in^2, $E = 29\,000$ ksi.

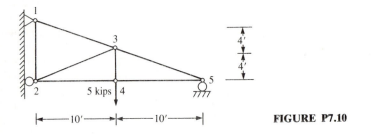

FIGURE P7.10

7.11. Find the bar forces for the truss of Fig. P7.11. The cross-sectional area of the vertical and diagonal bars is 20 cm^2 and that of the other bars 30 cm^2.

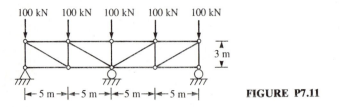

FIGURE P7.11

7.12. Using the three-moment equation find the moment at the middle support of the beam of Fig. P7.12. *EI* is constant.

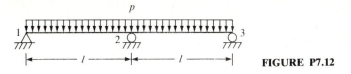

FIGURE P7.12

7.13. Determine the redundant moments and find the moments at the middle of the first and second spans of the beam of Fig. P7.13.

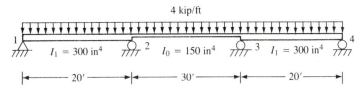

FIGURE P7.13

7.14. Find the moment at support 2 of the beam in Fig. 7.14.

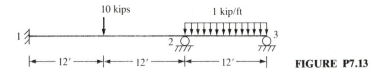

FIGURE P7.14

7.15. The frame of Fig. P7.15 is under the vertical and horizontal loads shown.
(a) Find the approximate moments in the beams under the vertical loads only.
(b) Find the moments in the beams and columns under the lateral loads using the portal or cantilever method.

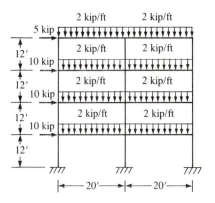

FIGURE P7.15

CHAPTER
8

INFLUENCE
LINES

Structures are often under moving loads or live loads that change their position. It is necessary to determine the location of the load that causes the maximum shear or moment. Consider, for example, the left reaction of a simply supported elastic beam, Fig. 8.1a, as a concentrated load traverses the beam from one end to the other.

If we construct a diagram giving the value of the left reaction as a unit load moves along the beam (Fig. 8.1b) then the reaction under a moving load P can be found by multiplying the ordinates of this diagram by P. The diagram of the reaction for a moving unit load is called the *influence line* of the left reaction. We will be able to find the critical location of the load causing the maximum reaction, as well as the value of the maximum reaction by using this diagram. With the influence line and superposition process we can also find the reaction for any number of loads acting at specific locations. Influence lines are commonly used to determine the pattern of live loads producing maximum effects.

8.1 DEFINITION OF INFLUENCE LINE

Influence lines for shear and moment represent the values of shear and moment at a fixed section of the beam when a unit load moves from one end of

204

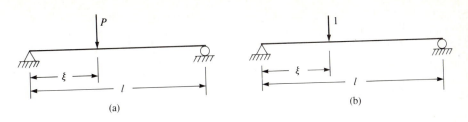

FIGURE 8.1

the beam to the other. In contrast, shear and moment diagrams for a beam represent the values of shear and moment at different sections of the beam for a fixed set of loads. The influence line of moment at a section of a simply supported beam (Fig. 8.3b) has a configuration similar to that of the moment diagram for a beam under a concentrated load. However, the two diagrams represent different things

Influence lines of other effects such as reaction or deflection can be obtained in a similar manner. Therefore, an influence line can be defined as follows.

> An influence line is a diagram whose ordinates are the values of some function of the structure (reaction, shear, moment, etc.) as a unit load moves across the structure.

8.2 CONSTRUCTION OF INFLUENCE LINES

Influence lines can be constructed in more than one way. In this section we discuss a direct procedure based on the definition of influence line. In Section 8.3 we introduce a short-cut procedure due to Muller–Breslau.

To find the influence line of the reaction of a simply supported beam, one possibility is to divide the beam into a number of divisions, say 10, and place a unit load at $0.1l$, $0.2l$, etc., along the beam (Fig. 8.2a), and calculate the reaction. The values of the reaction plotted at $0.1l$, $0.2l$, etc. in Fig. 8.2b

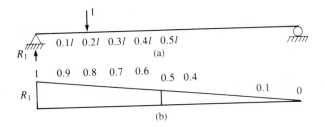

FIGURE 8.2

constitute the influence diagram for the left reaction. When the load is at the left-hand end, the left reaction is 1. This value is plotted at the left-hand end. For the load at 0.1*l*, the reaction is 0.9. This value is plotted at 0.1*l*, and so on.

Alternatively, we can place the unit load at a variable distance ξ from the left support, and write the equation for the left reaction. For the simply supported beam of Fig. 8.2a the reaction is

$$R_1 = 1 - \frac{\xi}{l}$$

The plot of this equation, which is a straight line, is the influence line of the left reaction (Fig. 8.2b). The variable ξ is used for the location of the load in order to differentiate it from the variable x, the location of an arbitrary section of the beam for shear and moment diagrams.

Similarly, influence lines for shear and moment at a section of a beam can be obtained by either placing a unit load at different intervals and calculating the values of shear and moment at that section, or by deriving equations for shear and moment in terms of the variable ξ, denoting the location of the unit load. For a unit load at the left of the section ($\xi < a$, Fig. 8.3a) the left reaction is

$$R_1 = 1 - \frac{\xi}{l}$$

The shear at *a* is

$$V = R_1 - 1 = 1 - \frac{\xi}{l} - 1 = -\frac{\xi}{l} \qquad \xi \le a$$

for $\xi = a$, $V = -a/l$. When the load is on the right of the section

$$V = R_1 = 1 - \frac{\xi}{l} \qquad \xi \ge a$$

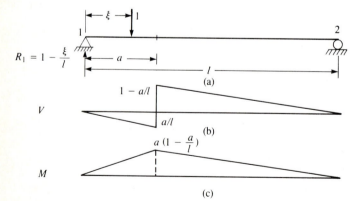

FIGURE 8.3

For $\xi = a$, $V = 1 - a/l$. A plot of the above equations for shear, that is the influence line of shear at a, is shown in Fig. 8.3b. Similarly, for the unit load on the left and right of the section the equations for moment at a are

$$M = a\left(1 - \frac{\xi}{l}\right) - (a - \xi) = \xi\left(1 - \frac{a}{l}\right) \qquad \text{for} \qquad \xi \leq a$$

and

$$M = a\left(1 - \frac{\xi}{l}\right) \qquad \text{for} \qquad \xi \geq a$$

For $\xi = a$, $M = a(1 - a/l)$. A plot of the latter equations, the influence line of moment at a, is shown in Fig. 8.3c.

For statically determinate beams, influence lines are composed of straight-line segments. For statically indeterminate beams, they are usually curves. Since the influence line of shear on the left of an interior support is different from that on the right of the support, we denote these by V^- and V^+, respectively.

Example 8.1. Find the influence lines of reaction R_2, shear on the right of support 2, V_2^+, and moment at point 3, the middle of the span, M_3 (Fig. 8.4a). The left span is 30 ft and the overhang is 10-ft long.

Solution. For the influence line of each function—reaction, shear, or moment at a section, we place a unit load at an arbitrary location ξ and calculate the value of that function at that section in terms of ξ.

(a) **Influence line of R_2.** For $0 \leq \xi \leq 30$ ft, we find R_2 by setting the sum of moments at the left support equal to zero:

$$R_2 \times 30 - 1 \times \xi = 0,$$

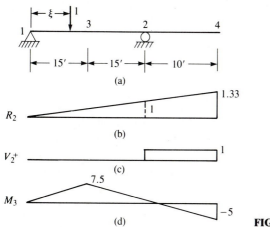

FIGURE 8.4

which gives

$$R_2 = \frac{\xi}{30}$$

Similarly for 30 ft $\leq \xi \leq 40$ ft,

$$R_2 \times 30 - 1 \times \xi = 0$$

or again

$$R_2 = \frac{\xi}{30}$$

A plot of the above equation for R_2, the influence line of the right reaction in terms of ξ, is shown in Fig. 8.4b.

(b) Influence line of shear V_2^+. For the unit load at the left of support 2, shear just to the right of support 2 can be found as sum of the forces on the right of the section, which is zero. For the unit load at the right of the section, shear at the section is again found from the right and is equal to 1. Thus, the plot of the influence line for shear is as shown in Fig. 8.4c.

(c) Influence line of moment M_3. For a unit load at distance ξ, the left reaction is found by setting the sum of moments at 2 equal to zero:

$$R_1 \times 30 - 1(30 - \xi) = 0$$

which gives

$$R_1 = 1 - \frac{\xi}{30}$$

The moment at section 3 is thus

$$M_3 = \left(1 - \frac{\xi}{30}\right) \times 15 - 1 \times (15 - \xi) = \tfrac{1}{2}\xi \qquad \text{for} \qquad \xi < 15 \text{ ft}$$

and

$$M_3 = \left(1 - \frac{\xi}{30}\right) \times 15 = 15 - \tfrac{1}{2}\xi \qquad \text{for} \qquad \xi > 15 \text{ ft}$$

A plot of these equations, i.e. influence line of moment at 3 is shown in Fig. 8.4d.

Example 8.2. Find the influence lines of reactions R_1, R_2, shear at point 5, and moment at support 2 for the beam of Fig. 8.5a. A hinge is located at point 4, the middle of the first span.

Solution.

(a) Influence line of reaction R_1. When $0 \leq \xi \leq 16$ m, the equation of condition, sum of the moments on the left 16-m segment with respect to the hinge equal zero, gives

$$R_1 \times 16 - 1(16 - \xi) = 0$$

or

$$R_1 = 1 - \frac{\xi}{16}$$

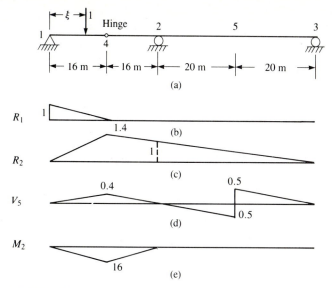

FIGURE 8.5

When the load is on the right of point 4, we can easily see from the equation of condition that $R_1 = 0$. Thus, the influence line of R_1 is as shown in Fig. 8.5b.

 (b) Influence line of reaction R_2. For $0 \le \xi \le 16$ m, again the equation of condition, as above, gives

$$R_1 = 1 - \frac{\xi}{16}$$

Sum of the moments at 3 gives

$$\left(1 - \frac{\xi}{16}\right) \times 72 - 1 \times (72 - \xi) + R_2 \times 40 = 0$$

or

$$R_2 = 0.0875\xi$$

At $\xi = 16$ m, $R_2 = 1.4$.

 When the unit load is on the right of the hinge, 16 m $\le \xi \le 72$ m, by the equation of condition the left reaction is zero. The reaction R_2 is found by setting the sum of moments at support 3 equal to zero,

$$-1(72 - \xi) + R_2 \times 40 = 0$$

or

$$R_2 = 1.8 - \frac{\xi}{40}$$

This gives $R_2 = 1.4$ for $\xi = 16$ m, $R_2 = 0$ for $\xi = 72$ m, point 3. We can plot the latter equations to obtain the influence line of reaction R_2 (Fig. 8.5c).

(c) Influence line of shear at 5. For $0 \le \xi \le 16$ m, the reactions are

$$R_1 = 1 - \frac{\xi}{16} \qquad R_2 = 0.0875\xi$$

Shear at section 5 is sum of the forces on the left of that section or $V_5 = R_1 - 1 + R_2 = 1 - (\xi/16) - 1 + 0.0875\xi = 0.025\xi$. For $\xi = 16$ m, $V_5 = 0.4$.
For 16 m $\le \xi \le 52$ m

$$R_1 = 0 \qquad R_2 = 1.8 - \frac{\xi}{40}$$

and

$$V_5 = -1 + 1.8 - \frac{\xi}{40} = 0.8 - \frac{\xi}{40}$$

At $\xi = 52$ m, $V_5 = -0.5$.
When the load is on the right of point 5, 52 m $\le \xi \le 72$ m, the reactions are the same as the last values, and shear at 5 is

$$V_5 = 1.8 - \frac{\xi}{40}$$

At $\xi = 52$, $V_5 = 0.5$ and at $\xi = 72$ m, $V_5 = 0$.
The plot of these equations, influence line of shear at point 5 is shown in Fig. 8.5d.

(d) Influence line of moment at support 2. As in part (c), for $0 \le \xi \le 16$ m, $R_1 = 1 - (\xi/16)$, $R_2 = 0.0875\xi$, and

$$M_2 = \left(1 - \frac{\xi}{16}\right) \times 32 - 1 \times (32 - \xi) = -\xi$$

For $\xi = 16$ m, $M_2 = -16$.
For 16 m $\le \xi \le 32$ m, with the corresponding reactions found in part (c),

$$M_2 = -1 \times (32 - \xi) = -32 + \xi$$

At $\xi = 16$, $M_2 = -16$, and for $\xi = 32$ m, $M_2 = 0$.
For 32 m $\le \xi \le 72$ m, with $R_1 = 0$, and $R_2 = 1.8 - (\xi/40)$ as found in part (c), $M_2 = 0$.

8.3 MULLER–BRESLAU PRINCIPLE

In some applications, for example, for finding the pattern of distributed live loads, it may be sufficient to have the shape of influence lines and the ordinates may not be necessary. The shape of the influence lines can be obtained by a simple procedure, due to Muller–Breslau. Since for statically determinate structures influence lines are composed of straight lines, the values of the influence line ordinates may be determined at a few locations, by applying unit loads at those locations and finding the values of the function in question. Alternatively, the values can be obtained from the geometry of the influence line as discussed below. The ordinates of influence lines can be also determined by the virtual work method. This procedure, which is the basis of the Muller–Breslau principle, will be discussed later.

The Muller–Breslau principle may be stated as follows.

The shape of the influence line for a function (moment, shear, reaction, etc.) can be obtained by removing the resistance of the structure to that function, at the section where the influence line is desired, and applying an internal force associated with that function at the section so as to produce a unit deformation at the section. The deformed shape that the structure will take represents the shape of the influence line.

It should be noted that for the shear influence line only a shear deformation must be induced at the section without any bending deformation. Thus, the slope of the shear diagram on both sides of the section will be the same.

Example 8.3. Find the shapes of influence lines for the left reaction, and shear and moment at section 4, 10 ft to the right of the first support of the beam in Fig. 8.6a. Calculate the ordinates of the influence lines.

Solution. To find the influence line of the left reaction we remove the support, apply a reaction force and induce a unit displacement in the direction of the reaction (Fig. 8.6b). The ordinates of the influence line at other points can be found by geometry. At the left end, point 3, the ordinate is found by similarity of triangles,

$$\frac{55}{40} \times 1 = 1.375 \text{ kips}$$

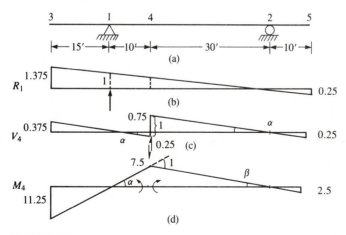

FIGURE 8.6

and at the right end, point 5,

$$\frac{10}{40} \times 1 = 0.25 \text{ kips}$$

To find the influence line of shear at section 4, we remove the resistance of the beam at that section against shear by introducing a cut at section 4. We then apply a positive internal shear at the cut (equal and opposite shear forces on both sides of the cut) to produce a unit shear deformation (Fig. 8.6c). The resulting deflected shape is the influence line of shear at section 4. The internal shear force is so chosen as to produce a positive shear deformation, that is, the shear deformation on the right of the cut minus that of the left is equal to +1. Note also that for a positive internal shear the two segments of the beam on the left and right are under positive shear. That is the left-hand side of each segment moves up relative to the right-hand side. Since the slope of the shear influence line on both sides of the cut is the same and the influence line for a statically determinate beam is composed of straight lines, we can find the ordinates by setting the slopes on both sides of the cut equal.

If shear at the right of section 4 is V_4^+, because the jump in shear is 1, shear on the left of the section would be $1 - V_4^+$. Setting the slope on the sides of the cut equal gives

$$\frac{V_4^+}{30} = \frac{1 - V_4^+}{10} \qquad \text{or} \qquad 10V_4^+ = 30 - 30V_4^+$$

or

$$V_4^+ = 0.75.$$

The ordinate of the shear influence line at point 3 is found by similarity of triangles,

$$\frac{V_3}{0.25} = \frac{15}{10} \qquad \text{or} \qquad V_3 = 0.375$$

and at 5

$$\frac{V_5}{10} = \frac{0.75}{30} \qquad \text{or} \qquad V_5 = 0.25$$

The moment influence line at section 4 is obtained by removing the resistance against bending, i.e. by introducing a hinge at that section and applying a positive internal moment (equal moments at both sides of the hinge) so as to produce a unit change of slope (Fig. 8.6d). If the slopes of the diagram on the left and right of the section are α and β, from geometry of the triangle (external angle is equal to the sum of the internal angles) and by assuming small deformations so that the angles in radians are equal to the tangents of the angles, we have $1 = \alpha + \beta$, or if the ordinate at section 4 is M_4

$$1 = \frac{M_4}{10} + \frac{M_4}{30} = \frac{4M_4}{30}$$

which gives $M_4 = 30/4 = 7.5$.

Other ordinates of the influence line can be found by similarity of triangles,

To find M_3 and M_5 we write

$$\frac{7.5}{M_3} = \frac{10}{15} \qquad \text{or} \qquad M_3 = 11.25$$

and

$$\frac{7.5}{M_5} = \frac{30}{10} \qquad \text{or} \qquad M_5 = 2.5$$

Example 8.4. Find the shear and moment influence lines just to the right of the left support of the beam in Fig. 8.7a.

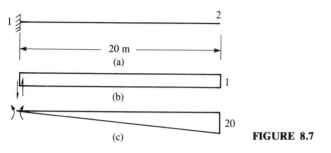

FIGURE 8.7

Solution. The influence line of shear is obtained by introducing a cut just to the right of the left-hand end, and applying a unit internal shear to induce a unit shear deformation (Fig. 8.7b). Since on the left of the cut the length of the beam segment is very short a very large force is needed to cause any change of slope there. Thus, this segment remains horizontal. Because the slope at the right of the cut must be the same as that on the left, the segment on the right of the cut will also remain horizontal but undergoes a unit displacement. To verify the shape of this influence line we observe that for a unit load acting anywhere on the right of the section shear is equal 1 at the section.

Similarly, the influence line of the moment just to the right of the left-hand end is obtained by introducing a hinge at that section and applying a positive internal moment to produce a unit change of slope. The ordinate of the influence line at the right-hand end is the change of slope times the length, i.e. $1 \times 20 = 20$. We see from Fig. 8.7c that this ordinate has a negative value.

Verification of the Muller–Breslau Principle

The Muller–Breslau principle can be verified by the virtual work method as follows. The principle of virtual work (virtual displacements) was discussed in Section 6.3. To apply the virtual work method a virtual displacement is introduced in the beam. This virtual displacement can take any shape as long as it is consistent with the support conditions. The external and internal works are then set equal. For beams the external work is done by the applied loads and the internal work by the bending moment through the change of slope.

For example, to find the influence line of the moment at a section of the simply supported beam of Fig. 8.8a, we proceed as follows. With the beam

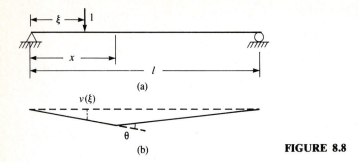

FIGURE 8.8

under a unit load at ξ, we introduce a virtual displacement such that internal work is done only by the moment at x. That is, we allow a change of slope at x only and nowhere else in the beam. This means that the beam segments on both sides of x displace as rigid bars hinged together at x. If the change of slope at x is θ then the virtual work equation, setting the internal and external works equal, gives

$$1 \cdot v(\xi) = M\theta$$

$v(\xi)$ is the virtual displacement at the location of the unit load and M is the moment at x. If we now choose the change of slope θ to be equal unity, then

$$M = v(\xi) \tag{8.1}$$

That is, the moment at section x due to a unit load at ξ is equal to the value of the virtual displacement at ξ. But the moment at a section x due to a unit load acting at an arbitrary distance ξ is by definition the influence line of moment at x. Thus, the Muller–Breslau principle for the moment influence line would state that the influence line of moment at a section can be obtained by introducing a hinge at that section and producing a unit change of angle by an internal moment applied at the hinge. The deflected shape gives the influence line.

The influence line for shear can be found similarly by introducing a virtual displacement such that work is done only by an internal shear force at that section. The virtual displacement in this case would be discontinuous at the section. However, as long as the displacement is consistent with the support conditions it can have any continuous or discontinuous form in the span. With the virtual displacement at the section taken equal to 1, we have $V \times 1 = 1 \times v(\xi)$. This proves the Muller–Breslau principle for shear influence line for beams. Determination of the ordinates of influence lines for determinate beams by the Muller–Breslau principle, was discussed in Section 8.3.

Influence line for a reaction can be obtained in a similar manner by choosing a virtual displacement so that the reaction can work through a unit displacement.

8.4 INFLUENCE LINES FOR CONTINUOUS BEAMS

The Muller–Breslau principle discussed above may be used to find either the shape or the shape and ordinates of influence lines for continuous beams. Determination of influence lines for determinate beams by the Muller–Breslau principle was discussed in Section 8.3.

Shape of Influence Lines

For many continuous-beam problems it is necessary to determine the critical pattern of the live loads. The shape of the influence line provides such information and can be found by the Muller–Breslau principle. Once the loading pattern producing the maximum effect is identified, the beam can be loaded accordingly to determine the forces and displacements. For statically indeterminate structures the shapes of influence lines are found in the same way as for determinate structures (Section 8.3). However, for indeterminate structures some segments of the influence line may be curves rather than straight lines.

Example 8.5. For the three-span beam of Fig. 8.9a, find the shape of the

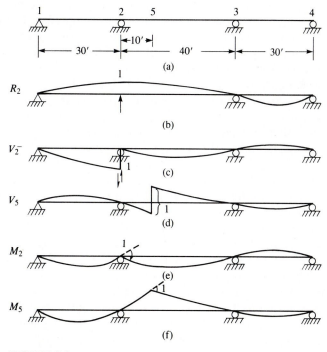

FIGURE 8.9

influence lines for the reaction R_2, shear V_2^- (on the left of support 2), and V_5 at point 5, as well as moments M_2 and M_5.

Solution. To find the influence line of reaction R_2, we remove the support 2 and apply a reaction force so as to produce a unit displacement at the support. The beam at the other supports does not deflect but can rotate. The shape of the influence line for R_2 is thus as shown in Fig. 8.9b.

The influence line for shear at the left of support 2 is obtained by making a cut just to the left of that support and applying a positive internal shear force causing a unit shear deformation. Note that the slope on both sides of the cut is the same. Again, at the supports the beam cannot deflect. For the slope on the right of the cut to be the same as that on the left, the beam must take the shape shown in Fig. 8.9c. The influence line of shear at section 5 is obtained in a similar fashion (Fig. 8.9d).

The influence line for the moment M_2 is found by introducing a hinge on the support 2, that is, by removing the resistance of the beam against bending. A positive internal moment is applied at the hinge, producing a unit change of slope. The shape of the influence line thus obtained is shown in Fig. 8.9e. Similarly, the influence line of M_5 is as shown in Fig. 8.9f.

Ordinates of Influence Lines

To find the ordinates of influence lines for continuous beams we can use the Muller–Breslau principle in conjunction with one of the methods of analysis for statically indeterminate beams and Maxwell–Betti's reciprocal theorem. The method of consistent deformations is used in the following discussions. The singularity method, which allows a direct derivation of formulas for influence lines, is discussed in Section 9.8.

Consider, for example, the two-span beam of Fig. 8.10a. To find the influence line of reaction R_2 we can use the method of consistent deformations. We remove the reaction R_2 to obtain a primary beam and find $\delta_{2\xi}$, the displacement at point 2 due to a unit load at ξ of the primary beam. Similarly, we must find v_{22}, the displacement at point 2 of the primary beam due to the unknown reaction R_2. As we saw in Section 7.1, this can be done by finding δ_{22}, the displacement at point 2 due to a unit load at 2. Then $v_{22} = -\delta_{22}R_2$. The negative sign is because in the design sign convention that we are using the directions of positive loads and positive reactions are opposite to each other. Noting that by Maxwell–Betti's theorem, Section 6.5, $\delta_{2\xi} = \delta_{\xi 2}$, we can use $\delta_{\xi 2}$, which is the deflection at ξ due to a unit load at 2. Thus, the primary beam under a unit load at 2 can be used to find both δ_{22} and $\delta_{\xi 2}$; in fact having the expressions for $\delta_{\xi 2}$, δ_{22} can be found by specifying ξ to be the location of the second reaction. Thus, only one beam with a unit load at the location of the support must be analyzed to find its deflection at an arbitrary point ξ. The conjugate beam method can be used for this purpose. By setting sum of the deflections at point 2, due to a unit load at ξ, $(\delta_{2\xi} = \delta_{\xi 2})$, and that due to the reaction, (v_{22}), equal to zero, we find $R_2 = \delta_{\xi 2}/\delta_{22}$. The reaction thus obtained is for a load acting at ξ and its equation is therefore the influence line of the reaction. The influence lines for other reactions as well as shear and moment can be found by the equations of statics.

Example 8.6. For the two-span beam of Fig. 8.10a, find the influence lines for the reaction R_2 and moment M_4, at the middle of the first span.

Solution. From the previous discussion $R_2 = \delta_{\xi 2}/\delta_{22}$, where $\delta_{\xi 2}$ is the deflection of the primary beam at ξ due to a unit load at 2, Fig. 8.10c and δ_{22} is deflection of the same beam at 2.

From the conjugate beam of Fig. 8.10d, the expression for the deflection $\delta_{\xi 2}$ is found as follows. By similarity of triangles the magnitude of the elastic load, depending on whether ξ is on the left or on the right of support 2, is

$$\frac{\xi}{20} \times \frac{10}{EI} = \frac{\xi}{2EI} \qquad \text{and} \qquad \frac{40 - \xi}{20} \times \frac{10}{EI} = \frac{1}{2} \frac{(40 - \xi)}{EI}$$

The left reaction for the conjugate beam is $100/EI$ and the deflection $\delta_{\xi 2}$ at ξ is

$$\delta_{\xi 2} = M'_\xi = \frac{100\xi}{EI} - \frac{\xi^3}{12EI} = \frac{\xi}{EI}\left(100 - \frac{\xi^2}{12}\right) \qquad \text{for} \qquad \xi \le 20\ \text{ft}$$

When ξ is on the right of support 2, calculating the moment in the conjugate beam from the right end gives

$$\delta_{\xi 2} = M'_\xi = \frac{40 - \xi}{EI}\left[100 - \frac{(40 - \xi)^2}{12}\right] \qquad \text{for} \qquad \xi \ge 20\ \text{ft}$$

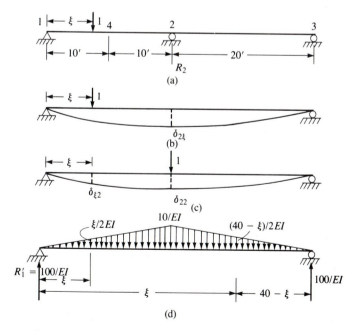

(a)

(b)

(c)

(d)

FIGURE 8.10

From the first equation for $\delta_{\xi 2}$,

$$\delta_{22} = \frac{20}{EI}\left(100 - \frac{20^2}{12}\right) = \frac{1\,333}{EI}$$

Therefore, the influence line for the middle reaction is

$$R_2 = \frac{\delta_{2\xi}}{\delta_{22}} = \frac{\delta_{\xi 2}}{\delta_{22}} = \frac{\xi}{1\,333}\left(100 - \frac{\xi^2}{12}\right) \qquad \text{for} \qquad \xi \le 20\,\text{ft}$$

and

$$R_2 = \frac{40 - \xi}{1\,333}\left[100 - \frac{(40 - \xi)^2}{12}\right] \qquad \text{for} \qquad \xi \ge 20\,\text{ft}$$

The plot of these equations gives the influence line of R_2.

The influence line of moment can be found from the reactions. To find the moment at the middle of the first span we find the first reaction by setting the sum of the moments with respect to the right-hand end equal to zero. This gives

$$R_1 \times 40 - 1 \times (40 - \xi) + R_2 \times 20 = 0$$

or

$$R_1 = 1 - \frac{\xi}{40} - 20R_2$$

The moment at the middle of the first span is $M_4 = 10R_1 - 1 \times (10 - \xi)$ for $\xi \le 10\,\text{ft}$ and $M_4 = 10R_1$ for $\xi \ge 10\,\text{ft}$, or

$$M_4 = 10 - \frac{\xi}{4} - 200R_2 - 10 + \xi = \frac{3}{4}\xi - 200R_2 \qquad \text{for} \qquad \xi \le 10\,\text{ft}$$

and

$$M_4 = 10 - \frac{\xi}{4} - 200R_2 \qquad \text{for} \qquad \xi \ge 10\,\text{ft}$$

Upon substitution for R_2 from influence line equations above, we get

$$M_4 = \frac{3}{4}\xi - 200\xi\left(100 - \frac{\xi^2}{12}\right) \qquad \text{for} \qquad \xi \le 10\,\text{ft}$$

and

$$M_4 = 10 - \frac{\xi}{4} - 200(40 - \xi)\left[100 - \frac{(40 - \xi)^2}{12}\right] \qquad \text{for} \qquad \xi \ge 20\,\text{ft}$$

8.5 INFLUENCE LINES FOR GIRDERS WITH FLOOR BEAMS

In some floor systems and bridge decks the loads are not applied to the girders (main beams) directly but are transferred from the stringers to the floor beams and then to the girders (Fig. 8.11a). In steel decks the stringers are often connected to floor beams by clip angles that transfer shear but not moment. Therefore the stringers can be assumed simply supported.

The influence lines for actions (reaction, shear, and moment) in a girder

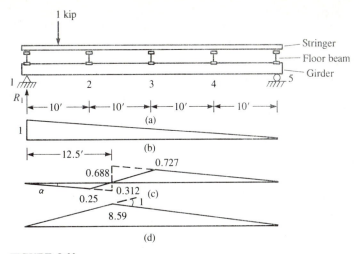

FIGURE 8.11

can be found by moving a unit load along the stringers and calculating the values of the action induced in the girder. For example, consider the case of a stringer that is placed at the end of the floor beams. This stringer is thus supported by the floor beams but is directly located above a girder. The loads of the stringer are transferred to the girder through the floor beams. We want to find the influence line of the reaction R_1 of a girder under a moving load applied on a stringer. We can find R_1 for the different positions of a unit load on the stringer as the load moves from the left-hand end to the right-hand end of the stringer and plot the results (Fig. 8.11b). When the load is above the first floor beam we get $R_1 = 1$. When it is above the second floor beam, $R_1 = 3/4$. Similarly, we can calculate R_1 when the load is at other floor beams. For the load at the last floor beam, $R_1 = 0$. If the laod is half-way between the first and second floor beams, the forces transferred to the girder by the first and the second floor beam reactions are equal to 1/2. Then the reaction R_1 is $(1/2) \times (1/2) + (3/4) \times (1/2) = 7/8$. For the load at other distances between the two floor beams, the reaction on each floor beam is proportional to the distance of the load from the other floor beam. Thus, the reaction influence line for the girder is the same (within a constant factor) as the influence line of the girder reaction if the load were directly applied to the girder.

Similarly, to find the shear influence line in panel 23 (Fig. 8.11c), we can first find the ordinates of the shear influence line when the unit load is at points 2 and 3. When the load is between the panel points we can calculate the forces transferred to the girder through floor beams 2 and 3 and then calculate the shear in panel 23. We will find that shear varies linearly for a unit load between panel points. The same is true for moment influence lines (Fig. 8.11d). A quicker way to construct the shear and moment influence lines in a panel is to draw the shear and moment influence lines for a section in the panel and connect the panel points. The moment influence line thus obtained is the same as moment influence lines for other beams. However, the shear influence line for the panel must be modified by connecting the values at the panel points (Fig. 8.11c).

8.6 INFLUENCE LINES FOR TRUSSES

Influence lines for truss bars provide the variation of the bar force as a unit load traverses the truss. Before we discuss influence lines for trusses, let us reconsider the method of sections (Section 3.8). With that method we can calculate the forces in the truss bars by passing a section through the members (Fig. 8.12a). The force in a chord member is then found by taking the moment of the forces acting on the free body on one side of the section. The moments are taken with respect to a point where all the unknown forces but one pass, yielding the unknown force in the chord member. The force in a diagonal member is found by setting the sum of the vertical forces on the free body equal to zero.

In writing the moment equation to find the force in a chord member, we note that the total moment is the sum of the chord force times its distance to the joint where the moment is calculated plus the moment of all the forces on the left of the joint. The latter is by definition the moment at that section of an equivalent beam. That is a beam with the same span length as the truss. For example, to find F_{57} in Fig. 8.12a we can take the moment of the forces on the free body on the left of section m–m at 6. The sum of moments at 6 is $M_6 - 8F_{57} = 0$. Where M_6 is the moment at a distance equal 30 ft from the left-hand end of a beam with an 80 ft span. This gives $F_{57} = M_6/8$.

Suppose M_6 is known for a unit load acting at a distance ξ; that is, we have the influence line diagram of the moment in an equivalent beam (with the

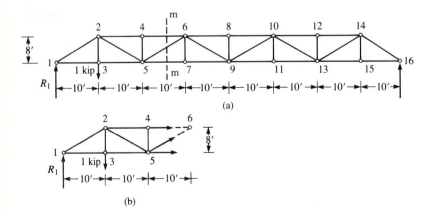

(a)

(b)

FIGURE 8.12

same span as the truss), at the same distance as joint 6. Then F_{57} can be found by dividing the ordinates of the diagram by 8 (the height of the truss).

Similarly, in calculating the sum of vertical forces acting on the free body diagram, Fig. 8.12b, we note that sum of the vertical forces is equal to the vertical component of F_{56} plus the sum of all the vertical forces on the left of bar 57. This is by definition shear V_{57} in the panel. Thus,

$$V_{57} + \frac{8}{\sqrt{8^2 + 10^2}} F_{56} = 0 \quad \text{which gives} \quad F_{56} = -\frac{\sqrt{8^2 + 10^2}}{8} V_{57} = -1.68 V_{57}$$

Therefore, the influence line for a diagonal member can be found by constructing the influence line of shear in the panel where the bar is located. The latter is the influence line for the vertical component of the bar force in question. The influence line of the bar force can therefore be found by multiplying the negative of the latter diagram by the length of the bar over the vertical projection of the length. The procedure for finding the influence line of shear in a panel was discussed in Section 8.5. This was done by finding the influence line of shear at a point in the panel (e.g., by the Muller–Breslau principle) and by connecting the points of the diagram at the panel end points.

Note also that the influence line of a lower chord member when the upper (non-load-carrying) joints are between the lower joints varies linearly between the two joints of that member.

Example 8.7. Find the influence lines for chord members 46 and 57 and diagonal 56 of Fig. 8.12.

Solution. To find the bar force F_{46} we can set sum of the moments at joint 5, 20 ft from the left support, equal to zero. This gives, $M_5 + 8F_{46} = 0$, where M_5 is the moment in an 80 ft span beam (same span as the truss) at $x = 20$ ft (same distance as joint 5). From the last equation $F_{46} = -M_5/8$. Therefore, we first draw the influence line of moment in the equivalent beam at $x = 20$ ft. This can be done by the Muller–Breslau principle (Section 8.3). We place a hinge at $x = 20$ ft and induce a unit change of slope at that section. The ordinate of the moment influence line at $x = 20$ ft can be found by setting the unit change of slope (external angle) equal to the sum of the internal angles (or their tangents, assuming small angles) (Fig. 8.13a). Thus,

$$1 = \frac{M_5}{20} + \frac{M_5}{60} \quad \text{which gives} \quad M_5 = 15$$

As mentioned above, the influence line of F_{46} is found by dividing the negative of the moment influence line by 8 (Fig. 8.13b).

Similarly, to find F_{57} we can calculate the moment of the free body diagram on the left of section m–m (Fig. 8.12a) at joint 6, which results in

$$F_{57} = \frac{M_6}{8}$$

Thus, we must find the influence line of moment in the equivalent beam at 30 ft,

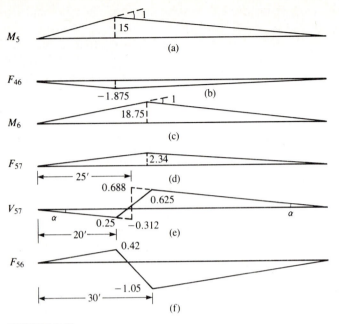

FIGURE 8.13

and divide the ordinates by 8. The influence line of M_6 is shown in Fig. 8.13c, and the influence line of F_{57} is given in Fig. 8.13d.

Finally, we found

$$V_{56} = -1.68V_{57}$$

To draw the influence line of shear in panel 57, we first draw the influence line of shear at a section in the panel. This gives the diagram of Fig. 8.13e including the dashed lines. The influence line of shear in the panel is then found by connecting the panel end points. Scaling this diagram by -1.68 gives the influence line of F_{56} (Fig. 8.13f).

8.7 USE OF INFLUENCE LINES

If we have the influence line of a function at a section of an elastic structure, we can find the value of the function at that section under any loading condition. By definition, the ordinate of the influence line at a point is the value of the function due to a unit load acting at that point. Therefore, the value of the function due to a load P is equal to the ordinate of the influence line at the point of application of the load times P. If more than one load is acting on the beam, then the value of the function is equal to the sum of each load times its corresponding ordinate of the influence line. For a uniformly distributed load, the value of the function is equal to the load intensity times the area of the influence line under the load.

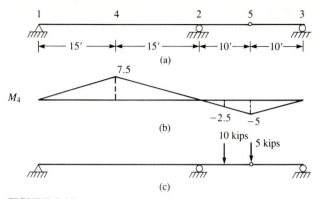

FIGURE 8.14

Example 8.8. Figure 8.14b shows the influence line of moment at section 4 of the beam of Fig. 8.14a. Find the moment at section 4 due to a 5 kip load acting at section 5, the loction of the hinge, and a 10 kip load at 5 ft to the left of the hinge (Fig. 8.14c). Also find the maximum positive and negative moments at section 4 due to a uniform live load of 2 kip/ft.

Solution. The ordinate of the influence line at any point of Fig. 8.14b gives the value of moment at section 4 due to a unit load applied at that point. For a load of 5 kips at point 5 the moment at section 4 is $5 \times (-5) = -25$. Thus, for the two loads of 5 kips and 10 kips from Fig. 8.14b the moment at section 4 is

$$M_4 = 5 \times (-5) + 10 \times (-2.5) = -50$$

For the uniform live load the maximum positive moment occurs when the load is applied on the portion of the beam where the ordinate of the influence line is positive. The magnitude of the positive moment is equal to the positive area of the influence line diagram times the magnitude of the load. Thus, $M_{4\,max}^+ = 2 \times (\frac{1}{2} \times 7.5 \times 30) = 225$ kip · ft.

Similarly, the value of the negative moment is equal to the load intensity times the negative area of the influence line,

$$M_{4\,max}^- = 2 \times [\tfrac{1}{2} \times (-5) \times 20] = -100 \text{ kip · ft.}$$

Example 8.9. Find the shear force at section 3 of the beam of Fig. 8.15a under a uniform live load 5 kN/m over the entire span, plus a concentrated load of 30 kN acting at the mid-span. Also calculate the maximum shear at section 3 for the uniform load alone. The influence line of shear at section 3 is shown in Fig. 8.15b.

Solution. The shear at section 3 due to the uniform load applied over the whole span is equal to the total area under the influence line times the load intensity. For the concentrated load the shear at section 3 is equal to the ordinate of the influence line at the mid-span times the magnitude of the load. The effect of both loads acting simultaneously is equal to the sum of the two values described. thus,

$$V_3 = 5 \times [\tfrac{1}{2} \times \tfrac{3}{4} \times 15 - \tfrac{1}{2} \times \tfrac{1}{4} \times 5] + 30 \times \tfrac{1}{2} = 40 \text{ kN}$$

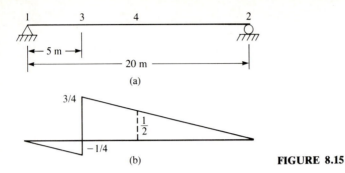

FIGURE 8.15

The maximum shear is produced under the uniform load when it is applied over the portion of the beam where the ordinate of the influence line is positive (since the value of the positive area is larger than that of the negative area). The value of the maximum shear is

$$V_{3,\,max} = 5 \times (\tfrac{1}{2} \times \tfrac{3}{4} \times 15) = 28.13 \text{ kN}$$

8.8 MAXIMUM EFFECTS UNDER MOVING LOADS

As can be seen from Example 8.9, maximum effects (for reaction, shear, moment, etc.) at a point can be found from influence lines. Uniform live loads are placed over portions of the beam corresponding to all positive or negative areas of influence line in order to produce maximum positive or negative values.

A set of concentrated moving loads produces maximum effects at a section when one of the loads is acting at that section. To find the maximum effect at a section under a set of concentrated loads, the location of the loads can be varied, placing the different loads at the section and calculating the value of the function from the influence line diagram. A better alternative is to find the change of the effect from the slope of the influence line. If we keep moving the loads from one end of the beam to the other as soon as the change becomes negative, the maximum has been reached (for the previous load position). With the position of the load causing the maximum value of the function known, the influence line can be used to calculate the maximum value of the function.

The following example demonstrates the application of this procedure.

Example 8.10. Find the maximum shear and moment at the mid-span of a simply supported beam with a 100 ft span under the truck load of Fig. 8.16a. The influence lines for shear and moment at the mid-span of the beam are shown in Figs. 8.16b, and 8.16c.

Solution. To find the position of the loads causing the maximum shear at the mid-span, we place the different loads at the mid-span and calculate the change of

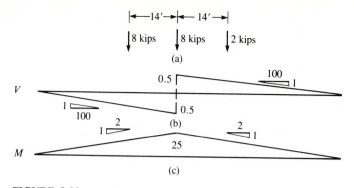

FIGURE 8.16

shear from the previous condition. As we move the loads from right to left, the left load approaches the section from the right, that is, it is just to the right of the mid-span. We need not perform any calculations for the first position of the loads since at this stage we are interested in the change of shear from one position of the load to another. Moving the loads to the left, when the left load is just to the left of the mid-span, the change in shear is $[-0.5 - (0.5)]8 = -8$. Going from the first position to the second the ordinate of the shear diagram changes from $+0.5$ to -0.5, Fig. 8.16b, resulting in a change of ordinate $-0.5 - (0.5) = -1$. Since the change of shear is negative, the maximum shear occurs under the previous loading condition, with the left load just to the right of the mid-span. The maximum shear at the mid-span is therefore,

$$V_{max} = 8 \times 0.5 + 8(0.5 - 14 \times 0.01) + 2(0.5 - 28 \times 0.01) = 7.44 \text{ kips}$$

The maximum moment at the mid-span occurs when one of the loads is acting there. To find the location of the loads causing the maximum moment at the mid-span, we proceed as above by moving the loads and calculating the change of moment. Again, when the left load is at the mid-span we need not perform any calculations. Since the slope of the moment influence line is 1 by 2 or 0.5, as the middle load is moved to the mid-span, the change in moment is

$$8 \times (-0.5 \times 14) + 8 \times (0.5 \times 14) + 2 \times (0.5 \times 14) = 14 \text{ kip} \cdot \text{ft}$$

As the right load is moved to the mid-span the change of moment from the previous position is

$$-8 \times 0.5 \times 14 - 8 \times 0.5 \times 14 + 2 \times 0.5 \times 14 = -98 \text{ kip} \cdot \text{ft}$$

The negative change indicates that the maximum moment has occurred under the previous position with the second load at the mid-span. Thus, the value of the maximum moment is

$$8 \times (25 - 0.5 \times 14) + 8 \times 25 + 2 \times (25 - 0.5 \times 14) = 380 \text{ kip} \cdot \text{ft}$$

Absolute Maximum Moment

For a beam with n loads on the span we can find the absolute maximum values of reaction, shear, and moment at any section by placing the first, second, . . . , loads at a given section and calculating the maximum value of the function at that section. By repeating this for different sections, we will be able to find the maximum value. If additional loads are moved on the span from the right as some of them are moved out from the left, we must include these loads in the subsequent calculations.

A better alternative is to find the change of the function as the next load moves to the section. This was done in the previous section by using influence lines. We can also find the maximum value of a function under a set of concentrated loads directly, without using influence lines. To achieve this end we can derive expressions for the change of reaction, shear, or moment in terms of the resultant force and its point of application.

Here we determine the absolute maximum moment in a simply supported beam under n concentrated loads, Fig. 8.17. If the distance of the resultant force, $\bar{P}$ from the first load is $\bar{x}$, then

$$\bar{x} = \frac{\sum\limits_{i}^{n} P_i a_i}{\sum\limits_{i=1}^{n} P_i} \tag{8.2}$$

When the load P_j is located at an arbitrary section x from the left end, the distance of the resultant from the right end of the beam is, (Fig. 8.17), $l - x - (\bar{x} - a_j)$. The moment at section x is

$$M = \bar{P}[l - x - (\bar{x} - a_j)]\frac{x}{l} - \sum_{i=1}^{j-1} P_i(a_j - a_i) \tag{8.3}$$

The last term in the above equation is the moment of the forces on the left of x; a_i is the distance of P_i from P_1. To find the absolute maximum moment, we can find the location of the section where the maximum moment occurs by setting the derivative of the moment, Eq. (8.3), with respect to x equal to zero.

$$\frac{dM}{dx} = \frac{\bar{P}(l - \bar{x} + a_j)}{l} - \frac{2\bar{P}x}{l} = 0$$

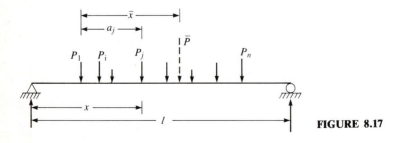

FIGURE 8.17

or

$$x = \frac{l}{2} - \frac{\bar{x} - a_j}{2} = \frac{l}{2} - \frac{d}{2} \tag{8.4}$$

where $d = \bar{x} - a_j$ is the distance between the load acting at the location of the maximum moment and the resultant force. The load acting at the location of the maximum moment is generally the load closest to the resultant force; $\bar{x}$ is the distance between the first load and the resultant force. Substitution of x from Eq. (8.4), into Eq. (8.3) gives the absolute maximum moment in the beam.

Example 8.11. Find the location and magnitude of the absolute maximum moment in a simply supported beam under a pair of equal moving loads (Fig. 8.18). The distance between the loads is b and their magnitude P.

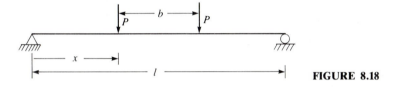

FIGURE 8.18

Solution. The critical section, where the maximum moment occurs, is under one of the loads, say the left load. By Eq. (8.4), with no load on the left of the critical section, the location of the maximum moment x is

$$x = \frac{l}{2} - \frac{b}{4}$$

Thus, the maximum moment occurs at $b/4$ from the mid-span. We observe that under the critical loading condition the resultant and one of the loads are symmetrically placed at both sides of the mid-span. The maximum moment is found by substituting the above value of x into Eq. (8.3),

$$M_{max} = 2P\left[l - \frac{l}{2} + \frac{b}{4} - \frac{b}{2}\right]\frac{(l/2) - (b/4)}{l} = \frac{P(2l - b)^2}{8l}$$

Example 8.12. Find the absolute maximum moment for the 100 ft span beam of Example 8.10, under the truck loads of Fig. 8.16a. The beam and the loads are shown in Fig. 8.19.

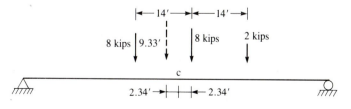

FIGURE 8.19

Solution. By Eq. (8.2), the distance from the first load to the resultant is

$$\bar{x} = \frac{8 \times 14 + 2 \times 28}{8 + 8 + 2} = 9.33 \text{ ft}$$

The load closest to the resultant is the middle load. The maximum moment occurs at sectiion x when the middle load is at section x. The distance between the latter load and the resultant is, $d = 9.33 - 14 = -4.67$ ft. therefore, by Eq. (8.4),

$$x = \frac{100}{2} + \frac{4.67}{2} = 52.34 \text{ ft}$$

The left reaction is $R_1 = 18 \times 52.34/100 = 9.42$ kips, and the maximum moment under the second load is

$$M_{max} = 9.42 \times 52.34 - 8 \times 14 = 381.04 \text{ kip} \cdot \text{ft}$$

LABORATORY EXPERIMENTS

Influence Lines–Direct Method

Select a simply supported beam with one or two strain gages mounted on it at a section (see Fig. L5.2). Apply the same load at different locations along the beam, measure the strain in the gage and the displacement at that or another point. Calculate the moments from the measured strains (see Bending Strain–Curvature in Laboratory Experiments of Chapter 5). If only one gage is used assume the strain to be the same magnitude with reverse sign on the other face of the beam. Plot the moment at the section with gage at the locations of the load. Compare this plot with the theoretical influence line of monent for the location of the gage. Verify that it is the same within a factor equal to the load used.

Similarly, plot the variation of the deflection at the measured point at the points of application of the load, and compare with the influence line of deflection at the measurement point.

Add another support and repeat the above steps for the two-span beam. Make sure that the beam does not lift off a support as the load moves along the beam.

Muller–Breslau Principle

Use a beam made of spring steel with a hinge at some distance from the end. Place the end points on supports (and other points for influence lines of statically indeterminate beams). By special levers, two pairs of pliers, or your hands, apply moments in the opposite directions at both sides of the hinge (Fig. L8.1a). Observe the shape and verify that it is the same as the influence line of a beam without a hinge at a section at the same distance as the hinge.

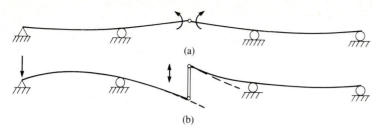

FIGURE L8.1

Remove the hinge and use a rigid link (a piece of wood with holes) and connect its ends to the two sides of the hinge so that it is perpendicular to the original axis of the beam. The left side of the link would be lower so as to produce a positive internal shear at the section. Move the link normal to the original axis of the beam until the ends of the beam on both sides of the link have parallel tangents (Fig. L8.1b). Verify that this is the shape of the influence line for shear at the location of the link for a beam that is continuous at that point.

PROBLEMS

8.1. Draw the influence lines for the following effects by moving a unit load along the beam and calculating the value of the functions (Fig. P8.1).
(a) Shear at left and right of point 2.
(b) Moment at points 2 and 3.

FIGURE P8.1

8.2. Derive the equations of the influence lines for the followng effects and plot the influence line diagrams (Fig. P8.2).
(a) Vertical reaction at points 1 and 4.
(b) Shear at points 2 and 3.
(c) Moment at points 1, 2 and 4.

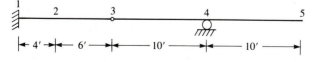

FIGURE P8.2

8.3. A unit load can travel from point 2 to 3 of the structure of Fig. P8.3. Find the

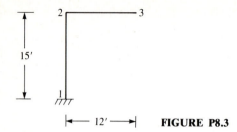

FIGURE P8.3

equation for M_1 (moment at the base, point 1) and plot the variation of this moment as a unit load moves from point 2 to 3.

8.4. Draw the influence lines for the following effects (Fig. P8.4).
 (*a*) The support reactions at points 3 and 5.
 (*b*) The shear on the left and right of point 3 and shear at points 2 and 4.
 (*c*) The moments at points 4 and 3.

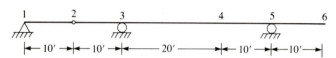

FIGURE P8.4

8.5. Draw the influence lines of the following effects (Fig. P8.5).
 (*a*) The reaction at supports 1, 2, and 3.
 (*b*) The shear at points 4, 5, and 6.
 (*c*) The moment at points 4, 5, and 6.

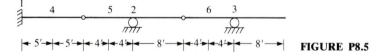

FIGURE P8.5

For Problems 8.6 to 8.8 draw the qualitative influence lines by the Muller–Breslau Principle.

8.6. (*a*) The reaction at supports 2 and 3.
 (*b*) The shear at point 5.
 (*c*) The moment at points 1 and 3.

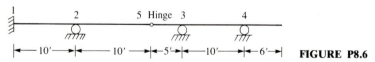

FIGURE P8.6

8.7. Shear at the left of point 2 and at point 3, moment at points 4 and 5.

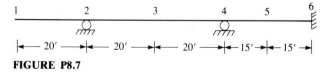

FIGURE P8.7

8.8. Shear and moment at point 2 in the beam.

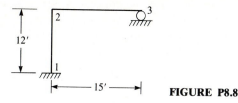

FIGURE P8.8

For Problems 8.9 and 8.10 draw the influence lines by the Muller–Breslau Principle and find their ordinates by geometry.

8.9. Shear on the right of points 2 and 4, moment at point 2.

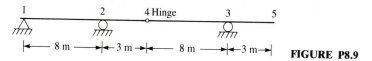

FIGURE P8.9

8.10. Moment and shear at point 5.

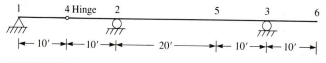

FIGURE P8.10

8.11. Find the influence line of shear in the second panel of Fig. P8.11 and moment at the third floor beam, 10 m from the left hand support.

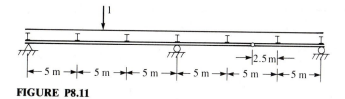

FIGURE P8.11

8.12. Find the influence lines for the bar forces 23, 35, 46 and 47 of the truss of Fig. P8.12, as a unit load moves along the bottom chord.

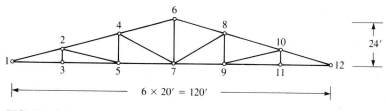

FIGURE P8.12

8.13. A travelling crane is supported by a pin-jointed plane truss that is pinned to a rigid wall at 2 and connected to the same wall by a tie-bar 15, as shown in Fig. P8.13. Construct influence lines for the axial forces in the members 56, 68, and 89 due to a unit load travelling along the lower chord of the truss. The axles of the crane trolley support vertical loads of 40 and 30 kN as shown in the figure. Find the greatest tensile and compressive forces in the members 56, 68 and 89, if the crane is operating in any position between the extreme ends of the truss. The axles of the trolley are 3 m apart.

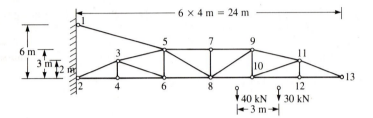

FIGURE P8.13

8.14. The influence lines of shear and moment at section 2 of the simply supported beam of Fig. P8.14a are shown in Figs. P8.14b and P8.14c. Find the maximum value of shear and moment due to the line loads shown in Figs. P8.14d and P8.14e, respectively.

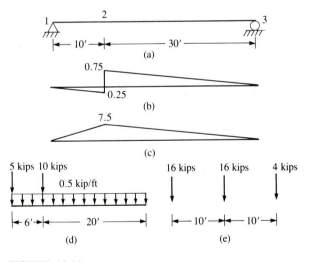

FIGURE P8.14

8.15. Using the influence lines of Example 8.3 (Fig. 8.6), find the maximum value of the reaction R_1, shear V_4, and moment M_4 due to the dead load of 1.5 kip/ft, a uniform live load of 3 kips/ft and a concentrated load of 15 kips. These loads are acting simultaneously.

8.16. A girder bridge has two girders supporting a concrete slab as shown in Fig. P8.16.

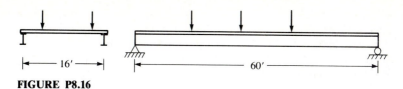

FIGURE P8.16

(*a*) Find the influence lines of shear at the left end of the girder and moment at the mid-span.

(*b*) If the load consists of an HS20 truck, described in Chapter 1 at the end of Section 1.8, find the critical lateral and longitudinal locations of the truck that produces the maximum shear and moment at the left-hand end and mid-span, respectively.

(*c*) Find the location and magnitude of the absolute maximum moment for the girder loads directly, without using the influence line.

CHAPTER
9

SINGULARITY METHOD

For beams under discontinuous loads, the shear and moment equations become cumbersome, requiring one equation for each segment of the beam between two loads and for each distributed-load region. As was seen in Chapter 5, Example 5.4, computation of deflection by integration in this case becomes rather tedious. This problem can be somewhat remedied by invoking the principle of superposition, that is, by finding the solution for each individual load and superimposing the results. On the other hand, in calculating the deflection by the moment–area method (Section 5.4), we do not treat the discontinuous loads any differently from continuous ones. The limitation of the analytical solution is in its inability to handle discontinuous functions. This limitation can be removed by using the singularity functions. The singularity method enables us to write a single equation for shear or moment irrespective of the number of discontinuous loads. With this moment equation, the differential equation of the beam can be integrated directly to find the deflection in much the same way as for a continuous load over the span, thus eliminating the matching conditions at the load discontinuities. That is to say, the singularity method involves only two unknown constants, which are found from the boundary conditions.

With the advent of microcomputers and the prominence of analytical techniques, the singularity method provides a simple way of formulating a

234

variety of problems. In this chapter we discuss the application of the singularity method to shear, moment, and deflection of beams under arbitrary loads. Analyses of variable-section beams and internal hinges are also presented. This is followed by the solution of continuous beams and application of the singularity method to influence lines. Solution of continuous beams by the singularity method is much the same as that for single-span beams. In the latter case the equation for moment is written in terms of the unknown redundant reactions. Application of the boundary conditions results in a system of equations in terms of the unknown redundant forces. The singularity method also provides an efficient method for influence lines. It allows for the derivation of a single expression for the influence line of any beam. The singularity method provides equally efficient procedures for deriving stiffness matrices and fixed-end forces of beams with variable cross-sections. This topic is discussed in Section 13.5.

9.1 SINGULARITY FUNCTIONS

The functions used in describing discontinuities are termed *singularity functions*. They may be expressed by Macaulay's brackets, the unit step function or Dirac delta function and its derivatives.

The singularity method can be postulated in the form of Macaulay's brackets or developed in terms of the unit step function. In this chapter we adopt the latter approach as it allows for the derivation of the pertinent equations and integrals and permits extensions to new classes of problems, while still maintaining the simplicity of the derivations. However, we also present Macaulay's brackets with an example of their application. A more rigorous approach can be taken by using the Dirac delta function, which is the derivative of the unit step function. The delta function and its derivatives are treated in texts on the theory of distributions.

Macaulay's Brackets

Macaulay has postulated a procedure for calculating the slope and defection of beams. In this method brackets are used to indicate a zero or nonzero value of an expression, depending on the sign of its value. Macaulay's brackets are defined as follows.

$$\langle x - a \rangle^n = \begin{cases} 0 & \text{for} \quad x - a \leq 0 \\ (x-a)^n & \text{for} \quad x - a > 0 \end{cases} \tag{9.1}$$

Macaulay also postulated the integral of the brackets as

$$\int \langle x - a \rangle^n \, dx = \frac{\langle x - a \rangle^{n+1}}{n+1} \tag{9.2}$$

For example, the expression $\langle x - 5 \rangle$ is equal zero when x is smaller than

5 and equals $(x - 5)$ when x is larger than 5. The integral of this expression is $\frac{1}{2}\langle x - 5 \rangle^2$.

For beams under concentrated loads, the above definitions provide a simple analytical solution technique. The following example shows how, using singularity functions, the shear or moment equations for beams can be written as single expressions. Using this equation for moment the differential equation of the beam can be easily integrated to obtain expressions for the slope and deflection.

Example 9.1. Write the expressions for shear and moment in a simply supported beam (Fig. 9.1) under a concentrated load P applied at a from the left end of the beam. Integrate the differential equation of the beam to obtain expressions for slope and deflection.

Solution. The expressions for shear and moment for this beam were found in Example 4.3, by dividing the beam up into two parts. The shear and moment were given as,

$$V = P\left(1 - \frac{a}{l}\right) \quad \text{and} \quad M = P\left(1 - \frac{a}{l}\right)x \quad \text{for} \quad x \leq a$$

and

$$V = P\left(1 - \frac{a}{l}\right) - P \quad \text{and} \quad M = P\left(1 - \frac{a}{l}\right)x - P(x - a) \quad \text{for} \quad x \geq a$$

We can combine the two shear equations into one using Macaulay's brackets. We note that the two shear equations are the same except for the additional term $-P$ in the expression for the beam segment on the right side of the load. If we write

$$V = P\left(1 - \frac{a}{l}\right) - P\langle x - a \rangle^0$$

Since by Eq. (9.1) the value of the bracket is zero on the left of the load, that is for $x < a$ or $x - a \leq 0$, we have

$$V = P\left(1 - \frac{a}{l}\right)$$

for the beam segment on the left side of the load. On the right of the load, $x > a$ or $x - a > 0$, the value in the brackets is positive and thus $\langle x - a \rangle^0 = 1$. Therefore, we get

$$V = P\left(1 - \frac{a}{l}\right) - P$$

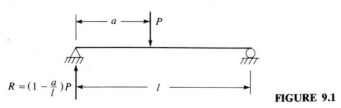

$$R = (1 - \tfrac{a}{l})P$$

FIGURE 9.1

In a similar manner we can write a single equation for the moment throughout the beam,

$$M = \frac{Pl}{2}x - P\langle x - a \rangle$$

Note that, in view of Eq. (9.2), the equation for moment is the integral of the shear equation. Again for $x \le a$, the value of the bracket is zero and $M = Plx/2$, while for $x > a$, the bracketed expression is positive, and $M = Plx/2 - P(x - a)$.

To find the expressions for the slope and deflection we can integrate the differential equation of the beam, Eq. (5.16). For this beam the differential equation is

$$EIv'' = -M = -\frac{Pl}{2}x + P\langle x - a \rangle$$

Integration of the above equation, using Eq. (9.2), yields

$$EIv' = -\frac{Pl}{4}x^2 + \frac{P}{2}\langle x - a \rangle^2 + C_1$$

and

$$EIv = -\frac{Pl}{12}x^3 + \frac{P}{6}\langle x - a \rangle^3 + C_1 x + C_2$$

The boundary conditions are $v = 0$ at $x = 0$ and $x = l$. At $x = 0$ the value in the brackets of the last equation is zero and at $x = l$ it is $(x - a)^3$. Thus, we get $C_2 = 0$ and

$$C_1 = \frac{Pl^3}{12} - \frac{P}{6l}(l - a)^3$$

The Unit Step Function

The unit step function or Heaviside function is a function of a variable x or more specifically a function of the argument $(x - a)$, where a is a fixed distance on the x axis.

By definition, the value of the function is zero when $(x - a)$ is negative and 1 when $(x - a)$ is positive (nonnegative). Thus,

$$H(x - a) = \begin{cases} 0 & \text{for} \quad x < a \\ 1 & \text{for} \quad x \ge a \end{cases} \tag{9.3}$$

A plot of this function is shown in Fig. 9.2. Instead of $H(x - a)$ we often use

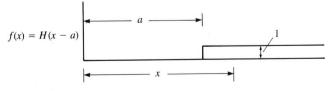

$f(x) = H(x - a)$

FIGURE 9.2

the abbreviated form H_a. The product of any function $f(x)$ with the step function will be either zero or the function itself depending on the argument of the step function. For example $f(x)H_a$ is equal to zero for $x < a$ and equal to $f(x)$ for $x \geq 0$.

9.2 INTEGRATION OF SINGULARITY FUNCTIONS

The integral of Macaulay's brackets was given by Eq. (9.2). In this section we derive integrals of functions that include the step function.

Consider the integral of the product of a continuous function $f(x)$ with the unit step function, that is,

$$\int_0^x f(x)H_a \, dx$$

In carrying out this integral we recognize two possibilities.

1. $x < a$. By definition (Eq. (9.3)) $H_a = 0$ everywhere and thus the integral is zero.
2. $x \geq a$. In this case

$$\int_0^x f(x)H_a \, dx = \int_0^a f(x) \cdot 0 \cdot dx + \int_a^x f(x) \cdot 1 \cdot dx$$

$$= \int_a^x f(x) \, dx$$

If we write

$$\boxed{\int_a^x f(x)H_a \, dx = H_a \int_a^x f(x) \, dx} \tag{9.4}$$

In view of the definition of the step function (Eq. (9.3)) we have incorporated both possibilities.

A function frequently encountered in calculating deflections of beams is $(x-a)^n H_a$. Referring to Eq. (9.4) the integral of this function is

$$\int_0^x (x-a)^n H_a \, dx = H_a \int_a^x (x-a)^n \, dx$$

To integrate $\int_a^x (x-a)^n \, dx$, we can change the variable as follows. Let $x - a = z$, then $dz = dx$. Thus, the integral becomes

$$\int_a^x (x-a)^n \, dx = \int_0^{x-a} z^n \, dz = \frac{z^{n+1}}{n+1} \Big|_0^{x-a}$$

$$\boxed{\int_a^x (x-a)^n H_a \, dx = \frac{(x-a)^{n+1}}{n+1} H_a} \tag{9.5}$$

Another function that appears in the formulation of variable section members is $(x - b)^n H_a$. Its integral can be found in a similar manner:

$$\int_0^x (x - b)^n H_a \, dx = H_a \int_a^x (x - b)^n \, dx$$

Let $x - b = z$, then $dz = dx$. The limits of the integral, a and x become $a - b$ and $x - b$, and

$$\int_0^x (x - b)^n \, dx = \int_{a-b}^{x-b} z^n \, dz = \frac{z^{n+1}}{n+1} \Big|_{a-b}^{x-b} = \frac{(x - b)^{n+1} - (a - b)^{n+1}}{n+1}$$

and thus,

$$\int_0^x (x - b)^n H_a \, dx = \frac{(x - b)^{n+1} - (a - b)^{n+1}}{n+1} H_a \qquad (9.6)$$

In the following sections, shear and moment equations are written for beams under concentrated and partially distributed loads, using the step function. Substituting the moment expressions in the differential equation of the beam, Eq. (5.16), and integrating using Eqs. (9.5) and (9.6) yields the expressions for slope and deflection.

9.3 CONCENTRATED LOADS

In the solution of beam problems the loads are involved in the equations for shear, moment, deflection, etc. The expressions for shear and moment can be written directly from the definitions given in Section 4.2. For example, shear can be found by summing the forces from the left of a beam to the section, and moment by summing the moments of the forces on the left of the section calculated at the section. As was seen in Example 9.1 using the singularity functions, one expression can be written for shear in a beam under a concentrated load. In that example we expressed the shear in terms of Macaulay's brackets. In writing the sum of forces for the shear at section x, the shear due to a concentrated load P acting at a, is

$$V_p = -PH_a \qquad (9.7a)$$

and its moment at x is

$$M_p = -P(x - a)H_a \qquad (9.7b)$$

Therefore, for the beam of Fig. 9.1, if R_1 is the left reaction, the shear force at x is $V = R_1 - PH_a$. From the definition of the step function, Eq. (9.3), $H_a = 0$ for $x < a$ (that is, at a section on the left of the load) and $V = R_1$. For a section on the right of the load, $x \geq a$, H_a is equal to 1 and $V = R_1 - P$. Therefore, a single expression can describe the shear throughout the beam. A similar expression can be found for moment using its definition. Alternatively, since moment is the integral of shear (Eq. (4.8)) it can be found by integrating the shear equation.

To write the expressions for shear and moment in a beam using singularity functions, we can take a section near the right end of the beam and write the equation for shear or moment in terms of the applied loads. To insure that the terms associated with each concentrated load drop out when the section is on the left of the load, we must multiply each of these terms by H_a, where a is the distance from the left hand end of the beam to the location of the concentrated load in question. Thus, for n concentrated loads P_i acting at a_i, the equation for shear would be $V = R_1 - \Sigma\, P_i H_{a_i}$, where the summation is for i from 1 to n, that is, for all the loads.

Example 9.2. Write the equations for shear and moment of the cantilever beam, Fig. 9.3, under two concentrated loads as shown, and find the expressions for slope and deflection.

Solution. The fixed end reactions are

$$R_0 = 15 + 10 = 25 \text{ kips}$$
$$M_0 = -15 \times 10 - (10 \times 20) = -350 \text{ kip} \cdot \text{ft}$$

The equation for shear is

$$V = 25 - 15H_{10} - 10H_{20}$$

and the moment equation is

$$M = -350 + 25x - 15(x - 10)H_{10} - 10(x - 20)H_{20}$$

The differential equation of the beam (Eq. (5.16)) becomes

$$EIv'' = 350 - 25x + 15(x - 10)H_{10} + 10(x - 20)H_{20}$$

By integration, using Eq. (9.5),

$$EIv' = 350x - 12.5x^2 + 7.5(x - 10)^2 H_{10} + 5(x - 20)^2 H_{20} + C_1$$

and

$$EIv = 175x^2 - 4.2x^3 + 2.5(x - 10)^3 H_{10} + 1.7(x - 20)^3 H_{20} + C_1 x + C_2$$

The boundary conditions; displacement and slope equal to zero at the left-hand end, give $C_1 = C_2 = 0$, since at the left-hand end of the beam the values of the step functions are zero.

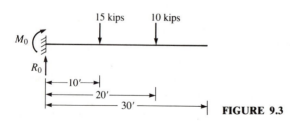

FIGURE 9.3

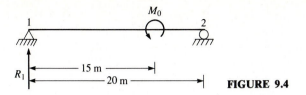

FIGURE 9.4

Example 9.3. Find the shear and moment equations for a simply supported beam with 20 m span (Fig. 9.4) under a concentrated moment of $M_0 = 500$ kN · m at 15 m from the left-hand end. Calculate the values of moment at the end points of the beam and at 15 m. Derive the equations for slope and deflection.

Solution

$$R_1 = \frac{500}{20} = 25 \text{ kN}$$

$$V = 25 \text{ kN}$$

$$M = 25x - 500H_{15}$$

At $x = 0$, $H_{15} = 0$ and thus $M = 0$.
At $x = 20$ m, $H_{15} = 1$ and $M = 25 \times 20 - 500 = 0$.
At point 3, just to the left of the concentrated moment, $x < 15$ and $H_{15} = 0$,

$$M = 25 \times 15 = 375 \text{ kN} \cdot \text{m}$$

and just to the right of the concentrated moment, $x > 15$ and $H_{15} = 1$,

$$M = 25 \times 15 - 500 = -125 \text{ kN} \cdot \text{m}$$

The differential equation of the beam is

$$EIv'' = -25x + 500H_{15}$$

Then, in view of Eq. (9.5),

$$EIv' = -12.5x^2 + 250(x - 15)H_{15} + C_1$$

and

$$EIv = -4.2x^3 + 125(x - 15)^2 H_{15} + C_1 x + C_2$$

At $x = 0$, $v = 0$ gives $C_2 = 0$, and at $x = 20$ m, $v = 0$, results in

$$C_1 = 4.2 \times 20^2 - 125(20 - 15)^2/20 = 1\,528$$

Example 9.4. Write the equations of shear and moment for a simply supported beam under n concentrated loads (Fig. 9.5) and derive the expressions for slope and deflection.

Solution. The left reaction is

$$R_1 = \frac{1}{l} \sum_{i=1}^{n} P_i(l - a_i)$$

The equation for shear is

$$V = R_1 - \sum_{i=1}^{n} P_i H_i$$

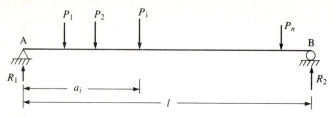

FIGURE 9.5

where $H_i = H(x - a_i)$ was defined by Eq. (9.3) and the equation for moment is

$$M = R_1 x - \sum_{i=1}^{n} P_i(x - a_i)H_i$$

The governing differential equation is

$$EIv'' = -M = -R_1 x + \sum_{i=1}^{n} P_i(x - a_i)H_i$$

Integration yields

$$EIv' = -\tfrac{1}{2}R_1 x^2 + \tfrac{1}{2}\sum_{i=1}^{n} P_i(x - a_i)^2 H_i + C_1$$

$$EIv = -\tfrac{1}{6}R_1 x^3 + \tfrac{1}{6}\sum_{i=1}^{n} P_i(x - a_i)^3 H_i + C_1 x + C_2$$

At $x = 0$, $v = 0$, $H_i = 0$ gives $C_2 = 0$.
At $x = l$, $v = 0$, $H_i = 1$ results in

$$C_1 = \tfrac{1}{6}R_1 l^2 - \frac{1}{6l}\sum_{i=1}^{n} P_i(l - a_i)^3$$

9.4 PARTIALLY DISTRIBUTED LOADS

For a partially distributed load p that starts at a and extends to the right-hand end of the beam (Fig. 9.6a), the shear and moment equations can be written in terms of the step function H_a. The shear due to this load is

$$V_p = -p(x - a)H_a \tag{9.8a}$$

and its moment at x is

$$M_p = -\tfrac{1}{2}p(x - a)^2 H_a \tag{9.8b}$$

For example, the shear and moment in the beam of Fig. 9.6a are

$$V = R_0 - p(x - a)H_a$$

$$M = M_0 + R_0 x - \frac{p}{2}(x - a)^2 H_a$$

where $R_0 = p(l - a)$ and $M_0 = -p(l - a)(a + l)/2$ are the reactions at the left-hand end of the beam.

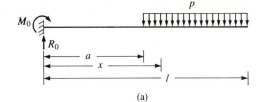

(a)

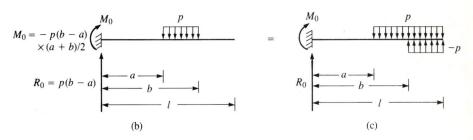

$M_0 = -p(b-a)$
$\times (a+b)/2$

$R_0 = p(b-a)$

(b)

R_0

(c)

FIGURE 9.6

For a load extending over a segment from a to b (Fig. 9.6b) it is easier to describe the load as the sum of two loads: p over a length starting at a, and $-p$ starting at b, both extending to the end of the beam (Fig. 9.6c). Thus, the magnitude of shear at an arbitrary section x due to this load is

$$V_p = -p(x-a)H_a + p(x-b)H_b \qquad (9.9a)$$

and the moment is

$$M_p = -\tfrac{1}{2}p(x-a)^2 H_a + \tfrac{1}{2}p(x-b)^2 H_b \qquad (9.9b)$$

Therefore, for the beam of Fig. 9.6 the shear and moment are

$$V = R_0 - p(x-a)H_a + p(x-b)H_b$$

$$M = M_0 + R_0 x - \frac{p}{2}(x-a)^2 H_a + \frac{p}{2}(x-b)^2 H_b$$

where

$$R_0 = p(b-a) \qquad \text{and} \qquad M_0 = -p(b-a)\frac{(a+b)}{2}$$

Example 9.5. Find the shear and moment equations for the simply supported beam of Fig. 9.7, under the distributed loads shown. Derive the equations of slope and deflection and calculate the deflection at $x = 20$ m.

Solution. The left reaction of the beam in Fig. 9.7 is

$$R_1 = \tfrac{1}{40}(8 \times 5 \times 27.5 + 15 \times 10 \times 10) = 65 \text{ kN}$$

Writing the shear at just before the right end of the beam, and including the

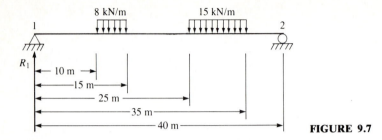

FIGURE 9.7

appropriate step functions to cancel the term when x is on the left of the load, we get the following shear equation. Note that each partially distributed load is expressed as the algebraic sum of two loads extending to the right end of the beam, as discussed above.

$$V = 65 - 8(x - 10)H_{10} + 8(x - 15)H_{15} - 15(x - 25)H_{25} + 15(x - 35)H_{35}$$

and

$$M = 65x - 4(x - 10)^2 H_{10} + 4(x - 15)^2 H_{15} - 7.5(x - 25)^2 H_{25} + 7.5(x - 35)^2 H_{35}$$

$$EIv'' = -65x + 4(x - 10)^2 H_{10} - 4(x - 15)^2 H_{15} + 7.5(x - 25)^2 H_{25} - 7.5(x - 35)^2 H_{35}$$

$$EIv' = -32.5x^2 + \tfrac{4}{3}(x - 10)^3 H_{10} - \tfrac{4}{3}(x - 15)^3 H_{15} + 2.5(x - 25)^3 H_{25}$$
$$- 2.5(x - 35)^3 H_{35} + C_1$$

$$EIv = -10.83x^3 + 0.33(x - 10)^4 H_{10} - 0.33(x - 15)^4 H_{15}$$
$$+ \tfrac{2.5}{4}(x - 25)^4 H_{25} - \tfrac{2.5}{4}(x - 35)^4 H_{35} + C_1 x + C_2$$

The condition of zero deflection at the left end gives $C_2 = 0$, since all the step functions are zero for $x = 0$. At $x = l$, $v = 0$, and the values of the step functions are 1. This gives

$$C_1 = \tfrac{1}{40}[10.83 \times 40^3 - 0.33(40 - 10)^4 + 0.33(40 - 15)^4$$
$$- \tfrac{2.5}{4}(40 - 25)^4 + \tfrac{2.5}{4}(40 - 35)^4]$$

$$C_1 = 13\,087$$

At $x = 20$ m, $H_{10} = H(x - 10) = H(20 - 10) = 1$, $H_{15} = 1$, $H_{25} = H_{35} = 0$.

$$EIv_{20} = -10.83 \times 20^3 + 0.33(20 - 10)^4 - 0.33(20 - 15)^4 + 13\,087 \times 20$$
$$= 178\,190$$

and $v_{20} = 178\,190/EI$.

9.5 VARIABLE SECTIONS

The singularity method can be used for problems with arbitrary variations of the section and properties. To demonstrate the procedure we start with the beam of Fig. 9.8.

The moment equation for the beam is

$$M = R_1 x - P(x - a)H_a$$

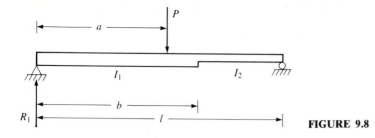

FIGURE 9.8

with

$$R_1 = P\left(1 - \frac{a}{l}\right)$$

We want to integrate the beam differential equation, that is,

$$v'' = -\frac{1}{EI} M$$

We can express the rigidity of the second segment of the beam in terms of that of the first one. Since the rigidity of the second segment can be expressed as a quotient of the rigidity of the first segment, we choose the proportional factor as $1 + r$, where r is used to express the change of $1/EI$ of the second segment with respect to that of the first segment. Thus,

$$\frac{1}{EI_2} = \frac{1}{EI_1}(1 + r) \qquad (9.10a)$$

From Eq. (9.10a) the constant r is

$$r = \frac{EI_1}{EI_2} - 1 = EI_1\left(\frac{1}{EI_2} - \frac{1}{EI_1}\right) \qquad (9.10b)$$

Therefore, a single equation can be used to express the rigidity at any point as

$$\frac{1}{EI} = \frac{1}{EI_1}(1 + rH_b) \qquad (9.10c)$$

where, as was defined in Eq. (9.3), the step function is

$$H_b = H(x - b) = \begin{cases} 0 & \text{for} \quad x < b \\ 1 & \text{for} \quad x \geq b \end{cases}$$

Upon substitution for the moment and rigidity in the differential equation of the beam we get

$$-v'' = [R_1 x - P(x - a)H_a]\left(\frac{1}{EI_1}(1 + rH_b)\right)$$

$$-EI_1 v'' = R_1 x - P(x - a)H_a + R_1 rxH_b - Pr(x - a)H_a H_b$$

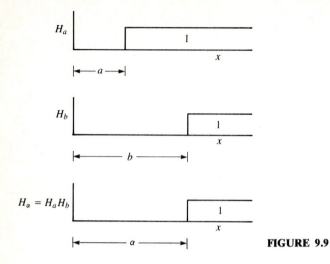

FIGURE 9.9

Referring to Fig. 9.9 we note that to obtain $H_a H_b$ we must take the ordinates of H_a and H_b at every distance along x and multiply them together. The result is a step function with a zero segment equal to the larger zero segment of H_a and H_b. Therefore, we have

$$H_a H_b = H_\alpha \qquad (9.11a)$$

where

$$\alpha = \max(a, b) \qquad (9.11b)$$

Therefore

$$-EI_1 v'' = R_1 x - P(x-a)H_a + R_1 rx H_b - Pr(x-a)H_\alpha$$

Upon integration we get

$$-EI v' = \tfrac{1}{2} R_1 x^2 - \tfrac{1}{2} P(x-a)^2 Ha + R_1 r \int_0^x x H_b \, dx - Pr \int_0^x (x-a) H_\alpha \, dx + C_1$$

The two integrals on the right can be found using Eqs. (9.4) and (9.6) as

$$\int_0^x x H_b \, dx = H_b \int_b^x x \, dx = \tfrac{1}{2}(x^2 - b^2) H_b$$

$$\int_0^x (x-a) H_\alpha \, dx = \tfrac{1}{2}[(x-a)^2 - (\alpha-a)^2] H_\alpha$$

Thus

$$-EI v' = \tfrac{1}{2} R_1 x^2 - \tfrac{1}{2} P(x-a)^2 H_a + \tfrac{1}{2} R_1 r(x^2 - b^2) H_b - \tfrac{1}{2} Pr$$
$$\times [(x-a)^2 - (\alpha-a)^2] H_\alpha + C_1$$

Integration of the last equation gives the deflection as

$$-EI v = \tfrac{1}{6} R_1 x^3 - \tfrac{1}{6} P(x-a)^3 H_a + \tfrac{1}{2} R_1 r[\tfrac{1}{3}(x^3 - b^3) - b^2(x-b)] H_b$$
$$- \tfrac{1}{2} Pr[\tfrac{1}{3}(x-a)^3 - \tfrac{1}{3}(\alpha-a)^3 - (\alpha-a)^2(x-\alpha)] H_\alpha + C_1 x + C_2$$

Application of the boundary conditions, that is, $v = 0$ at $x = 0$ and $x = l$ gives the values of C_1 and C_2 as

$$C_1 = -\tfrac{1}{6}R_1l^2 + \frac{P}{6l}(l-a)^3 - \frac{R_1r}{2l}[\tfrac{1}{3}(l^3 - b^3) - b^2(l-b)]$$

$$+ \frac{Pr}{2l}[\tfrac{1}{3}(l-a)^3 - \tfrac{1}{3}(\alpha - a)^3 - (\alpha - a)^2(l - \alpha)]$$

and $C_2 = 0$.

Note that at $x = 0$ all the step functions are zero, and at $x = l$ they are equal to unity. Furthermore, if we calculate the values of rotation and deflection at specific points on the beam then the step functions take on either 0 or 1 and therefore disappear from the equations.

Example 9.6. Find the slope and deflection equations for a beam with m-step changes of the section and n concentrated loads (Fig. 9.10).

Solution. In this case there are m different segments. Following the procedure described above, we can write the flexibility of any segment as

$$\frac{1}{EI} = \frac{1}{EI_1}\left(1 + \sum_{j=1}^{m-1} r_j H_j\right)$$

$$r_j = EI_1\left(\frac{1}{EI_{j+1}} - \frac{1}{EI_j}\right)$$

with $H_j = H(x - b_j)$.

The equation for moment from Example 9.4 is

$$M = R_1x - \sum_{i=1}^{n} P_i(x - a_i)H_i$$

with

$$R_1 = \frac{1}{l}\sum_{i=1}^{n} P_i(l - a_i) \qquad \text{and} \qquad H_i = H(x - a_i)$$

The governing equation of the beam is

$$v'' = -\frac{M}{EI}$$

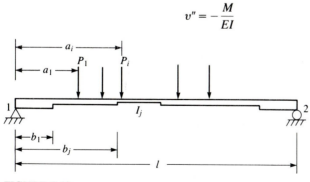

FIGURE 9.10

becomes, upon substitution for $1/EI$ and M,

$$-EI_1v'' = \left(1 + \sum_{j=1}^{m-1} r_j H_j\right)\left(R_1 x - \sum_{i=1}^{n} P_i(x - a_i)H_i\right)$$

$$-EI_1v'' = R_1 x - \sum_{i=1}^{n} P_i(x - a_i)H_i + R_1 \sum_{j=1}^{m-1} x r_j H_j$$

$$- \sum_{i=1}^{n} \sum_{j=1}^{m-1} P_i r_j(x - a_i)H_i H_j$$

As in Eqs. 9.11 we can let

$$H_i H_j = H_k = H(x - c_k) \qquad \text{where} \qquad c_k = \max(a_i, b_j)$$

Then

$$-EI_1v'' = R_1 x - \sum_{i=1}^{n} P_i(x - a_i)H_i + R_1 \sum_{j=1}^{m-1} x r_j H_j$$

$$- \sum_{i=1}^{n} \sum_{j=1}^{m-1} P_i r_j(x - a_i)H_k$$

Upon integration using Eqs. (9.5) and (9.6),

$$-EI_1v' = \tfrac{1}{2}R_1 x^2 - \tfrac{1}{2}\sum_{i=1}^{n} P_i(x - a_i)^2 H_i + \tfrac{1}{2}R_1 \sum_{j=1}^{m-1} r_j(x^2 - b_j^2)H_j$$

$$- \sum_{i=1}^{n} \sum_{j=1}^{m-1} P_i r_j[\tfrac{1}{2}(x^2 - c_k^2) - a_i(x - c_k)]H_k + C_1$$

$$-EI_1v = \tfrac{1}{6}R_1 x^3 - \tfrac{1}{6}\sum_{i=1}^{n} P_i(x - a_i)^3 H_i + \tfrac{1}{2}R_1 \sum_{j=1}^{m-1} r_j[\tfrac{1}{3}(x^3 - b_j^3) - b_j^2(x - b_j)]H_j$$

$$- \tfrac{1}{2}\sum_{i=1}^{n} \sum_{j=1}^{m-1} P_i r_j[\tfrac{1}{3}(x^3 - c_k^3) - c_k^2(x - c_k) - a_i(x - c_k)^2]H_k + C_1 x + C_2$$

The boundary conditions are $v = 0$ at $x = 0$, which yields $C_2 = 0$, and $v = 0$ at $x = l$, with $H_i = H_j = H_k = 1$, which yield C_1 as follows

$$C_1 l = -\tfrac{1}{6}R_1 l^3 + \tfrac{1}{6}\sum_{i=1}^{n} P_i(l - a_i)^3 - \tfrac{1}{2}R_1 \sum_{j=1}^{m-1} r_j[\tfrac{1}{3}(l^3 - b_j^3) - b_j^2(l - b_j)]$$

$$+ \tfrac{1}{2}\sum_{i=1}^{n} \sum_{j=1}^{m-1} P_i r_j[\tfrac{1}{3}(l^3 - c_k^3) - c_k^2(l - c_k) - a_i(l - c_k)^2]$$

Example 9.7. Find the mid-span deflection of the beam of Fig. 9.11, with $E = 29\,000$ ksi. The load 5 kips is applied at the mid-span.

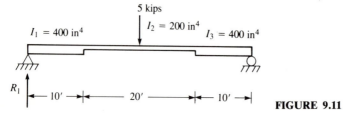

FIGURE 9.11

Solution. Using the procedure of the previous example here we have

$$r_1 = EI_1\left(\frac{1}{EI_2} - \frac{1}{EI_1}\right) = 400\left(\frac{1}{200} - \frac{1}{400}\right) = 1$$

and

$$r_2 = EI_1\left(\frac{1}{EI_3} - \frac{1}{EI_2}\right) = 400\left(\frac{1}{400} - \frac{1}{200}\right) = -1$$

$$\begin{cases} a_1 = 20 & c_1 = \max(a_1, b_1) = 20 & i = 1, \quad j = 1 \\ b_1 = 10 & c_2 = \max(a_1, b_2) = 30 & i = 1, \quad j = 2 \\ b_2 = 30 \end{cases}$$

$$R_1 = \tfrac{1}{40} \times 5 \times (40 - 20) = 2.5 \text{ kips}$$

$$40C_1 = -\tfrac{1}{6} \times 2.5 \times 40^3 + \tfrac{1}{6} \times 5 \times (40 - 20)^3 - \tfrac{1}{2} \times 2.5$$
$$\times \{[\tfrac{1}{3}(40^3 - 10^3) - 10^2(40 - 10)] - \tfrac{1}{3}(40^3 - 30^3) + 30^2(40 - 30)]\}$$
$$+ \tfrac{5}{2}(-1)[\tfrac{1}{3}(40^3 - 20^3) - 20^2(40 - 20) - 20(40 - 20)^2]$$
$$+ \tfrac{5}{2}(-1)[\tfrac{1}{3}(40^3 - 30^3) - 30^2(40 - 30) - 20(40 - 30)^2]$$

or $C_1 = -875$, which gives upon substitution into the equation for v with $x = 20$,

$$-400Ev_3 = \tfrac{1}{6} \times 2.5 \times 20^3 + \tfrac{1}{2} \times 2.5 \times [\tfrac{1}{3}(20^3 - 10^3) - 10^2(20 - 10)] - 875 \times 20$$

or

$$v_3 = 31.25/E$$

9.6 INTERNAL HINGES

Application of the singularity method to beams with internal hinges requires additional considerations. For instance, let us examine the beam of Fig. 9.12. The beam has a hinge at b and a concentrated load P is acting at a distance a from the left-hand end.

Setting the moments of the forces on the left of the hinge, calculated at the hinge, equal to zero, we can find the left reaction as

$$R = \frac{P(b - a)}{b}$$

The moment equation for the beam is

$$M = Rx - P(x - a)H_a$$

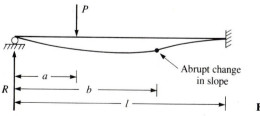

Abrupt change in slope

FIGURE 9.12

This equation is the same as the moment equation for the same beam without the hinge. Thus, if we substitute this expression in the differential equation (Eq. (5.16)) and integrate to obtain the deflection, we will end up with the deflection of a beam without the hinge, that is, the wrong results. This is because we are not considering the discontinuity of the slope at the hinge. Furthermore, since the deflection at the left-hand end and the deflection and rotation at the right-hand end of the beam are specified, we will end up with two constants of integration and three conditions for finding their values.

One way to handle this problem is to integrate the equation to get a relation for the slope. Noting that the slope is discontinuous at the hinge, we can add a term to reflect this discontinuity. This is done by adding the product of a constant C and a step function that changes its value at b. This constant plus the two constants of integration are then evaluated from the three boundary conditions. Another way to handle this problem is by using the Dirac delta function. This alternative provides a relationship between the change of slope and moment that is valid for discontinuous slopes as well. Here we stay with the first alternative, using the familiar step function.

The governing differential equation is

$$EIv'' = -Rx + P(x - a)H_a$$

Integration of this equation gives

$$EIv' = -\tfrac{1}{2}Rx^2 + \tfrac{1}{2}P(x - a)^2 H_a + C_1$$

As mentioned before, this equation is not correct for the beam with a hinge. Thus, we add the term

$$CH_b \tag{9.12}$$

to the slope equation. In the expression (9.12), C is an additional constant to be found, which is actually equal to the change of slope at the hinge. Thus, the equation for slope of the beam with a hinge has the form,

$$EIv' = -\tfrac{1}{2}Rx^2 + \tfrac{1}{2}P(x - a)^2 H_a + CH_b + C_1$$

Upon integration,

$$EIv = -\tfrac{1}{6}Rx^3 + \tfrac{1}{6}P(x - a)^3 H_a + C(x - b)H_b + C_1x + C_2$$

The problem is now well posed with the constants of integration equal in number to the boundary conditions and a change of slope taking place at the hinge. Applying the boundary conditions, $v = 0$ at $x = 0$ and $v = v' = 0$ at $x = l$, we find the unknown constants C, C_1, and C_2 as follows.

$$C_1 = \tfrac{1}{2}Rl^2 - \tfrac{1}{2}P(l - a)^2 - \frac{Rl^3}{3b} - \frac{P}{6b}(l - a)^3 + \frac{Pl}{2b}(l - a)^2$$

$$C_2 = 0 \quad \text{and} \quad C = \frac{Rl^3}{3b} + \frac{P}{6b}(l - a)^3 - \frac{Pl}{2b}(l - a)^2$$

9.7 CONTINUOUS BEAMS BY THE SINGULARITY METHOD

The solution of continuous beams by the singularity method follows the same steps as that of statically determinate beams. The unknown reactions are treated as concentrated forces. An expression is written for the moment throughout the beam (in terms of the unknown reactions). The differential equation of the beam is integrated and the boundary conditions are applied. The resulting equations along with the equations of statics will provide a sufficient number of equations for the solution of the problem. The process just outlined is the method of consistent deformations presented in Chapter 7. However, the singularity method allows a more direct solution.

Note that in the singularity method the origin of the x coordinate is always taken at the left-hand end of the beam.

Example 9.8. Find the reactions of the two-span beam of Fig. 9.13 by the singularity method.

Solution. The moment equation for this beam is

$$M = R_1 x + R_2(x - 15)H_{15} - 30(x - 6)H_6$$

The differential equation of the beam is

$$EIv'' = -M = -R_1 x - R_2(x - 15)H_{15} + 30(x - 6)H_6$$

Upon integration, using Eq. (9.5), we get

$$EIv' = -\tfrac{1}{2}R_1 x^2 - \tfrac{1}{2}R_2(x - 15)^2 H_{15} + 15(x - 6)^2 H_6 + C_1$$

and

$$EIv = -\tfrac{1}{6}R_1 x^3 - \tfrac{1}{6}R_2(x - 15)^3 H_{15} + 5(x - 6)^3 H_6 + C_1 x + C_2$$

The unknowns consist of the reactions and the two constants of integration. There are three boundary conditions, $v = 0$ at $x = 0$, 15 ft, and 33 ft. These along with two equilibrium equations provide a sufficient number of equations for the solution of the problem. Applying the boundary conditions and noting that $H_{15} = H(x - 15) = 0$ for $x < 15$ and $H_{15} = 1$ for $x \ge 15$, we get

$$C_2 = 0$$

$$-\tfrac{1}{6}R_1(15)^3 + 5(15 - 6)^3 + 15C_1 + C_2 = 0$$

$$-\tfrac{1}{6}R_1(33)^3 - \tfrac{1}{6}R_2(33 - 15)^3 + 5(33 - 6)^3 + 33C_1 + C_2 = 0$$

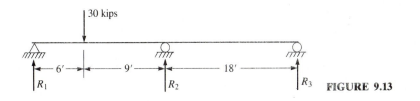

FIGURE 9.13

or

$$C_2 = 0 \qquad (a)$$

$$-562.5R_1 + 15C_1 + C_2 = -3\,645 \qquad (b)$$

$$-5989.5R_1 - 972R_2 + 33C_1 + C_2 = -98\,415 \qquad (c)$$

The two equilibrium equations are the sum of the vertical forces equal to zero and sum of the moment with respect to the right-hand end equal to zero. The best way is to use the moment equation along with the above equations to find the unknowns. The force equilibrium equation then gives the value of the right reaction, R_3. However, this reaction is not necessary for shear and moment diagrams. The moment equation gives

$$R_1 \times 33 + R_2 \times 18 = 30 \times 27 \qquad (d)$$

The solution of Eqs. (a) to (d) is

$$R_1 = 15.71 \text{ kips} \qquad R_2 = 16.2 \text{ kips} \qquad C_1 = 346.1 \qquad C_2 = 0$$

Equilibrium of the vertical forces then gives $R_3 = -1.91$ kips.

Substitution of R_1 and R_2 into the moment equation results in the expression for the moment throughout the beam as

$$M = 15.71x + 16.2(x - 15)H_{15} + 30(x - 6)H_6$$

The derivative of the moment is shear, and by Eq. (9.5) we get,

$$V = 15.71 + 16.2H_{15} + 30H_6$$

Similarly, substitution of the reactions and constants C_1 and C_2 into the equations for the slope and deflection results in expressions that can be used to calculate the slope and deflection at any point.

Example 9.9. Find the reactions of the beam of Fig. 9.14.

Solution. Here again there is one redundant force. However, the solution of continuous beams with the singularity method can always be carried out in the same manner as for Example 9.8. The equation for the moment in this case is

$$M = R_1x + R_2(x - 8)H_8 - 10\frac{(x - 8)^2}{2}H_8$$

or

$$M = R_1x + R_2(x - 8)H_8 - 5(x - 8)^2H_8$$

The differential equation for the beam is $v'' = -M/EI$.

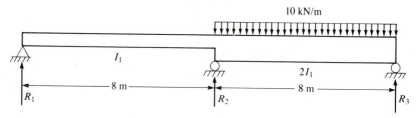

FIGURE 9.14

As discussed in Section 9.5, for variable sections, we can express the rigidity throughout the beam as

$$\frac{1}{EI} = \frac{1}{EI_1}(1 + r_1 H_8)$$

where

$$r_1 = I_1\left(\frac{1}{I_2} - \frac{1}{I_1}\right) = \frac{I_1}{I_2} - 1 = \frac{1}{2} - 1 = -\frac{1}{2}$$

Thus,

$$\frac{1}{EI} = \frac{1}{EI_1}(1 - \frac{1}{2}H_8)$$

The differential equation becomes

$$EI_1 v'' = -(1 - \tfrac{1}{2}H_8)[R_1 x + R_2(x - 8)H_8 - 5(x - 8)^2 H_8]$$

Expanding the expression and noting (Eq. (9.11a)) that $H_8 H_8 = H_8$, we get

$$EI_1 v'' = -R_1 x - R_2(x - 8)H_8 + 5(x - 8)^2 H_8$$
$$+ \tfrac{1}{2}R_1 x H_8 + \tfrac{1}{2}R_2(x - 8)H_8 - \tfrac{5}{2}(x - 8)^2 H_8$$

Integration of this equation gives

$$EI_1 v' = -\tfrac{1}{2}R_1 x^2 - \tfrac{1}{2}R_2(x - 8)^2 H_8 + \tfrac{5}{3}(x - 8)^3 H_8 + \tfrac{1}{4}R_1(x^2 - 8^2)H_8$$
$$+ \tfrac{1}{4}R_2(x - 8)^2 H_8 - \tfrac{5}{6}(x - 8)^3 H_8 + C_1$$

and

$$EI_1 v = -\tfrac{1}{6}R_1 x^3 - \tfrac{1}{6}R_2(x - 8)^3 H_8 + \tfrac{5}{12}(x - 8)^4 H_8$$
$$+ \tfrac{1}{12}R_1(x^3 - 8^3)H_8 - \frac{8^2}{4}R_1(x - 8)H_8 + \tfrac{1}{12}R_2(x - 8)^3 H_8$$
$$- \tfrac{5}{24}(x - 8)^4 H_8 + C_1 x + C_2$$

The boundary conditions are $v = 0$ at $x = 0$, 8, 16 m, and the equilibrium equation is the sum of moments with respect to the right-hand end equal to zero, leading to

$$C_2 = 0$$

$$-\frac{8^3}{6}R_1 + 8C_1 + C_2 = 0$$

$$-\frac{16^3}{6}R_1 - \tfrac{1}{6}R_2(16 - 8)^3 + \tfrac{5}{12}(16 - 8)^4$$

$$+ \tfrac{1}{12}R_1(16^3 - 8^3) - \frac{8^2}{4}R_1(16 - 8)^2 + \tfrac{1}{12}R_2(16 - 8)^3$$

$$- \tfrac{5}{24}(16 - 8)^4 + 16C_1 + C_2 = 0$$

or

$$-85.33R_1 + 8C_1 = 0$$

and

$$-512R_1 - 42.67R_2 + 16C_1 = -853.33$$

These two equations plus the equilibrium of the moments with respect to the right end,

$$16R_1 + 8R_2 - 10 \times \frac{8^2}{2} = 0$$

provide three equations whose solution gives

$$R_1 = -3.33 \text{ kN}$$
$$R_2 = 46.67 \text{ kN}$$
$$C_1 = -35.56$$

Upon substitution of these values into the moment, slope, and deflection equations, we can find the values of the latter at any point.

9.8   INFLUENCE LINES BY THE SINGULARITY METHOD

The sinularity method provides an efficient method for obtaining expressions for influence lines, especially for continuous beams with variable cross-sections. The advantage of this procedure is that a single equation can be derived for the influence line.

Determinate Beams

For determinate beams the influence lines of reactions, shear, and moment are obtained by the equations of statics (plus equations of condition if applicable). The process is similar to finding shear and moment in beams. That is, a unit load is applied at an arbitrary distance ξ, and the shear and moment at any section is obtained in terms of the variable ξ.

Example 9.10. Find the influence line of the left reaction R_1, and shear and moment at x for the beam of Fig. 9.15a.

Solution. For a unit load at variable location, the reaction at the left support is

$$R_1 = 1 - \frac{\xi}{l}$$

This equation in terms of the variable ξ provides the influence line for the left reaction. Since this is a linear relation in terms of ξ, it can be plotted by finding two points at $\xi = 0$, and $\xi = l$. A plot of R_1 is shown in Fig. 9.15b.
    Shear at x under the unit load at ξ is

$$V = R_1 - H_\xi$$

$$= 1 - \frac{\xi}{l} - H_\xi$$

where

$$H_\xi = H(x - \xi) = \begin{cases} 0 & x < \xi \\ 1 & x \geq \xi \end{cases}$$

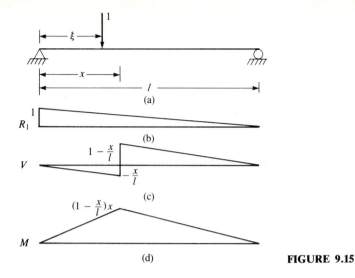

FIGURE 9.15

H_ξ is the unit step function discussed in Sections 9.1 and 9.2. For a fixed value of x, the above equation for V gives the influence line of shear in terms of ξ. Similarly, the moment at x is

$$M = R_1x - (x - \xi)H_\xi$$

$$= \left(1 - \frac{\xi}{l}\right)x - (x - \xi)H_\xi$$

Again for a fixed value of x, the above equation gives the influence line of the moment at x in terms of ξ.

The influence lines for shear and moment, V and M, are shown in Figs. 9.15c, d.

Example 9.11. Find the influence line of deflection at x for the beam of Fig. 9.16a.

Solution. Although less common than moment and shear, influence lines for rotation and deflection can be found at a section. With the singularity method this is done by integrating the differential equation of the beam. Here

$$EIv'' = -M = -\left(1 - \frac{\xi}{l}\right)x + (x - \xi)H_\xi$$

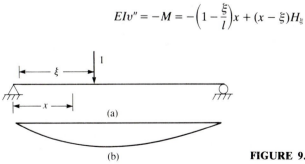

FIGURE 9.16

Upon integration in terms of x,

$$EIv' = -\tfrac{1}{2}\left(1 - \frac{\xi}{l}\right)x^2 + \tfrac{1}{2}(x - \xi)^2 H_\xi + C_1$$

and

$$EIv = -\tfrac{1}{6}\left(1 - \frac{\xi}{l}\right)x^3 + \tfrac{1}{6}(x - \xi)^3 H_\xi + C_1 x + C_2$$

Application of the boundary conditions ($v = 0$ for $x = 0$ and l) results in

$$C_1 = \frac{1}{6}\left(1 - \frac{\xi}{l}\right)l^2 - \frac{1}{6l}(l - \xi)^3 \qquad \text{and} \qquad C_2 = 0$$

The equation for v gives the influence line of deflection at section x, in terms of ξ, the variable location of the unit load. Note that, as ξ varies, $H_\xi = H(x - \xi)$ is equal to zero when $x < \xi$ and equal to 1 when $x \geq \xi$. The influence line of deflection is shown in Fig. 9.16b.

Indeterminate Beams

The singularity method can be applied to influence lines of indeterminate beams in a manner similar to that for determinate beams. Thus, for a unit load applied at an arbitrary distance ξ from the first support of the beam, an equation can be written for moment throughout the beam in terms of the unknown reactions. Upon integration of the differential equation for the beam an expression is obtained for the deflection of the whole beam. Application of the boundary conditions at the supports plus equilibrium equations provides a sufficient number of equations for determining the constants of integration and the unknown reactions. The equilibrium equations are obtained by setting the sum of the vertical forces and sum of the moments at the right-hand end of the beam equal to zero. Variable-section beams can be handled in the same manner as described in Section 9.5.

Example 9.12. For the three span beam of Fig. 9.17, find the influence lines for the shear at the middle of the first span and moment at the middle of the second span.

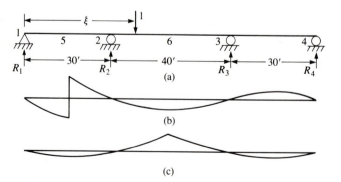

(a)

(b)

(c)

FIGURE 9.17

Solution. The equations for shear and moment are

$$V = R_1 + R_2 H_{30} + R_3 H_{70} - H(x - \xi)$$

$$M = R_1 x + R_2(x - 30)H_{30} + R_3(x - 70)H_{70} - (x - \xi)H(x - \xi)$$

$$-EIv'' = M = R_1 x + R_2(x - 30)H_{30} + R_3(x - 70)H_{70} - (x - \xi)H(x - \xi)$$

$$-2EIv' = R_1 x^2 + R_2(x - 30)^2 H_{30} + R_3(x - 70)^2 H_{70}$$
$$- (x - \xi)^2 H(x - \xi) + 2C_1$$

$$-6EIv = R_1 x^3 + R_2(x - 30)^3 H_{30} + R_3(x - 70)^3 H_{70}$$
$$- (x - \xi)^3 H(x - \xi) + 6C_1 x + 3C_2$$

$v = 0$ at $x = 0$ gives $C_2 = 0$.

Also, $v = 0$ at $x = 30$, 70, 100 ft gives

$$30^3 R_1 - (30 - \xi)^3 H(30 - \xi) + 180C_1 = 0$$

$$70^3 R_1 + 40^3 R_2 - (70 - \xi)^3 H(70 - \xi) + 420C_1 = 0$$

$$100^3 R_1 + 70^3 R_2 + 30^3 R_3 - (100 - \xi)H(100 - \xi) + 600C_1 = 0$$

The equilibrium equation, $\Sigma M_4 = 0$, gives $100R_1 + 70R_2 + 30R_3 - (100 - \xi) = 0$. Thus, we get

$$27\,000R_1 + 180C_1 = (30 - \xi)^3 H(30 - \xi)$$

$$343\,000R_1 + 64\,000R_2 + 420C_1 = (70 - \xi)^3 H(70 - \xi)$$

$$1\,000\,000R_1 + 343\,000R_2 + 27\,000R_3 + 600C_1 = (100 - \xi)^3 H(100 - \xi)$$

$$100R_1 + 70R_2 + 30R_3 = 100 - \xi$$

or

$$
\begin{bmatrix}
27\,000 & 0 & 0 & 130 \\
343\,000 & 64\,000 & 0 & 420 \\
1\,000\,000 & 343\,000 & 2\,700 & 600 \\
100 & 70 & 30 & 0
\end{bmatrix}
\begin{bmatrix}
R_1 \\ R_2 \\ R_3 \\ C_1
\end{bmatrix}
=
\begin{bmatrix}
(30 - \xi)^3 H(30 - \xi) \\
(70 - \xi)^3 H(70 - \xi) \\
(100 - \xi)^3 H(100 - \xi) \\
100 - \xi
\end{bmatrix}
$$

The inverse of the matrix on the left-hand side is

$$
10^{-6}
\begin{bmatrix}
-28 & 13 & -3 & 2\,698 \\
63 & -37 & 12 & -10\,781 \\
-53 & 42 & -18 & 49\,497 \\
13\,619 & -2\,324 & 623 & -560\,311
\end{bmatrix}
$$

By multiplying this matrix by the vector of the right-hand side, we get the unknowns as

$$R_1 = 10^{-6}[-28(30 - \xi)^3 H(30 - \xi) + 13(70 - \xi)^3 H(70 - \xi)$$
$$- 3(100 - \xi)^3 H(100 - \xi) + 2698(100 - \xi)]$$

$$R_2 = 10^{-6}[63(30 - \xi)^3 H(30 - \xi) - 37(70 - \xi)^3 H(70 - \xi)$$
$$+ 12(100 - \xi)^3 H(100 - \xi) - 10\,781(100 - \xi)]$$

$$R_3 = 10^{-6}[-53(30 - \xi)^3 H(30 - \xi) + 42(100 - \xi)^3 H(100 - \xi)$$
$$- 18(100 - \xi)H(100 - \xi) + 49\,497(100 - \xi)]$$

Substituting the R_i in the expressions for V and M results in moment and shear at section x in terms of the location of the load ξ.

For section 5, $x = 15$, $V_5 = R_1 - H(15 - \xi)$ since $H_{30} = H_{70} = 0$ at this point, leading to

$$10^6 V_5 = -28(30 - \xi)^3 H(30 - \xi) + 13(70 - \xi)^3 H(70 - \xi)$$
$$- 3(100 - \xi)^3 H(100 - \xi) + 2\,698(100 - \xi) - 10^{-6} H(15 - \xi)$$

Note that $H(a - \xi) = 0$ for $a < \xi$ and 1 for $a > \xi$. At point 6,

$$M_6 = 50R_1 + 20R_2 - (50 - \xi)H(50 - \xi)$$

or

$$10^6 M_6 = -140(30 - \xi)^3 H(30 - \xi) - 90(70 - \xi)^3 H(70 - \xi)$$
$$+ 90(100 - \xi)^3 H(100 - \xi) - 10^{-6}(50 - \xi)H(50 - \xi)$$

Plots of these equations, the shear and moment influence lines at 5 and 6, are shown in Figs. 9.17b and 9.17c. For variable-section members and those with internal hinges, the process is the same as those discussed in Sections 9.5 and 9.6.

PROBLEMS

9.1. For the beam of Fig. P9.1 find the equations for slope and deflection in terms of EI using Macaulay's brackets and calculate the deflection at the mid-span.

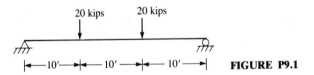

FIGURE P9.1

Solve the following problems in terms of EI using the singularity method.

9.2. Find the deflection at the left end and at the mid-span of the beam in Fig. P9.2.

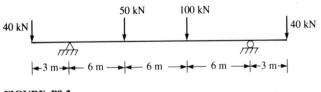

FIGURE P9.2

9.3. Find the deflection at point 3 and at the mid-span of the beam in Fig. P9.3.

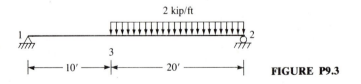

FIGURE P9.3

9.4. Calculate the deflection and rotation at the free end of the cantilever in Fig. P9.4.

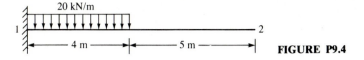

FIGURE P9.4

9.5. Find the slope and deflection of the beam in Fig. P9.5 at the mid-span.

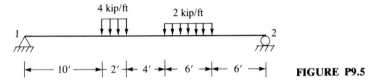

FIGURE P9.5

9.6. Find the deflection and slope of the beam in Fig. P9.6 at point 3.

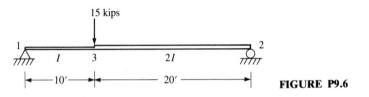

FIGURE P9.6

9.7. Find the deflection at the end of the overhang in Fig. P9.7.

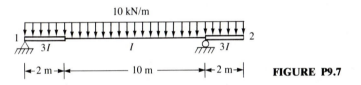

FIGURE P9.7

9.8. Calculate the deflection of the beam of Fig. P9.8 at the hinge.

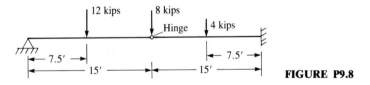

FIGURE P9.8

9.9. Determine the deflection at the mid-span of the beam in Fig. P9.9.

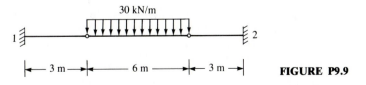

FIGURE P9.9

9.10. For the beam of Fig. P9.10 calculate the slope on both sides of the hinge and the deflection at the hinge.

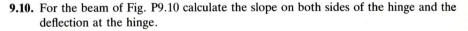

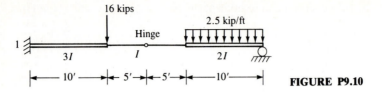

FIGURE P9.10

9.11. Calculate the mid-span deflection of the beam in Fig. P9.11.

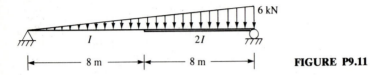

FIGURE P9.11

For the following problems find the redundant reactions using the singularity method.

9.12.

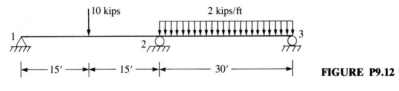

FIGURE P9.12

9.13.

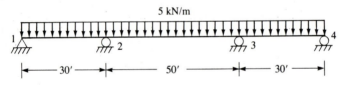

FIGURE P9.13

9.14

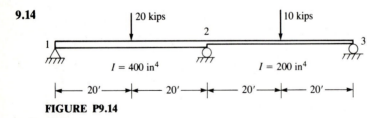

FIGURE P9.14

9.15. The beam of Fig. P9.15 is supported by a vertical spring at point 2. Calculate the reaction of this elastic support. The stiffness of the vertical spring is $K = 200$ kip/in, $E = 29\,000$ ksi.

    Hint: The vertical spring provides a reaction R_2 against displacement. Find the deflection equation in terms of R_2. The boundary condition at the spring is $v_2 = R_2/K$. This and other boundary conditions plus the equibrium equations allow the complete solution.

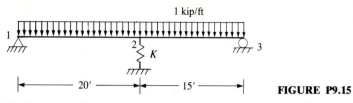

FIGURE P9.15

9.16. Find the equations for moment and shear for beam 234 of the structure in Fig. P9.16. I is constant for all the members and equal 200 in^4 and $E = 29\,000$ ksi.

> *Hint:* The resistance of beam 12 at point 2 can be replaced by a torsional spring, the stiffness of which is equal to the rotational stiffness at the right of member 12. The solution procedure is similar to that of Problem 9.15.

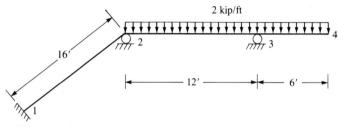

FIGURE P9.16

Find the expressions for influence lines of the following structures

9.17. Influence line of the left reaction, shear on the left of the second support, and moment at the middle of the first span in Fig. P9.17.

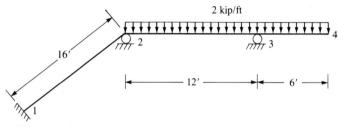

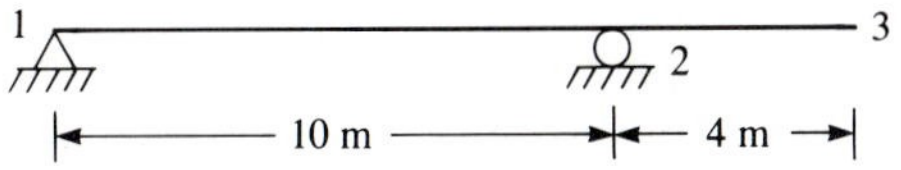

FIGURE P9.17

9.18. Influence line of shear on the left and right of support 2 and that of moment at support 2 in Fig. P9.18.

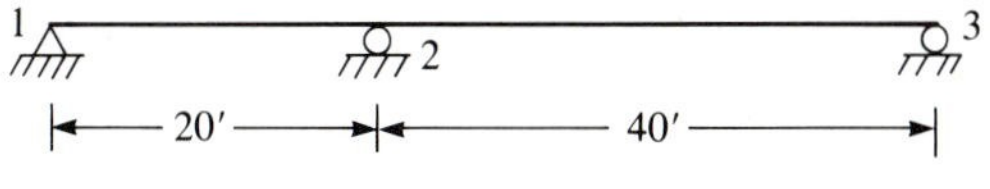

FIGURE P9.18

9.19. Influence line of moment at the middle of the second span in Fig. P9.19.

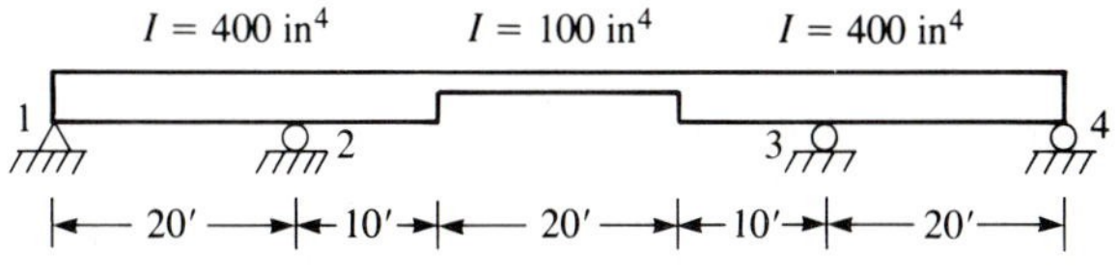

FIGURE P9.19

9.20. Influence line of shear at the middle of the second span and moment at the middle of the third span in Fig. P9.20.

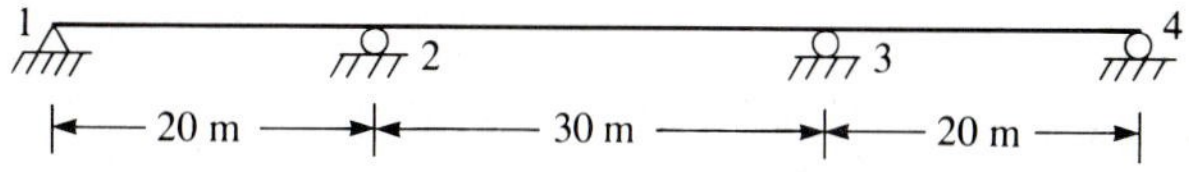

FIGURE P9.20

CHAPTER
10

FORCE-
DISPLACEMENT
RELATIONS,
MOMENT
DISTRIBUTION,
AND SLOPE-
DEFLECTION
METHODS

Before starting the discussion of general methods of analysis we must establish the sign convention to be used. In addition we derive the force–displacement relations and fixed-end forces. These are needed for the general methods of analysis for continuous beams and frames such as moment distribution, slope deflection, and stiffness methods. The analysis sign convention discussed in Section 10.1 will be used for all the methods mentioned above as well as for the stiffness formulation of arches, beam–columns and torsion problems. Similarly, the force–displacement relations and fixed-end forces derived in Sections 10.2 and 10.3 will be used in the moment distribution and slope deflection as well as stiffness methods.

10.1 ANALYSIS SIGN CONVENTION

The design sign convention, discussed in Section 4.1 and used so far, is appropriate for simple structures and especially for design purposes. During the design it is important to be able to easily identify which side of the beam is in compression and which side in tension. In reinforced concrete structures, for example, this facilitates the installation of reinforcing bars, which must be placed in the tension zones, to compensate for the weakness of concrete in carrying tension. In steel beams the compression flange may have to be

262

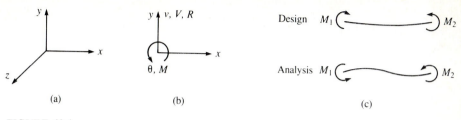

FIGURE 10.1

checked for buckling. In the design sign convention some of the positive values were so chosen as to facilitate the identification of the sense of stresses in the members. Other signs were selected so that the commonly encountered values are positive. This sign convention is not easy to use for the analysis of complex structures. For example, when analyzing structures with members having different orientations it would not be easy to identify the top of the members, needed in the design convention, in a consistent manner throughout the structure. We must therefore use a sign convention by which the senses of forces and displacements can be readily identified, so that the equilibrium and compatibility equations can be written easily. The analysis sign convention discussed below can be used for this purpose.

In the analysis sign convention all entities are considered positive when they act in the positive directions of the coordinate system chosen. In this book, a right-handed coordinate system (Fig. 10.1a) is used. The axes of this coordinate system are such that if we turn a corkscrew by rotating the x axis toward the y axis, the cork screw moves in the positive direction of z. Thus, for two-dimensional structures, the signs of distances, displacements, rotations, loads, shear forces, reactions, and moments are positive if they act in the directions shown in Fig. 10.1b.

The analysis sign convention is quite suitable for general methods of analysis. However, the moment and shear forces found by these methods are often used for design purposes. Thus, before drawing moment diagrams, the moments in the members must be converted to the design sign convention. Comparing the sign of moments acting at the two ends of a member in the two conventions (Fig. 10.1c), we observe that to convert the moments from the analysis to the design sign convention (and vice versa) the sign of moments at the left ends of the members must be changed.

It can be easily verified that in the analysis sign convention a positive moment causes a negative change of slope from a section to an adjacent one. Therefore, the differential equation of the beam (Eq. (5.16)) still has the form $EIv'' = -M$ in this sign convention.

10.2 FORCE–DISPLACEMENT RELATIONS

Consider the beam of Fig. 10.2 under the end forces and moments and a distributed load over the span. We want to find expressions relating the end

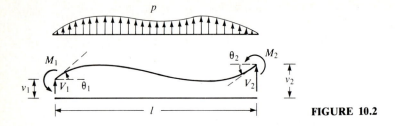

FIGURE 10.2

forces and moments to the end displacements and rotations. For brevity we will use the term "force" to mean both force and moment, and "displacement" to mean both displacement and rotation. In Fig. 10.2 the applied loads as well as the end forces and displacements of the beam are shown in the positive senses of the analysis sign convention.

To find the force–displacement relationship for this element we integrate the differential equation of the beam,

$$-EIv'' = M = M_1 - V_1x + m(x) \tag{10.1}$$

where $m(x)$ is the moment due to the applied loads only and is positive when counterclockwise. For example for a uniformly distributed load p, $m(x) = -\frac{1}{2}px^2$. Equation (10.1) integrates to

$$-EIv' = M_1x - \frac{1}{2}V_1x^2 + f(x) + C_1 \tag{10.2}$$

where $f(x) = \int m(x)\,dx$ and

$$-EIv = \frac{1}{2}M_1x^2 - \frac{1}{6}V_1x^3 + g(x) + C_1x + C_2 \tag{10.3}$$

where $g(x) = \int f(x)\,dx$.

The boundary conditions are, $v' = \theta_1$, $v = v_1$ at $x = 0$, and $v' = \theta_2$, $v = v_2$ at $x = l$. The first two boundary conditions give

$$-EI\theta_1 = C_1 \quad \text{and} \quad -EIv_1 = C_2$$

$f(0)$ and $g(0)$ are zero for all loads. The last two boundary conditions, upon substituting for C_1 and C_2 from the above equations, give

$$-EI\theta_2 = M_1l - \frac{1}{2}V_1l^2 + f(l) - EI\theta_1$$

$$-EIv_2 = \frac{1}{2}M_1l^2 - \frac{1}{6}V_1l^3 + g(l) - EI\theta_1l - EIv_1$$

By the equilibrium equations, sum of the moments at the right end and sum of the vertical forces are equal to zero. Thus,

$$M_1 - V_1l + m(l) + M_2 = 0 \qquad V_1 + q + V_2 = 0$$

where $q = \int_0^l p(x)\,dx$. We find

$$V_1 = \frac{(M_1 + M_2)}{l} + \frac{1}{l}m(l) \qquad V_2 = -(V_1 + q) \tag{10.4}{10.5}$$

Substitution of V_1 in the equations for θ_2 and v_2 gives upon rearranging

$$M_1 - M_2 = \frac{2EI}{l} \theta_1 - \frac{2EI}{l} \theta_2 + m(l) - \frac{2}{l} f(l)$$

$$2M_1 - M_2 = \frac{6EI}{l} \theta_1 - \frac{6EU}{l^2} v_1 - \frac{6EI}{l^2} v_2 + m(l) - \frac{6}{l^2} g(l)$$

Solving for M_1 and M_2 we get

$$M_1 = \frac{2EI}{l} (2\theta_1 + \theta_2) + \frac{6EI}{l^2} (v_1 - v_2) + M_1^F \qquad (10.6)$$

$$M_2 = \frac{2EI}{l} (\theta_1 + 2\theta_2) + \frac{6EI}{l^2} (v_1 - v_2) + M_2^F \qquad (10.7)$$

where

$$M_1^F = \frac{2}{l^2} [lf(l) - 3g(l)] \qquad (10.8)$$

$$M_2^F = -\frac{1}{l^2} [l^2 m(l) - 4lf(l) + 6g(l)] \qquad (10.9)$$

M_1^F and M_2^F are the end moments for $\theta_1 = \theta_2 = 0$ and $v_1 = v_2 = 0$, that is, fixed-end moments. They can be calculated from a fixed-end beam with the same in-span loads or more readily by Eqs. (10.8) and (10.9). Equations (10.6) and (10.7) show that the moments developed at the ends of a member are equal to the moments due to the end notations and displacements plus the fixed end moments.

The end shear forces can be found from Eqs. (10.4) and (10.5) as

$$V_1 = \frac{6EI}{l^2} (\theta_1 + \theta_2) + \frac{12EI}{l^3} (v_1 - v_2) + V_1^F \qquad (10.10)$$

$$V_2 = -\frac{6EI}{l^2} (\theta_1 + \theta_2) - \frac{6EI}{l^3} (v_1 - v_2) + V_2^F \qquad (10.11)$$

where fixed-end shear forces are

$$V_1^F = \frac{6}{l^3} [lf(l) - 2g(l)] \qquad (10.12)$$

$$V_2^F = -\left[\frac{6}{l^3} [lf(l) - 2g(l)] + q \right] \qquad (10.13)$$

We can see from the above equations that the end shear forces are equal to the sum of the forces due to the end displacements, plus the fixed-end shear forces. Force–displacement relations for variable section members can be derived in a similar manner. This is done in Section 13.5.

10.3 FIXED-END FORCES

As mentioned in Section 10.2, the end forces developed in a member involve the end displacements, rotations, and the in-span loads. In-span loads exhibit themselves in the form of fixed-end forces.

The fixed-end forces can be found by the equations derived above, or by analyzing a fixed-end beam under the applied loads. The equations derived for calculating the fixed-end forces can be summarized as follows. We recall that, with the x axis directed to the right, positive loads and shear forces act upward and positive moments are counterclockwise. To calculate the fixed end forces the only thing we need is an expression for the moment of the applied loads (without the end reactions) in the analysis sign convention. Thus

$$m(x) = \text{moment due to the applied loads at section } x$$

$$f(x) = \int m(x)\, dx \tag{10.14a}$$

$$g(x) = \int f(x)\, dx \tag{10.14b}$$

$$q = \int p(x)\, dx = \text{total load on the span} \tag{10.15}$$

$$M_1^F = \frac{2}{l^2}[lf(l) - 3g(l)] \tag{10.16a}$$

$$M_2^F = -\frac{1}{l^2}[l^2 m(l) - 4lf(l) + 6g(l)] \tag{10.16b}$$

$$V_1^F = \frac{6}{l^3}[lf(l) - 2g(l)] \tag{10.17a}$$

$$V_2^F = -\frac{6}{l^3}[lf(l) - 2g(l)] - q \tag{10.17b}$$

Uniformly Distributed Loads

For a uniformly distributed load p over the entire span,

$$m(x) = -\tfrac{1}{2}px^2 \qquad f(x) = -\tfrac{1}{6}px^3 \qquad g(x) = -\tfrac{1}{24}px^4 \qquad q = pl$$

$$M_1^F = \frac{2}{l^2}[l(-\tfrac{1}{6}pl^3) - 3(-\tfrac{1}{24}pl^4)] = -\frac{pl^2}{12}$$

$$M_2^F = -\frac{1}{l^2}[l^2(-\tfrac{1}{2}pl^2) - 4l(-\tfrac{1}{6}pl^3) + 6(-\tfrac{1}{24}pl^4)] = \frac{pl^2}{12}$$

$$V_1^F = \frac{6}{l^3}[l(-\tfrac{1}{6}pl^3) - 2(-\tfrac{1}{24}pl^4)] = -\frac{pl}{2}$$

$$V_2^F = -\left(\frac{6}{l^3}[l(-\tfrac{1}{6}pl^3) - 2(-\tfrac{1}{24}pl^4)] + pl\right) = -\frac{pl}{2}$$

Concentrated Loads

For a concentrated load we can use the unit step function to find $m(x)$ as discussed in Section 9.3.

For a concentrated load P acting at a from the left-hand end with $b = l - a$,

$$m(x) = -P(x - a)H_a \qquad \text{gives} \qquad m(l) = -Pb$$
$$f(x) = -\tfrac{1}{2}P(x - a)^2 H_a \qquad\qquad f(l) = -\tfrac{1}{2}Pb^2$$
$$g(x) = -\tfrac{1}{6}P(x - a)^3 H_a \qquad\qquad g(l) = -\tfrac{1}{6}Pb^3$$

and

$$q = P$$

$$M_1^F = \frac{2}{l^2}[l(-\tfrac{1}{2}Pb^2) - 3(-\tfrac{1}{6}Pb^3)] = -\frac{Pb^2}{l^2}(l - b) = -\frac{Pb^2 a}{l^2}$$

$$M_2^F = -\frac{1}{l^2}[l^2(-Pb) - 4l(-\tfrac{1}{2}Pb^2) + 6(-\tfrac{1}{6}Pb^3)]$$

$$= \frac{Pb}{l^2}(l^2 - 2lb + b^2) = \frac{Pb}{l^2}(l - b^2) = \frac{Pba^2}{l^2}$$

$$V_1^F = \frac{6}{l^3}[l(-\tfrac{1}{2}Pb^2) - 2(-\tfrac{1}{6}Pb^3)] = -\frac{Pb^2}{l^3}(3l - 2b)$$

$$= -\frac{Pb^2}{l^3}(3a + b)$$

$$V_2^F = -\left(\frac{6}{l^3}[l(-\tfrac{1}{2}Pb^2) - 2(-\tfrac{1}{6}Pb^3)] + P\right) = -\frac{Pa^2}{l^3}(a + 3b)$$

For partially distributed loads the fixed-end forces can be found using the representation discussed in Section 9.4.

Analysis of structures with a number of methods, including moment distribution, slope deflection, and stiffness method involves fixed-end forces. Fixed-end forces for some commonly encountered loads are compiled in Tables 10.1a, and 10.1b in the analysis sign convention. The moment diagrams for these cases are also shown in the design conventions to simplify the construction of moment diagrams for indeterminate structures. For gravity (downward) loads, the moment diagrams must be inverted.

TABLE 10.1a
Fixed-end Forces

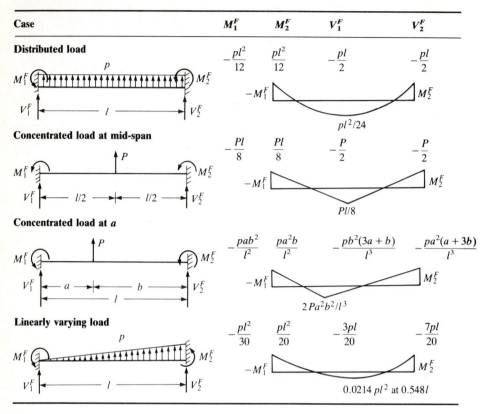

Case	M_1^F	M_2^F	V_1^F	V_2^F
Distributed load	$-\dfrac{pl^2}{12}$	$\dfrac{pl^2}{12}$	$-\dfrac{pl}{2}$	$-\dfrac{pl}{2}$
Concentrated load at mid-span	$-\dfrac{Pl}{8}$	$\dfrac{Pl}{8}$	$-\dfrac{P}{2}$	$-\dfrac{P}{2}$
Concentrated load at a	$-\dfrac{pab^2}{l^2}$	$\dfrac{pa^2b}{l^2}$	$-\dfrac{pb^2(3a+b)}{l^3}$	$-\dfrac{pa^2(a+3b)}{l^3}$
Linearly varying load	$-\dfrac{pl^2}{30}$	$\dfrac{pl^2}{20}$	$-\dfrac{3pl}{20}$	$-\dfrac{7pl}{20}$

For distributed load, mid-span diagram value $pl^2/24$. For concentrated load at mid-span, $Pl/8$. For concentrated load at a, $2Pa^2b^2/l^3$. For linearly varying load, $0.0214\,pl^2$ at $0.548l$.

10.4 MOMENT DISTRIBUTION METHOD

The moment distribution method provides an iterative solution technique for continuous beams and frames. It is especially suitable for hand calculations. Consider the beam of Fig. 10.3a. Upon application of the loads, the beam deforms, producing joint rotations at 1, 2, and 3. If before application of the loads we prevent all the joints from rotating (Fig. 10.3b) and then apply the loads and release joints 1, 2, and 3, the beam will have the same final displacements, rotations, and moments as in Fig. 10.3a. Alternatively, we can prevent the rotation of all the joints, apply the loads, and then release the joint restraints one at a time, allow the joint to undergo rotation, restrain the joint again in its new position, and proceed to the next joint (Figs. 10.7c,d). If we proceed in this manner to release and fix all the joints several times, (except for the fixed end) at some point the release of the joints will not cause any additional deformation and the deformed beam will be the same as the one

TABLE 10.1b
Fixed-end forces (Con'd)

Case	M_1^F	M_2^F	V_1^F	V_2^F
Partially distributed load	$-\dfrac{pl}{12l^2}[a^2(6l^2 - 8al + 3a^2)$ $- c^2(6l^2 - 8cl + 3c^2)]$	$\dfrac{pb}{2}(2l - 2c + b) + V_1^F l - M_1^F$	$-\dfrac{p}{2l^3}[-a(2l^3 - 2a^2 l$ $+ a^3) + c(2l^3 - 2c^2 l + c^3)]$	$-(V_1^F + pb)$
Partially linear load	$-\dfrac{P}{60bl^2}[(l-a)^4(2l + 3a)$ $+ d^3(3d^2 - 5dl)$ $+ 15bd - 20bl]$	$V_1 l - M_1^F$ $- Pb(3c + b)/6$	$-\dfrac{P}{20bl^3}[(l - a)^4(3l + 2a)$ $+ d^3(2d^2 - 5dl$ $+ 10db - 20bl)]$	$-(V_1^F + pb/2)$
Triangular load	$-\dfrac{5pl^2}{96}$	$\dfrac{5pl^2}{96}$	$-\dfrac{pl}{4}$ $\quad -3pl^2/96$	$-\dfrac{pl}{4}$
End displacement	$-\dfrac{6EI\Delta}{l^2}$	$-\dfrac{6EI\Delta}{l^2}$	$-\dfrac{12EI\Delta}{l^3}$	$\dfrac{12EI\Delta}{l^3}$

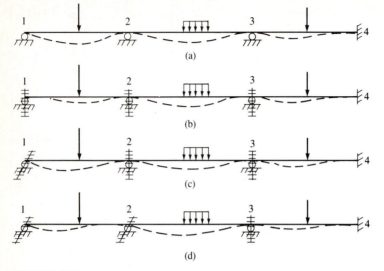

FIGURE 10.3

under the loads without restraining any joints. The latter process, called the moment distribution, is iterative and at each iteration involves the solution of a two-span beam with fixed ends. Thus, the solution at each step can be conveniently obtained, as we will see below. In practice, convergence is fast and is usually achieved after a few iterations.

Before we discuss the process of analyzing the beam during each iteration, we must define the parameters that are used in the solution process. These include stiffness and distribution factors and fixed-end moments. The sign convention used in the moment distribution method is the analysis sign convention discussed in Section 10.1. As mentioned before, for moments, conversion from the design to analysis sign convention (and vice versa) involves changing the sign of the moments on the left of each member.

Stiffness Factor

Consider the supported cantilever of Fig. 10.4. If we apply a moment M_1 at support 1, a rotation θ_1 takes place there such that

$$M_1 = K\theta_1 \qquad (10.18a)$$

Letting θ_1 equal 1, we can see that K is the moment producing a unit rotation. K is called the rotational stiffness.

FIGURE 10.4

We can find an expression for the rotational stiffness by integrating the differential equation of the beam of Fig. 10.4. Alternatively, we can use the expressions derived in Section 10.2 and set the in-span loads and the end displacements as well as the right rotation, θ_2, equal to zero. Thus, from Eq. (10.6),

$$M_1 = \frac{4EI}{l}\,\theta_1 \qquad\qquad (10.18b)$$

Comparison of this equation with Eq. (10.18a) reveals that the rotational stiffness is

$$K = \frac{4EI}{l} \qquad\qquad (10.19a)$$

The moment distribution method involves the ratio of stiffness of the members. Thus, the stiffness factor, K_{ij}, is often used instead of K, where i and j refer to the end points of the member and

---

The stiffness factor is $K_{ij} = I/l$

Therefore, for member 12 we have

$$M_1 = 4EK_{12}\theta_1 \qquad\qquad (10.19b)$$

Carry-Over Factor

Consider the supported cantilever again with a moment M_1 acting at the left-hand end (Fig. 10.5). A moment M_2 is developed at end 2 of the elastic beam, which is proportional to M_1, that is,

$$M_2 = CM_1 \qquad\qquad (10.20a)$$

where C is called the carry-over factor. It is the ratio of the moments at ends 2 and 1.

To find an expression for the carry-over factor we can calculate the moment developed at the right-hand end of the beam and find its ratio to the moment at the left-hand end. The right-end moment can be calculated by Eq. (10.7). With no in-span loads, no end displacements, and $\theta_2 = 0$, this gives

$$M_2 = \frac{2EI}{l}\,\theta_1 \qquad\qquad (10.20b)$$

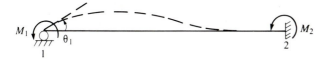

FIGURE 10.5

Comparing this with the expression in Eq. (10.18b), we find that

$$M_2 = \tfrac{1}{2}M_1 \tag{10.20c}$$

Thus,

> for members with a constant cross-section, the carry-over factor is $\tfrac{1}{2}$.

The carry-over factor for members with variable cross-sections is not $\tfrac{1}{2}$. Expressions for finding carry-over factors of variable section members are derived in Section 13.5.

Distribution Factor

In the moment distribution process we release the joints one at a time. Therefore, each step involves a situation where two or more members with one end fixed meet at a joint with a moment applied at that joint. For example, for continuous beams each step would consist of the solution of a two-span beam with fixed ends such as that of Fig. 10.6a.

If M_2 is the sum of the fixed-end moments at end 2 of the members meeting at joint 2, as we will see in the next section, upon release of joint 2 an unbalanced moment $M_u = -M_2$ will act at that joint, producing rotation. Because of this rotation the two members develop internal moments at end 2. If M_{21} is the moment at end 2 in member 12 and M_{23} is the moment at end 2 in member 23, then these two moments must balance the unbalanced moment M_u at joint 2, that is, each one of these moments has to balance a portion of the applied moment M_u. Thus,

$$M_{21} = D_{21}M_u = -D_{21}M_2 \tag{10.21a}$$

$$M_{23} = D_{23}M_u = -D_{23}M_2 \tag{10.21b}$$

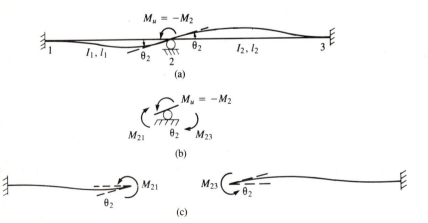

(a)

(b)

(c)

FIGURE 10.6

D_{21} and D_{23} are called the distribution factors. They are the proportions in which moment M_u gets distributed among the members. Equilibrium of joint 2 requires that

$$M_{21} + M_{23} - M_u = 0 \quad \text{or} \quad M_{21} + M_{23} + M_2 = 0. \tag{10.22}$$

Because of the compatibility of rotations end 2 of members 12 and 23 would have a rotation θ_2. From Eq. (10.19b),

$$M_{21} = 4EK_{21}\theta_2 \tag{10.23a}$$

$$M_{23} = 4EK_{23}\theta_2 \tag{10.23b}$$

From the latter equations,

$$\theta_2 = \frac{M_{21}}{4EK_{21}} = \frac{M_{23}}{4EK_{23}}$$

which gives $M_{23} = (K_{23}/K_{21})M_{21}$. Upon substitution into Eq. (10.22), we find

$$M_{21} = -\frac{K_{21}}{K_{21} + K_{23}} M_2 = -\frac{K_{21}}{\Sigma K_{ij}} M_2$$

and the equation before last gives

$$M_{23} = -\frac{K_{23}}{\Sigma K_{ij}} M_2$$

where $\Sigma K_{ij} = K_{21} + K_{23}$ is the sum of the stiffness factors of the members meeting at joint 2. In view of the definition (Eqs. (10.21a,b)) the distribution factors are

$$D_{21} = \frac{K_{21}}{\Sigma K_{ij}} \tag{10.24a}$$

$$D_{23} = \frac{K_{23}}{\Sigma K_{ij}} \tag{10.24b}$$

The distribution factor for each member connecting to a joint is equal to the stiffness factor of that member divided by the sum of the stiffness factors for all the members connecting to that joint.

Fixed-End Moments

The fixed-end moments are required in the moment distribution method. Fixed-end moments can be found by considering each member with its loading as a fixed-end beam. Any method such as the method of consistent deformations, integration, singularity method, etc., can be used for finding the fixed-end moments. For some frequently encountered loads, the fixed-end forces are given in Tables 10.1a and 10.1b. For other cases the expressions given in Section 10.3 may be used to find the fixed end moments.

Distribution of Moments

Consider the two-span beam of Fig. 10.7a, under the loads shown, with the ends 1 and 3 fixed. We want to find the moments M_{21} and M_{23} developed in the two members near joint 2. To achieve this end we use the free body diagrams of Figs. 10.7b and 10.7c. Equilibrium of joint 2 requires that

$$M_{21} + M_{23} = 0 \qquad (10.25)$$

Because of the compatibility of rotations, the slopes at end 2 of the two spans must be the same and equal to θ_2. From the force-displacement relations, Eq. (10.6), for the members in Figs. 10.7b and 10.7c, with no deflections and no rotations at the far ends, we get

$$M_{21} = 4EK_{21}\theta_2 + M_{21}^F \qquad (10.26a)$$

$$M_{23} = 4EK_{23}\theta_2 + M_{23}^F \qquad (10.26b)$$

Here K_{21} and K_{23} are I/l for members 21 and 23. Thus, the moment developed at the end of each member is the sum of a term due to the rotation plus the fixed-end moment at that end. Substitution of Eqs. (10.26a) and (10.26b) into the equilibrium equation (10.25) gives

$$4E(K_{21} + K_{23})\theta_2 = -(M_{21}^F + M_{23}^F) \qquad (10.27)$$

Equation (10.27) indicates that the moment causing the rotation at joint 2 in Fig. 10.7a is the negative of the algebraic sum of the fixed-end moments at

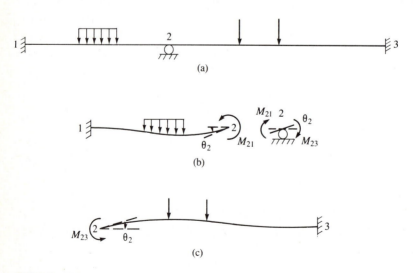

(a)

(b)

(c)

FIGURE 10.7

joint 2, acting as an external moment. This is called the *unbalanced* or *equivalent joint moment.*

Letting $M_u = -M_2 = -(M_{21}^F + M_{23}^F)$ and solving for θ_2 from Eq. (10.27), we get

$$\theta_2 = -\frac{M_2}{4E(K_{21} + K_{23})} = -\frac{M_2}{4E \sum K_{ij}}$$

Where M_2 is the sum of the fixed end moments at joint 2. Then, by Eq. (10.26a),

$$M_{21} = -\frac{K_{21}}{\sum K_{ij}} M_2 + M_{21}^F = -D_{21}M_2 + M_{21}^F = -D_{21}(M_{21}^F + M_{23}^F) + M_{21}^F$$

and by Eq. (10.26b),

$$M_{23} = -\frac{K_{23}}{\sum K_{ij}} M_2 + M_{23}^F = -D_{23}M_2 + M_{23}^F = -D_{23}(M_{21}^F + M_{23}^F) + M_{23}^F$$

Thus, the moment at the end of a member after releasing a joint is the algebraic sum of the fixed-end and distributed moments. From the above discussion of carry-over factor, we see that the internal moment induced by the rotation of a joint produces a carry-over moment, at the other end of the members attached to that joint. For end 3 of member 23, this carry over moment is

$$\tfrac{1}{2}D_{23}M_u = -\tfrac{1}{2}D_{23}M_2 = -\tfrac{1}{2}D_{23}(M_{21}^F + M_{23}^F).$$

Hence, upon release of a joint the moment developed at the end of a member is equal to the fixed-end moment at that end minus the distribution factor times the negative of sum of the fixed-end moments at that joint. The carry-over moments at the far ends of the members are equal to half of the distributed moments at the near ends. In the subsequent releases of the joint, the fixed-end moments come from the moments carried over to that joint, which are not as yet distributed.

Steps in the Moment Distribution Method

The following steps are involved in the moment distribution method.

1. Calculate the stiffness factor for all the members and the distribution factors at all the joints. The carry-over factor is $\tfrac{1}{2}$ for members with constant cross-section

2. Find the fixed-end moments for all the members.

3. Start out by fixing all the joints, and release them one at a time. Distribute the unbalanced moment at the released joint and carry over the moments to the far ends of the members, then fix the joint again.

4. Continue releasing joints until the distributed moments are insignificant. If

the last moments carried over are small and cannot be distributed, it is better to discard them so that the joints remain in equilibrium.

5. Sum up the moments at each end of the members to obtain the final moments.

Example 10.1. Find the member end moments for the beam of Fig. 10.8 by the moment distribution method. The moments of inertia for both spans are constant and equal to I.

Solution. The stiffness factors, $K_{ij} = I_{ij}/l_{ij}$, for the two members are

$$K_{12} = \frac{I}{30} \quad \text{and} \quad K_{23} = \frac{I}{40}$$

The distribution factors are $D_{ij} = K_{ij}/\sum K_{ij}$ and are thus equal to 1 for joints 1 and 3. For joint 2,

$$D_{21} = \frac{K_{21}}{\sum K_{ij}} = \frac{K_{21}}{K_{21} + K_{23}} = \frac{I/30}{(I/30) + (I/40)} = 0.57$$

$$D_{23} = \frac{K_{23}}{\sum K_{ij}} = \frac{K_{23}}{K_{21} + K_{23}} = \frac{I/40}{(I/30) + (I/40)} = 0.43$$

As expected, the sum of the distribution factors at each joint is equal to 1. The carry-over factor for members with constant cross-section is $\frac{1}{2}$. The fixed-end moments for the two members are, from Table 10.1a,

$$M_{12}^F = -\frac{pl^2}{12} = -\frac{-2 \times 30^2}{12} = 150 \text{ kip} \cdot \text{ft}$$

$$M_{21}^F = \frac{pl^2}{12} = \frac{-2 \times 30^2}{12} = -150 \text{ kip} \cdot \text{ft}$$

$$M_{23}^F = -\frac{Pl}{8} = -\frac{-8 \times 40}{8} = 40 \text{ kip} \cdot \text{ft}$$

$$M_{32}^F = \frac{Pl}{8} = \frac{-8 \times 40}{8} = -40 \text{ kip} \cdot \text{ft}$$

We record the distribution factors and the fixed-end moments on the sketch of the beam. Here joint 2 is the only one that is allowed to rotate. We release this joint, distribute the unbalanced moment and carry over half of the distributed moments to the other ends of the members.

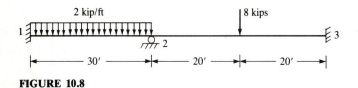

FIGURE 10.8

Since joints 1 and 3 are fixed, they cannot be released. The sum of fixed-end moments at joint 2 is $-150 + 40 = -110$. The negative of this moment multiplied by the distribution factors gives the distributed moments recorded on the second line of the sketch at joint 2. An underline indicates that the moments above the line are distributed. After the distribution, half of the distributed moments are carried over to the other ends of the members. Thus, half of 62.7, or 31.4 (rounded), is carried over to end 1, and half of 47.3, or 23.7, is carried over to end 3. Since joint 2 is already released and is in equilibrium and no other joint can be released, the distribution process is complete. The sum of moments at each end is the actual moment in the beam at that point. These are the values shown in the last line of the sketch.

Example 10.2. Find the member-end moments for the beam of Fig. 10.9. The left-hand end of the beam is fixed, while all the other supports are rollers. A uniform load of -5 kN/m is applied over the entire length of the beam, as shown in Fig. 10.9. The middle span is 30 m long and the other two spans are 20 m. The moment of inertia of the middle span is $1.2I$ and that of the other spans is I.

Solution. The stiffness factors are

$$K_{12} = \frac{I}{20} \qquad K_{23} = \frac{1.2I}{30} = \frac{I}{25} \qquad \text{and} \qquad K_{34} = \frac{I}{20}$$

The distribution factors are

$$D_{21} = \frac{K_{12}}{K_{12} + K_{23}} = \frac{I/20}{(I/20) + (I/25)} = 0.56$$

$$D_{23} = \frac{K_{23}}{K_{12} + K_{23}} = \frac{I/25}{(I/20) + (I/25)} = 0.44$$

$$D_{32} = \frac{K_{32}}{K_{32} + K_{34}} = \frac{I/25}{(I/20) + (I/25)} = 0.44$$

$$D_{34} = \frac{K_{34}}{K_{32} + K_{34}} = \frac{I/20}{(I/20) + (I/25)} = 0.56$$

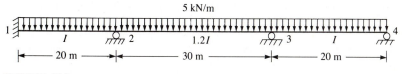

FIGURE 10.9

Again because each member (span) has a constant cross-section, the carry over factor is $\frac{1}{2}$. The fixed-end moments from Table 10.1a, for $p = -5 \text{ kN/m}$, are

$$M_{12} = -M_{21} = -\frac{-5 \times 20^2}{12} = 167$$

$$M_{23} = -M_{32} = -\frac{-5 \times 30^2}{12} = 375$$

$$M_{34} = -M_{43} = -\frac{-5 \times 20^2}{12} = 167$$

The values of the moments are rounded to the closest integers. These fixed-end moments are recorded on the sketch of the beam below. Joint 1 of the beam is fixed. Therefore, joints 2, 3, and 4 are released in turn, their unbalanced moments are distributed and carried over, and then the joint is locked again. After each distribution a horizontal line is drawn under the moments to indicate that the moments above the line are already distributed.

Starting at joint 2, the unbalanced moment is $-(-167 + 375) = -208$. This moment is multiplied by 0.56 and recorded on the left of joint 2 and multiplied by 0.44 and recorded on the right of joint 2. Half of the latter values are carried over to joint 1 and to the left of joint 3, respectively. A line under the numbers on the left and right of 2 indicates that the moments at joint 2 are balanced. We then proceed with joint 3. The unbalanced moment at this joint is the negative of sum of the fixed-end moments plus the moments carried over. Thus, the unbalanced moment is $-(-375 + 167 - 46) = 254$. The products of this value with 0.44 and 0.56 are recorded on the left and on the right of joint 3. Again half of these distributed moments are carried over to the other ends of the members. A line is drawn under the values at joint 3 to indicate that the moments there are all distributed and the joint moments are balanced for the new rotated position. Joint 3 is now locked back and we proceed with joint 4. The unbalanced moment at this joint is $-(-167 + 71) = 96$. Since there is only one member at this joint, the distributed moment at 4 is 96. Half of the latter value is carried over to the right of joint 3, and a line is drawn under the values at joint 4 to indicate that the moments at this joint are balanced. Joint 4 is then locked back and we can proceed to either joint 2 or 3. As mentioned before, the order of the joint release is immaterial. However, it is preferable to proceed to the joint with the largest unbalanced moment in order to accelerate the convergence. Therefore, we continue at joint 2, which has the largest unbalanced moment of 56. Continuing in this manner we stop the process when the distributed moments are less than 1, the tolerance used for this problem. The order of joint release for this problem has been 2, 3, 4, 3, 2, 3, and 4.

The final moments, the sums of the columns, are shown below the double lines, indicating the end of the process.

Example 10.3. Find the moments in the beams shown in Figs. 10.10a, 10.10b and 10.10c. All the spans have the length l and moment of inertia I. The loads are uniformly distributed with magnitude p.

Solution. The details of the moment distribution are shown below for the three beams.

Beam (a):

$pl^2/12$	$-pl^2/12$
$pl^2/24$	$pl^2/12$
$pl^2/8$	0

Beam (b):

$pl^2/12$	$-pl^2/12$	$pl^2/12$	$-pl^2/12$
$-pl^2/12$	$-pl^2/24$	$pl^2/24$	$pl^2/12$
0	$-pl^2/8$	$pl^2/8$	0

Beam (c):

$pl^2/12$	$-pl^2/12$	$pl^2/12$	$-pl^2/12$	$pl^2/12$	$-pl^2/12$	$pl^2/12$	$-pl^2/12$

For the beam of Fig. 10.10a, the hinge at the right-hand end is released and the moment is distributed and carried over, we complete the moment distribution process by finding the final moments.

For the beam of Fig. 10.10b, we release the two ends of the beam and distribute and carry over the moments. When we try to balance the moments at the middle joint we find that they are already balanced and thus the solution is complete. The sum of the columns give the moments in the beam.

For the beam of Fig. 10.10c, after the fixed-end moments are applied, we observe that all the joints are balanced and therefore, no further distribution is necessary.

From the above results we can see that under a uniform load the fixed-end moment for a supported cantilever and the moment at the middle support of a

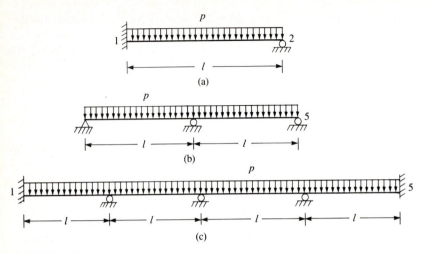

FIGURE 10.10

two span beam, with equal spans, and hinged ends, is $-pl^2/8$. For an n-span beam, with equal spans and fixed ends the moment at all the supports is $-pl^2/12$.

Beams with Overhangs

For beams with overhangs we can apply the fixed-end moment at the overhang along with the other fixed-end moments and treat the joint at the continuous end of the overhang as a hinge. In other words, the distribution and carry over factors are zero for an overhang.

Example 10.4. Find the end moments in the beam of Fig. 10.11a and draw the shear and moment diagrams.

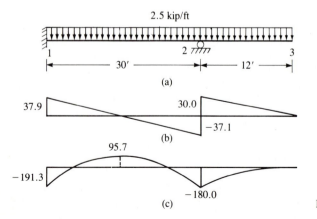

FIGURE 10.11

Solution. The fixed-end moments with the analysis sign convention are

$$M_{12} = -M_{21} = -\frac{pl^2}{12} = -\frac{-2.5 \times 30^2}{12} = 187.5$$

$$M_{23} = \frac{2.5 \times 12^2}{2} = 180.0$$

The moments are distributed as follows

The only joint to be released is 2. Upon releasing this joint, the unbalanced moment is $-(-187.5 + 180) = 7.5$. Since the distribution factor for the overhang is zero, the unbalanced moment must be taken by member 21. After the moments are balanced at joint 2, half of the distributed moment, 3.8, is carried to joint 1. Since there is no other joint to be released the distribution is complete. As can be seen, the final moments at joint 2 are in equilibrium.

To convert the sign of moments to that of the design sign convention, we must change the sign of the moments on the left of the members. Thus, the moment at joint 1 will become -191.3 kip · ft, and at the right of joint 2, -180 kip · ft. To draw the shear diagram we first find the reactions of the beam. The moment on the left of support 2 is -180. Setting this moment equal to that calculated from the loads and the end forces (Fig. 10.11a),

$$-191.3 + R_1 \times 30 - 2.5 \times \frac{30^2}{2} = -180$$

gives $R_1 = 37.9$ kips. Then sum of the vertical forces gives

$$R_1 + R_2 - 2.5 \times 42 = 0,$$

$$R_2 = -R_1 + 2.5 \times 42 = -37.9 + 2.5 \times 42 = 67.1 \text{ kips}.$$

With these reactions the shear force at any section can be calculated. For uniformly distributed load the shear diagram is linear. Therefore, we only need to calculate the values of shear at the ends of the members. At the left of support 2 the shear is $37.9 - 2.5 \times 30 = -37.1$ kips, and at the right of support 2 it is $-37.1 + 67.1 = 30.0$ kips.

Therefore, the shear diagram is as shown in Fig. 10.11b. The moment diagram is drawn by superimposing the moment diagram due to the load and that due to the end moments (Section 7.2). The moment diagram is shown in Fig. 10.11c.

Reduced Stiffness for Hinged Ends

In the moment distribution process, hinged-end members can be treated as we have done so far, the same way as other members. However, since the final

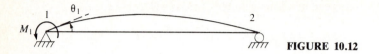

FIGURE 10.12

moments at hinged ends are zero, we can expedite the process by using a reduced stiffness factor for hinged end members. We can derive the stiffness factor and fixed-end moments for members with one end hinged and the other fixed and use these values without ever fixing the hinged ends. This would require the development of new tables of fixed-end moments for supported cantilevers under different loads, similar to Tables 10.1a and 10.1b. To eliminate the need for such tables, we can use reduced stiffness factors for hinged-end members, but start out the process by fixing all the joints, including the hinges. However, once released, the hinged ends are not to be fixed again and no more carry-over moments are transferred to them.

The reduced stiffness factor for a member with one end hinged and the other fixed can be found by referring to Fig. 10.12. Setting the end displacements and fixed-end moments equal to zero, Eqs. (10.6) and (10.7) give

$$M_1 = \frac{2EI}{l}(2\theta_1 + \theta_2) \qquad M_2 = \frac{2EI}{l}(\theta_1 + 2\theta_2)$$

Since $M_2 = 0$, the second equation gives $\theta_2 = -\theta_1/2$. Substitution in the first equation results is

$$M_1 = \frac{3EI}{l}\theta_1$$

Setting $M_1 = 4EK'_{ij}\theta_1$ as before, we find

$$K'_{ij} = \frac{3}{4}\frac{I}{l} = \tfrac{3}{4}K_{ij} \qquad (10.28)$$

Thus,

the reduced stiffness factor is $\dfrac{3}{4}\left(\dfrac{I}{l}\right)$.

Example 10.5. Find the end moments for the beam of Example 10.2 (Fig. 10.9) using the reduced stiffness factor for member 34.

Solution. The stiffness factors for the first two members are the same as before, that is,

$$K_{12} = \frac{I}{20} \qquad K_{23} = \frac{I}{25}$$

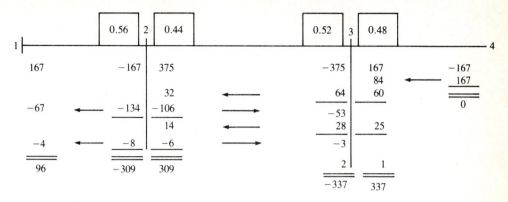

FIGURE 10.13

For member 34, the reduced stiffness factor is

$$K'_{34} = \frac{3}{4}\frac{I}{l} = \frac{3}{4}\frac{I}{20} = \frac{I}{26.67}$$

Again the distribution factors at joint 2 are the same as before, that is,

$$D_{21} = 0.56 \quad \text{and} \quad D_{23} = 0.44$$

However,

$$D_{32} = \frac{I/25}{(I/25) + (I/26.67)} = 0.52$$

and

$$D_{34} = \frac{I/26.67}{(I/25) + (I/26.67)} = 0.48$$

As discussed earlier, we start out with all the joints fixed, including the hinge at 4. In this example we start at joint 4. Since we have used the reduced stiffness factor for member 34, once joint 4 is released it must not be fixed again. Furthermore, after it is released no carry-over moment should be transferred to joint 4. Note that if we do not start with joint 4, carry-over moments will be transferred there until this joint is released. The order of joint releases for this problem has been, 4, 3, 2, 3, 2, 3. The details of the calculations are given in Fig. 10.13.

10.5 MOMENT DISTRIBUTION METHOD FOR FRAMES

The moment distribution method is especially suitable for hand calculation of moments in frames. In applying the moment distribution method to frames, we must distinguish between frames with sidesway (lateral deflection) and those without. This is because sidesway induces additional moments in the frame. If a frame is prevented from lateral deflection, for example, by lateral supports, or if both the frame geometry and the loads are symmetric, then no sidesway will take place. It should be noted that in applying the moment distribution method to frames the axial deformations of the members are neglected.

Frames Without Sidesway

The moment distribution method for frames without sidesway is identical to that for beams (Section 10.4). If a frame has more than two members connecting at a joint, then the unbalanced moment at that joint must be distributed among all the members. Similarly, if a frame member has a hinged end, then the reduced stiffness can be used for that member.

As for beams, the solution process consists of determining the stiffness factors for all the members, the distribution factors at all the joints, where more than one member is present, and the fixed-end moments for all the members with in-span loads. At the start of the process all the joints are fixed. The joints that can incur rotation are released one at a time. Once a joint is released, the unbalanced moment (the negative of sum of the fixed-end moments at that joint) is distributed among the members connecting to that joint, in proportion to their distribution factors. The joint is then fixed in its deformed configuration and the process is continued by releasing the other joints one at a time. Convergence is reached when the distributed moments at all the joints have negligible values as compared to the final joint moments, for example, when they are less than 1% of the final moments.

Example 10.6. Find the member-end moments for the frame of Fig. 10.14a and draw the moment diagram. The geometry and loads are shown in the figure.

Solution. Since there is no sidesway, because the frame is restrained at point 4 and the axial deformations are neglected in the moment distribution method, the solution is similar to that of continuous beams. The stiffness factors are

$$K_{12} = \frac{I}{10} \qquad K_{23} = \frac{2I}{20} = \frac{I}{10}$$

$$K_{34} = \frac{2I}{20} = \frac{I}{10} \qquad K_{35} = \frac{I}{10}$$

The distribution factors are

$$D_{21} = D_{23} = \frac{I/10}{(I/10) + (I/10)} = 0.5$$

and

$$D_{32} = D_{34} = D_{35} = \frac{I/10}{(I/10) + (I/10) + (I/10)} = 0.33$$

The fixed-end moments are

$$M_{23}^F = -M_{32}^F = -\frac{Pl}{8} = -\frac{-20 \times 20}{8} = 50$$

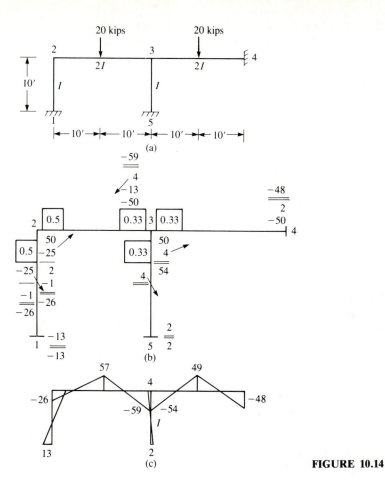

FIGURE 10.14

and

$$M_{34}^F = -M_{43}^F = -\frac{Pl}{8} = -\frac{-20 \times 20}{8} = 50$$

The moment distribution is performed on the sketch of the frame (Fig. 10.14b). For clarity, at the left-hand ends of the members the distributed moments are recorded below the member (on the right of vertical members) and at the right-hand ends above the members (on the left of vertical members).

Starting at joint 2, which has the largest unbalanced moment, we distribute the moment between members 21 and 23. This is done by multiplying the unbalanced moment by the distribution factors of each member connecting to the joint. The values thus obtained are recorded on the sketch. Half of the distributed moments (that is, −13) is carried over to the other ends of members 21 and 23. We then restrain joint 2 and proceed to unlock joint 3. The unbalanced moment at that joint is −(−13 − 50 + 50) = 13, which is distributed between the three members meeting at joint 3. Half of the distributed moments (2) is carried over to the far ends of the members. Upon releasing joint 2 again, the distributed moments are -1, a small value compared to the moment at the joint (~ 60). Thus, the distribution process is stopped at this point. Note that the ends 2 and 4 are fixed and cannot be released during the distribution process.

The moments at the ends of the members are then summed up to give the member-end moments. Note that at each joint the sum of the final moments of all the member ends must add up to zero. This is necessary for equilibrium of the joints.

Before drawing the moment diagram, we must convert the sign of the moments to the design convention. This is done by changing the signs of the moments at the left-hand ends of the members. Therefore, in the design sign convention, $M_{12} = 16$, $M_{21} = -26$, $M_{23} = -26$, $M_{32} = -59$, $M_{34} = -54$, $M_{35} = 4$, and $M_{23} = -2$. The moment diagram is shown in Fig. 10.14c. The positive moments in the two spans are found by superposition of the moment diagrams for simply supported beams with a load at the center and the moment diagram due to the end moments.

Example 10.7. Find the end moments for the frame of Fig. 10.15a and calculate the shear forces at the bases of the vertical members, points 1 and 4.

Solution. Because the frame is symmetric and under symmetric loads, there is no sidesway. The stiffness factors, distribution factors, and fixed-end moments are as follows.

$$K_{12} = K_{34} = \frac{I}{6} \qquad K_{23} = \frac{1.5I}{15} = \frac{I}{10}$$

$$D_{21} = \frac{I/6}{(I/6) + (I/10)} = 0.64 \qquad D_{23} = \frac{I/10}{(I/6) + (I/10)} = 0.36$$

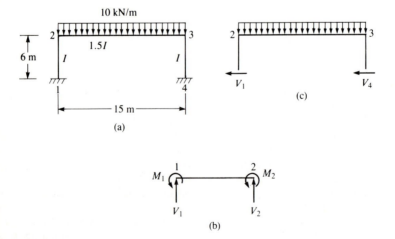

(a)

(b)

(c)

FIGURE 10.15

$$D_{32} = \frac{I/10}{(I/6) + (I/10)} = 0.36 \qquad D_{34} = \frac{I/6}{(I/6) + (I/10)} = 0.64$$

$$M_{23}^F = -M_{32}^F = -\frac{pl^2}{12} = -\frac{-10 \times 15^2}{12} = 188$$

We start the moment distribution at joint 2 and continue by alternating between joints 2 and 3. The distributed moments are shown in Fig. 10.16.

To find the shear force at the base, we consider the columns 12 and 34. As shown in Fig. 10.15b, with the analysis sign convention, shear at end 1 is

$$V_1 = \frac{M_1 + M_2}{l} = \frac{-73 - 146}{6} = -36.5$$

and at end 4 of member 34 is

$$V_4 = \frac{73 + 146}{6} = 36.5$$

As expected, these forces are equal and opposite. This is because, for the free body diagram of the frame separated at the bases of the columns, the horizontal forces must be in equilibrium. With the base shears as the only horizontal forces, their algebraic sum must be zero.

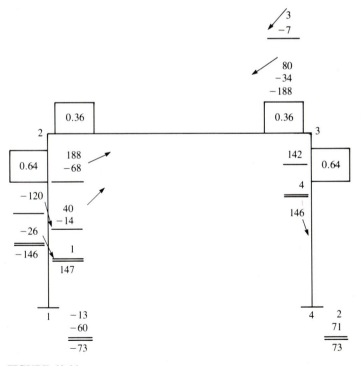

FIGURE 10.16

Frames With Sidesway

When a frame is not laterally restrained or the structure and the loads are not symmetric, it undergoes a lateral displacement (sidesway). To find the solution of such frames, a superposition procedure is used. The moments developed in the members are due to the rotations of the joints and the lateral displacement of the joints. In the first step the sidesway is ignored and the moments due to the joint rotations are found. This step involves the same moment distribution process as for frames without sidesway. If the structure is separated at the base, the sum of horizontal forces, V_0, can be calculated from the shear forces at the base of the columns along with the applied horizontal forces. V_0 will not be zero because of the sidesway. The second step considers the effect of sidesway on the structure. In this step the moments developed in the members due to the sidesway are found.

From Eq. (10.6) or Table 10.1b the magnitude of the moments at the ends of a member due to relative end displacements or sidesway $\Delta = v_2 - v_1$ with no joint rotations and loads, that is, the fixed end moments due to sidesway, are $M_1 = M_2 = -(6EI/l^2)\Delta$. If the magnitude of sidesway were known we could calculate the fixed-end moments due to sidesway and find the internal moments by the moment distribution. However, Δ is not known, what we know is that for equilibrium the sum of the horizontal forces must be zero. This condition can be used to calculate Δ. Instead of calculating the horizontal forces in terms of Δ, since the sidesway fixed-end moments are proportional to Δ, we can calculate the horizontal forces in terms of the sidesway fixed-end moments for one of the members. The sidesway moments for other members with the same end displacements are proportional to their I/l^2. For inclined members undergoing the same horizontal end displacement, the fixed-end moments due to the sidesway are in addition proportional to the ratio of the displacement normal to the axis of the inclined member and the sidesway displacement.

In the solution process, a value (usually a round number, say 100 units) is assumed for the sidesway fixed-end moments at the ends of one of the members. The fixed-end moments at the ends of other members with the same displacement are in proportion to the I/l^2 values of the members. After distributing these fixed-end moments, the base shear V_1 can be calculated for the assumed sidesway fixed-end moments. Since the moment was arbitrarily assumed, the sidesway shear thus found is not the correct value. If the correct sidesway moment is α times the assumed value, then the correct sidesway shear would be αV_1. By equilibrium of the horizontal forces we must have

$$V_0 + \alpha V_1 = 0 \tag{10.29}$$

Note that in calculating V_0 and V_1 the positive direction of shear must be the same. This equation gives the value of α. The internal moments due to the sidesway can be corrected by multiplying them by α. The final values of the member-end moments are the sum of the moments without sidesway and the corrected internal moments due to the sidesway.

For an n-story frame the shear due to sidesway of each floor is found, using assumed (usually the same) fixed-end moments, while the sidesway of all the other floors is prevented. The sum of shear forces at the base of each story is set equal to zero, leading to n equation for the determination of n parameters α_i.

Example 10.8. Find the member-end moments for the frame of Fig. 10.17.

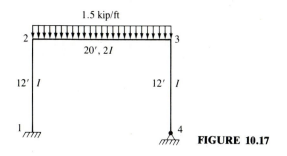

FIGURE 10.17

Solution. The frame is not symmetric because of the hinge at joint 4. Thus, it will experience sidesway. The stiffness and distribution factors are,

$$K_{12} = \frac{I}{12} \qquad K_{23} = \frac{2I}{20} = \frac{I}{10} \qquad K'_{34} = \frac{3}{4} \times \frac{I}{12} = \frac{I}{16}$$

$$D_{21} = \frac{I/12}{I/12 + I/10} = 0.45 \qquad D_{23} = \frac{I/10}{I/12 + I/10} = 0.55$$

$$D_{32} = \frac{I/10}{I/10 + I/16} = 0.62$$

$$D_{34} = \frac{I/16}{I/10 + I/16} = 0.38$$

In the first step, ignoring the sidesway, the moments are calculated (Fig. 10.18), from which the base shear can be determined.

The base shear, the sum of the horizontal forces, in this step is

$$V_0 = -4.1 + 1.9 = -2.2$$

This value is not zero because we have ignored the sidesway. To find the effect of the sidesway we assume a fixed-end moment of $100 \text{ kip} \cdot \text{ft}$ at the ends of member 21 due to the sidesway. We note that member 34 has the same sidesway. With the rotations restricted, the fixed-end moments at the ends of member 34 are also $100 \text{ kip} \cdot \text{ft}$. This is because when the end rotations are restricted (and with axial deformations ignored) the top of members 12 and 43 displace the same amount. Since I/l^2 is the same for both members, the moments developed at the ends of

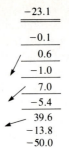

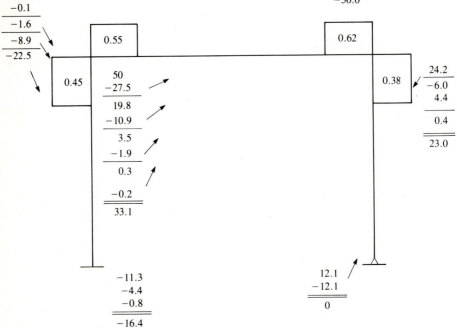

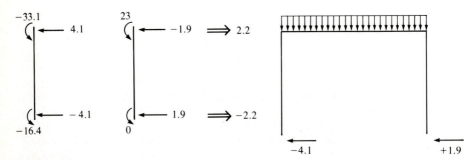

FIGURE 10.18

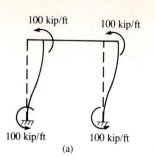

100 kip/ft   100 kip/ft

100 kip/ft   100 kip/ft

(a)   **FIGURE 10.19a**

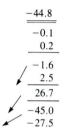

−44.8
−0.1
0.2
−1.6
2.5
26.7
−45.0
−27.5

59.6

0.1
0.4
↑4.1
−45.0
100.0

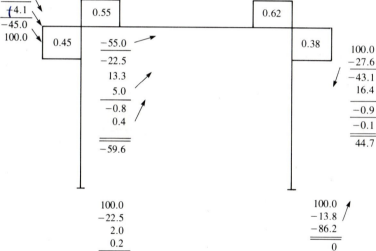

| 0.45 | 0.55 | | | 0.62 | 0.38 |

−55.0
−22.5
13.3
5.0
−0.8
0.4
―――
−59.6

100.0
−27.6
−43.1
16.4
−0.9
−0.1
―――
44.7

100.0
−22.5
2.0
0.2
―――
79.7

100.0
−13.8
−86.2
―――
0

FIGURE 10.19b   (b)

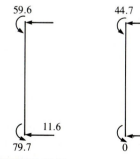

59.6   44.7

11.6   3.7
79.7   0

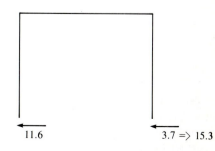

11.6   3.7 ⟹ 15.3

FIGURE 10.19c

both members are the same. Distribution of the fixed-end moments shown in Fig. 10.19a is done in Fig. 10.19b.

The shear forces at the bases of the columns are found from the moments in the columns, giving a total force $V_1 = 15.3$ kips, as indicated in Fig. 10.19c. Equilibrium of the horizontal force without sidesway and that due to sidesway gives,

$$V_0 + \alpha V_1 = 0 \qquad \text{or} \qquad -2.2 + 15.3\alpha = 0$$

which results in

$$\alpha = 0.144$$

Thus, the moments in the members are found as the sum of the values without sidesway and those due to sidesway as follows.

$$M_{12} = -16.4 + 0.144 \times 79.7 = -4.92 \text{ kip} \cdot \text{ft}$$

$$M_{21} = -M_{23} = -33.1 + 0.144 \times 59.6 = -24.52 \text{ kip} \cdot \text{ft}$$

$$M_{34} = -M_{32} = -23.0 - 0.144 \times 44.7 = -29.44 \text{ kip} \cdot \text{ft}$$

$$M_{43} = 0$$

10.6 SLOPE–DEFLECTION METHOD

The slope–deflection method is an analytical method for the determination of end moments in statically indeterminate beams and frames. However, it is not as commonly used as the moment distribution and stiffness methods. The slope–deflection method involves the solution of a system of simultaneous equations.

Slope–Deflection Equations

If in Eqs. (10.6) and (10.7) we let $\Delta = v_2 - v_1$ and $\psi = \Delta/l$, we will get the slope–deflection equations as

$$M_1 = 2EK(2\theta_1 + \theta_2 - 3\psi) + M_1^F \qquad (10.30a)$$

$$M_2 = 2EK(\theta_1 + 2\theta_2 - 3\psi) + M_2^F \qquad (10.30b)$$

Here again, $K = I/l$ is the stiffness factor, E is the modulus of elasticity, Δ the relative displacement of the ends, and ψ the rotation of the chord of the member. The above relations give the end moments of a member due to end displacements and rotations as well as the applied loads.

Slope–Deflection Procedure

Consider the two-span beam of Fig. 10.20a. Under the applied loads the rotations θ_1, θ_2, and θ_3 take place at the three supports. Moments are also developed throughout the beam. Separating the spans from the supports, we can write the equilibrium of the moments for the free body of each support

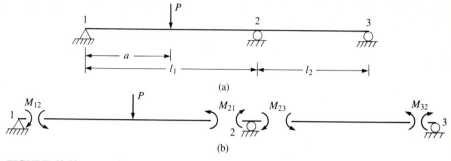

FIGURE 10.20

(Fig. 10.20b). Thus, the equilibrium equations are

$$M_{12} = 0 \qquad M_{21} + M_{23} = 0 \qquad M_{32} = 0$$

There are three equilibrium equations. They can be expressed in terms of the three unknown rotations, θ_1, θ_2, and θ_3. This is done by using the slope–deflection equations (Eqs., (10.30a) and (10.30b)). Since there are no deflections at the supports, we get

$$M_{12} = 2EK_{12}(2\theta_1 + \theta_2) + M_{12}^F \qquad M_{21} = 2EK_{12}(\theta_1 + 2\theta_2) + M_{21}^F$$

$$M_{23} = 2EK_{23}(2\theta_2 + \theta_3) \qquad M_{32} = 2EK_{23}(\theta_2 + 2\theta_3)$$

The fixed-end moments for the first span can be found from Table 10.1a. Substitution of the above relations in the equilibrium equations results in

$$2\theta_1 + \theta_2 = -M_{12}^F/(2EK_{12})$$

$$K_{12}\theta_1 + 2(K_{12} + K_{23})\theta_2 + K_{23}\theta_3 = -M_{21}^F/(2E)$$

$$\theta_2 + 2\theta_3 = 0$$

The solution of this system of simultaneous equations gives the values of the rotations. With the rotations calculated, the above slope–deflection equations give the values of the end moments.

Steps in the Slope–Deflection Method

Application of the slope–deflection method involves the following steps.

1. Identify the unknown support rotations and deflections.
2. Write the equilibrium equations at all the supports in terms of the end moments.
3. Express the end moments in terms of the support rotations and deflections as well as the fixed-end moments. The fixed-end moments can be calculated from Tables 10.1a and 10.1b.

4. Substitute the expressions obtained in step 3 in the equilibrium equations.
5. Solve the equilibrium equations to find the unknown support deflections and rotations.
6. Using the slope-deflection equations calculate the end moments.

Example 10.9. Find the member-end moments for the two-span beam of Fig. 10.21 by the slope deflection method.

Solution. The fixed-end moments for the two spans are, from Table 10.1a,

$$M_{12}^F = -M_{21}^F = -\frac{pl^2}{12} = \frac{2 \times 20^2}{12} = 66.67 \text{ kip} \cdot \text{ft}$$

$$M_{23}^F = -M_{32}^F = -\frac{Pl}{8} = \frac{5 \times 30}{8} = 18.75 \text{ kip} \cdot \text{ft}$$

The member-end moments in terms of the end rotations are, by Eqs. (10.30a) and (10.30b) with $\psi = \theta_1 = 0$,

$$M_{12} = 2EK_{12}(\theta_2) + M_{12}^F = \frac{2EI_1}{l_1}\theta_2 + M_{12}^F = \frac{EI}{10}\theta_2 + 66.67$$

$$M_{21} = 2EK_{12}(2\theta_2) + M_{21}^F = \frac{4EI_1}{l_1}\theta_2 + M_{21}^F = \frac{EI}{5}\theta_2 - 66.67$$

$$M_{23} = 2EK_{23}(2\theta_2 + \theta_3) + M_{23}^F = \frac{2EI_2}{l_2}(2\theta_2 + \theta_3) + M_{23}^F = \frac{EI}{7.5}\theta_2 + \frac{EI}{15}\theta_3 + 18.75$$

$$M_{32} = 2EK_{23}(\theta_2 + 2\theta_3) + M_{32}^F = \frac{2EI_2}{l_2}(\theta_2 + 2\theta_3) + M_{23}^F = \frac{EI}{15}\theta_2 + \frac{EI}{7.5}\theta_3 - 18.75$$

Equilibrium of moments at joints 2 and 3 gives

$$M_{21} + M_{23} = 0 \qquad M_{32} = 0$$

or

$$\frac{EI}{5}\theta_2 - 66.67 + \frac{EI}{7.5}\theta_2 + \frac{EI}{15}\theta_3 + 18.75 = 0$$

$$\frac{EI}{15}\theta_2 + \frac{EI}{7.5}\theta_3 - 18.75 = 0$$

Thus

$$5EI\theta_2 + EI\theta_3 = 718.8$$

$$EI\theta_2 + 2EI\theta_3 = 281.25$$

giving

$$EI\theta_2 = 128.48 \qquad EI\theta_3 = 76.38$$

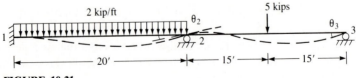

FIGURE 10.21

Then
$$M_{12} = 12.85 + 66.67 = 79.52$$

$$M_{21} = \frac{128.48}{5} - 66.67 = -40.97$$

$$M_{23} = \frac{128.48}{7.5} + \frac{76.38}{15} + 18.75 = 40.97$$

$$M_{32} = \frac{128.48}{15} + \frac{76.38}{7.5} - 18.75 = 0$$

Example 10.10. Find the end moments for the previous problem if the middle support is settled 6 in (Fig. 10.22).

Solution. Since this is a linear elastic structure, we can find the solution by superposition. Thus, we will find the moments due to the support settlement here. These can be added to the moments found in the previous problem due to the applied loads. Fixed-end moments in the slope–deflection equations are due just to the in-span loads. The slope–deflection equations (Eqs. 10.30a and 10.30b), with $\Delta = -0.5$ ft, $\theta_1 = 0$ and no loads, give

$$M_{12} = 2EK_{12}\left(2\theta_1 + \theta_2 - 3\frac{\Delta}{l_1}\right) = \frac{1}{10} EI\theta_2 + \frac{3}{400} EI$$

$$M_{21} = 2EK_{12}\left(\theta_1 + 2\theta_2 - 3\frac{\Delta}{l_1}\right) = \frac{1}{5} EI\theta_2 + \frac{3}{400} EI$$

$$M_{23} = 2K_{23}\left(2\theta_2 + \theta_3 - 3\frac{\Delta}{l_{23}}\right)\frac{1}{7.5} EI\theta_2 + \frac{1}{15} EI\theta_3 + \frac{1}{300} EI$$

$$M_{32} = 2K_{23}\left(\theta_2 + 2\theta_3 - 3\frac{\Delta}{l_{23}}\right)\frac{1}{15} EI\theta_2 + \frac{1}{7.5} EI\theta_3 + \frac{1}{300} EI$$

The equilibrium equations are the same as those in Example 10.9, that is,

$$M_{21} + M_{23} = 0 \qquad M_{32} = 0$$

or
$$\frac{1}{5} EI\theta_2 + \frac{3}{400} EI + \frac{1}{7.5} EI\theta_2 + \frac{1}{15} EI\theta_3 + \frac{1}{300} EI = 0$$

$$\frac{1}{15} EI\theta_2 + \frac{1}{7.5} EI\theta_3 + \frac{1}{300} EI = 0$$

or
$$100\theta_2 + 20\theta_3 = -\frac{13}{4}$$

$$20\theta_2 + 40\theta_3 = -1$$

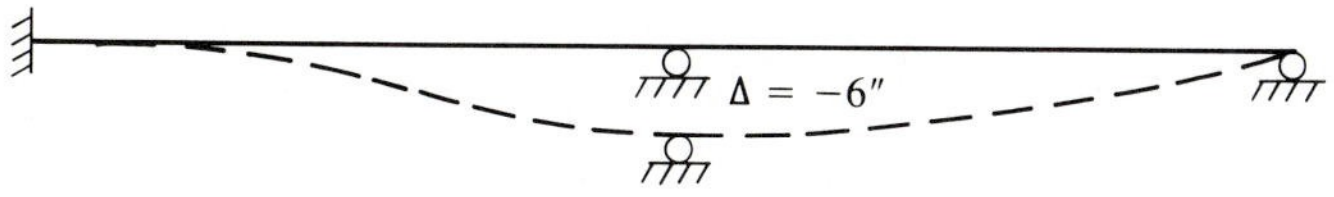

FIGURE 10.22

The solution of these two simultaneous equations gives

$$\theta_2 = -\frac{5.5}{180} = -0.031 \text{ rad}$$

$$\theta_3 = \frac{-1 + (5.5/9)}{40} = -0.0097 \text{ rad}$$

Thus, the additional moments due to support settlement are

$$M_{12} = \frac{1}{10} EI(-0.031) + \frac{3}{400} EI = 0.0044 EI$$

$$M_{21} = \frac{1}{5} EI(-0.031) + \frac{3}{400} EI = 0.0013 EI$$

$$M_{23} = \frac{1}{7.5} EI(-0.031) + \frac{1}{15} EI(0.0097) + \frac{1}{300} EI = 0.0015 EI$$

$$M_{32} = \frac{1}{15} EI(-0.031) + \frac{1}{7.5} EI(-0.0097) + \frac{1}{300} EI = 0$$

Example 10.11. Find the end moments for the beam of Fig. 10.23 using the slope–deflection method.

Solution. Since this beam is statically indeterminate to the first degree, we need to calculate one unknown only. The unknown can be taken as the rotation of the beam at the second support, θ_2. The effect of the overhang is to produce a fixed-end moment equal to $100 \text{ kN} \cdot \text{m}$ at support 2. Therefore, the equivalent joint moment at 2 is $-100 \text{ kN} \cdot \text{m}$. Once the moments M_{12} and M_{21} are found, all the other forces can be found by statics. The slope deflection equations for this beam are

$$M_{12} = 2EK_{12}(2\theta_1 + \theta_2) = \frac{2}{10} EI\theta_2$$

$$M_{21} = 2EK_{12}(\theta_1 + 2\theta_2) = \frac{4}{10} EI\theta_2$$

The equilibrium equation at joint 2 is

$$M_{21} - 100 = 0 \qquad \text{or} \qquad M_{21} = 100 \text{ KN} \cdot \text{m}$$

which gives

$$\theta_2 = \frac{10}{4EI} M_{21} = \frac{250}{EI}$$

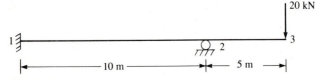

FIGURE 10.23

and then

$$M_{12} = \frac{2}{10} EI\theta_2 = \frac{2EI \times 50}{10EI} = 50 \text{ kN} \cdot \text{m}$$

10.7 SLOPE–DEFLECTION METHOD FOR FRAMES

Application of the slope–deflection method to frames is similar to that of beams. In frames, often the ends of the members undergo lateral displacements in addition to rotations.

Example 10.12. Find the member-end moments for the frame of Fig. 10.24 with the slope–deflection method. $I = 60$ in^4 for the vertical members and 24 in^4 for the horizontal members.

Solution. The effect of the cantilever 35 can be included by replacing it with its end moment. This will reduce the number of equations to solve. The fixed end moment at the end of the cantilever is

$$M_{35}^F = -\frac{pl^2}{2} = \frac{3 \times 6^2}{2} = 54 \text{ kip} \cdot \text{ft}$$

The fixed end moments for member 23 are, from Table 10.1a,

$$M_{21}^F = -\frac{Pl}{8} = -\frac{-10 \times 10}{8} = 12.5 \text{ kip} \cdot \text{ft}$$

$$M_{23}^F = -M_{32}^F = -\frac{pl^2}{12} = -\frac{-3 \times 20^2}{12} = 100 \text{ kip} \cdot \text{ft}$$

The unknown displacements and rotations are the lateral displacement and rotation of joint 2, denoted by Δ_2 and θ_2, and the rotations of joints 3 and 4, θ_3, θ_4. Since the axial deformations are neglected, the lateral displacement of joint 3 is the same as that of joint 2, that is Δ_2.

The equilibrium equations consist of equilibrium of forces at the joints undergoing displacements. That is, equilibrium of moments at joints 2, 3, and 4 and equilibrium of the horizontal forces. Thus,

$$M_{21} + M_{23} = 0 \qquad M_{32} + M_{34} + 54 = 0 \qquad M_{43} = 0$$

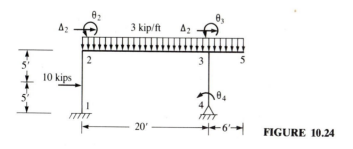

FIGURE 10.24

and
$$V_{12} + V_{43} - 10 = 0$$

The shear forces V_{21} and V_{34} can be expressed in terms of the end moments by considering the equilibrium of members 21 and 34, which gives

$$V_{12} = \frac{M_{12} + M_{21}}{l_{12}} + 5 \qquad V_{43} = \frac{M_{34} + M_{43}}{l_{34}}$$

From the slope–deflection equations (Eqs. 10.30a and 10.30b),

$$M_{12} = 2EK_{12}\left(\theta_2 - \frac{3\Delta_2}{l_{12}}\right) + M_{12}^F = 0.2EI(\theta_2 - 0.3\Delta_2) + 12.5$$

$$M_{21} = 2EK_{12}\left(2\theta_2 - \frac{3\Delta_2}{l_{12}}\right) + M_{21}^F = 0.2EI(2\theta_2 - 0.3\Delta_2) - 12.5$$

$$M_{23} = 2EK_{23}(2\theta_2 + \theta_3) + M_{23}^F = 0.1EI(2\theta_2 + \theta_3) + 100$$

$$M_{32} = 2EK_{32}(\theta_2 + 2\theta_3) + M_{32}^F = 0.1EI(\theta_2 + 2\theta_3) - 100$$

$$M_{34} = 2EK_{34}\left(2\theta_3 + \theta_4 - \frac{3\Delta_2}{l_{34}}\right) = 0.2EI(2\theta_3 + \theta_4 - 0.3\Delta_2)$$

$$M_{43} = 2EK_{34}\left(\theta_3 + 2\theta_4 - \frac{3\Delta_2}{l_{34}}\right) = 0.2EI(\theta_3 + 2\theta_4 - 0.3\Delta_2)$$

Substituting the latter relations in the equilibrium equations and dividing by EI yields

$$6\theta_2 + \theta_3 - 0.6\Delta_2 = -\frac{875}{EI}$$

$$\theta_2 + 6\theta_3 + 2\theta_4 - 0.6\Delta_2 = \frac{460}{EI}$$

$$\theta_3 + 2\theta_4 - 0.3\Delta_2 = 0$$

and the last equilibrium equation upon substitution for V_{12} and V_{43} and multiplying by $10/EI$ gives

$$\theta_2 + \theta_3 + \theta_4 - 0.4\Delta_2 = -\frac{83.3}{EI}$$

The solution of these equations yields the joint rotations and displacement as

$$\theta_2 = -294.8/EI$$

$$\theta_3 = 68.4/EI$$

$$\theta_4 = -240.6/EI$$

$$\Delta_2 = -1375.7/EI$$

Substitution of the above displacements back into the slope–deflection equations gives the member-end moments in kip · ft as

$$M_{12} = 36.0 \qquad M_{21} = -47.88$$

$$M_{23} = 47.88 \qquad M_{32} = -115.80$$

$$M_{34} = 61.78 \qquad M_{43} = 0$$

LABORATORY EXPERIMENTS

Moment Distribution

A moment distribution system can be built for demonstrating the moment distribution for beams and frames. Such a system has been designed by the author and is shown in Fig. L10.1. It can also be used for experimentation with stiffness properties of the structures, influence lines, Laboratory Experiments in Chapter 8, etc. This system provides roller, hinge, and fixed supports. A mechanism allows the supports to rotate and then be made fixed in any rotated position. Dials indicate the amount of rotations taking place at the supports.

Using this moment distribution system with continuous beams or frames made of spring steel, we can follow the physical interpretation of the moment distribution method as an iterative process.

Let us first produce the deformed conditions of a beam or frame, by allowing rotations of the joints to take place, except of course at the joints that are fixed. When we apply the loads, the supports rotate and the dials indicate the deformed configuration of the structure. This is the way beams and frames deform under loads in "real life."

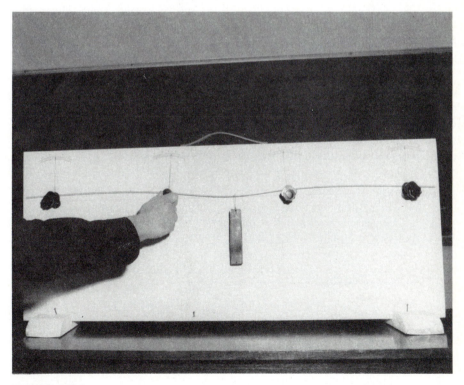

FIGURE L10.1
Moment distribution system

In proceeding with the moment distribution procedure, we reach the same final deformed configuration under the loads in a different way. At the start we lock all the joints then apply the loads. This causes equivalent joint moments (the negative of the sum of the fixed-end moments) at the joints. Starting at a support that is not permanently fixed, we release the joint and allow it to rotate under the equivalent joint moments. We then lock the joint with the beam in the deformed position by tightening a knob and proceed with releasing another joint. It is easy to see that after releasing a joint we have a two-span beam, with the two far ends fixed and under an unbalanced moment at the middle support, Fig. 10.6. As discussed under Distribution Factor (Section 10.4), the internal moments can be easily found at the three joints of the two-span beam by distributing the unbalanced moment and calculating the carry over moments. After a few rounds of releasing and fixing the joints, the release of the joints does not cause any additional rotation in the system. That is, the joints have reached their final rotated positions. The dials show that this is the same deformed position as in the first experiment.

Frame Sidesway

Using reduced-scale frame models, made out of spring steel, show that symmetric frames under symmetrical loads do not undergo any lateral displacements or sidesway. On the other hand, if either the frame or the load is unsymmetrical the frame undergoes sidesway.

If the frame has strain gages we can prevent the sidesway first, and measure the strain induced by the rotation of the joints. We then remove the restraint against the sidesway and measure the new strains. We can show that the total strains (or moments) are equal to the sum of those due to the joint rotations plus those due to sidesway displacements, thus verifying the validity of the moment distribution method for frames with sidesway.

PROBLEMS

10.1. If a moment $M_2 = 100$ kip · ft is applied to the right-hand end of a beam, what is the moment M_1 developed at the left, fixed-end (Fig. P10.1).

FIGURE P10.1

10.2. The beams of Figs. P10.2a have their left and right ends fixed, respectively.

(a) Calculate the right end rotation of the left beam, θ_L, and the left end rotation of the right beam θ_R, in terms of M_L and M_R.

(b) The moment M_0 is applied at the middle support of the two-span beam of Fig. 10.2b. Using the results of part (a) and the equilibrium equation

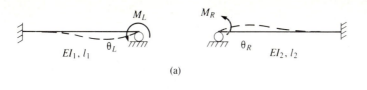

(a)

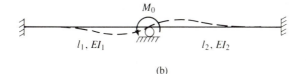

(b)

FIGURE P10.2

$M_L + M_R = M_0$ and the compatibility equation $\theta_L = \theta_R = \theta$, find the internal moments M_L and M_R on the left and right of support 2, in terms of M_0.

Solve the following problems by the moment distribution method.

10.3. Determine the support moments M_1 to M_5 in Fig. P10.3.

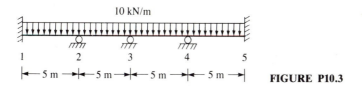

FIGURE P10.3

10.4. Find the support moments for the beam of Fig. P10.4 and draw the moment diagram.

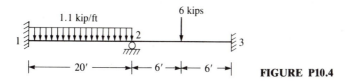

FIGURE P10.4

10.5. Find the support moments for the beam of Fig. P10.5 and draw the moment diagram.

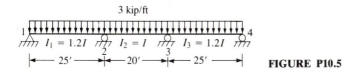

FIGURE P10.5

For Problems P10.6 through P10.9 find the moments at all the supports by the moment distribution method.

10.6.

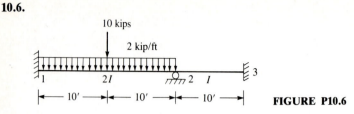

FIGURE P10.6

10.7.

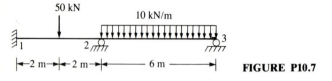

FIGURE P10.7

10.8.

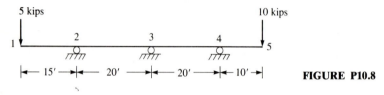

FIGURE P10.8

10.9.

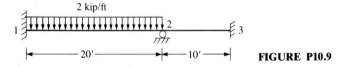

FIGURE P10.9

10.10. Find the moment diagram and the reactions of the three-span beam with uniform load of Fig. P10.10 using the moment distribution method. Use the reduced stiffness for the end members.

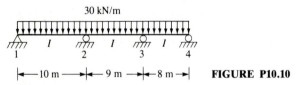

FIGURE P10.10

10.11. Find the moment and shear diagrams for the three-span continuous beam in Fig. P10.11. Use the moment distribution method to find the moments at the supports.

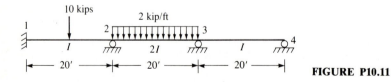

FIGURE P10.11

10.12. Find the support moments for the two-span beam in Fig. P10.12.

(*a*) Not using reduced stiffness.
(*b*) Using reduced stiffness.

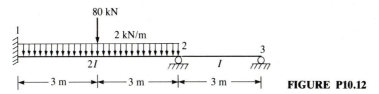

FIGURE P10.12

10.13. The middle support of the two-span beam of Fig. P10.13 settles by 3 in. Find the moment developed at the middle support due to this settlement. $I = 900$ in^4 and $E = 3\,100$ ksi.

Hint: Find the fixed-end moments due to the settlement from Table 10.1b and distribute.

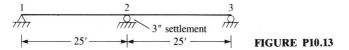

FIGURE P10.13

10.14. The three members of the frame shown in Fig. P10.14 are connected together at point 2 and are hinged at the other ends. The moment of inertia for all the members is *I*.

(*a*) Find the member-end moments by the moment distribution method.
(*b*) Draw the moment diagram.
(*c*) Find the shear force at point 3.

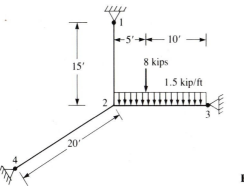

FIGURE P10.14

Find the member-end moments for the following frames by the moment distribution method.

10.15.

Hint: Add the negative of the joint moment to one side of the joint, distribute

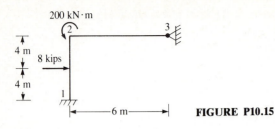

FIGURE P10.15

this moment along with others, but exclude the original value when summing up to find the member-end moments.

10.16.

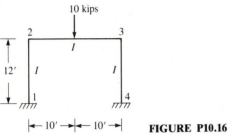

FIGURE P10.16

10.17.

See Hint in problem 10.15.

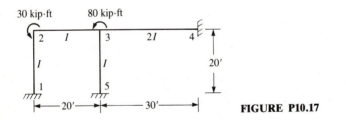

FIGURE P10.17

10.18.

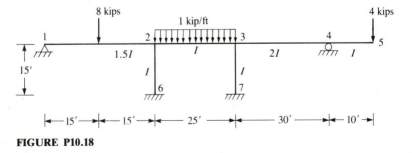

FIGURE P10.18

10.19. Find the moment diagram and the base shear at point 4 in Fig. P10.19.

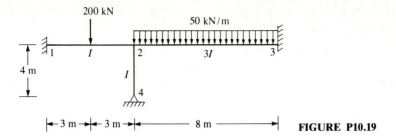

200 kN
50 kN/m

FIGURE P10.19

10.20. Find the moment diagram and the reactions at point 4 in Fig. P10.20.

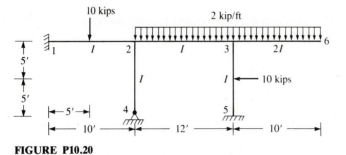

10 kips
2 kip/ft
10 kips

FIGURE P10.20

10.21. Calculate the moments by the moment distribution method and draw the moment diagram for the frame of Fig. P10.21. *EI* is constant.

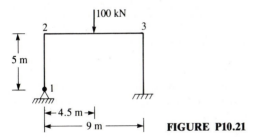

100 kN

FIGURE P10.21

10.22. Find the moment diagram and the base shear forces. *EI* is constant.

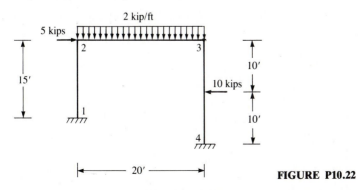

2 kip/ft
5 kips
10 kips

FIGURE P10.22

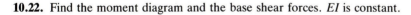
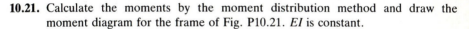
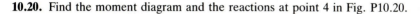

10.23. Find the moments at the supports and draw the bending moment diagram for the frame of Fig. P10.23.

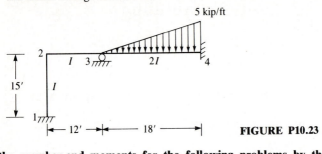

FIGURE P10.23

Find the member-end moments for the following problems by the slope–deflection method.

10.24. *EI* is constant (Fig. P10.24).

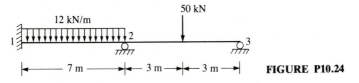

FIGURE P10.24

10.25. *EI* is constant (Fig. P10.25).

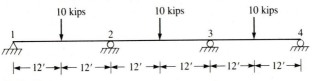

FIGURE P10.25

10.26. The right end support, 3, of the beam of Fig. P10.26 settles by 5 cm. Find the moments developed in the members due to this settlement in terms of *EI*. *EI* is constant.

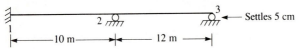

FIGURE P10.26

10.27. Find the reactions, moment, and shear diagrams, and the horizontal deflection at 3 in Fig. P10.27.

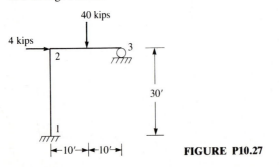

FIGURE P10.27

10.28. *EI* is constant (Fig. P10.28).

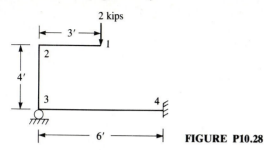

FIGURE P10.28

10.29. For the frame shown in Fig. P10.29, with constant *EI,* solve for all the reactions and draw the shear and moment diagrams.

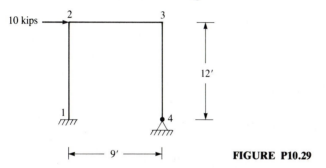

FIGURE P10.29

For Problems 10.30 and 10.31 find the member-end moments by the slope-deflection method.

10.30.

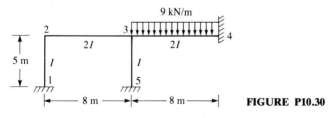

FIGURE P10.30

10.31.

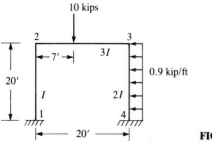

FIGURE P10.31

CHAPTER
11

STIFFNESS
METHOD
FOR
BEAMS

11.1 STIFFNESS METHOD

The stiffness method presents a general procedure for the analysis of trusses, beams, and frames. Its counterpart, the finite-element method, provides a versatile technique for the solution of two- and three-dimensional solids. These methods along with the flexibility method discussed in Section 13.6 are sometimes referred to as matrix methods because matrix notation is the natural notation for the formulation and solution of problems with these methods. The two common matrix methods for trusses, beams, and frames are the stiffness and flexibility methods. The stiffness and flexibility matrices have an important role in these solution methods. They are also known as the displacement and force methods, respectively, because they use the joint displacements or forces as the basic variables of the problem. The stiffness method can be formulated in many different ways. The direct stiffness method seems to be the easiest for both programming on the computer and hand calculation. The direct stiffness method has now become the commonest method of structural analysis. In the following discussions as well as in the subsequent chapters the term stiffness method is used to mean the direct stiffness method. The stiffness method is covered in detail in this book.

The first step in the development of the stiffness method is establishing the force–displacement relationship. This is a relationship between a set of independent forces and the corresponding displacements. As mentioned in Section 10.2, the term displacement is used to signify both displacement and rotation, and the term force is used for both force and moment. The force–displacement relation can be developed for a member, a structure, or a portion thereof. The set of independent displacements are called the *degrees of freedom*. They are the displacements of the joints or the nodes where the elements are connected together. A segment with its end joints is called an *element* or *member*. The relationship between the nodal forces and the corresponding displacements is called the *force–displacement* or *stiffness relation*.

The relationship between the member-end forces and displacements is the first step in the solution of any problem with the stiffness method. Equilibrium of the joint forces and compatibility of the joint displacements, at the two sides of the joints, lead to the equilibrium equations in terms of the unknown displacements. The joint displacements are found by solving the equilibrium equations. The member-end forces are subsequently determined from the force–displacement relations for the elements.

In the next section we establish the basis of the direct stiffness method. As we will see, in the stiffness method discussed here the structure stiffness matrix can be obtained directly by adding the terms of the element stiffness matrices. That is why the procedure is called the *direct* stiffness method. Regardless of the complexity of the structure and the number of degrees of freedom per joint, the following steps must be taken in the direct stiffness method.

1. Find the force–displacement relation for each element. This will result in the element stiffness matrices. Such relations are found by substituting the element parameters in the force–displacement equations derived for the type of element used. Force–displacement equations are derived in this book for different types of elements and can be used for solving problems.
2. Establish the equilibrium of the forces acting at each joint.
3. Satisfy the compatibility of the displacements on the sides of each joint.
4. Express the equilibrium equations in terms of the joint displacements. This leads to the equilibrium equations in terms of the structure stiffness matrix and the load vector.
5. Solve the equilibrium equations to find the joint displacements.
6. Determine the member-end displacement vectors using the compatibility of displacements, and calculate the member-end forces.

It should be mentioned that for complex problems with a large number of elements, application of the stiffness method by hand becomes rather time consuming. Other methods such as moment distribution may be easier to apply in such cases. With a computer program, however, application of the stiffness method becomes simple and efficient. A discussion of the procedures for programming the stiffness method is included in Section 19.2.

11.2 BEAM ELEMENTS WITH ROTATIONAL DEGREES OF FREEDOM

In this section we discuss the stiffness method for continuous beams, using beam elements with rotational degrees of freedom only. Thus, for continuous beams the support points are used as joints. The discussion of elements with both displacement and rotational degrees of freedom is deferred to Section 11.3. In the latter case more than one element may be used for each span of a continuous beam, allowing the direct calculation of the displacements and rotations within the span.

To establish the procedure for the application of the direct stiffness method to continuous beams, we use a beam such as the one shown in Fig. 11.1. However, the derivations are general and can be applied to any other beam. The beam of Fig. 11.1 has four degrees of freedom, the rotations of supports 1, 2, 3, and 4. The derivation of the solution procedure is done in the following steps.

The analysis sign convention discussed in Section 10.1 is used for the stiffness method. A right-handed coordinate system (defined in Section 10.1) is used as shown in Fig. 11.2. All entities such as displacements, rotations, loads, shear forces, reactions, and bending moments are positive when they act in the positive directions of the coordinate system. For drawing moment diagrams, conversion from analysis to design sign convention is made by changing the sign of the moments on the left end of the members. The positive sense of variables for beams in the analysis sign convention are shown in Fig. 11.2. Thus, for beams with the x coordinate pointing from left to right, positive loads, shear forces, and reactions, as well as deflections, are upward and positive rotations and moments are in the counterclockwise sense.

In this section we show the basis of the direct stiffness method and establish the procedure for solving problems using this method.

Force–Displacement Relations

For beam elements with rotational degrees of freedom only, the force–displacement equations consist of relations between the end moments and rotations. Consider a typical element of a continuous beam, for example, the middle span of the beam of Fig. 11.1, with span length l, modulus of elasticity E, and moment of inertia I. Moments M_1 and M_2 are applied by the left and right portions of the structure to this element, Fig. 11.3. The rotations θ_1 and θ_2 are the slopes of the continuous beam at the second and third supports. The loads on the element are the same as those applied on the second span of the

FIGURE 11.1

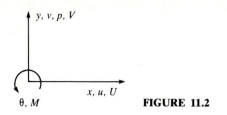

y, v, p, V

x, u, U

θ, M **FIGURE 11.2**

continuous beam. Note that instead of support numbers 2 and 3 we have used numbers 1 and 2. We will always use 1 and 2 to refer to the left and right ends of a typical element. Alternatively one can use i and j for this purpose.

In Section 10.3 we found the relationship between the end forces and displacements for a beam under end forces and external loads. The general force–displacement expressions were given by Eqs. (10.6), (10.7), (10.10), and (10.11). For the element of Fig. 11.3, the end displacements are zero. The relations between the end moments M_1 and M_2 and rotations θ_1 and θ_2, Eqs. (10.6) and (10.7), become

$$M_1 = \frac{4EI}{l}\,\theta_1 + \frac{2EI}{l}\,\theta_2 + M_1^F \tag{11.1a}$$

$$M_2 = \frac{2EI}{l}\,\theta_1 + \frac{4EI}{l}\,\theta_2 + M_2^F \tag{11.1b}$$

where M_1^F and M_2^F are the fixed-end moments under the loads applied over the span. In matrix form the above equations become

$$
\begin{array}{cc}
 & \theta_1 \quad \theta_2 \\
\begin{bmatrix} M_1 \\ M_2 \end{bmatrix} = \frac{2EI}{l} \begin{bmatrix} 2 & 1 \\ 1 & 2 \end{bmatrix} \begin{bmatrix} \theta_1 \\ \theta_2 \end{bmatrix} + \begin{bmatrix} M_1^F \\ M_2^F \end{bmatrix}
\end{array} \tag{11.2a}
$$

θ_1 and θ_2 above the matrix indicate that the terms in the first and second columns are associated with the left and right rotations, respectively. This expression can be written in compact form as

$$\mathbf{S} = \mathbf{Ks} + \mathbf{S}^F \tag{11.2b}$$

where $\mathbf{S}$ is the vector of the end moments, $\mathbf{s}$ the vector of the end rotations, $\mathbf{S}^F$ is the vector of the fixed-end moments, $\mathbf{K}$ is the rotational stiffness matrix for a

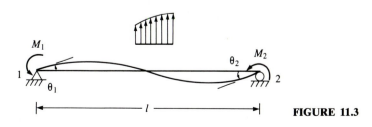

M_1

M_2

θ_2

1

θ_1

2

l

FIGURE 11.3

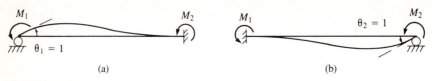

FIGURE 11.4

beam element,

$$\mathbf{K} = \frac{2EI}{l} \begin{bmatrix} 2 & 1 \\ 1 & 2 \end{bmatrix} \qquad (11.3)$$

The terms in the first column of the stiffness matrix are the values of the left and right moments required to maintain a unit rotation at the left-hand end of the member while the rotation at the right-hand end is kept at zero and no loads are acting on the member. The terms in the second column are the left and right moments needed to maintain a unit rotation at the right-hand end while the rotation at the left-hand end is zero and no loads are acting on the member.

Thus, an alternative procedure for finding the terms of the stiffness matrix is to set the rotation at one end of the beam, without any external loads, equal to 1, while the rotation at the other end is kept at 0, and find the end moments required to maintain this deflected configuration. The solution of the beams shown in Fig. 11.4a and 11.4b for M_1 and M_2 yields the two columns of the stiffness matrix.

Note that we can rearrange Eqs. (11.1a) and (11.1b) to express the end rotations in terms of the end moments. In such a relationship the vector of the end rotations would be equal to a matrix times the vector of end moments. This matrix is called the *flexibility matrix* and is the inverse of the stiffness matrix. The flexibility matrix is used in the flexibility method (Section 13.9).

Equilibrium of Joints

Consider the three-span beam of Fig. 11.5a under the applied loads with the external moments M_1 to M_4 acting at joints 1 to 4. These loads and moments cause the joint rotations r_1 to r_4. To find these rotations we can use the method of consistent deformations discussed in Chapter 7. We must find the internal and external moments acting at each joint, establish the equilibrium of the joints, and invoke the compatibility of the rotations.

Figure 11.5b shows the spans separated at the joints. The internal member-end moments on the left and right of each member are indicated by subscripts 1 and 2, while a superscript denotes the element (span) number. Thus, instead of using the two ends to denote a member, as was done in the other methods such as the moment distribution and slope–deflection methods,

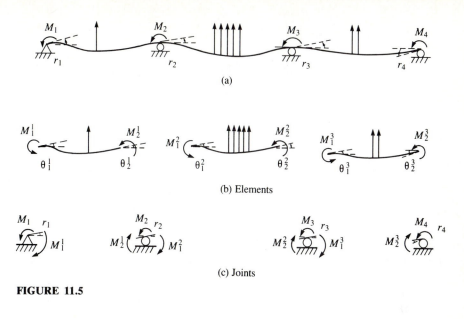

FIGURE 11.5

we use the element numbers. As we will see later, the notation can be simplified by dropping the superscripts without creating any confusion. However, in this derivation we keep the superscripts for clarity of the discussion. Figure 11.5b shows the internal member-end moments with positive signs along with the member-end rotations. Figure 11.5c shows the moments acting on the joints. M_i are the applied joint moments. By the principle of action–reaction, the internal moments acting on the joints must be equal and opposite to the member-end moments, as shown in Fig. 11.5c. Figure 11.5c also shows the joint rotations r_i. Equilibrium of the moments for joints 1 through 4 gives

$$M_1^1 - M_1 = 0$$

$$M_2^1 + M_1^2 - M_2 = 0$$

$$M_2^2 + M_1^3 - M_3 = 0 \qquad (11.4)$$

$$M_2^3 - M_4 = 0$$

The end moments for each member can be expressed in terms of the end rotations using Eqs. (11.1a, b). Hence, for members 1, 2, and 3 we have

$$M_1^1 = \frac{4EI_1}{l_1}\,\theta_1^1 + \frac{2EI_1}{l_1}\,\theta_2^1 + M_1^{1F}$$

$$M_2^1 = \frac{2EI_1}{l_1}\,\theta_1^1 + \frac{4EI_1}{l_1}\,\theta_2^1 + M_2^{1F}$$

$$M_1^2 = \frac{4EI_2}{l_2}\,\theta_1^2 + \frac{2EI_2}{l_2}\,\theta_2^2 + M_1^{2F}$$

$$(11.5)$$

$$M_2^2 = \frac{2EI_2}{l_2} \theta_1^2 + \frac{4EI_2}{l_2} \theta_2^2 + M_2^{2F}$$

$$M_1^3 = \frac{4EI_3}{l_3} \theta_1^3 + \frac{2EI_3}{l_3} \theta_2^3 + M_1^{3F}$$

$$M_2^3 = \frac{2EI_3}{l_3} \theta_1^3 + \frac{4EI_3}{l_3} \theta_2^3 + M_2^{3F}$$

where I_i and l_i are the moment of inertia and length of the ith member.

Compatibility of Rotations

Because of the continuity of the beam at the supports, compatibility of the rotations requires that

$$\theta_1^1 = r_1 \qquad \theta_2^1 = \theta_1^2 = r_2 \qquad \theta_2^2 = \theta_1^3 = r_3 \qquad \theta_2^3 = r_4 \qquad (11.6)$$

Equilibrium of the Joints in Terms of Joint Rotations

Substitution of these compatibility relations into the member-end moments, Eqs. (11.5), and in turn substitution of the member-end moments found into the equilibrium equations, Eqs. (11.4), gives

$$\frac{4EI_1}{l_1} r_1 + \frac{2EI_1}{l_1} r_2 = M_1 - M_1^{1F}$$

$$\frac{2EI_1}{l_1} r_1 + \left(\frac{4EI_1}{l_1} + \frac{4EI_2}{l_2}\right) r_2 + \frac{2EI_2}{l_2} r_3 = M_2 - (M_2^{1F} + M_1^{2F})$$

$$\frac{2EI_3}{l_3} r_2 + \left(\frac{4EI_3}{l_3} + \frac{4EI_2}{l_2}\right) r_3 + \frac{2EI_2}{l_2} r_4 = M_3 - (M_2^{2F} + M_1^{3F})$$

$$\frac{2EI_3}{l_3} r_3 + \frac{4EI_3}{l_3} r_4 = M_4 - M_2^{3F}$$

$$(11.7)$$

We can see from the above equations that, if two members are connected at a given joint, then their stiffness terms relating to the shared degree of freedom are added in the structure stiffness matrix. Defining the negative of the sum of the fixed-end moments as equivalent joint moments, we can also see that the terms on the right-hand side of the equilibrium equations are sums of the applied moments at the joints and the equivalent joint moments. The above equilibrium equations in matrix form become

$$
\begin{bmatrix}
\dfrac{4EI_1}{l_1} & \dfrac{2EI_1}{l_1} & 0 & 0 \\[2ex]
\dfrac{2EI_1}{l_1} & \left(\dfrac{4EI_1}{l_1}+\dfrac{4EI_2}{l_2}\right) & \dfrac{2EI_2}{l_2} & 0 \\[2ex]
0 & \dfrac{2EI_2}{l_2} & \left(\dfrac{4EI_2}{l_2}+\dfrac{4EI_3}{l_3}\right) & \dfrac{2EI_3}{l_3} \\[2ex]
0 & 0 & \dfrac{2EI_3}{l_3} & \dfrac{4EI_3}{l_3}
\end{bmatrix}
\begin{bmatrix} r_1 \\ r_2 \\ r_3 \\ r_4 \end{bmatrix}
=
\begin{bmatrix}
M_1 - M_1^{1F} \\
M_2 - (M_2^{1F} + M_1^{2F}) \\
M_3 - (M_2^{2F} + M_1^{3F}) \\
M_4 - M_2^{3F}
\end{bmatrix}
$$

(11.8a)

or

$$\mathbf{K}_s \mathbf{r} = \mathbf{R} \tag{11.8b}$$

where $\mathbf{K}_s$ is the structure stiffness matrix, $\mathbf{r}$ is the vector of the joint rotations, and $\mathbf{R}$ is the vector of joint moments also called the external force vector or load vector.

The terms of the load vector on the right-hand side of Eq. (11.8a) have the form

$$R_i = M_i - \sum M_i^F \tag{11.8c}$$

where M_i is the external moment applied at the joint and $-\sum M_i^F$ is the equivalent joint moment, which is the negative of sum of the fixed-end moments at the joint. Note that, in the abence of in-span loads, the equivalent joint moments are zero and the load vector consists of the applied joint moments.

Note also that the stiffness matrix is always symmetric, whether for an element or a structure (Eqs. (11.2a) and (11.8a)). This can be proved by Maxwell–Betti's reciprocal theorem, section 6.5.

Direct Stiffness Method

We observe that in the equilibrium equations (Eq. (11.8a)) when two elements contribute to the same degree of freedom, their pertinent stiffness terms are added together to form the terms in the structure stiffness matrix associated with that degree of freedom. Similarly, each term of the external force vector in Eq. (11.8a) is the sum of the applied joint load associated with that degree of freedom plus the equivalent joint load (negative of sum of the fixed-end moments at that degree of freedom). This observation enables us to assemble the structure stiffness matrix and the load vector directly without having to go through the lengthy steps taken to obtain Eq. (11.8a). Thus, to find the structure stiffness matrix and load vector needed to write the equilibrium equation for the structure, we first find the moment-rotation relations Eq. ((11.2a)), for all the elements. We then construct the structure stiffness matrix by summing the terms, from each element stiffness matrix, contributing to each degree of freedom. Similarly, we assemble the external force vector by summing the applied external moment and the equivalent joint moments for each degree of freedom. The mechanics of this process will be discussed below, under steps in the direct stiffness method for beams. In the assembly process the terms associated with zero joint rotations can be left out in order to reduce the size of the structure stiffness matrix and the number of equilibrium equations. This in effect amounts to applying the boundary conditions. Thus, the size of the structure stiffness matrix and the external force vector will be the same as the number of degrees of freedom, with each term consisting of contributions from all the elements sharing the same degree of freedom.

Joint Rotations and Member-End Moments

The equilibrium equations (Eq. (11.8a)) can be solved for the joint rotations r_i. Once the joint rotations are known, the compatibility equations (Eq. (11.6)) can be used to relate the joint rotations to the member-end rotations. Then using the moment-rotation relations (Eq. (11.5)), or their general form (Eq. (11.2a)), the member-end moments can be determined. For example, for member 1 of Fig. 11.5 we have

$$\begin{bmatrix} \theta_1^1 \\ \theta_2^1 \end{bmatrix} = \begin{bmatrix} r_1 \\ r_2 \end{bmatrix}$$

and

$$\begin{bmatrix} M_1^1 \\ M_2^1 \end{bmatrix} = \frac{2EI_1}{l_1} \begin{bmatrix} 2 & 1 \\ 1 & 2 \end{bmatrix} \begin{bmatrix} r_1 \\ r_2 \end{bmatrix} + \begin{bmatrix} M_1^F \\ M_2^F \end{bmatrix}^1$$

Steps in the Direct Stiffness Method for Beams

The relations obtained in the previous sections are used in applying the direct stiffness method. The procedure described below is for hand calculations. However, the use of matrix manipulation computer programs or spread sheets would expedite the process. Special considerations for developing computer programs using the stiffness method are discussed in Section 19.2. The steps in the solution of beams by the stiffness method are as follows.

   Step 1. Number the elements (spans) 1 to m and denote the degrees of freedom by r_1 to r_n on the joints that can undergo rotations. The joints must also be numbered using a separate sequence of numbers. The first and second joints associated with each element establish the axis of the element and the positive direction of the member end moments.

Step 2. Using Eq. (11.2a), reproduced below, calculate the moment–rotation relations for each element.

$$\begin{matrix} \theta_1 & \theta_2 \end{matrix}$$

$$\begin{bmatrix} M_1 \\ M_2 \end{bmatrix} = \frac{2EI}{l} \begin{bmatrix} 2 & 1 \\ 1 & 2 \end{bmatrix} \begin{bmatrix} \theta_1 \\ \theta_2 \end{bmatrix} + \begin{bmatrix} M_1^F \\ M_2^F \end{bmatrix}$$

The fixed-end moments in the above equation can be found by the formulas in Tables 10.1a, 10.1b, by Eqs. (10.14) to (10.16), or by any other method. For each element, replace θ_1 and θ_2 in the above equation by r_i, the rotation degrees of freedom of the joints associated with that element. If an end of a member is fixed, then a zero is used for the degree of freedom. For example, if the left end rotation of element 3 corresponds to degree of freedom r_2 and the right end is fixed and no load is acting over the element, then the force–displacement relation for this element would be

$$\begin{matrix} r_2 & 0 \end{matrix}$$

$$\begin{bmatrix} M_1 \\ M_2 \end{bmatrix}^3 = \frac{2EI_2}{l_2} \begin{bmatrix} 2 & 1 \\ 1 & 2 \end{bmatrix} \begin{bmatrix} r_2 \\ 0 \end{bmatrix}$$

Step 3. The equilibrium equations are in terms of an $n \times n$ structure stiffness matrix $\mathbf{K}_s$, the joint rotation vector $\mathbf{r}$, and an $n \times 1$ external force vector $\mathbf{R}$. Set up the equilibrium in matrix form as follows. Indicate the n degrees of freedom r_1 to r_n above the structure stiffness matrix. Assemble the structure stiffness matrix by adding to the location (r_i, r_j) of the structure stiffness matrix the terms of element stiffness matrices calculated in the element moment–rotation relations, developed in step 2. Note that the location where each term gets added in the structure stiffness is indicated by r_i above the element stiffness matrices and to their right (in the vector $\mathbf{r}$). Thus, by going through all the element stiffness matrices, assembly of the structure stiffness matrix is completed. The elements of the external force vector for each degree of freedom r_i is equal to the applied moment in that degree of freedom, plus the equivalent joint moments (negatives of sum of the fixed-end moments at the joint). Note that the fixed-end moments are calculated in step 2.

$$\begin{matrix} r_1 & r_2 & \cdots & r_n \end{matrix}$$

$$\begin{bmatrix} & & \end{bmatrix} \begin{bmatrix} r_1 \\ r_2 \\ \vdots \\ r_n \end{bmatrix} = \begin{bmatrix} M_1 - \sum M_1^F \\ \vdots \\ M_n - \sum M_n^F \end{bmatrix}$$

or

$$\mathbf{K}_s \mathbf{r} = \mathbf{R}$$

Step 4. Solve the equilibrium equations to find the joint rotations $\mathbf{r}$.

Step 5. Using the moment–rotation relations in step 2, and substituting for the rotations r_i found in step 4, calculate the end moments for each element.

Example 11.1. In this example we will use the direct stiffness procedure outlined above to assemble the structure stiffness matrix and external force vector in terms of the parameters of the beam in Fig. 11.5a, to show that we get the same result as we found by the method of consistent deformations leading to Eq. (11.8a). For the three-span beam of Fig. 11.5a, with the moments M_1 to M_4 applied at joints 1 to 4, and the applied loads, find the equilibrium equations from which the joint rotations can be found. Write the equilibrium equations in terms of parameters I_i and l_i for each span.

Solution. There are three elements (spans) numbered 1 to 3 from left to right. The degrees of freedom, rotations of joints 1 to 4, are indicated by r_1 through r_4 in Fig. 11.5a.

Following the steps outlined above, the moment rotation relations for the three elements are as follows. In these equations the left and right angles θ_1 and θ_2 for each element are replaced by the rotations of the pertinent joints r_i. The superscripts indicate the element numbers.

$$\begin{matrix} & r_1 & r_2 & \\ \begin{bmatrix} M_1^1 \\ M_2^1 \end{bmatrix} = \frac{2EI_1}{l_1} \begin{bmatrix} 2 & 1 \\ 1 & 2 \end{bmatrix} \begin{bmatrix} r_1 \\ r_2 \end{bmatrix} + \begin{bmatrix} M_1^{1F} \\ M_2^{1F} \end{bmatrix} \end{matrix}$$

$$\begin{matrix} & r_2 & r_3 & \\ \begin{bmatrix} M_1^2 \\ M_2^2 \end{bmatrix} = \frac{2EI_2}{l_2} \begin{bmatrix} 2 & 1 \\ 1 & 2 \end{bmatrix} \begin{bmatrix} r_2 \\ r_3 \end{bmatrix} + \begin{bmatrix} M_1^{2F} \\ M_2^{2F} \end{bmatrix} \end{matrix}$$

$$\begin{matrix} & r_3 & r_4 & \\ \begin{bmatrix} M_1^3 \\ M_2^3 \end{bmatrix} = \frac{2EI_3}{l_3} \begin{bmatrix} 2 & 1 \\ 1 & 2 \end{bmatrix} \begin{bmatrix} r_3 \\ r_4 \end{bmatrix} + \begin{bmatrix} M_1^{3F} \\ M_2^{3F} \end{bmatrix} \end{matrix}$$

Since there are four degrees of freedom (possible joint rotations), the structure stiffness matrix is 4×4, and the load vector is 4×1. Starting from element 1, we add the terms of this element stiffness matrix to the locations of the structure stiffness matrix indicated on the latter matrix. Thus, the term in the first row identified by r_1 and the column identified by r_1 is written at the location (r_1, r_1) of the structure stiffness matrix. Note that in assembling the structure stiffness matrix, if a value already exists at that location the new value will be added to it. Continuing this process for all the element moment–rotation relations, that is assembling all the element stiffness matrices in the structure stiffness matrix results in the left matrix below.

Similarly, each element of the external force vector is equal to the applied moment minus the sum of the fixed-end moments acting at that joint. The first term of the force vector is thus $M_1 - M_1^{1F}$; the second term, at joint 2, is

$M_2 - (M_2^{1M} + M_1^{2F})$, etc. Thus, the equilibrium equations are

$$
\begin{array}{cccc}
r_1 & r_2 & r_3 & r_4
\end{array}
$$

$$
\begin{bmatrix}
\dfrac{4EI_1}{l_1} & \dfrac{2EI_1}{l_1} & 0 & 0 \\[2mm]
\dfrac{2EI_1}{l_1} & \left(\dfrac{4EI_1}{l_1} + \dfrac{4EI_2}{l_2}\right) & \dfrac{2EI_2}{l_2} & 0 \\[2mm]
0 & \dfrac{2EI_2}{l_2} & \left(\dfrac{4EI_2}{l_2} + \dfrac{4EI_3}{l_3}\right) & \dfrac{2EI_3}{l_3} \\[2mm]
0 & 0 & \dfrac{2EI_3}{l_3} & \dfrac{4EI_3}{l_3}
\end{bmatrix}
\begin{bmatrix} r_1 \\[2mm] r_2 \\[2mm] r_3 \\[2mm] r_4 \end{bmatrix}
=
\begin{bmatrix}
M_1 - M_1^{1F} \\[2mm]
M_2 - (M_2^{1F} + M_1^{2F}) \\[2mm]
M_3 - (M_3^{2F} + M_1^{3F}) \\[2mm]
M_4 - M_2^{3F}
\end{bmatrix}
$$

This equation is identical to Eq. 11.8a found by the method of consistent deformations.

Example 11.2. Find the solution of the supported cantilever in Fig. 11.6a using the stiffness method.

Solution. In this case we have one element with only one degree of freedom, r_1, the rotation of the right end. There is no externally applied moment at this joint. Thus, the only joint load is the equivalent joint load, the negative of the fixed-end moment at the right end. The fixed-end moments are found from Table 10.1a. The moment–rotation equation for this element is

$$
\begin{array}{cc} 0 & r_1 \end{array}
$$

$$
\begin{bmatrix} M_1 \\ M_2 \end{bmatrix} = \frac{2EI}{l} \begin{bmatrix} 2 & 1 \\ 1 & 2 \end{bmatrix} \begin{bmatrix} 0 \\ r_1 \end{bmatrix} + \begin{bmatrix} -\dfrac{pl^2}{12} \\[2mm] \dfrac{pl^2}{12} \end{bmatrix}
$$

The zero in the above equation indicates that the left joint is fixed and thus its rotation is zero. The rotation of the right end is not zero, but equal to r_1; r_1 will be found in the solution process.

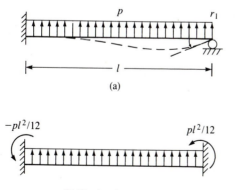

(a)

$-pl^2/12$ $pl^2/12$

(b) Fixed-end moments   **FIGURE 11.6**

Since there is only one degree of freedom, r_1, as shown in Fig. 11.6a, the structure stiffness matrix is 1×1. Referring to the moment–rotation relation for the element, we can see that the structure stiffness term (r_1, r_1) is $4EI/l$, and the equivalent joint load associated with r_1 is $-(pl^2/12)$. Thus, the equilibrium equation is

$$\frac{4EI}{l} r_1 = -\frac{pl^2}{12}$$

From the above equation the rotation of the right end is $r_1 = -pl^3/48EI$. The member-end moments are found by using the above moment–rotation relation upon substitution for r_1. This gives

$$\begin{bmatrix} M_1 \\ M_2 \end{bmatrix} = \frac{2EI}{l} \begin{bmatrix} 2 & 1 \\ 1 & 2 \end{bmatrix} \begin{bmatrix} 0 \\ -pl^3/48EI \end{bmatrix} + \begin{bmatrix} -pl^2/12 \\ pl^2/12 \end{bmatrix} = \begin{bmatrix} -pl^2/8 \\ 0 \end{bmatrix}$$

Thus, the left end moment is $-pl^2/8$, the right end rotation is $-pl^3/48EI$ and as expected, the right end moment is zero.

Example 11.3. Find the joint rotations and member end moments for the beam of Fig. 11.7a under two 20 kip loads and draw the moment diagram.

Solution. The element numbers and degrees of freedom are indicated in Fig. 11.7a. The magnitude of the fixed-end moments from Table 10.1a is $Pl/8$. Thus, the fixed-end moments for member 1 with the analysis sign convention are

$$M_1^F = -M_2^F = \frac{-Pl}{8} = \frac{20 \times 40}{8} = 100 \text{ kip} \cdot \text{ft}$$

and for member 2,

$$M_1^F = -M_2^F = \frac{20 \times 20}{8} = 50 \text{ kip} \cdot \text{ft}$$

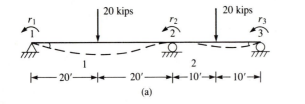

(a)

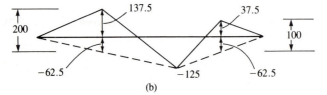

(b)

FIGURE 11.7

Using Eq. (11.2a), the moment–rotation relations for the elements are

$$\begin{bmatrix} M_1 \\ M_2 \end{bmatrix}^1 = EI \begin{matrix} r_1 \quad\quad r_2 \\ \begin{bmatrix} 0.1 & 0.05 \\ 0.05 & 0.1 \end{bmatrix} \end{matrix} \begin{bmatrix} r_1 \\ r_2 \end{bmatrix} + \begin{bmatrix} 100 \\ -100 \end{bmatrix}$$

$$\begin{bmatrix} M_1 \\ M_2 \end{bmatrix}^2 = EI \begin{matrix} r_2 \quad\quad r_3 \\ \begin{bmatrix} 0.2 & 0.1 \\ 0.1 & 0.2 \end{bmatrix} \end{matrix} \begin{bmatrix} r_2 \\ r_3 \end{bmatrix} + \begin{bmatrix} 50 \\ -50 \end{bmatrix}$$

Note that the superscripts on the moment vectors denote the element numbers.

The structure stiffness matrix is assembled from the terms of the element stiffness matrices in the above relations. The load vector consists of the equivalent joint moments only. From the above relations, for degree of freedom r_1, the equivalent joint moment is -100 kip · ft. For degree of freedom r_2, joint 2, the equivalent joint moment is $-(-100 + 50) = 50$ kip · ft; and for r_3 it is 50 kip · ft. Thus, the equilibrium equations are

$$EI \begin{matrix} r_1 \quad\quad\quad r_2 \quad\quad r_3 \\ \begin{bmatrix} 0.1 & 0.05 & 0 \\ 0.05 & (0.1 + 0.2) & 0.1 \\ 0 & 0.1 & 0.2 \end{bmatrix} \end{matrix} \begin{bmatrix} r_1 \\ r_2 \\ r_3 \end{bmatrix} = \begin{bmatrix} -100 \\ 50 \\ 50 \end{bmatrix}$$

Solving the above equations we find the joint rotations,

$$\begin{bmatrix} r_1 \\ r_2 \\ r_3 \end{bmatrix} = \frac{1}{EI} \begin{bmatrix} -1\,166.66 \\ 333.33 \\ 83.33 \end{bmatrix}$$

Using these values in the element moment-rotation relations we find the member-end moments,

$$\begin{bmatrix} M_1 \\ M_2 \end{bmatrix}^1 = EI \begin{bmatrix} 0.1 & 0.05 \\ 0.05 & 0.1 \end{bmatrix} \frac{1}{EI} \begin{bmatrix} 1\,166.66 \\ 333.33 \end{bmatrix} + \begin{bmatrix} 100 \\ -100 \end{bmatrix} = \begin{bmatrix} 0 \\ -125 \end{bmatrix}$$

$$\begin{bmatrix} M_1 \\ M_2 \end{bmatrix}^2 = EI \begin{bmatrix} 0.2 & 0.1 \\ 0.1 & 0.2 \end{bmatrix} \frac{1}{EI} \begin{bmatrix} 333.33 \\ -83.33 \end{bmatrix} + \begin{bmatrix} 50 \\ -50 \end{bmatrix} = \begin{bmatrix} 125 \\ 0 \end{bmatrix}$$

Before drawing the moment diagram, the sign of the moments must be converted to that of design convention. For the latter, moments will be positive when they cause compression at the top of the beam. Thus, the sign of moments at the left of each member must be changed. Therefore, the moment on the left end of the second span becomes -125 kip · ft. With these end moments and the applied loads we can draw the moment diagram by superimposing the diagrams due to the support moments and that of the applied loads (Section 7.2). The moment diagram is shown in Fig. 11.7b.

Example 11.4. Find the moments at the supports of the beam in Fig. 11.8a and draw the moment diagram.

Solution. In dealing with beams with overhangs, considering rotational degrees

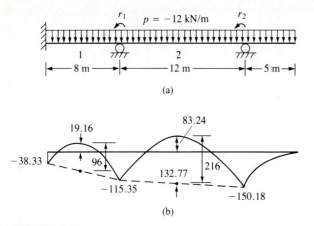

FIGURE 11.8

of freedom only, we can represent the effect of the overhang by including its fixed-end moment (as a cantilever) in the equivalent joint moment. After the stiffness solution is completed and the member-end moments are found, we must include the overhang and its load in drawing the moment diagram.

The element numbers and degrees of freedom are indicated in Fig. 11.8a. In this case there are only two possible joint rotations, that is, two degrees of freedom. The element moment–rotation relations are (Eq. (11.2a))

$$\begin{bmatrix} M_1 \\ M_2 \end{bmatrix}^1 = EI\begin{bmatrix} 0.50 & 0.25 \\ 0.25 & 0.50 \end{bmatrix}\begin{bmatrix} 0 \\ r_1 \end{bmatrix} + \begin{bmatrix} 64 \\ -64 \end{bmatrix}$$

$$\begin{bmatrix} M_1 \\ M_2 \end{bmatrix}^2 = EI\begin{bmatrix} 0.333 & 0.167 \\ 0.167 & 0.333 \end{bmatrix}\begin{bmatrix} r_1 \\ r_2 \end{bmatrix} + \begin{bmatrix} 144 \\ -144 \end{bmatrix}$$

The degrees of freedom associated with the terms of the element stiffness matrices are shown above the matrices. The structure stiffness matrix is 2×2 and is found by adding the pertinent terms from the element stiffness matrices. The elements of the external force vector consist of the equivalent joint moments, the negatives of the fixed-end moments. Note that the fixed-end moment for the cantilever is included with the fixed-end forces at joint 3. For a uniformly distributed load this moment is $pl^2/2 = 12 \times 5^2/2 = 150 \text{ kN} \cdot \text{m}$. Thus, the load vector is as shown in the following equilibrium equations,

$$EI\begin{bmatrix} (0.5 + 0.333) & 0.167 \\ 0.167 & 0.333 \end{bmatrix}\begin{bmatrix} r_1 \\ r_2 \end{bmatrix} = \begin{bmatrix} -80 \\ -6 \end{bmatrix}$$

The solution of the equilibrium equations gives

$$\begin{bmatrix} r_1 \\ r_2 \end{bmatrix} = \frac{1}{EI}\begin{bmatrix} -102.7 \\ 33.4 \end{bmatrix}$$

The member-end moments are found by using the above moment–rotation relations upon substituting the values for the joint rotations just found. Hence, the member-end moments for the two elements are

$$\begin{bmatrix} M_1 \\ M_2 \end{bmatrix}^1 = EI \begin{bmatrix} 0.5 & 0.25 \\ 0.25 & 0.5 \end{bmatrix} \begin{bmatrix} 0 \\ -102.7/EI \end{bmatrix} + \begin{bmatrix} 64 \\ -64 \end{bmatrix} = \begin{bmatrix} 38.33 \\ -115.35 \end{bmatrix}$$

$$\begin{bmatrix} M_1 \\ M_2 \end{bmatrix}^2 = EI \begin{bmatrix} 0.333 & 0.167 \\ 0.167 & 0.333 \end{bmatrix} \frac{1}{EI} \begin{bmatrix} -102.7 \\ 33.4 \end{bmatrix} + \begin{bmatrix} 144 \\ -144 \end{bmatrix}$$

$$= \begin{bmatrix} 115.35 \\ -150.00 \end{bmatrix}$$

To draw the moment diagram we convert the sign of the moments to that of the design convention by changing the sign of the moments on the left ends of the elements. The moment diagram is drawn in Fig. 11.8b. The moment at the mid-spans is equal to the difference between the mid-span moment of a simply supported beam with the applied load ($pl^2/8$ here) and the negative moment at the mid-span, that is, the average of the negative member-end moments.

11.3 BEAM ELEMENTS WITH DISPLACEMENT AND ROTATIONAL DEGREES OF FREEDOM

The stiffness method using beam elements with rotational degrees of freedom only can be applied to most beam problems. However, in some cases, such as beams with overhangs, it is more convenient to use beam elements that incorporate displacement as well as rotational degrees of freedom. With this element the deflections at the ends of cantilevers are found directly. In addition, forces and deflections at other points in the span can be found by using more than one element per span.

The 4 × 4 Beam Stiffness Matrix

The force–displacement relations, which relate the end forces to the end displacements, were derived in Section 10.2 (Eqs. (10.6), (10.7), (10.10), and (10.11)). These are

$$V_1 = \frac{12EI}{l^3} v_1 + \frac{6EI}{l^2} \theta_1 - \frac{12EI}{l^3} v_2 + \frac{6EI}{l^2} \theta_2 + V_1^F \qquad (11.9a)$$

$$M_1 = \frac{6EI}{l^2} v_1 + \frac{4EI}{l} \theta_1 - \frac{6EI}{l^2} v_2 + \frac{2EI}{l} \theta_2 + M_1^F \qquad (11.9b)$$

$$V_2 = -\frac{12EI}{l^3} v_1 - \frac{6EI}{l^2} \theta_1 + \frac{12EI}{l^3} v_2 - \frac{6EI}{l^2} \theta_2 + V_2^F \qquad (11.9c)$$

$$M_2 = \frac{6EI}{l^2} v_1 + \frac{2EI}{l} \theta_1 - \frac{6EI}{l^2} v_2 + \frac{4EI}{l} \theta_2 + M_2^F \qquad (11.9d)$$

Writing Eqs. (11.9) in matrix form gives

$$
\begin{bmatrix} V_1 \\ M_1 \\ V_2 \\ M_2 \end{bmatrix} = \begin{bmatrix} \dfrac{12EI}{l^3} & \dfrac{6EI}{l^2} & -\dfrac{12EI}{l^3} & \dfrac{6EI}{l} \\[2mm] \dfrac{6EI}{l^2} & \dfrac{4EI}{l} & -\dfrac{6EI}{l^2} & \dfrac{2EI}{l} \\[2mm] -\dfrac{12EI}{l^3} & -\dfrac{6EI}{l^2} & \dfrac{12EI}{l^3} & -\dfrac{6EI}{l^2} \\[2mm] \dfrac{6EI}{l^2} & \dfrac{2EI}{l} & -\dfrac{6EI}{l^2} & \dfrac{4EI}{l} \end{bmatrix} \begin{bmatrix} v_1 \\ \theta_1 \\ v_2 \\ \theta_2 \end{bmatrix} + \begin{bmatrix} V_1^F \\ M_1^F \\ V_2^F \\ M_2^F \end{bmatrix} \qquad (11.10a)
$$

or, in compact form,

$$ \mathbf{S} = \mathbf{Ks} + \mathbf{S}^F \qquad (11.10b) $$

where $\mathbf{S}$ and $\mathbf{s}$ are the force and displacement vectors. $\mathbf{K}$ is the 4×4 stiffness matrix for a beam element including displacement and rotational degrees of freedom, and $\mathbf{S}^F$ is the vector of fixed-end forces. Note that the compact form of the force–displacement relations is the same for beam elements with two and four degrees of freedom (Eqs. (11.2b) and (11.10b)). In fact, this form is the same for every element, although the size of the matrices and vectors and their terms will differ for the different elements.

The degrees of freedom in the above equations are the displacements and rotations of the left and right ends of the element and in that order. Note that each column of the stiffness matrix in Eq. (11.10a) may be obtained by setting one of the degrees of freedom (end displacement or rotation) equal to unity while all the other degrees of freedom are kept at zero. The end forces found by solving for such a beam with its boundary conditions will give the terms in the column of the stiffness matrix related to that degree of freedom. Each term of the stiffness matrix is the force in the corresponding row of the force vector due to a unit value of the corresponding displacement while all the other end displacements are kept at zero. For example, the term K_{24} (stiffness term in row 2, column 4) is the moment M_1 due to a unit rotation θ_2, while all the other end displacements are zero.

Solution Procedure

The solution procedure is the same as with the 2×2 stiffness matrix outlined in Section 11.2, except for the addition of displacement degrees of freedom. In numbering the degrees of freedom, all the degrees of freedom for a given joint must be completed before going on to the next joint. At each joint the displacement degree of freedom is numbered before the rotational degree of freedom.

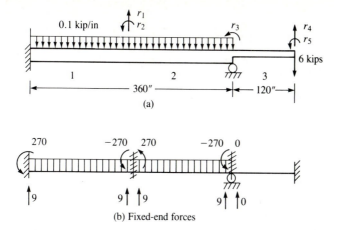

(a)

(b) Fixed-end forces

FIGURE 11.9

Example 11.4. Using three elements as shown in Fig. 11.9a, find the displacements and rotations of the joints and the member-end forces. The length of elements 1 and 2 is 180 in. The moment of inertia for the first span is 200 in⁴ and for the overhang 100 in⁴. The modulus of elasticity is 29 000 ksi.

Solution. The element numbers and the degrees of freedom are shown in Fig. 11.9a. As mentioned before, when there are both displacement and rotational degrees of freedom at a joint, the displacement is numbered before the rotation. For this beam there are five degrees of freedom. Since there are in-span loads, the solution involves the fixed-end forces. The fixed-end forces are calculated using Table 10.1a and shown in Fig. 11.9b.

The force–displacement relations for the elements are found by Eq. (11.10a). The terms in columns 1 and 2 of the stiffness matrices are associated with the displacement and rotation of the left end, and those in columns 3 and 4 are associated with the displacement and rotation at the right end of the element. The structure degrees of freedom associated with the displacements and rotations of the left and right ends of the elements are also shown above the stiffness matrices.

$$
\begin{bmatrix} V_1 \\ M_1 \\ V_2 \\ M_2 \end{bmatrix}^1 = \begin{bmatrix} 11.93 & 1\,074.07 & -11.93 & 1\,074.07 \\ 1\,074.07 & 128\,888.8 & -1\,074.07 & 64\,444.44 \\ -11.93 & -1\,074.07 & 11.934\,25 & -1\,074.07 \\ 1\,074.07 & 64\,444.44 & -1\,074.07 & 128\,888.8 \end{bmatrix} \begin{bmatrix} 0 \\ 0 \\ r_1 \\ r_2 \end{bmatrix} + \begin{bmatrix} 9 \\ 270 \\ 9 \\ -270 \end{bmatrix}
$$

$$
\begin{bmatrix} V_1 \\ M_1 \\ V_2 \\ M_2 \end{bmatrix}^2 = \begin{bmatrix} 11.93 & 1\,074.07 & -11.93 & 1\,074.07 \\ 1\,074.07 & 128\,888.8 & -1\,074.07 & 64\,444.44 \\ -11.93 & -1\,074.07 & 11.934\,15 & -1\,074.07 \\ 1\,074.07 & 64\,444.44 & -1\,074.07 & 128\,888.8 \end{bmatrix} \begin{bmatrix} r_1 \\ r_2 \\ 0 \\ r_3 \end{bmatrix} + \begin{bmatrix} 9 \\ 270 \\ 9 \\ -270 \end{bmatrix}
$$

$$
\begin{bmatrix} V_1 \\ M_1 \\ V_2 \\ M_2 \end{bmatrix}^3 = \begin{bmatrix} \overset{0}{20.13} & \overset{r_3}{1\,208.33} & \overset{r_4}{-20.13} & \overset{r_5}{1\,208.33} \\ 1\,208.33 & 96\,666.66 & -1\,208.33 & 48\,333.33 \\ -20.13 & -1\,208.33 & 20.13 & -1\,208.33 \\ 1\,208.33 & 48\,333.33 & -1\,208.33 & 96\,666.66 \end{bmatrix} \begin{bmatrix} 0 \\ r_3 \\ r_4 \\ r_5 \end{bmatrix}
$$

The structure stiffness matrix for this problem is 5×5, because there are 5 degrees of freedom. The terms of the structure stiffness matrix are found by adding the terms from the element stiffness matrices. The location to which each term of the element stiffness matrix is added in the structure stiffness matrix is indicated by the degree of freedom above and to the right of the element and structure stiffness matrices. Thus, the term $1, 1$ (indicated by r_1, r_1) in the structure stiffness matrix is the sum of the terms $1, 1$ (indicated by r_1, r_1) in the first element stiffness matrix, $\mathbf{K}_1$, plus the term $1, 1$ in $\mathbf{K}_2$. There is no term $1, 1$ in $\mathbf{K}_3$. The other terms of the structure stiffness matrix are found in a similar manner. The elements of the external force vector consist of the equivalent joint forces (the negatives of sum of the fixed-end forces at that joint), except for the -6 kip vertical joint load which is the applied load at joint 3. The equilibrium equations are

$$
\begin{bmatrix}
\overset{r_1}{(11.93+11.93)} & \overset{r_2}{(-1\,074.07+1\,074.07)} & \overset{r_3}{1\,074.07} & \overset{r_4}{0} & \overset{r_5}{0} \\
(-1\,074.07+1\,074.07) & (128\,888.8+128\,888.8) & 64\,444.44 & 0 & 0 \\
1\,074.07 & 64\,444.44 & (128\,888.8+96\,666.66) & -1\,208.33 & 48\,333.33 \\
0 & 0 & -1\,208.33 & 20.13 & -1\,208.33 \\
0 & 0 & 48\,333.33 & -1\,208.33 & 96.666.66
\end{bmatrix}
\times
\begin{bmatrix} r_1 \\ r_2 \\ r_3 \\ r_4 \\ r_5 \end{bmatrix}
=
\begin{bmatrix} -(9+9) \\ -(270+270) \\ 270 \\ -6 \\ 0 \end{bmatrix}
$$

The solution of the above system of simultaneous equations gives the joint displacements as follows.

$$
\begin{bmatrix} r_1 \\ r_2 \\ r_3 \\ r_4 \\ r_5 \end{bmatrix} = \begin{bmatrix} -1.0055 \\ 0.0013 \\ 0.0056 \\ -0.5214 \\ -0.0093 \end{bmatrix}
$$

With these values, referring to Fig. 11.9a, we get the following element displacement vectors.

$$
\begin{array}{cccc}
\text{Element} \rightarrow & 1 & 2 & 3 \\
\begin{bmatrix} v_1 \\ \theta_1 \\ v_2 \\ \theta_2 \end{bmatrix} : &
\begin{bmatrix} 0 \\ 0 \\ -1.0055 \\ 0.0013 \end{bmatrix} &
\begin{bmatrix} -1.0055 \\ 0.0013 \\ 0 \\ 0.0056 \end{bmatrix} &
\begin{bmatrix} 0 \\ 0.0056 \\ -0.5214 \\ -0.0093 \end{bmatrix}
\end{array}
$$

Using the latter vectors in the element force–displacement relations results in the member end forces as

$$
\begin{array}{cccc}
\text{Element} \rightarrow & 1 & 2 & 3 \\
\begin{bmatrix} V_1 \\ M_1 \\ V_2 \\ M_2 \end{bmatrix} : &
\begin{bmatrix} 19.5 \\ 1\,260 \\ -1.5 \\ 630 \end{bmatrix} &
\begin{bmatrix} 1.5 \\ -630 \\ 16.5 \\ -720 \end{bmatrix} &
\begin{bmatrix} 6 \\ 720 \\ -6 \\ 0 \end{bmatrix}
\end{array}
$$

   To express the above forces in the design sign convention, for drawing shear and moment diagrams, we must change the sign of moments on the left of each member.

LABORATORY EXPERIMENTS

Element Stiffness Matrix

Using a single span beam element made of spring steel or a thicker strip, install strain gages near the two ends. Then using a lever with a gap that fits on the strip apply a moment at one end of the element and measure the rotation at that end without allowing displacement at that end or displacement and rotation at the other end. Measure the strains in the gages on both ends and calculate the moments (see Bending Strain–Curvature Relation in Laboratory Experiments of Chapter 5). Scale all the values so that the end rotation is 1 unit. Calculate the terms in the rotational stiffness matrix (Eq. (11.3) or with terms (2, 2) and (4, 2) in the stiffness matrix of Eq. (11.10a)) and compare with the measured moments.

   Next induce a small displacement at one end without allowing any rotation there, with no displacement and rotation at the other end. Find the moments from the measured strains and compare these values to terms (2, 1) and (4, 1) of the stiffness matrix in Eq. (11.10a).

Structure Stiffness Matrix

The equation relating any set of forces and moments to the associated displacements and rotations is in terms of a stiffness matrix.

Using a two-span beam, apply a load P at a point of one of the spans and measure the displacement v in the direction of the load. Calculate the stiffness $K = P/v$ using the values of P and v. Then using the stiffness method find the deflection due to the same load P. Calculate the value of K as the ratio of the load and the calculated deflection and compare it to the previous value.

PROBLEMS

11.1. A beam element is 10 ft long with rigidity EI and under a uniform load 2 kip/ft (Fig. P11.1). With kips and inches used as the units, its left and right end rotations are $6000/EI$ and $4500/EI$. Find the end moments M_1 and M_2 and then the end shear forces V_1 and V_2.

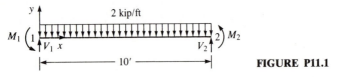

FIGURE P11.1

11.2. Show the degrees of freedom for the beam of Fig. P11.2 with the elements shown.

FIGURE P11.2

11.3. The expressions relating the end moments and rotations for the three elements of a structure are given below. The degrees of freedom are indicated on the member stiffness matrices. No concentrated loads are applied at the joints. Find the structure stiffness matrix and external force vector.

$$\begin{bmatrix} M_1 \\ M_1 \end{bmatrix}^1 = \begin{bmatrix} 120 & 60 \\ 60 & 120 \end{bmatrix}\begin{bmatrix} 0 \\ r_1 \end{bmatrix} + \begin{bmatrix} 100 \\ -100 \end{bmatrix} \qquad \begin{bmatrix} M_1 \\ M_1 \end{bmatrix}^2 = \begin{bmatrix} 200 & 100 \\ 100 & 200 \end{bmatrix}\begin{bmatrix} r_1 \\ r_2 \end{bmatrix} + \begin{bmatrix} 150 \\ -150 \end{bmatrix}$$

$$\begin{bmatrix} M_1 \\ M_1 \end{bmatrix}^3 = \begin{bmatrix} 200 & 100 \\ 100 & 200 \end{bmatrix}\begin{bmatrix} r_2 \\ 0 \end{bmatrix}$$

11.4. Part of the stiffness matrix for a structure is shown below. Fill in the terms of this matrix that are left blank.

$$K = EI \begin{bmatrix} 120 & 60 & -45 \\ & 400 & -10 \\ & & 280 \end{bmatrix}$$

11.5. The member-end moments calculated from the stiffness matrix for a beam are shown in Fig. P11.5. The first and second joints for each element are on the left and right of the span. Write the values of these moments in the design sign convention for drawing the moment diagram.

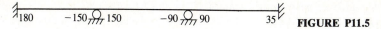

180   −150 150   −90 90   35   **FIGURE P11.5**

11.6. Find the moment at the middle support of the beam of Fig. 11.6 using the stiffness method. *EI* is constant.

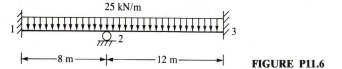

|←——8 m——→|←——12 m——→| **FIGURE P11.6**

11.7. Using the 2×2 beam stiffness matrix find the joint rotations and member-end moments for the beam of Fig. P11.7. The modulus of elasticity is $E = 29\,000$ ksi.

−200 kip·in   200 kip·in

$I = 300$ in^4 $I = 200$ in^4

|←———— 200″ ————→|←—— 120″ ——→|

FIGURE P11.7

11.8. Calculate the rotations of the beam in Fig. P11.8 at the supports and the moment at the middle support. Note that in addition to the loads a moment of 100 kip · ft is acting at the first support.

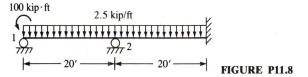

100 kip·ft   2.5 kip/ft

|←——— 20′ ———→|←——— 20′ ———→| **FIGURE P11.8**

11.9. For the beam of Fig. P11.9 with $E = 29\,000$ ksi and $I = 150$ in^4, using the 2×2 stiffness matrix:

(a) Find the joint rotations and member end moments.
(b) Draw the moment diagram.

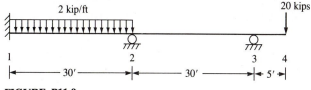

2 kip/ft   20 kips

|←———— 30′ ————→|←———— 30′ ————→|← 5′→|

FIGURE P11.9

11.10. Using the stiffness method, find the moments at the support of the beam in Fig. P11.10. *EI* is constant.

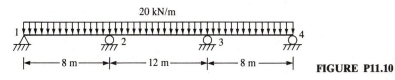

20 kN/m

|←——8 m——→|←——12 m——→|←—— 8 m ——→| **FIGURE P11.10**

11.11. A cantilever beam with a span of 150 in and $EI = 10^7$ kip · in² carries a uniform gravity load of 0.1 kip/in. The left end of this beam is fixed and the right end is free. Using the stiffness method, find the displacement and rotation of the free end and the moment and shear at the fixed end.

11.12. Find the vertical deflection of point 3 and the member-end forces for the beam of Fig. P11.12. $E = 3\,000$ ksi and $I = 400$ in⁴.

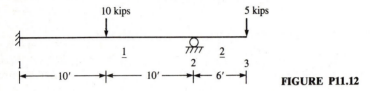

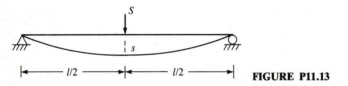

10 kips   5 kips

1   2   2   3

1

|← 10' →|← 10' →|← 6' →| **FIGURE P11.12**

11.13. The mid-span deflection s of the beam in Fig. P11.13 under a concentrated load S applied at the mid-span is $Sl^3/24EI = S/3.50$. What is the stiffness relation, $S = Ks$, for the force and deflection at the mid-span of the beam.

S

s

|← $l/2$ →|← $l/2$ →| **FIGURE P11.13**

11.14. A load P_1 and moment M_1 are applied at the free end of a cantilever causing a deflection v_1 and rotation θ_1. Find the force–displacement relation for this structure in terms of l, E, and I.

    Hint: Using stiffness or any other method for calculating the deflection, find the end deflection v_1 and rotation θ_1 in terms of P_1 and M_1. Solve these equations to find P_1 and M_1 in terms of v_1 and θ_1 and cast in matrix form.

11.15. For a beam element with constant cross-sections the shear forces V_1 and $V_2 = -V_1$ are applied at the two ends. Equation (6.24) gives $v_2 - v_1 = V_2l/GA_s$, where the shear area $A_s = A/\kappa$. Express the vector of the end shear forces in terms of the end displacements to find the stiffness matrix for shear deformation.

CHAPTER
12

STIFFNESS METHOD FOR TRUSSES AND FRAMES

The stiffness method provides a general procedure for the solution of trusses and frames. Small problems can be done by hand. More complex problems are best handled by computer. Computer programs using the stiffness method for two-dimensional trusses and frames are readily available. The steps in the development of such programs are delineated in Section 19.2. The discussion of frames in this chapter is divided into frames without sidesway and those with sidesway. Frames without sidesway, and with rotational degrees of freedom only, are treated in a manner similar to continuous beams with rotational degrees of freedom only. Frames with sidesway are also divided into square frames, for which a simple solution procedure can be worked out, and general frames with inclined members and axial deformations. Although the procedure for general frames can be used in all cases, the simpler solution techniques are included here because they are easier to use and help understand the behavior of frames.

The sign convention used in the application of the stiffness method for trusses and frames is the analysis convention discussed in Section 10.1. The positive directions of the coordinate axes and forces are to the right and upward, and positive moments and rotations are counterclockwise.

12.1 STIFFNESS METHOD FOR TRUSSES

Application of the stiffness method to trusses is similar to that for beams. However, in this case all the loads are applied at the joints. The arbitrary orientation of truss members requires transformation of the forces to global coordinates.

Development of the stiffness method for trusses involves the derivation of the truss element stiffness matrix, transformation of the coordinates, and equilibrium of the joints. The solution of equilibrium equations gives the joint displacements, which in turn lead to the member forces.

Truss Element Stiffness Matrix

Consider a bar under axial end forces U_1 and U_2 and end displacements u_1 and u_2. We found in Section 5.1 (Eq. (5.4c)) that

$$u_2 = \frac{U_2 l}{EA} + u_1$$

From Fig. 12.1 we see that by equilibrium $U_1 = -U_2$. Substituting this in the above equation and solving for U_1, we get

$$U_1 = \frac{EA}{l}(u_1 - u_2)$$

Also

$$U_2 = -\frac{EA}{l}(u_1 - u_2)$$

Expressing the above equations in matrix form, we find

$$\begin{bmatrix} U_1 \\ U_2 \end{bmatrix} = \frac{EA}{l}\begin{bmatrix} 1 & -1 \\ -1 & 1 \end{bmatrix}\begin{bmatrix} u_1 \\ u_2 \end{bmatrix} \tag{12.1a}$$

or in compact notation

$$\mathbf{S} = \mathbf{K}\mathbf{s} \tag{12.1b}$$

This is the force–displacement relation for a truss element. The matrix

$$\mathbf{K} = \frac{EA}{l}\begin{bmatrix} 1 & -1 \\ -1 & 1 \end{bmatrix} \tag{12.1c}$$

is the stiffness matrix for the truss element.

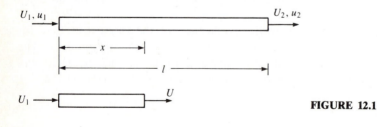

FIGURE 12.1

Transformation of Coordinates

In a truss the bars have different orientations. Thus, before we write the equilibrium of the joints, the bar forces must be decomposed into their components in a global coordinate system that is common to all the members. As mentioned before, the global axes are chosen in the horizontal and vertical directions, directed to the right and upward.

For a typical truss bar making an angle α with the horizontal direction (Fig. 12.2a), the axial deformations at the two ends can be expressed in terms of their components in the global coordinates. The components of the displacements and forces in the global coordinates are denoted by an overbar. The end displacements along the axis of the bar can be expressed in terms of their global displacements as

$$u_1 = \bar{u}_1 \cos \alpha + \bar{v}_1 \sin \alpha$$
$$u_2 = \bar{u}_2 \cos \alpha + \bar{v}_2 \sin \alpha$$

(12.2a)

Letting $c = \cos \alpha$, $s = \sin \alpha$, and casting Eq. (12.2a) in matrix form, we have

$$\begin{bmatrix} u_1 \\ u_2 \end{bmatrix} = \begin{bmatrix} c & s & 0 & 0 \\ 0 & 0 & c & s \end{bmatrix} \begin{bmatrix} \bar{u}_1 \\ \bar{v}_1 \\ \bar{u}_2 \\ \bar{v}_2 \end{bmatrix}$$

(12.2b)

or

$$\mathbf{s} = \mathbf{T}\bar{\mathbf{s}}$$

(12.2c)

The components of the end forces in the global coordinates are (Fig. 12.2b)

$$\bar{U}_1 = U_1 \cos \alpha \qquad \bar{V}_1 = U_1 \sin \alpha$$
$$\bar{U}_2 = U_2 \cos \alpha \qquad \bar{V}_2 = U_2 \sin \alpha$$

(12.3a)

$$\begin{bmatrix} \bar{U}_1 \\ \bar{V}_1 \\ \bar{U}_2 \\ \bar{V}_2 \end{bmatrix} = \begin{bmatrix} c & 0 \\ s & 0 \\ 0 & c \\ 0 & s \end{bmatrix} \begin{bmatrix} U_1 \\ U_2 \end{bmatrix}$$

(12.3b)

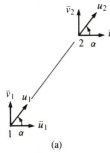

(a)

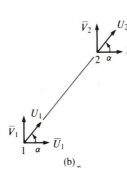

(b)

FIGURE 12.2

or

$$\bar{S} = T^T S \tag{12.3c}$$

Substituting Eq. (12.2b) into Eq. (12.1a) we have

$$\begin{bmatrix} U_1 \\ U_2 \end{bmatrix} = \frac{EA}{l} \begin{bmatrix} 1 & -1 \\ -1 & 1 \end{bmatrix} \begin{bmatrix} c & s & 0 & 0 \\ 0 & 0 & c & s \end{bmatrix} \begin{bmatrix} \bar{u}_1 \\ \bar{v}_1 \\ \bar{u}_2 \\ \bar{v}_2 \end{bmatrix} \tag{12.4a}$$

or

$$\begin{bmatrix} U_1 \\ U_2 \end{bmatrix} = \frac{EA}{l} \begin{bmatrix} c & s & -c & -s \\ -c & -s & c & s \end{bmatrix} \begin{bmatrix} \bar{u}_1 \\ \bar{v}_1 \\ \bar{u}_2 \\ \bar{v}_1 \end{bmatrix} \tag{12.4b}$$

and in compact form

$$S = B\bar{s} \tag{12.4c}$$

where

$$B = KT \tag{12.4d}$$

Equation (12.4b) is used to calculate the bar forces from the joint displacements in the global coordinates. Premultiplying both sides of Eq. (12.4b) by the transpose of the transformation matrix yields

$$\begin{bmatrix} c & 0 \\ s & 0 \\ 0 & c \\ 0 & s \end{bmatrix} \begin{bmatrix} U_1 \\ U_2 \end{bmatrix} = \frac{EA}{l} \begin{bmatrix} c & 0 \\ s & 0 \\ 0 & c \\ 0 & s \end{bmatrix} \begin{bmatrix} c & s & -c & -s \\ -c & -s & c & s \end{bmatrix} \begin{bmatrix} \bar{u}_1 \\ \bar{v}_1 \\ \bar{u}_2 \\ \bar{v}_2 \end{bmatrix}$$

Substituting from Eq. (12.3b) on the left-hand side of the above equation, and carrying the multiplication of the transformation matrices on the right-hand side, we obtain

$$\begin{bmatrix} \bar{U}_1 \\ \bar{V}_1 \\ \bar{U}_2 \\ \bar{V}_2 \end{bmatrix} = \frac{EA}{l} \begin{bmatrix} c^2 & cs & -c^2 & -cs \\ cs & s^2 & -cs & -s^2 \\ -c^2 & -cs & c^2 & cs \\ -cs & -s^2 & cs & s^2 \end{bmatrix} \begin{bmatrix} \bar{u}_1 \\ \bar{v}_1 \\ \bar{u}_2 \\ \bar{v}_2 \end{bmatrix} \tag{12.5a}$$

or

$$\bar{S} = \bar{K}\bar{s} \tag{12.5b}$$

This is the force–displacement relation in the global coordinates. The truss stiffness matrix in the global coordinate system is

$$K = \frac{EA}{l} \begin{bmatrix} c^2 & cs & -c^2 & -cs \\ cs & s^2 & -cs & -s^2 \\ -c^2 & -cs & c^2 & cs \\ -cs & -s^2 & cs & s^2 \end{bmatrix} \tag{12.5c}$$

In the above stiffness matrix c and s are the sine and cosine of the angle between the global and local x axes. The local x axis is directed from the first to the second end of the member. The first and second ends can be chosen arbitrarily. However, once selected, their order must be kept throughout the analysis. The angle is measured from the global to the local x axis in a counterclockwise sense. If X_1, Y_1 and X_2, Y_2 are the coordinates of the first and second ends in the global coordinates, then

$$c = \frac{X_2 - X_1}{l} \tag{12.6a}$$

and

$$s = \frac{Y_2 - Y_1}{l} \tag{12.6b}$$

where the length of the member is

$$l = \sqrt{(X_2 - X_1)^2 + (Y_2 - Y_1)^2} \tag{12.6c}$$

Assembly of Elements and Solution Procedure

Assembly of the elements and establishment of the equilibrium equations is done in the same way as for beams, Chapter 11. The steps in the process are as follows.

1. Number the elements, joints and degrees of freedom, that is, the possible horizontal and vertical displacements for each joint, in the global coordinate system. The horizontal degree of freedom is numbered before the vertical degree of freedom.
2. Find the element force–displacement relationships in the global coordinates, Eq. 12.5a, and write the degrees of freedom r_i in the element displacement vectors and above the columns of the stiffness matrices.
3. Set up the equilibrium equations in terms of the structure stiffness matrix and external force or load vector. The rank of the structure stiffness matrix is equal to the number of degrees of freedom. The elements of the structure stiffness matrix are found by adding the terms of the element stiffness matrices associated with that degree of freedom. The elements of the load vector consist of the joint loads acting in the degree of freedom associated with each term.
4. Solve the equilibrium equations to find the joint displacements.
5. Determine the element displacement vectors in the global coordinates from the above joint displacements, by adding zeros for the fixed joints.
6. Find the member-end forces by Eq. (12.4b).

> **Example 12.1.** Find the bar forces for the truss of Fig. 12.3 under a 10 kip vertical load at joint 2. The cross-sectional areas are $10\ \text{in}^2$ for the inclined bars and $20\ \text{in}^2$ for the horizontal bar. $E = 29\,000$ ksi.

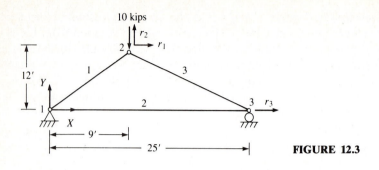

FIGURE 12.3

Solution. The element numbers are shown in Fig. 12.3. The degrees of freedom consist of the horizontal and vertical displacements of joint 2 and horizontal displacement of joint 3. These displacements are denoted by r_1, r_2, and r_3 in the global coordinate system. The lengths of the three bars are calculated from their projections. They are $l_1 = 15$ ft, $l_2 = 25$ ft, and $l_3 = 16$ ft.

The element force–displacement relations, in the global coordinates, are found from Eq. (12.5a). For element 1, by Eq. (12.6), $c = (9-0)/15 = 0.6$, $s = (12-0)/15 = 0.8$. Thus, the first element of the stiffness matrix, with units of kips and inches, is

$$\frac{EA_1}{l_1} c^2 = \frac{29\,000 \times 10}{15 \times 12} \times 0.6^2 = 580$$

The other terms of the element stiffness matrix are calculated in a similar manner. The force–displacement relations, in the global coordinate system, for the three elements are given below. The degrees of freedom are shown above the columns of the element stiffness matrices and in the element displacement vectors.

$$\begin{bmatrix} \bar{U}_1 \\ \bar{V}_1 \\ \bar{U}_2 \\ \bar{V}_2 \end{bmatrix}^1 = \begin{bmatrix} \overset{0}{580} & \overset{0}{733.3} & \overset{r_1}{-580} & \overset{r_2}{-773.3} \\ 773.3 & 1\,031.1 & -773.3 & -1\,031.3 \\ -580 & -773.3 & 580 & 773.3 \\ -773.3 & -1\,031.1 & 773.3 & 1\,031.1 \end{bmatrix} \begin{bmatrix} 0 \\ 0 \\ r_1 \\ r_2 \end{bmatrix}$$

$$\begin{bmatrix} \bar{U}_1 \\ \bar{V}_1 \\ \bar{U}_2 \\ \bar{V}_2 \end{bmatrix}^2 = \begin{bmatrix} \overset{0}{1\,933.3} & \overset{0}{0} & \overset{r_3}{-1\,933.3} & \overset{0}{0} \\ 0 & 0 & 0 & 0 \\ -1\,933.3 & 0 & 1\,933.3 & 0 \\ 0 & 0 & 0 & 0 \end{bmatrix} \begin{bmatrix} 0 \\ 0 \\ r_3 \\ 0 \end{bmatrix}$$

$$\begin{bmatrix} \bar{U}_1 \\ \bar{V}_1 \\ \bar{U}_2 \\ \bar{V}_2 \end{bmatrix}^3 = \begin{bmatrix} \overset{r_1}{733.3} & \overset{r_2}{-580} & \overset{r_3}{-773.3} & \overset{0}{580} \\ -580 & 435 & 580 & -435 \\ -773.3 & 580 & 773.3 & -580 \\ 580 & -435 & -580 & 435 \end{bmatrix} \begin{bmatrix} r_1 \\ r_2 \\ r_3 \\ 0 \end{bmatrix}$$

The structure stiffness matrix is 3×3 because there are three degrees of freedom. The contributing terms from the element stiffness matrices are added to each element of the structure stiffness matrix.

There is only one load, -10 kips, acting in the second degree of freedom. Therefore, the only nonzero term in the load vector is -10. The equilibrium equations are

$$
\begin{array}{c}
 \\
1 \\
2 \\
3
\end{array}
\begin{array}{ccc}
\quad r_1 & \quad r_2 & \quad r_3
\end{array}
\left[\begin{array}{ccc}
(580 + 773.3) & (773.3 - 580) & -773.3 \\
(773.3 - 580) & (1\,031.1 + 435) & 580 \\
-773.3 & 580 & (1\,933.3 + 773.3)
\end{array}\right]
\left[\begin{array}{c}
r_1 \\
r_2 \\
r_3
\end{array}\right]
=
\left[\begin{array}{c}
0 \\
-10 \\
0
\end{array}\right]
$$

The solution of the above system of equations gives the joint displacements as

$$
\left[\begin{array}{c}
r_1 \\
r_2 \\
r_3
\end{array}\right]
=
\left[\begin{array}{c}
0.00026 \\
-0.008 \\
0.0025
\end{array}\right]
$$

The bar forces are found by Eq. (12.4b), using the element displacement vectors in the force-displacement relations in the global coordinates with the values just found. The member-end forces for element 1 are

$$
\left[\begin{array}{c}
U_1 \\
U_2
\end{array}\right]^1
=
\frac{29\,000 \times 10}{15 \times 12}
\left[\begin{array}{cccc}
0.6 & 0.8 & -0.6 & -0.8 \\
-0.6 & -0.8 & 0.6 & 0.8
\end{array}\right]
\left[\begin{array}{c}
0 \\
0 \\
0.0026 \\
-0.008
\end{array}\right]
=
\left[\begin{array}{c}
8 \\
-8
\end{array}\right]
$$

For element 2,

$$
\left[\begin{array}{c}
U_1 \\
U_2
\end{array}\right]^2
=
\frac{29\,000 \times 20}{25 \times 12}
\left[\begin{array}{cccc}
1 & 0 & -1 & 0 \\
0 & -1 & 0 & -1
\end{array}\right]
\left[\begin{array}{c}
0 \\
0 \\
0.0025 \\
0
\end{array}\right]
=
\left[\begin{array}{c}
-4.8 \\
4.8
\end{array}\right]
$$

and for element 3,

$$
\left[\begin{array}{c}
U_1 \\
U_2
\end{array}\right]^3
=
\frac{29\,000 \times 10}{20 \times 12}
\left[\begin{array}{cccc}
0.8 & 0.6 & -0.8 & -0.6 \\
-0.8 & -0.6 & 0.8 & 0.6
\end{array}\right]
\left[\begin{array}{c}
0.0026 \\
-0.008 \\
0.0025 \\
0
\end{array}\right]
=
\left[\begin{array}{c}
6 \\
-6
\end{array}\right]
$$

Note that the positive directions of the bar forces are from the first to the second joint of the element.

From the direction and the sign of the end forces it is seen that members 1 and 3 are in compression, while member 2 is in tension.

Example 12.2. The truss of Fig. 12.4 has 4 joints and 6 members. The horizontal and vertical members are 10 ft long with 6 in^2 cross-sectional area. The diagonal members have a length of 12.14 ft and cross-sectional area of 8 in^2. A vertical load of 30 kips is applied at joint 2 and a horizontal load of 15 kips is applied at joint 3.

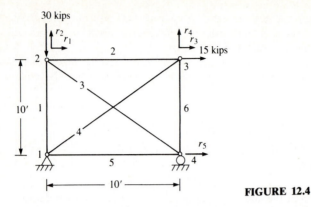

FIGURE 12.4

The modulus of elasticity of the bars is 29 000 ksi. Find the joint displacements and member end forces for this truss.

Solution. This truss is statically indeterminate internally. However, the stiffness method applies in the same manner to determinate and indeterminate structures. The degrees of freedom consist of the horizontal and vertical displacements of joints 2 and 3 and the horizontal displacement of joint 4. Therefore, there are 5 degrees of freedom, as indicated in Fig. 12.4. The element numbers are also shown in the figure. The element force–displacement relations in the global coordinates are as follows.

$$
\begin{bmatrix} U_1 \\ V_1 \\ U_2 \\ V_2 \end{bmatrix}^1 = \begin{bmatrix} 0 & 0 & 0 & 0 \\ 0 & 1\,450 & 0 & -1\,450 \\ 0 & 0 & 0 & 0 \\ 0 & -1\,450 & 0 & 1\,450 \end{bmatrix} \begin{bmatrix} 0 \\ 0 \\ r_1 \\ r_2 \end{bmatrix}
$$

$$
\begin{bmatrix} U_1 \\ V_1 \\ U_2 \\ V_2 \end{bmatrix}^2 = \begin{bmatrix} 1\,450 & 0 & -1\,450 & 0 \\ 0 & 0 & 0 & 0 \\ -1\,450 & 0 & 1\,450 & 0 \\ 0 & 0 & 0 & 0 \end{bmatrix} \begin{bmatrix} r_1 \\ r_2 \\ r_3 \\ r_4 \end{bmatrix}
$$

$$
\begin{bmatrix} U_1 \\ V_1 \\ U_2 \\ V_2 \end{bmatrix}^3 = \begin{bmatrix} 683.53 & -683.53 & -683.53 & 683.53 \\ -683.53 & 683.53 & 683.53 & -683.53 \\ -683.53 & 683.53 & 683.53 & -683.53 \\ 683.53 & -683.53 & -683.53 & 683.53 \end{bmatrix} \begin{bmatrix} r_1 \\ r_2 \\ r_5 \\ 0 \end{bmatrix}
$$

$$
\begin{bmatrix} U_1 \\ V_1 \\ U_2 \\ V_2 \end{bmatrix}^4 = \begin{bmatrix} 683.53 & 683.53 & -683.53 & -683.53 \\ 683.53 & 683.53 & -683.53 & -683.53 \\ -683.53 & -683.53 & 683.53 & 683.53 \\ -683.53 & -683.53 & 683.53 & 683.53 \end{bmatrix} \begin{bmatrix} 0 \\ 0 \\ r_3 \\ r_4 \end{bmatrix}
$$

$$
\begin{bmatrix} U_1 \\ V_1 \\ U_2 \\ V_2 \end{bmatrix}^5 =
\begin{array}{cccc} 0 & 0 & r_5 & 0 \end{array}
\begin{bmatrix} 1\,450 & 0 & -1\,450 & 0 \\ 0 & 0 & 0 & 0 \\ -1\,450 & 0 & 1\,450 & 0 \\ 0 & 0 & 0 & 0 \end{bmatrix}
\begin{bmatrix} 0 \\ 0 \\ r_5 \\ 0 \end{bmatrix}
$$

$$
\begin{bmatrix} U_1 \\ V_1 \\ U_2 \\ V_2 \end{bmatrix}^6 =
\begin{array}{cccc} r_3 & r_4 & r_5 & 0 \end{array}
\begin{bmatrix} 0 & 0 & 0 & 0 \\ 0 & 1\,450 & 0 & -1\,450 \\ 0 & 0 & 0 & 0 \\ 0 & -1\,450 & 0 & 1\,450 \end{bmatrix}
\begin{bmatrix} r_3 \\ r_4 \\ r_5 \\ 0 \end{bmatrix}
$$

The number of degree of freedom, which determines the location of each term in the structure stiffness matrix, is shown on top and on the right of each term of the element stiffness matrices. Using these numbers we can assemble the 5×5 structure stiffness matrix as shown in the following equilibrium equations. The load vector is shown on the right of these equations.

$$
\begin{array}{ccccc} r_1 & r_2 & r_3 & r_4 & r_5 \end{array}
$$
$$
\begin{bmatrix}
2\,133.53 & -683.53 & -1\,450 & 0 & -683.53 \\
-683.53 & 2\,133.53 & 0 & 0 & 683.53 \\
-1\,450 & 0 & 2\,133.53 & 683.53 & 0 \\
0 & 0 & 683.53 & 2\,133.53 & 0 \\
-683.53 & 683.53 & 0 & 0 & 2\,133.53
\end{bmatrix}
\begin{bmatrix} r_1 \\ r_2 \\ r_3 \\ r_4 \\ r_5 \end{bmatrix} =
\begin{bmatrix} 0 \\ -30 \\ 15 \\ 0 \\ 0 \end{bmatrix}
$$

The joint displacements obtained by solving the above equations are

$$
\begin{bmatrix} r_1 \\ r_2 \\ r_3 \\ r_4 \\ r_5 \end{bmatrix} =
\begin{bmatrix} 0.0057 \\ -0.0142 \\ 0.0122 \\ -0.0039 \\ 0.0064 \end{bmatrix}
$$

With these values the displacement vectors on the right of the force-displacement relations in the global coordinates are used in Eq. (12.4b) to find the member forces. For member 1,

$$
\begin{bmatrix} U_1 \\ U_2 \end{bmatrix} = \frac{29\,000 \times 6}{120}
\begin{bmatrix} 1 & 0 & -1 & 0 \\ -1 & 0 & 1 & 0 \end{bmatrix}
\begin{bmatrix} 0 \\ 0 \\ 0.0057 \\ -0.0142 \end{bmatrix} =
\begin{bmatrix} 20.68 \\ -20.68 \end{bmatrix}
$$

The force vectors for all the members, found similarly, are given below.

Element→	1	2	3	4	5	6
$\begin{bmatrix} U_1 \\ U_2 \end{bmatrix}$:	$\begin{bmatrix} 20.68 \\ -20.68 \end{bmatrix}$	$\begin{bmatrix} -9.310 \\ 9.31 \end{bmatrix}$	$\begin{bmatrix} 13.18 \\ -13.18 \end{bmatrix}$	$\begin{bmatrix} -8.03 \\ 8.03 \end{bmatrix}$	$\begin{bmatrix} -9.31 \\ 9.31 \end{bmatrix}$	$\begin{bmatrix} 5.68 \\ -5.68 \end{bmatrix}$

12.2 STIFFNESS METHOD FOR FRAMES WITH ROTATIONAL DEGREES OF FREEDOM

As mentioned earlier, a general method of solution for two-dimensional frames using an element stiffness matrix including end displacements and rotations as well as axial deformations is not difficult to implement in a computer program. For hand calculation it may be preferable to use the 2×2 or 4×4 element stiffness matrices in order to reduce the amount of algebra involved and to gain insight into the behavior of frames.

Therefore, in this section we consider frames with joint rotations only, using the 2×2 stiffness matrix developed for beams in Section 11.2. We then proceed with frames with sidesway in Section 12.3, using the 4×4 stiffness matrix derived in Section 11.3 including coordinate transformations, followed by a discussion of the stiffness method for general two-dimensional frames in Section 12.4.

Frames with Rotational Degrees of Freedom Only

For frames without joint displacements (no sidesway) the 2×2 beam stiffness matrix (Eq. 11.2a), can be used with a solution process similar to that of the stiffness method for continuous beams.

Example 12.3. Find the joint rotations and member end moments for the steel frame of Fig. 12.5 with the geometry and loads shown in the figure. The lengths of the members 1, 2, and 3 are 18, 15, and 10 ft, respectively, and the moments of inertia are 200, 100, and 200 in⁴, respectively. The modulus of elasticity is $E = 29\,000$ ksi. The external loads consist of a concentrated 8-kip load applied at the mid-span of the inclined member and normal to it, and a 3-kip/ft gravity load on the horizontal member.

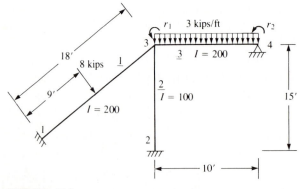

FIGURE 12.5

Solution. Since there is no sidesway, the solution is similar to that of continuous beams outlined in Section 11.2. There are two degrees of freedom for this problem, the rotations of joints 3 and 4. We denote these degrees of freedom by r_1 and r_2.

For joints 1 and 3 of the inclined member, the fixed-end moments are (Table 10.1a) $\pm Pl/8 = \pm 8 \times 18 \times 12/8 = \pm 216$ kip · in. The member-end moments at joints 3 and 4 of the horizontal member are $\pm (3/12)(10 \times 12)^2/12 = \pm 300$ kip · in.

The equivalent joint moments for joints with rotation are $-(-216 + 300) = -84$ kip · in at joint 3, and 300 kip · in at joint 4.

The force–displacement relation for the members,

$$\begin{bmatrix} M_1 \\ M_2 \end{bmatrix} = \frac{2EI}{l} \begin{bmatrix} 2 & 1 \\ 1 & 2 \end{bmatrix} \begin{bmatrix} \theta_1 \\ \theta_2 \end{bmatrix} + \begin{bmatrix} M_1^F \\ M_2^F \end{bmatrix}$$

applied to the three elements with the associated degrees of freedom, gives

$$\begin{bmatrix} M_1 \\ M_2 \end{bmatrix}^1 = \begin{bmatrix} \overset{0}{107\,407} & \overset{r_1}{53\,704} \\ 53\,704 & 107\,407 \end{bmatrix} \begin{bmatrix} 0 \\ r_1 \end{bmatrix} + \begin{bmatrix} 216 \\ -216 \end{bmatrix}$$

$$\begin{bmatrix} M_1 \\ M_2 \end{bmatrix}^2 = \begin{bmatrix} \overset{0}{64\,444} & \overset{r_1}{32\,222} \\ 32\,222 & 64\,444 \end{bmatrix} \begin{bmatrix} 0 \\ r_1 \end{bmatrix} + \begin{bmatrix} 0 \\ 0 \end{bmatrix}$$

$$\begin{bmatrix} M_1 \\ M_2 \end{bmatrix}^3 = \begin{bmatrix} \overset{r_1}{193\,333} & \overset{r_2}{96\,667} \\ 96\,667 & 193\,333 \end{bmatrix} \begin{bmatrix} r_1 \\ r_2 \end{bmatrix} + \begin{bmatrix} 300 \\ -300 \end{bmatrix}$$

The degrees of freedom are indicated above the columns of the element stiffness matrices and in the element displacement vectors. The structure stiffness matrix is 2×2. Each element of this matrix consists of the sum of the pertinent terms of the element stiffness matrices. The terms of the load vector consist of the equivalent joint moments for the joints calculated above. The structure stiffness matrix and the load vector **R** are shown in the following equilibrium equations:

$$\begin{matrix} r_1 & r_2 & \mathbf{R} \end{matrix}$$
$$\begin{bmatrix} 36\,585 & 96\,667 \\ 96\,667 & 193\,333 \end{bmatrix} \begin{bmatrix} r_1 \\ r_2 \end{bmatrix} = \begin{bmatrix} -84 \\ 300 \end{bmatrix}$$

The solution of this equation gives

$$\begin{bmatrix} r_1 \\ r_2 \end{bmatrix} = \begin{bmatrix} -0.000\,73 \\ 0.001\,92 \end{bmatrix}$$

Using the member-end rotations in the force–displacement equations, we get the end moments. For member 1,

$$\begin{bmatrix} M_1 \\ M_2 \end{bmatrix}^1 = \begin{bmatrix} 107\,242.0 & 53\,621.01 \\ 53\,621.01 & 107\,242.0 \end{bmatrix} \begin{bmatrix} 0 \\ -0.000\,73 \end{bmatrix} + \begin{bmatrix} 216 \\ -216 \end{bmatrix} = \begin{bmatrix} 176.34 \\ -295.32 \end{bmatrix}$$

The other member-end forces are found in a similar manner. The force vectors are thus as follows.

$$\text{Member} \rightarrow \quad 1 \qquad\qquad 2 \qquad\qquad 3$$

$$\begin{bmatrix} M_1 \\ M_2 \end{bmatrix}: \quad \begin{bmatrix} 176.34 \\ -295.32 \end{bmatrix} \quad \begin{bmatrix} -23.80 \\ -47.59 \end{bmatrix} \quad \begin{bmatrix} 342.92 \\ 0 \end{bmatrix}$$

12.3 SQUARE FRAMES WITH DISPLACEMENT AND ROTATIONAL DEGREES OF FREEDOM

When frame joints undergo displacements, the 2×2 stiffness matrix involving only joint rotations is no longer appropriate and a stiffness matrix that includes the joint displacements must be used. The general solution of frames, discussed in the following section, with 6×6 stiffness and transformation matrices, can of course be used for square frames. Here the 4×4 stiffness matrix derived in Section 11.3 (Eq. (11.10a)) is used, which may provide a simpler solution procedure. This matrix does not include the axial deformation of the member. Thus, the displacements of the end joints of a member, along the axis of the member, are equal and must thus be denoted by the same degree of freedom. The element force–displacement relation in this case is the same as Eq. (11.10a), that is,

$$\begin{bmatrix} V_1 \\ M_1 \\ V_2 \\ M_2 \end{bmatrix} = \begin{bmatrix} \dfrac{12EI}{l^3} & \dfrac{6EI}{l^2} & -\dfrac{12EI}{l^3} & \dfrac{6EI}{l^2} \\[2mm] \dfrac{6EI}{l^2} & \dfrac{4EI}{l} & -\dfrac{6EI}{l^2} & \dfrac{2EI}{l} \\[2mm] -\dfrac{12EI}{l^3} & -\dfrac{6EI}{l^2} & \dfrac{12EI}{l^3} & -\dfrac{6EI}{l^2} \\[2mm] \dfrac{6EI}{l^2} & \dfrac{2EI}{l} & -\dfrac{6EI}{l^2} & \dfrac{4EI}{l} \end{bmatrix} \begin{bmatrix} v_1 \\ \theta_1 \\ v_2 \\ \theta_2 \end{bmatrix} + \begin{bmatrix} V_1^F \\ M_1^F \\ V_2^F \\ M_2^F \end{bmatrix}$$

or

$$\mathbf{S} = \mathbf{Ks} + \mathbf{S}^F$$

where $\mathbf{S}^F$ is the vector of the fixed-end forces.

We recall that the above stiffness matrix applies to all continuous beam elements because they have the same orientations. In the case of square frames the stiffness terms associated with the moments remain the same, because moments do not change with the orientation of the member.

In the above stiffness matrix the first two columns are associated with the displacement and rotation of the first joint of the element and the last two columns with the displacement and rotation of the second joint. This is important when assigning the structure degrees of freedom to the stiffness matrices in the member force–displacement relations. The terms associated with shear force would not require transformation only if the shear forces in the local and global coordinates have the same positive sense. The local x axis is directed from the first joint to the second one.

> If the global coordinate axes are directed to the right and upward, and the first joints are chosen on the left of horizontal members and the upper joints of vertical members, then the above force–displacement equation holds true in the global coordinates, without any need for transformation.

If, on the other hand, the right or the lower joints are chosen as the first joints of the members, then to obtain the global force–displacement relation the sign of all the shear terms $(6EI/l^2)$ in the stiffness matrix and the sign of the fixed-end shears must be changed. The latter statement can be proved by using the transformation matrix derived in the next section.

Example 12.4. Find the joint displacements and member end forces for the portal frame of Fig. 12.6. The moment of inertia of the columns is 200 in^4 and their length is 12 ft. The moment of inertia of the girder is 400 in^4 and its length is 20 ft. A uniformly distributed load of 3 kip/ft is applied on the girder and a concentrated horizontal load of 10 kips is applied at joint 2. The columns are fixed at their base and the axial deformations of the members are negligible. $E = 29\,000$ ksi.

Solution. If the horizontal load did not exist, then, because of symmetry of the structure and load, no sidesway would occur and a solution with 2×2 stiffness matrices could be used. However, because of the horizontal load, a lateral displacement takes place and the stiffness matrix must include displacement degrees of freedom. Therefore, we use 4×4 stiffness matrices. The degrees of freedom consist of the lateral displacement and rotation of joint 2, and the rotation of joint 3. Note that, since axial deformations are neglected, the lateral displacement of joint 3 is the same as that of joint 2 and is therefore denoted by r_1.

The force–displacement relations for the three members are given below.

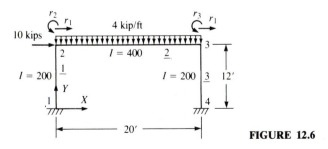

FIGURE 12.6

Note that the global coordinates are to the right and upward, while the local or member coordinates are to the right for the horizontal member and downward for the vertical members. That is, the first joint for the horizontal member is the left joint while the left joints for the vertical members are the upper joints, 2 and 3.

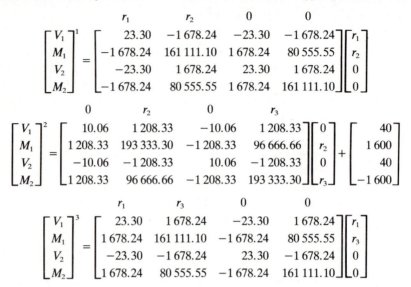

$$\begin{bmatrix} V_1 \\ M_1 \\ V_2 \\ M_2 \end{bmatrix}^1 = \begin{matrix} r_1 & r_2 & 0 & 0 \\ \begin{bmatrix} 23.30 & -1\,678.24 & -23.30 & -1\,678.24 \\ -1\,678.24 & 161\,111.10 & 1\,678.24 & 80\,555.55 \\ -23.30 & 1\,678.24 & 23.30 & 1\,678.24 \\ -1\,678.24 & 80\,555.55 & 1\,678.24 & 161\,111.10 \end{bmatrix} \end{matrix} \begin{bmatrix} r_1 \\ r_2 \\ 0 \\ 0 \end{bmatrix}$$

$$\begin{bmatrix} V_1 \\ M_1 \\ V_2 \\ M_2 \end{bmatrix}^2 = \begin{matrix} 0 & r_2 & 0 & r_3 \\ \begin{bmatrix} 10.06 & 1\,208.33 & -10.06 & 1\,208.33 \\ 1\,208.33 & 193\,333.30 & -1\,208.33 & 96\,666.66 \\ -10.06 & -1\,208.33 & 10.06 & -1\,208.33 \\ 1\,208.33 & 96\,666.66 & -1\,208.33 & 193\,333.30 \end{bmatrix} \end{matrix} \begin{bmatrix} 0 \\ r_2 \\ 0 \\ r_3 \end{bmatrix} + \begin{bmatrix} 40 \\ 1\,600 \\ 40 \\ -1\,600 \end{bmatrix}$$

$$\begin{bmatrix} V_1 \\ M_1 \\ V_2 \\ M_2 \end{bmatrix}^3 = \begin{matrix} r_1 & r_3 & 0 & 0 \\ \begin{bmatrix} 23.30 & 1\,678.24 & -23.30 & 1\,678.24 \\ 1\,678.24 & 161\,111.10 & -1\,678.24 & 80\,555.55 \\ -23.30 & -1\,678.24 & 23.30 & -1\,678.24 \\ 1\,678.24 & 80\,555.55 & -1\,678.24 & 161\,111.10 \end{bmatrix} \end{matrix} \begin{bmatrix} r_1 \\ r_3 \\ 0 \\ 0 \end{bmatrix}$$

The structure degrees of freedom associated with the terms of the member stiffness matrices are indicated above the stiffness matrices and in the displacement vectors. Since there are no transformations involved, the above relations are also valid in the global coordinates.

The structure stiffness matrix is obtained by adding the terms from the element stiffness matrices to the pertinent locations of the structure stiffness matrix. The elements of the load vector are the sums of the applied joint loads and the equivalent joint forces, due to in-span loads, in each degree of freedom, in the global coordinates. In degree of freedom 1 there is only a 10-kip horizontal load. In degree of freedom 2 the equivalent joint moment is the negative of the fixed-end moment, or $-1\,600$ kip · ft. Similarly, in degree of freedom 3 the moment is $-(-1\,600) = 1\,600$ kip · ft. Since the transformation matrix for this element is the identity matrix, the fixed-end force vectors in the local and global coordinates are the same. The equilibrium equations are thus as follows

$$\begin{matrix} r_1 & r_2 & r_3 & & R \\ \begin{bmatrix} 46.60 & -1\,678.24 & 1\,678.24 \\ -1\,678.24 & 354\,444.40 & 96\,666.66 \\ 1\,678.24 & 96\,666.66 & 354\,444.40 \end{bmatrix} \begin{bmatrix} r_1 \\ r_2 \\ r_3 \end{bmatrix} & = & \begin{bmatrix} 10 \\ -1\,600 \\ 1\,600 \end{bmatrix} \end{matrix}$$

The solution of the above equations gives the joint displacements as

$$\begin{bmatrix} r_1 \\ r_2 \\ r_3 \end{bmatrix} = \begin{bmatrix} 0.293\,142 \\ -0.007\,29 \\ 0.005\,116 \end{bmatrix}$$

The element displacement vectors are found by identifying the member-end displacements with the structure degrees of freedom. The restrained degrees of freedom will have a zero value.

Using the element displacement vectors, the member-end forces are found by the element force–displacement relations. For example, for member 1,

$$\begin{bmatrix} V_1 \\ M_1 \\ V_2 \\ M_2 \end{bmatrix}^1 = \begin{bmatrix} 23.30 & -1\,678.24 & -23.30 & -1\,678.24 \\ -1\,678.24 & 161\,111.10 & 1\,678.24 & 80\,555.55 \\ -23.30 & 1\,678.24 & 23.30 & 1\,678.24 \\ -1\,678.24 & 80\,555.55 & 1\,678.24 & 161\,111.10 \end{bmatrix}$$

$$\times \begin{bmatrix} 0.293\,142 \\ -0.007\,29 \\ 0 \\ 0 \end{bmatrix} = \begin{bmatrix} -5.41 \\ -683.90 \\ 5.41 \\ -96.10 \end{bmatrix}$$

The other end forces are found in a similar manner, leading to the following force vectors.

Member→	1	2	3
$\begin{bmatrix} V_1 \\ M_1 \\ V_2 \\ M_2 \end{bmatrix}$:	$\begin{bmatrix} -5.41 \\ -683.90 \\ 5.41 \\ -96.10 \end{bmatrix}$	$\begin{bmatrix} 37.36 \\ 683.90 \\ 42.63 \\ -1\,316.10 \end{bmatrix}$	$\begin{bmatrix} 15.41 \\ 1\,316.10 \\ -15.41 \\ 903.90 \end{bmatrix}$

The validity of these results can be verified by checking the equilibrium of the forces at each joint.

12.4 GENERAL TWO-DIMENSIONAL FRAMES

In applying the stiffness method to frames, especially when using computers, it is easier to include three degrees of freedom per joint (two displacements and one rotation). This is true even when axial deformations of the members are negligible. Although this increases the size of the stiffness matrices and the number of simultaneous equations to be solved, it makes the identification of the degrees of freedom easier and the application of the stiffness method straightforward. The increased size does not usually present any problems when the solution is carried out on a computer. The steps and procedures for implementing the stiffness method in a computer program are given in Section 19.2.

Stiffness Matrix for Beam Elements with Axial Deformations

Since axial deformations are independent of bending, the force–displacement relationships for beams with axial deformations can be developed separately for the axial and bending deformations.

FIGURE 12.7

The stiffness matrix, relating the member-end forces and displacements (Fig. 12.7), is a combination of the truss (Section 12.1) and beam (Section 12.3) stiffness matrices. The force–displacement relations thus take the following form.

$$
\begin{bmatrix} U_1 \\ V_1 \\ M_1 \\ U_2 \\ V_2 \\ M_2 \end{bmatrix} =
\begin{bmatrix}
\dfrac{EA}{l} & 0 & 0 & -\dfrac{EA}{l} & 0 & 0 \\
0 & \dfrac{12EI}{l^3} & \dfrac{6EI}{l^2} & 0 & -\dfrac{12EI}{l^3} & \dfrac{6EI}{l^2} \\
0 & \dfrac{6EI}{l^2} & \dfrac{4EI}{l} & 0 & -\dfrac{6EI}{l^2} & \dfrac{2EI}{l} \\
-\dfrac{EA}{l} & 0 & 0 & \dfrac{EA}{l} & 0 & 0 \\
0 & -\dfrac{12EI}{l^3} & -\dfrac{6EI}{l^2} & 0 & \dfrac{12EI}{l^3} & -\dfrac{6EI}{l^2} \\
0 & \dfrac{6EI}{l^2} & \dfrac{2EI}{l} & 0 & -\dfrac{6EI}{l^2} & \dfrac{4EI}{l}
\end{bmatrix}
\begin{bmatrix} u_1 \\ v_1 \\ \theta_1 \\ u_2 \\ v_2 \\ \theta_2 \end{bmatrix}
+
\begin{bmatrix} U_1^F \\ V_1^F \\ M_1^F \\ U_2^F \\ V_2^F \\ M_2^F \end{bmatrix}
$$

$$(12.7a)$$

or, again,

$$\mathbf{S} = \mathbf{Ks} + \mathbf{S}^F \qquad (12.7b)$$

where **S** is the member force vector, consisting of the axial and shear forces plus moments; **K** is the 6×6 stiffness matrix and **s** is the vector of end displacements; $\mathbf{S}^F$ is the vector of fixed-end forces. The fixed-end shear forces and moments for common loads are given in Tables 10.1a and 10.1b. Fixed-end axial forces are found in a manner similar to shear.

Transformation of Coordinates

In general, frame members may be oriented in any direction. When two members with different orientations are connected at a joint (Fig. 12.8), it is necessary to decompose the forces in a common coordinate system so that the equilibrium equations can be written at the joint by summing up the force components in two perpendicular directions.

Instead of decomposing the forces at each joint, it is better to express the

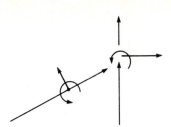

FIGURE 12.8

force–displacement relations in the global coordinates providing the force components in the global coordinate system. To achieve this we must find the relationship between the member-end forces and their components in the global coordinate system. The member-end forces and deformations and their components in the global coordinates are shown in Fig. 12.9 at the first end of a member. The forces at the other end of the member would be similar.

It is seen that the forces and displacements transform in a similar manner. In the local or member coordinate system, the x axis is along the axis of the member directed from the first joint to the second. The angle between the local and global x axes is α.

To relate the force components in the local coordinates, U_1, V_1, M_1, to those in the global coordinates, $\bar{U}_1$, $\bar{V}_1$, $\bar{M}_1$ (Fig. 12.9a), we can write the following expressions

$$U_1 = \bar{U}_1 \cos \alpha + \bar{V}_1 \sin \alpha$$

$$V_1 = -\bar{U}_1 \sin \alpha + \bar{V}_1 \cos \alpha$$

$$M_1 = \bar{M}_1$$

Similar expressions can be written for forces at end 2 of the member. Thus, the vector of forces in the local coordinates, $\mathbf{S}$, and that in the global coordinates, $\bar{\mathbf{S}}$, are related through the transformation matrix $\mathbf{T}$ as follows:

$$\mathbf{S} = \mathbf{T}\bar{\mathbf{S}} \qquad\qquad (12.8a)$$

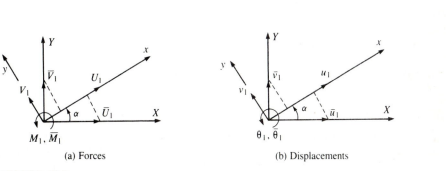

(a) Forces   (b) Displacements

FIGURE 12.9

The expanded form of the above equation is

$$
\begin{bmatrix} U_1 \\ V_1 \\ M_1 \\ U_2 \\ V_2 \\ M_2 \end{bmatrix} = \begin{bmatrix} c & s & 0 & 0 & 0 & 0 \\ -s & c & 0 & 0 & 0 & 0 \\ 0 & 0 & 1 & 0 & 0 & 0 \\ 0 & 0 & 0 & c & s & 0 \\ 0 & 0 & 0 & -s & c & 0 \\ 0 & 0 & 0 & 0 & 0 & 1 \end{bmatrix} \begin{bmatrix} \bar{U}_1 \\ \bar{V}_1 \\ \bar{M}_1 \\ \bar{U}_2 \\ \bar{V}_2 \\ \bar{M}_2 \end{bmatrix} \tag{12.8b}
$$

In the above equations the cosine and sine terms are calculated from the coordinates of the joints using the following equations:

$$
c = \cos \alpha = \frac{X_2 - X_1}{l} \tag{12.9a}
$$

$$
s = \sin \alpha = \frac{Y_2 - Y_1}{l} \tag{12.9b}
$$

$$
l = \sqrt{(X_2 - X_1)^2 + (Y_2 - Y_1)^2} \tag{12.9c}
$$

We can find the forces in the global coordinates in terms of those in the local coordinates as $\bar{\mathbf{S}} = \mathbf{T}^{-1}\mathbf{S}$. It can be easily verified that the transformation matrix $\mathbf{T}$ is an orthogonal matrix. That is,

$$
\mathbf{T}^{-1} = \mathbf{T}^T
$$

Thus,

$$
\bar{\mathbf{S}} = \mathbf{T}^T \mathbf{S} \tag{12.10}
$$

The displacement vectors in the local and global coordinates are related (Fig. 12.9b) in a similar manner as the forces:

$$
\mathbf{s} = \mathbf{T}\bar{\mathbf{s}} \tag{12.11a}
$$

and

$$
\bar{\mathbf{s}} = \mathbf{T}^T \mathbf{s} \tag{12.11b}
$$

where

$$
\mathbf{s}^T = [u_1 \quad v_1 \quad \theta_1 \quad u_2 \quad v_2 \quad \theta_2]^T \tag{12.11c}
$$

and

$$
\bar{\mathbf{s}}^T = [\bar{u}_1 \quad \bar{v}_1 \quad \bar{\theta}_1 \quad \bar{u}_2 \quad \bar{v}_2 \quad \bar{\theta}_2]^T \tag{12.11d}
$$

Referring to Eq. (12.7b), which relates the member-end forces and displacements in the local coordinates,

$$
\mathbf{S} = \mathbf{K}\mathbf{s} + \mathbf{S}^F
$$

Substituting for $\mathbf{s}$ from Eq. (12.11a) into the last equation, we get the forces in the local coordinates in terms of the displacements in the global coordinates as

$$
\mathbf{S} = \mathbf{K}\mathbf{T}\bar{\mathbf{s}} + \mathbf{S}^F = \mathbf{B}\bar{\mathbf{s}} + \mathbf{S}^F \tag{12.12a}
$$

$$
\begin{bmatrix} U_1 \\ V_1 \\ M_1 \\ U_2 \\ V_2 \\ M_2 \end{bmatrix}
=
\begin{bmatrix}
\dfrac{EA}{l}c & \dfrac{EA}{l}s & 0 & -\dfrac{EA}{l}c & -\dfrac{EA}{l}s & 0 \\[2ex]
-\dfrac{12EI}{l^3}s & \dfrac{12EI}{l^3}c & \dfrac{6EI}{l^2} & -\dfrac{12EI}{l^3}s & -\dfrac{12EI}{l^3}c & \dfrac{6EI}{l^2} \\[2ex]
-\dfrac{6EI}{l^2}s & \dfrac{6EI}{l^2}c & \dfrac{4EI}{l} & \dfrac{6EI}{l^2}s & -\dfrac{6EI}{l^2}c & \dfrac{2EI}{l} \\[2ex]
-\dfrac{EA}{l}c & -\dfrac{EA}{l}s & 0 & \dfrac{EA}{l}c & \dfrac{EA}{l}s & 0 \\[2ex]
\dfrac{12EI}{l^3}s & -\dfrac{12EI}{l^3}c & -\dfrac{6EI}{l^2} & -\dfrac{12EI}{l^3}s & \dfrac{12EI}{l^3}c & -\dfrac{6EI}{l^2} \\[2ex]
-\dfrac{6EI}{l^2}s & \dfrac{6EI}{l^2}c & \dfrac{2EI}{l} & \dfrac{6EI}{l^2}s & -\dfrac{6EI}{l^2}c & \dfrac{4EI}{l}
\end{bmatrix}
\begin{bmatrix} \bar{u}_1 \\ \bar{v}_1 \\ \bar{\theta}_1 \\ \bar{u}_2 \\ \bar{v}_2 \\ \bar{\theta}_2 \end{bmatrix}
+
\begin{bmatrix} U_1^F \\ V_1^F \\ M_1^F \\ U_2^F \\ V_2^F \\ M_2^F \end{bmatrix}
$$

$$(12.12b)$$

Premultiplying both sides of Eq. (12.12a) by $\mathbf{T}^T$ we get

$$\mathbf{T}^T\mathbf{S} = \mathbf{T}^T\mathbf{KT\bar{s}} + \mathbf{T}^T\mathbf{S}^F$$

Referring to Eq. (12.10) we can write

$$\bar{\mathbf{S}} = \bar{\mathbf{K}}\bar{\mathbf{s}} + \bar{\mathbf{S}}^F \qquad (12.13a)$$

where $\bar{\mathbf{K}}$ is the element stiffness matrix in the global coordinate system given by

$$\bar{\mathbf{K}} = \mathbf{T}^T\mathbf{KT} \qquad (12.13b)$$

and the fixed-end force vector in the global coordinates is

$$\bar{\mathbf{S}}^F = \mathbf{T}^T\mathbf{S}^F \qquad (12.13c)$$

Upon carrying out the matrix multiplication of Eq. (12.13b) we get

$$
\bar{\mathbf{K}} = \dfrac{EI}{l^3}
\begin{bmatrix}
\dfrac{Al^2c^2}{I}+12s^2 & & & & & \\[2ex]
\left(\dfrac{Al^2}{I}-12\right)cs & \dfrac{Al^2s^2}{I}+12c^2 & & \text{Symmetric} & & \\[2ex]
-6ls & 6lc & 4l^2 & & & \\[2ex]
-\left(\dfrac{Al^2c^2}{I}+12s^2\right) & -\left(\dfrac{Al^2}{I}-12\right)cs & 6ls & \dfrac{Al^2c^2}{I}+12s^2 & & \\[2ex]
-\left(\dfrac{Al^2}{I}-12\right)cs & -\left(\dfrac{Al^2s^2}{I}+12c^2\right) & -6lc & \left(\dfrac{Al^2}{I}-12\right)cs & \dfrac{Al^2s^2}{I}+12c^2 & \\[2ex]
-6ls & 6lc & 2l^2 & 6ls & -6lc & 4l^2
\end{bmatrix}
$$

$$(12.13d)$$

This is the stiffness matrix of a beam element with 6 degrees of freedom in the global coordinate system. In Eqs. (12.12b) and (12.13d), $c = \cos \alpha$, $s = \sin \alpha$, calculated by Eqs. (12.9); A is the area of the cross-section; I is the moment of inertia; E is the modulus of elasticity; and l the length of the element.

Note that the equivalent joint loads, included in the load vector (of equilibrium equations) due to span loads, is the negative of sum of the fixed-end forces at the joint after they are transformed to global coordinates by Eq. (12.13c). After finding the displacements in the global coordinates, the member force vectors are computed in the local coordinates by Eq. (12.12b), so that they can be used directly in design.

Solution Process

The steps in the solution of general two-dimensional frames are as follows.

1. Number the members, joints and structure degrees of freedom.
2. Find the force–displacement relations in the global coordinates for all the elements, using the global stiffness matrix (Eq. (12.13d)) and indicate the degrees of freedom in the displacement vectors and above the stiffness matrices.
3. Find the equilibrium equations in terms of an $n \times n$ structure stiffness matrix, where n is the number of degrees of freedom. Indicate the degrees of freedom r_1 to r_n above the structure stiffness matrix and in the displacement vector. Calculate the elements of the structure stiffness matrix by adding the terms from the member stiffness matrices in the global coordinates denoted by the same degrees of freedom.

   Also calculate the elements of the load vector, each term being equal to the applied load in the associated degree of freedom plus the equivalent joint force. The latter is the negative of the sum of the fixed-end forces in that degree of freedom.
4. Solve the equilibrium equations thus obtained to find the joint displacements.
5. Determine the member-end displacement vectors from the joint displacements, by adding zeros for the restrained displacements. Find the member end forces by Eq. 12.12b.

   The shear and moment diagrams can be drawn using the member-end forces and applied in-span loads. However, the sign of the moments on the left of each member must be reversed first to express the values in the design sign convention.

> **Example 12.5.** Find the joint displacements, and member-end forces for the frame in Fig. 12.10. In addition to a distributed gravity load of 2.2 kip/ft on element 2, a horizontal load of 15 kips is applied at joint 2 and an upward load of 41 kips is acting at joint 3. The geometry of the frame is shown in Fig. 12.10. The

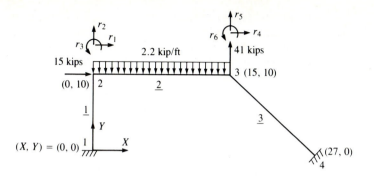

FIGURE 12.10

origin of the global coordinates is at the lower left joint. The global coordinates of joints 1, 2, 3, and 4 are $(0, 0)$, $(0, 10)$, $(15, 10)$, and $(27, 0)$ feet, respectively. The moments of inertia for all the members is $250\,\text{in}^4$ and the area of the cross-section is $10\,\text{in}^2$. The modulus of elasticity is $29\,000$ ksi.

Solution. The lengths of elements 1, 2, and 3 are 10, 15, and 15.62 ft, respectively. The cosine and sine of the angles between the member axes and the global X axis are $(0, 1)$, $(1, 0)$, and $(0.76, -0.64)$ for members 1, 2 and, 3. The degrees of freedom are shown in Fig. 12.10.

The force-displacement relations in the global coordinates are for the three elements (Eqs. (12.13a) and (12.13d))

$$
\begin{bmatrix} \bar{U}_1 \\ \bar{V}_1 \\ \bar{M}_1 \\ \bar{U}_2 \\ \bar{V}_2 \\ \bar{M}_2 \end{bmatrix}^1 =
\begin{array}{c}
\begin{matrix} 0 \quad\quad 0 \quad\quad\quad 0 \quad\quad\quad r_1 \quad\quad\quad r_2 \quad\quad\quad r_3 \end{matrix} \\
\begin{bmatrix}
50.34 & 0 & -3\,020.83 & -50.34 & 0 & -3\,020.83 \\
0 & 2\,416.66 & 0 & 0 & -2\,416.66 & 0 \\
-3\,020.83 & 0 & 241\,666.60 & 3\,020.83 & 0 & 120\,833.30 \\
-50.34 & 0 & 3\,020.83 & 50.34 & 0 & 3\,020.83 \\
0 & -2\,416.66 & 0 & 0 & 2\,416.66 & 0 \\
-3\,020.83 & 0 & 120\,833.30 & 3\,020.83 & 0 & 241\,666.60
\end{bmatrix}
\end{array}
\begin{bmatrix} 0 \\ 0 \\ 0 \\ r_1 \\ r_2 \\ r_3 \end{bmatrix}
$$

$$
\begin{bmatrix} \bar{U}_1 \\ \bar{V}_1 \\ \bar{M}_1 \\ \bar{U}_2 \\ \bar{V}_2 \\ \bar{M}_2 \end{bmatrix}^2 =
\begin{array}{c}
\begin{matrix} r_1 \quad\quad\; r_2 \quad\quad\; r_3 \quad\quad\; r_4 \quad\quad\; r_5 \quad\quad\; r_6 \end{matrix} \\
\begin{bmatrix}
1\,611.11 & 0 & 0 & -1\,611.11 & 0 & 0 \\
0 & 14.91 & 1\,342.59 & 0 & -14.91 & 1\,342.59 \\
0 & 1\,342.59 & 161\,111.10 & 0 & -1\,342.59 & 80\,555.55 \\
-1\,611.11 & 0 & 0 & 1\,611.11 & 0 & 0 \\
0 & -14.91 & -1\,342.59 & 0 & 14.91 & -1\,342.59 \\
0 & 1\,342.59 & 80\,555.55 & 0 & -1\,342.59 & 161\,111.10
\end{bmatrix}
\end{array}
\begin{bmatrix} r_1 \\ r_2 \\ r_3 \\ r_4 \\ r_5 \\ r_6 \end{bmatrix} +
\begin{bmatrix} 0 \\ 16.5 \\ 495 \\ 0 \\ 16.5 \\ -495 \end{bmatrix}
$$

$$
\begin{bmatrix} \bar{U}_1 \\ \bar{V}_1 \\ \bar{M}_1 \\ \bar{U}_2 \\ \bar{V}_2 \\ \bar{M}_2 \end{bmatrix}^3 =
\begin{array}{c}
\begin{matrix} r_4 \quad\quad\quad r_5 \quad\quad\quad r_6 \quad\quad\quad 0 \quad\quad\quad 0 \quad\quad\quad 0 \end{matrix} \\
\begin{bmatrix}
918.46 & -754.37 & 792.57 & -918.46 & 754.37 & 792.57 \\
-754.37 & 641.85 & 951.09 & 754.37 & -641.85 & 951.09 \\
792.57 & 951.09 & 154\,711.20 & -792.57 & -951.09 & 77\,355.61 \\
-918.46 & 754.37 & -792.57 & 918.46 & -754.37 & -792.57 \\
754.37 & -641.85 & -951.09 & -754.37 & 641.85 & -951.09 \\
792.57 & 951.09 & 77\,355.61 & -792.57 & -951.09 & 154\,711.20
\end{bmatrix}
\end{array}
\begin{bmatrix} r_4 \\ r_5 \\ r_6 \\ 0 \\ 0 \\ 0 \end{bmatrix}
$$

Since there are six degrees of freedom, the structure stiffness matrix is 6×6. Adding each term of the element stiffness matrices in the global coordinates to the appropriate location of the structure stiffness matrix indicated by the degrees of freedom, we obtain the structure stiffness matrix and the equilibrium equations given below.

The fixed-end shear forces for member 2 (Table 10.1a) are $2.2 \times 15/2 = 16.5$, and the fixed-end moments are $\pm 2.2 \times 15^2/12 = \pm 41.25 \text{ kip} \cdot \text{ft}$ or $\pm 495 \text{ kip} \cdot \text{in}$ at joints 2 and 3.

The elements of the load vector consist of the sum of the applied joint loads and equivalent joint loads in the global coordinates. Only member 2 has fixed-end forces, with the force vector $[0, 16.5, 495, 0, 16.5, -495]^T$. Since the transformation matrix for element 2 is the identity matrix, this vector remains the same in the global coordinates. The first term of the load vector that is associated with degree of freedom 1 is the 15-kip lateral load. The second term is the equivalent joint shear acting at joint 2 in the vertical direction (in degree of freedom 2). This is -16.5 kips. The third and the sixth terms are the equivalent joint moments at joints 2 and 3. There is no lateral force at joint 3, therefore the force associated with the fourth degree of freedom is zero. The vertical force at joint 3 consists of the equivalent joint shear of -16.5 plus the applied load of 41 kips or 24.5 kips. These values are used in the load vector below.

$$
\begin{array}{cccccc}
r_1 & r_2 & r_3 & r_4 & r_5 & r_6
\end{array}
\qquad \mathbf{R}
$$

$$
\begin{bmatrix}
1\,661.45 & 0 & 3\,020.83 & -1\,611.11 & 0 & 0 \\
0 & 2\,431.58 & 1\,342.59 & 0 & -14.91 & 1\,342.59 \\
3\,020.83 & 1\,342.59 & 402\,777.70 & 0 & -1\,342.59 & 80\,555.55 \\
-1\,611.11 & 0 & 0 & 2\,529.57 & -754.37 & 792.57 \\
0 & -14.91 & -1\,342.59 & -754.37 & 656.77 & -391.49 \\
0 & 1\,342.59 & 80\,555.55 & 792.57 & -391.49 & 315\,822.30
\end{bmatrix}
\begin{bmatrix}
r_1 \\ r_2 \\ r_3 \\ r_4 \\ r_5 \\ r_6
\end{bmatrix}
=
\begin{bmatrix}
15 \\ -16.5 \\ -495 \\ 0 \\ 24.5 \\ 495
\end{bmatrix}
$$

The solution of the above equilibrium equations gives the joint displacements:

$$
\begin{bmatrix}
r_1 \\ r_2 \\ r_3 \\ r_4 \\ r_5 \\ r_6
\end{bmatrix}
=
\begin{bmatrix}
0.461\,34 \\ -0.002\,64 \\ -0.003\,19 \\ 0.460\,45 \\ 0.560\,74 \\ 0.001\,93
\end{bmatrix}
$$

The displacement vectors for the three elements are found from the above displacements by adding zeros for the restrained-end displacements.

The element force vectors in the local coordinates are found by Eq. (12.12b), that is, $\mathbf{S} = \mathbf{B\bar{s}} + \mathbf{S}^F$, where $\bar{s}$ is the element displacement vector in the global coordiantes and $\mathbf{B} = \mathbf{KT}$. The member force vectors are found as follows. For element 1,

$$
\begin{bmatrix}
U_1 \\ V_1 \\ M_1 \\ U_2 \\ V_2 \\ M_2
\end{bmatrix}^1
=
\begin{bmatrix}
0 & 2\,416.66 & 0 & 0 & -2\,416.66 \\
-50.340 & 3020.83 & 50.34 & 0 & 3\,020.83 \\
-3\,020.830 & 241\,666.66 & 3\,020.83 & 0 & 120\,833.33 \\
0 & -2\,416.66 & 0 & 2\,416.66 & 0 \\
50.340 & -3\,020.83 & -50.34 & 0 & -3\,020.83 \\
-3\,020.830 & 120\,833.33 & 3\,020.83 & 0 & 241\,666.66
\end{bmatrix}
\begin{bmatrix}
0 \\ 0 \\ 0 \\ 0.461\,34 \\ -0.002\,64 \\ -0.003\,19
\end{bmatrix}
$$

For element 2,

$$
\begin{bmatrix} U_1 \\ V_1 \\ M_1 \\ U_2 \\ V_2 \\ M_2 \end{bmatrix}^2 =
\begin{bmatrix}
1.611.11 & 0 & 0 & -1\,611.11 & 0 & 0 \\
0 & 14.91 & 1\,342.59 & 0 & -14.91 & 1\,342.59 \\
0 & 1\,342.59 & 161\,111.10 & 0 & -1\,342.59 & 80\,555.55 \\
-1\,611.11 & 0 & 0 & 1\,611.11 & 0 & 0 \\
0 & -14.91 & -1\,342.59 & 0 & 14.91 & -1\,342.59 \\
0 & 1\,342.59 & 80\,555.55 & 0 & -1\,342.59 & 161\,111.10
\end{bmatrix}
$$
$$
\times
\begin{bmatrix}
0.461\,34 \\
-0.002\,64 \\
-0.003\,19 \\
0.460\,45 \\
0.560\,74 \\
0.001\,93
\end{bmatrix}
+
\begin{bmatrix}
0 \\
16.5 \\
495 \\
0 \\
16.5 \\
-495
\end{bmatrix}
$$

For element 3,

$$
\begin{bmatrix} U_1 \\ V_1 \\ M_1 \\ U_2 \\ V_2 \\ M_2 \end{bmatrix}^3 =
\begin{bmatrix}
1\,188.52 & -990.43 & 0 & -1\,188.52 & 990.43 & 0 \\
8.45 & 10.14 & 1\,238.04 & -8.45 & -10.14 & 1\,238.04 \\
792.57 & 951.09 & 154\,711.22 & -792.57 & -951.09 & 77\,355.61 \\
-1\,188.52 & 990.43 & 0 & 1\,188.52 & -990.43 & 0 \\
-8.45 & -10.14 & -1\,238.04 & 8.45 & 10.14 & -1\,238.04 \\
792.57 & 951.09 & 77\,355.61 & -792.57 & -951.09 & 154\,711.22
\end{bmatrix}
\begin{bmatrix}
0.460\,45 \\
0.560\,74 \\
0.001\,93 \\
0 \\
0 \\
0
\end{bmatrix}
$$

Carrying out the above matrix operations we get the following member-end force vectors for the three elements:

Element →	1	2	3
U_1	6.39	1.43	−8.11
V_1	13.56	6.39	11.97
M_1	1\,007.23	−620.83	1\,197.44
U_2	−6.39	−1.43	8.11
V_2	−13.56	26.60	−11.97
M_2	620.83	−1\,197.44	1\,047.85

LABORATORY EXPERIMENTS

Truss Element Stiffness

Apply a load U at the end of a wire and measure the end deflection u. Show that $P = Ku$ with $K = EA/l$.

Frame Structure Stiffness

For the frame of Fig. L12.1 a horizontal and vertical (upward) load can be applied by a bucket and pulley system as shown. Apply a horizontal load and a gradually increasing vertical load until the end point remains at the same vertical position as the unloaded structure. Measure the horizontal deflection and scale the loads so that this horizontal deflection is 1. Call the horizontal

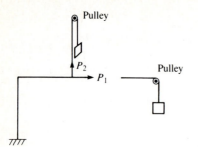

FIGURE L12.1

and vertical loads thus obtained K_{11} and K_{21}. Then remove the loads and apply a vertical load and a gradually increasing horizontal load so that the horizontal position of the end is the same as that of the unloaded structure. Scale these loads so that the vertical displacement is 1. Call these loads K_{12} and K_{22}.

Using the stiffness method and the dimensions and properties of the structure find the structure stiffness matrix relating P_1 and P_2 to the corresponding end displacements r_1 and r_2. Compare this stiffness matrix with the terms K_{ij} found above.

PROBLEMS

12.1. The origin of the global coordinate system for the truss of Fig. P12.1 is at joint 1. The x and y axes are to the right and upward. The element and joint numbers are indicated on the figure. The first and second joints for element 1 are joints 2 and 1, while that of elements 2 and 3 are 1, 3 and 4, 1 respectively. $A = 5$ in² for all bars; $E = 29\,000$ ksi.
  (a) Find the element stiffness matrices in the local coordinates and the transformation matrices for these three elements.
  (b) If the displacements of joints 1 and 3 are found to be -0.03 and -0.02 in in the vertical direction and zero in the horizontal direction, find the member-end forces for the three elements.

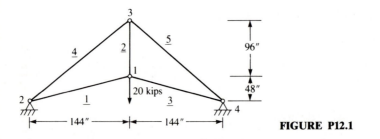

FIGURE P12.1

12.2. Determine the displacements of joints 2 and 3 and the bar forces of the truss of Fig. P12.2 using the stiffness method. The cross-sectional areas of the bars is 3 cm² and $E = 200\,000$ MPa.

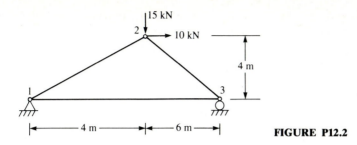

FIGURE P12.2

12.3. Calculate the joint displacements and bar forces of the truss of Fig. P12.3 with the stiffness method. The cross-sectional areas of the bars are 10 in^2 for bars 13 and 34 and 5 in^2 for the others, $E = 29\,000$ ksi.

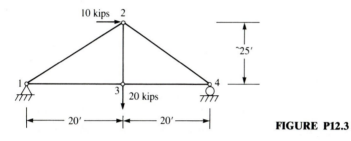

FIGURE P12.3

12.4. Find the bar forces for the members of the truss in Fig. P12.4. The area of the cross-section for all the bars is 4 in^2 and $E = 29\,000$ ksi.

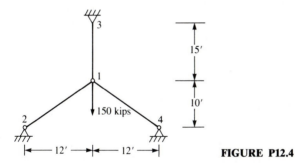

FIGURE P12.4

12.5. Determine the bar forces for the truss of Fig. P12.5. The area of the cross-section of the bars is 2 cm^2, $E = 200,000$ MPa.

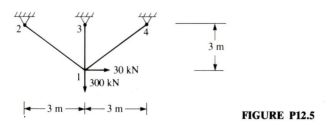

FIGURE P12.5

12.6. Determine the bar forces for the members of the truss in Fig. P12.6. E and A are constant and the same for all the members.

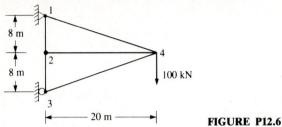

FIGURE P12.6

12.7. The area of the cross-section of bars 13 and 35 is 15 in^2 and that of the other members 10 in^2. Calculate the bar forces for the members. $E = 2\,000$ ksi.

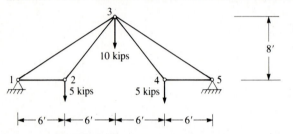

FIGURE P12.7

12.8. For the structure of Fig. P12.8 the cross-sectional area of all the bars is 2 in^2 and $E = 29\,000$ ksi. Calculate the bar forces.

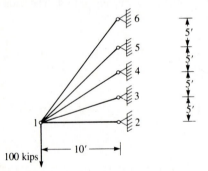

FIGURE P12.8

12.9. Calculate the bar forces of the truss of Fig. P12.9 for the cross-sectional area of the bars equal 3 in^2 and $E = 29\,000$ ksi.

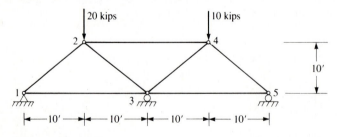

FIGURE P12.9

12.10. Determine the bar forces for the truss of Fig. P12.10. The area of cross-section of the bars is $5\,\text{cm}^2$ and $E = 200\,000$ MPa.

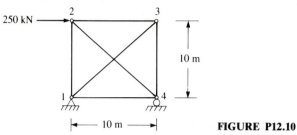

250 kN

10 m

10 m

FIGURE P12.10

Find the member-end forces for the following frames using the 2×2 beam element stiffness matrix.

12.11. $E = 29\,000$ ksi.   **12.12.** $E = 200\,000$ MPa.

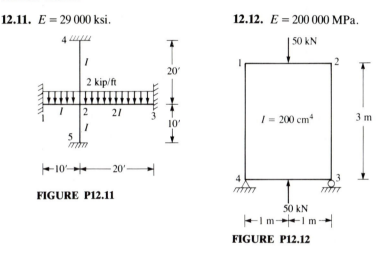

I

20'

2 kip/ft

I   2   2*I*   3

10'

5

|←10'→|←──20'──→|

FIGURE P12.11

50 kN

1   2

$I = 200\,\text{cm}^4$ 3 m

4   3

50 kN

|←1 m→|←1 m→|

FIGURE P12.12

Find the member-end forces for the following frames using the 4×4 beam element stiffness matrix.

12.13. The moment of inertia of the beam is $400\,\text{in}^4$ and that of the column $300\,\text{in}^4$. $E = 29\,000$ ksi. Use two 7.5 ft elements for the column and two 5 ft elements for the beam.

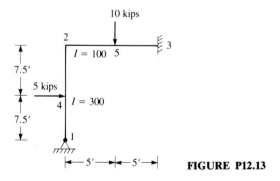

10 kips

2

$I = 100$ 5   3

7.5'

5 kips

4    $I = 300$

7.5'

1

|←5'→|←5'→| **FIGURE P12.13**

12.14. $I = 8\,000\,\text{cm}^4$; $E = 200\,000\,\text{MPa}$.

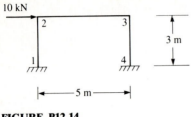

10 kN

FIGURE P12.14

12.15. $E = 29\,000\,\text{ksi}$.

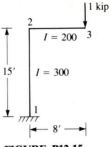

FIGURE P12.15

12.16. $I = 100\,\text{in}^4$; $E = 29\,000\,\text{ksi}$.

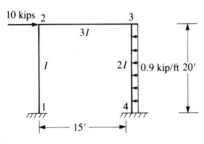

FIGURE P12.16

12.17. $E = 29\,000\,\text{ksi}$.

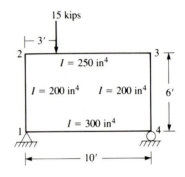

FIGURE P12.17

12.18. $E = 29\,000\,\text{ksi}$.

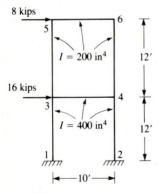

FIGURE P12.18

12.19. $E = 200\,000\,\text{MPa}$.

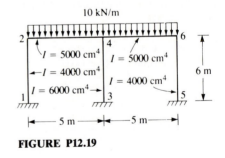

FIGURE P12.19

Find the member-end forces for the following frames using the 6×6 beam element stiffness matrix.

12.20. $I = 2\,000$ in^4; $A = 250$ in^2; $E = 3\,000$ ksi.

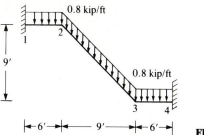

0.8 kip/ft

0.8 kip/ft

9′

|←6′→|← 9′ →|←6′→|

FIGURE P12.20

12.21. $I = 1\,000$ cm^4; $A = 100$ cm^2; $E = 200\,000$ MPa.

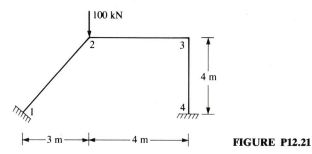

100 kN

4 m

|← 3 m →|← 4 m →|

FIGURE P12.21

12.22. $I = 8\,000$ cm^4; $E = 20\,000$ MPa; $A = 750$ cm^2.

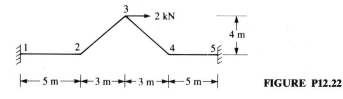

2 kN

4 m

|← 5 m →|← 3 m →|← 3 m →|← 5 m →|

FIGURE P12.22

12.23. $I = 200$ in^4; $E = 29\,000$ ksi; $A = 12$ in^2.

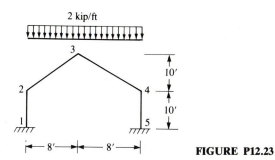

2 kip/ft

10′

10′

|← 8′ →|← 8′ →|

FIGURE P12.23

The following structures are composed of beam and truss elements. Using the 6×6 stiffness matrices for the beams and 4×4 stiffness matrices in the global coordinates for the truss elements find the forces in the members. The truss elements are indicated by the hinges at both ends of the elements.

12.24. $E = 29\,000$ ksi. For the beam $A = 10$ in^2, $I = 200$ in^4, for the cable $A = 1$ in^2.

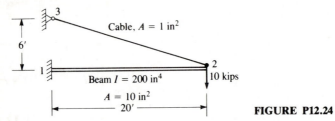

FIGURE P12.24

12.25. For the beam members 24, 34, 45, $I = 4\,000$ cm^4, $A = 40$ cm^2; $A = 30$ cm^2 for truss elements 12, 65. $E = 200\,000$ MPa. For the cables 13, 36 (also modeled as truss elements when in tension) $A = 4$ cm^2.

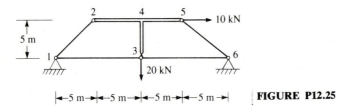

FIGURE P12.25

12.26. A cable-stayed bridge is shown in Fig. P12.26. The middle pier can be modeled by a roller, point 3. Using beam and truss elements find the joint displacements and bar forces under the gravity load of 50 kN/m. For the beam elements, $A = 0.7$ m^2, $I = 0.15$ m^4, and $E = 20\,000$ MPa. For the cables, $A = 20$ cm^2 and $E = 200\,000$ MPa.

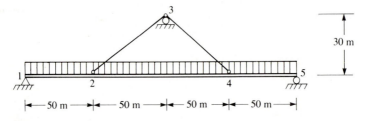

FIGURE P12.26

12.27. Figure P12.27 shows the model of a bow-string. For the beam elements, $I = 150$ in^4, $A = 18$ in^2. For the cables, $A = 0.5$ in^2, $E = 29\,000$ ksi. Find the joint displacements and the member-end forces.

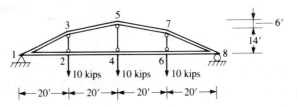

FIGURE P12.27

CHAPTER 13

ADDITIONAL TOPICS IN MATRIX METHODS

In this chapter we discuss some additional topics in the stiffness method. These include support settlements, internal hinges, elastic supports, flexible connections, variable sections, temperature effects, inclined rollers and influence lines by the stiffness method. The last section is an introduction to the flexibility method.

13.1 SUPPORT SETTLEMENT

Support settlements may induce significant stresses in structures. If the support settlement is significant, then it must be included in the analysis. The amount of settlement can be estimated by the methods of soil mechanics and depends on the type and depth of the compressible materials on which the foundation rests. If settlement is possible but, because of lack of data, cannot be estimated, then it would be better to use a statically determinate structure. Statically determinate structures may be less economical than indeterminate ones. However, their stresses are not affected by support settlements.

The effect of settlement in structures may be considered in two ways. The first is by adding the fixed-end forces induced by the settlement to the fixed-end forces due to loads. If this procedure is used, the settlement displacements must be set equal to zero when calculating the member forces, since their effects are already included as fixed-end forces. The member forces due to support settlement may also be calculated separately and added to those due to the loads.

An alternative procedure is to specify the settlement displacements in the equilibrium equations. In this case no fixed-end forces due to settlement must be included. If the alternative procedure is used, the terms in the load vector associated with support displacements are equal to the sum of equivalent joint forces due to loads and the unknown joint forces due to the settlement. These unknown forces, in general, need not be determined. The equilibrium equations can be solved for the unknown displacements. Using these displacements and the values of support settlements in the element force–displacement relation, the member forces can be determined. Although the former alternative is easier to apply the latter procedure is used in the solution of the following example for the purpose of demonstration.

With the latter procedure, the steps in the solution of structures with support settlements are much the same as those without settlement, discussed in Chapters 11 and 12. The only difference is that the displacement degrees of freedom must be considered, and that the equilibrium equations must be rearranged to place the support displacements together, so that the unknown displacements can be found by partitioning the matrices.

Example 13.1. Find the end forces for a three-span beam with equal span lengths of 10 m (Fig. 13.1a). Supports 2 and 3 settle by 1.0 and 8.0 cm, respectively. The beam is under a uniform load of 12 kN/m; $E = 200\,000$ MPa, and $I = 0.04$ m^4.

Solution. The degrees of freedom are the displacements and rotations of supports 2 and 3. These are denoted by r_1, r_2, r_3, and r_4 (Fig. 13.1a). r_1 and r_3 are specified as -0.01 and -0.08 m, respectively. The force–displacement relations for the members with displacement and rotation degrees of freedom (Eq.

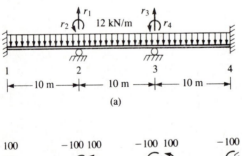

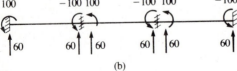

(b)   FIGURE 13.1

(11.10a)), are

$$
\begin{bmatrix} V_1 \\ M_1 \\ V_2 \\ M_2 \end{bmatrix} = \begin{matrix} \begin{matrix} v_1 & \theta_1 & v_2 & \theta_2 \end{matrix} \\ \begin{bmatrix} \dfrac{12EI}{l^3} & \dfrac{6EI}{l^2} & -\dfrac{12EI}{l^3} & \dfrac{6EI}{l^2} \\[2mm] \dfrac{6EI}{l^2} & \dfrac{4EI}{l} & -\dfrac{6EI}{l^2} & \dfrac{2EI}{l} \\[2mm] -\dfrac{12EI}{l^3} & -\dfrac{6EI}{l^2} & \dfrac{12EI}{l^3} & -\dfrac{6EI}{l^2} \\[2mm] \dfrac{6EI}{l^2} & \dfrac{2EI}{l} & -\dfrac{6EI}{l^2} & \dfrac{4EI}{l} \end{bmatrix} \end{matrix} \begin{bmatrix} v_1 \\ \theta_1 \\ v_2 \\ \theta_2 \end{bmatrix} + \begin{bmatrix} V_1^F \\ M_1^F \\ V_2^F \\ M_2^F \end{bmatrix}
$$

Since I and l are the same for the three spans, the stiffness matrices will be the same for the three elements. The force–displacement relations, along with the structure degrees of freedom for the three elements, are

$$
\begin{bmatrix} V_1 \\ M_1 \\ V_2 \\ M_2 \end{bmatrix}^1 = \begin{matrix} \begin{matrix} 0 & \quad 0 & \quad r_1 & \quad r_2 \end{matrix} \\ \begin{bmatrix} 96 & 480 & -96 & 480 \\ 480 & 3200 & -480 & 1\,600 \\ -96 & -480 & 96 & -480 \\ 480 & 1\,600 & -480 & 3\,200 \end{bmatrix} \end{matrix} \begin{bmatrix} 0 \\ 0 \\ r_1 \\ r_2 \end{bmatrix} + \begin{bmatrix} 60 \\ 100 \\ 60 \\ -100 \end{bmatrix}
$$

$$
\begin{bmatrix} V_1 \\ M_1 \\ V_2 \\ M_2 \end{bmatrix}^2 = \begin{matrix} \begin{matrix} r_1 & \quad r_2 & \quad r_3 & \quad r_4 \end{matrix} \\ \begin{bmatrix} 96 & 480 & -96 & 480 \\ 480 & 3\,200 & -480 & 1\,600 \\ -96 & -480 & 96 & -480 \\ 480 & 1\,600 & -480 & 3\,200 \end{bmatrix} \end{matrix} \begin{bmatrix} r_1 \\ r_2 \\ r_3 \\ r_4 \end{bmatrix} = \begin{bmatrix} 60 \\ 100 \\ 60 \\ -100 \end{bmatrix}
$$

$$
\begin{bmatrix} V_1 \\ M_1 \\ V_2 \\ M_2 \end{bmatrix}^3 = \begin{matrix} \begin{matrix} r_3 & \quad r_4 & \quad 0 & \quad 0 \end{matrix} \\ \begin{bmatrix} 96 & 480 & -96 & 480 \\ 480 & 3\,200 & -480 & 1\,600 \\ -96 & -480 & 96 & -480 \\ 480 & 1\,600 & -480 & 3\,200 \end{bmatrix} \end{matrix} \begin{bmatrix} r_3 \\ r_4 \\ 0 \\ 0 \end{bmatrix} + \begin{bmatrix} 60 \\ 100 \\ 60 \\ -100 \end{bmatrix}
$$

The fixed-end forces, calculated using Table 10.1a, are shown in Fig. 13.1b.

The structure stiffness matrix is 4×4 and is found by adding the terms of the element stiffness matrices to the pertinent locations in the structure stiffness matrix. The elements of the load vector are the equivalent joint forces (negative of sum of the fixed-end forces for each degree of freedom). Thus, the equilibrium equations are

$$
\begin{matrix} \begin{matrix} r_1 & \quad r_2 & \quad r_3 & \quad r_4 \end{matrix} \\ \begin{bmatrix} 192 & 0 & -96 & 480 \\ 0 & 6\,400 & -480 & 1\,600 \\ -96 & -480 & 192 & 0 \\ 480 & 1\,600 & 0 & 6\,400 \end{bmatrix} \end{matrix} \begin{bmatrix} r_1 \\ r_2 \\ r_3 \\ r_4 \end{bmatrix} = \begin{bmatrix} -120 \\ 0 \\ -120 \\ 0 \end{bmatrix}
$$

Since r_1 and r_3 have known values of -0.01 and $-0.08\,\text{m}$, by rearranging the equilibrium equations we get the following matrix equation. Note that when switching rows i and j of the structure stiffness matrix, to place the unknown displacements near each other, columns i and j must also be switched.

$$\begin{bmatrix} 6\,400 & 1\,600 & 0 & -480 \\ 1\,600 & 6\,400 & 480 & 0 \\ 0 & 480 & 192 & -96 \\ -480 & 0 & -96 & 192 \end{bmatrix} \begin{bmatrix} r_2 \\ r_4 \\ -0.01 \\ -0.08 \end{bmatrix} = \begin{bmatrix} 0 \\ 0 \\ F_1 - 120 \\ F_2 - 120 \end{bmatrix}$$

or

$$\begin{bmatrix} \mathbf{K}_{11} & \mathbf{K}_{12} \\ \mathbf{K}_{21} & \mathbf{K}_{22} \end{bmatrix} \begin{bmatrix} r_2 \\ r_4 \\ -0.01 \\ -0.08 \end{bmatrix} = \begin{bmatrix} 0 \\ 0 \\ F_1 - 120 \\ F_2 - 120 \end{bmatrix}$$

In the above equations, F_1 and F_2 are the joint forces due to settlement. Although their values can be calculated from the above equations, in general this is not necessary. Expanding the above equations in terms of submatrices, we get

$$\mathbf{K}_{11}\begin{bmatrix} r_2 \\ r_4 \end{bmatrix} + \mathbf{K}_{12}\begin{bmatrix} -0.01 \\ -0.08 \end{bmatrix} = \begin{bmatrix} 0 \\ 0 \end{bmatrix}$$

or

$$\begin{bmatrix} r_2 \\ r_4 \end{bmatrix} = \mathbf{K}_{11}^{-1}\left(\begin{bmatrix} 0 \\ 0 \end{bmatrix} - \mathbf{K}_{12}\begin{bmatrix} -0.01 \\ -0.08 \end{bmatrix}\right) = \begin{bmatrix} 166.7 & -41.7 \\ -41.7 & 166.7 \end{bmatrix}\begin{bmatrix} -38.4 \\ 4.8 \end{bmatrix} = \begin{bmatrix} -0.0066 \\ 0.0024 \end{bmatrix}$$

Thus, the structure displacements, including the support settlement, are

$$\begin{bmatrix} r_1 \\ r_2 \\ r_3 \\ r_4 \end{bmatrix} = \begin{bmatrix} -0.01 \\ -0.0066 \\ -0.08 \\ 0.0024 \end{bmatrix}$$

The member forces can be found from $\mathbf{S} = \mathbf{K}\mathbf{s} + \mathbf{S}^F$ given above for the three elements. $\mathbf{s}$ is the element displacement vector which is now determined for each element using the values just found. Thus, the member-end forces are

$$\begin{bmatrix} V_1 \\ M_1 \\ V_2 \\ M_2 \end{bmatrix}^1 = \begin{bmatrix} 96 & 480 & -96 & 480 \\ 480 & 3\,200 & -480 & 1\,600 \\ -96 & -480 & 96 & -480 \\ 480 & 1\,600 & -480 & 3\,200 \end{bmatrix}\begin{bmatrix} 0 \\ 0 \\ -0.01 \\ -0.0066 \end{bmatrix} + \begin{bmatrix} 60 \\ 100 \\ 60 \\ -100 \end{bmatrix} = \begin{bmatrix} 57.79 \\ 94.24 \\ 62.21 \\ -116.32 \end{bmatrix}$$

$$\begin{bmatrix} V^1 \\ M_1 \\ V_2 \\ M_2 \end{bmatrix}^2 = \begin{bmatrix} 96 & 480 & -96 & 480 \\ 480 & 3\,200 & -480 & 1\,600 \\ -96 & -480 & 96 & -480 \\ 480 & 1\,600 & -480 & 3\,200 \end{bmatrix}\begin{bmatrix} -0.01 \\ -0.0066 \\ -0.08 \\ 0.0024 \end{bmatrix} + \begin{bmatrix} 60 \\ 100 \\ 60 \\ -100 \end{bmatrix} = \begin{bmatrix} 64.7 \\ 116.32 \\ 55.30 \\ -69.28 \end{bmatrix}$$

$$\begin{bmatrix} V_1 \\ M_1 \\ V_2 \\ M_2 \end{bmatrix}^3 = \begin{bmatrix} 96 & 480 & -96 & 480 \\ 480 & 3\,200 & -480 & 1\,600 \\ -96 & -480 & 96 & -480 \\ 480 & 1\,600 & -480 & 3\,200 \end{bmatrix}\begin{bmatrix} -0.08 \\ 0.0024 \\ 0 \\ 0 \end{bmatrix} + \begin{bmatrix} 60 \\ 100 \\ 60 \\ -100 \end{bmatrix} = \begin{bmatrix} 53.47 \\ 69.28 \\ 66.53 \\ -134.56 \end{bmatrix}$$

13.2 INTERNAL HINGES

In the stiffness method, internal hinges can be handled in two ways. One way is to derive special force–displacement relations by specifying that the moment is zero at the hinge in Eq. (11.10a). The rotation at the hinge can be found and the end forces expressed in terms of the other joint displacements and rotation. This would lead to a reduced stiffness matrix and modified fixed-end forces. Although not as common in practice, connections that do not allow transfer of shear or axial force can also be handled in a similar manner.

Another alternative is to use two different degrees of freedom for the rotations on the two sides of the hinge. This would increase the number of degrees of freedom by one for each hinge. However, since this procedure uses the standard force–displacement relations, and gives the rotations on both sides of the hinge directly, it is easier to apply. The latter procedure is used in the following example.

Example 13.2. Find the member-end forces for the beam of Fig. 13.2 with member 2 hinged to members 1 and 3 at its ends. The right end of member 2 is on support 2. $E = 29\,000$ ksi and $I = 100$ in^4.

Solution. The degrees of freedom are shown in Fig. 13.2. Member 1 is hinged at its right end, while member 2 is hinged at both ends, and member 3 at its left end. The rotations at the left and right of the first hinge are denoted by r_2 and r_3 and those on the left and right of the second hinge by r_4 and r_5. r_1 is the vertical displacement of the first hinge.

The force–displacement relations for the elements (Eq. 11.10a)) are shown below. The fixed-end forces are calculated using Table 10.1a.

$$
\begin{bmatrix} V_1 \\ M_1 \\ V_2 \\ M_2 \end{bmatrix}^1 =
\begin{matrix}
& 0 & 0 & r_1 & r_2 \\
& \begin{bmatrix} 20 & 1\,208 & -20 & 1\,208 \\ 1\,208 & 96\,667 & -1\,208 & 48\,333 \\ -20 & -1\,208 & 1\,20 & -1\,208 \\ 1\,208 & 48\,333 & -1\,208 & 96\,667 \end{bmatrix}
\end{matrix}
\begin{bmatrix} 0 \\ 0 \\ r_1 \\ r_2 \end{bmatrix} +
\begin{bmatrix} 15 \\ 300 \\ 15 \\ -300 \end{bmatrix}
$$

$$
\begin{bmatrix} V_1 \\ M_1 \\ V_2 \\ M_2 \end{bmatrix}^2 =
\begin{matrix}
& r_1 & r_3 & 0 & r_4 \\
& \begin{bmatrix} 20 & 1\,208 & -20 & 1\,208 \\ 1\,208 & 96\,667 & -1\,208 & 48\,333 \\ -20 & -1\,208 & 20 & -1\,208 \\ 1\,208 & 48\,333 & -1\,208 & 96\,667 \end{bmatrix}
\end{matrix}
\begin{bmatrix} r_1 \\ r_3 \\ 0 \\ r_4 \end{bmatrix} +
\begin{bmatrix} 6 \\ 180 \\ 6 \\ -180 \end{bmatrix}
$$

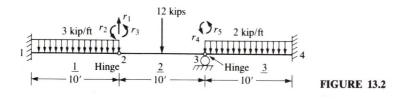

FIGURE 13.2

$$
\begin{bmatrix} V_1 \\ M_1 \\ V_2 \\ M_2 \end{bmatrix}^3 = \begin{matrix} 0 & r_5 & 0 & 0 \\ \begin{bmatrix} 20 & 1\,208 & -20 & 1\,208 \\ 1\,208 & 96\,667 & -1\,208 & 48\,333 \\ -20 & -1\,208 & 20 & -1\,208 \\ 1\,208 & 48\,333 & -1\,208 & 96\,667 \end{bmatrix} \end{matrix} \begin{bmatrix} 0 \\ 0 \\ r_5 \\ 0 \end{bmatrix} + \begin{bmatrix} 10 \\ 200 \\ 10 \\ -200 \end{bmatrix}
$$

The structure stiffness matrix is 5×5 and the load vector is 5×1. The elements of the structures stiffness matrix are found by assembling the element stiffness matrices shown in the above force–displacement relations. The elements of the load vector are the equivalent joint forces, that is, the negative of sum of the fixed-end forces acting at each degree of freedom. These result in the following equilibrium equations

$$
\begin{matrix} r_1 & r_2 & r_3 & r_4 & r_5 \end{matrix}
$$
$$
\begin{bmatrix} 40 & -1\,208 & 1\,208 & 1\,208 & 0 \\ -1\,208 & 96\,667 & 0 & 0 & 0 \\ 1\,208 & 0 & 96\,667 & 48\,333 & 0 \\ 1\,208 & 0 & 48\,333 & 96\,667 & 0 \\ 0 & 0 & 0 & 0 & 96\,667 \end{bmatrix} \begin{bmatrix} r_1 \\ r_2 \\ r_3 \\ r_4 \\ r_5 \end{bmatrix} = \begin{bmatrix} -21 \\ 300 \\ -180 \\ 180 \\ -200 \end{bmatrix}
$$

The unknown displacements and rotations are found from the solution of the above equilibrium equations, yielding

$$
\begin{bmatrix} r_1 \\ r_2 \\ r_3 \\ r_4 \\ r_5 \end{bmatrix} = \begin{bmatrix} -3.4262 \\ -0.0397 \\ 0.0248 \\ 0.0323 \\ -0.0021 \end{bmatrix}
$$

Upon substitution of the values for displacements and rotations, the element force–displacement relations given above yield the member-end force as

$$
\begin{matrix} \text{Element} \rightarrow & 1 & 2 & 3 \end{matrix}
$$
$$
\mathbf{S} = \begin{bmatrix} V_1 \\ M_1 \\ V_2 \\ M_2 \end{bmatrix} : \begin{bmatrix} 36 \\ 2\,520 \\ -6 \\ 0 \end{bmatrix} \begin{bmatrix} 6 \\ 0 \\ 6 \\ 0 \end{bmatrix} \begin{bmatrix} 7.5 \\ 0 \\ 12.5 \\ -300 \end{bmatrix}
$$

13.3   ELASTIC SUPPORTS

In some cases the supports are not unyielding but deform under the applied loads. Elastic supports, or supports provided for a structure by other elastic members, can be modeled by elastic springs. To determine the stiffness of the support spring, the force–displacement relation must be determined for the elastic support. This is done by applying an arbitrary force at that point and finding the corresponding displacement there. The ratio of the applied load to the displacement is the stiffness of the elastic support. If the supporting element provides resistance against rotation, then a torsional spring is used to model the support stiffness of the members. Torsional stiffness is discussed in the Section 17.2.

Structure Stiffness Matrix Including Elastic Supports

Analysis of structures with elastic supports by the stiffness method is quite similar to that of structures with rigid supports. The only difference is the treatment of the elastic supports.

To find a procedure for solving elastically supported structures with the stiffness method, consider the structure of Fig. 13.3a. This is a two-span beam with span lengths l_1 and l_2, moments of inertia I_1 and I_2, and modulus of elasticity E. The two ends of the beam are fixed with a load P_1 and moment M_2 acting at joint 2, and arbitrarily loads applied over the two spans. At joint 2 the beam is on an elastic support with vertical and rotational stiffness K_v and K_θ, respectively. We want to find the displacement r_1 and rotation r_2 at joint 2 and the forces at the ends of the two elements.

As was done in Section 11.2, the direct stiffness procedure can be derived by writing the equilibrium of the joints in terms of the joint displacements. For joint 2 of the elastically supported beam in Fig. 13.3a, the external and internal forces are shown in Fig. 13.3b. Equilibrium of the vertical forces and moments at this joint give

$$F_v + V_2^1 + V_1^2 = P_1 \tag{13.1a}$$

$$M_\theta + M_2^1 + M_1^2 = M_2 \tag{13.1b}$$

where F_v and M_θ are the force and moment induced in the vertical and torsional springs. V_2^1 and M_2^1 are the shear and moment at the right end of

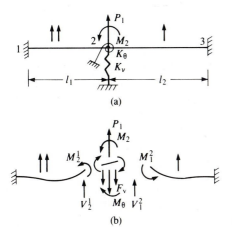

(a)

(b)   **FIGURE 13.3**

element 1, and V_1^2, M_1^2 are the shear and moment at the left end of element 2. If r_1 and r_2 are the displacement and rotation of joint 2, then the elastic forces in the springs are,

$$F_v = -K_v r_1 \qquad (13.2a)$$

$$M_\theta = -K_\theta r_2 \qquad (13.2b)$$

The member-end forces in Eqs. (13.1) can be expressed in terms of the member-end displacements by Eqs. (10.6), (10.7), (10.10) and (10.11). Since the displacement and rotation at the left end of the first element and at the right end of the second element are zero, the forces in terms of r_1 and r_2 become

$$V_2^1 = \frac{12EI_1}{l_1^3} r_1 - \frac{6EI_1}{l_1^2} r_2 + V_2^{F_1} \qquad (13.3a)$$

$$M_2^1 = -\frac{6EI_1}{l_1^2} r_1 + \frac{4EI_1}{l_1} r_2 + M_2^{F_1} \qquad (13.3b)$$

$$V_1^2 = \frac{12EI_2}{l_2^3} r_1 - \frac{6EI_2}{l_2^2} r_2 + V_1^{F_2} \qquad (13.3c)$$

$$M_1^2 = \frac{6EI_2}{l_2^2} r_1 + \frac{4EI_2}{l_2} r_2 + M_1^{F_2} \qquad (13.3d)$$

Here the superscripts for shear and moment denote the element numbers. Substituting Eqs. (13.2) and (13.3) in the equilibrium equations, Eqs. (13.1a) and (13.1b), and casting the latter two equations in matrix form, we get

$$\begin{bmatrix} \dfrac{12EI_1}{l_1^3} + \dfrac{12EI_2}{l_2^3} + K_v & -\dfrac{6EI_1}{l_1^2} + \dfrac{6EI_2}{l_2^2} \\[2ex] -\dfrac{6EI_1}{l_1^2} + \dfrac{6EI_2}{l_2^2} & \dfrac{4EI_1}{l_1} + \dfrac{4EI_2}{l_2} + K\theta \end{bmatrix} \begin{bmatrix} r_1 \\ r_2 \end{bmatrix} = \begin{bmatrix} P_1 - (V_2^{F_1} + V_1^{F_2}) \\ M_2 - (M_2^{F_1} + M_1^{F_2}) \end{bmatrix} \qquad (13.4)$$

Examination of the above equations reveals that the equilibrium equations for structures with elastic supports can be found in a manner similar to the direct stiffness method for structures with rigid supports (Chapters 11 and 12). The only difference is that after assembling the structure stiffness matrix, the stiffness of the support springs must be added to the diagonal terms of this matrix associated with the degrees of freedom of the support deformations. The member-end forces are found in exactly the same way as before, using the force–displacement relations for the elements. The forces in the support springs can be found by Eqs. (13.2a) and (13.2b).

Example 13.3. Find the displacements and member forces for the beam of Fig. 13.3a if the two spans are equal and 30 ft long. The beam is supported at point 2 by a spring with axial stiffness $K_v = 10$ kips/in ($K_\theta = 0$). The moment of inertia for both spans is 300 in⁴. The only load is a 9-kip concentrated gravity load acting at point 2. $E = 29\,000$ ksi.

Solution. There are two degrees of freedom, the displacement and rotation of the beam at point 2. The stiffness matrices for the two elements are identical. They are calculated by Eq. (11.10a) and given below with the associated structure degrees of freedom. Since there is vertical displacement at point 2, the 4×4 element stiffness matrices that include displacement and rotation degrees of freedom must be used. The element force–displacement relations for the two elements are,

$$\begin{bmatrix} V_1 \\ M_1 \\ V_2 \\ M_2 \end{bmatrix}^1 = \begin{bmatrix} 2.24 & 402.78 & -24 & 402.78 \\ 402.78 & 96\,666.67 & -402.78 & 48\,333.33 \\ -2.24 & -402.78 & 2.24 & -402.78 \\ 402.78 & 48\,333.33 & -402.78 & 96\,666.67 \end{bmatrix} \begin{matrix} 0 \\ 0 \\ r_1 \\ r_2 \end{matrix}$$

$$\begin{matrix} 0 & 0 & r_1 & r_2 \end{matrix}$$

$$\begin{bmatrix} V_1 \\ M_1 \\ V_2 \\ M_2 \end{bmatrix}^2 = \begin{bmatrix} 2.24 & 402.78 & -2.24 & 402.78 \\ 402.78 & 96\,666.67 & -402.78 & 48\,333.33 \\ -2.24 & -402.78 & 2.24 & -402.78 \\ 402.78 & 48\,333.33 & -402.78 & 96\,666.67 \end{bmatrix} \begin{matrix} r_1 \\ r_2 \\ 0 \\ 0 \end{matrix}$$

$$\begin{matrix} r_1 & r_2 & 0 & 0 \end{matrix}$$

The structure stiffness matrix is 2×2 and is found by assembling the element stiffness matrices as discussed in Section 11.2. Without considering the elastic support, the structure stiffness matrix is

$$\mathbf{K}'_s = \begin{bmatrix} 4.48 & 0 \\ 0 & 193\,333 \end{bmatrix} \begin{matrix} r_1 \\ r_2 \end{matrix}$$

$$\begin{matrix} r_1 & r_2 \end{matrix}$$

The spring provides resistance in the vertical direction, that is, against displacement r_1. Thus, we must add the stiffness of the spring, 10 kip/in, to location 1, 1 of the structure stiffness matrix. The loads in degrees of freedom 1 and 2 are -9 and 0. The equilibrium equations are thus

$$\begin{bmatrix} 14.48 & 0 \\ 0 & 193\,333 \end{bmatrix} \begin{bmatrix} r_1 \\ r_2 \end{bmatrix} = \begin{bmatrix} -9.00 \\ 0 \end{bmatrix}$$

$$\begin{matrix} r_1 & r_2 \end{matrix}$$

or

$$\mathbf{K}_s \mathbf{r} = \mathbf{R}$$

The rest of the solution process is the same as for the other structures, as outlined at the end of Section 11.2. The displacements are found from the equilibrium equations as

$$\begin{bmatrix} r_1 \\ r_2 \end{bmatrix} = \begin{bmatrix} -0.62 \\ 0 \end{bmatrix}$$

The member displacement vectors are

$$\text{Element} \rightarrow \qquad 1 \qquad\qquad 2$$

$$\mathbf{s} = \begin{bmatrix} v_1 \\ \theta_1 \\ v_2 \\ \theta_2 \end{bmatrix} : \quad \begin{bmatrix} 0 \\ 0.00 \\ -0.62 \\ 0 \end{bmatrix} \quad \begin{bmatrix} -0.62 \\ 0 \\ 0 \\ 0 \end{bmatrix}$$

The products of the element stiffness matrices and displacement vectors give the member force vectors, $S = Ks$. These are

$$
\text{Element} \rightarrow \qquad 1 \qquad\qquad 2
$$

$$
S = \begin{bmatrix} V_1 \\ M_1 \\ V_2 \\ M_2 \end{bmatrix} : \qquad \begin{bmatrix} 1.39 \\ 250.35 \\ -1.39 \\ 250.35 \end{bmatrix} \qquad \begin{bmatrix} -1.39 \\ -250.35 \\ 1.39 \\ -250.35 \end{bmatrix}
$$

The force in the spring is the product of the spring stiffness and the associated displacement, that is, $10 \times (-0.62) = -6.2$ kips (compression).

13.4 FLEXIBLE CONNECTIONS

In beams and frames it is usually assumed that the moment connections are rigid, that is, the rotations of the members at the two sides of the connection are the same. In Section 13.2 we discussed internal hinges, where the rotation on the two sides of the hinge are different. Of course in that case no moment is transferred from one side of the hinge to the other. In some cases, such as for some steel structures with bolted connections, the connections are not rigid but semi-rigid or flexible. In these cases the connections transfer moments but the rotation on the two sides of the connections are not the same. The internal forces are somewhere between those of hinged and rigid connections. Accurate evaluation of the internal forces requires the inclusion of the connection stiffness in the analysis. This can be done in two ways. The first one is to develop a member with springs at the ends. The end of the member is not directly connected to the other members but through the spring, representing the flexible connection.

Another procedure is to use a spring element for the connection and to represent the rotations of the two sides of the connection by two different degrees of freedom, as was done for internal hinges in Section 13.2.

A flexible bending connection is represented by a torsional spring (Figs. 13.4a and 13.4b). If the moments acting at the left and right of the spring are M_1 and M_2 and the rotations are θ_1 and θ_2, then equilibrium of moments requires that

$$
M_1 + M_2 = 0 \qquad \text{or} \qquad M_2 = -M_1 \tag{13.5a}
$$

The relative rotation of the right side with respect to the left is $\theta_2 - \theta_1$, and the moment causing this rotation calculated from the right is M_2. If the torsional stiffness of the spring is K_θ, then

$$
M_2 = K_\theta(\theta_2 - \theta_1) \tag{13.5b}
$$

In view of Eqs. (13.5a) and (13.5b) we can write M_1 and M_2 in terms r_1 and r_2. In matrix form we get

$$
\begin{matrix} \quad\theta_1 \quad\ \theta_2 \quad \end{matrix}
$$
$$
\begin{bmatrix} M_1 \\ M_2 \end{bmatrix} = K_\theta \begin{bmatrix} 1 & -1 \\ -1 & 1 \end{bmatrix} \begin{bmatrix} \theta_1 \\ \theta_2 \end{bmatrix} \tag{13.6}
$$

This is the moment–rotation relationship for a flexible bending connection. The stiffness matrix in Eq. (13.6) will be assembled along with the element stiffness matrices to form the structure stiffness m' rix. Similar expressions can be obtained for flexible shear and axial member connections.

Example 13.4. The frame of Fig. 13.4b has a flexible connection at joint 2 with stiffness $10^4 \, \text{kN} \cdot \text{m/rad}$. For both members the length is 6 m. $E = 200\,000$ MPa, and $I = 60\,000$ cm^4. A uniform gravity load of 20 kN/m is applied on the beam. Find the rotations at both sides of the connection and the member-end forces.

Solution. Since the rotations at the two sides of the connection are different, two different degrees of freedom, r_1 and r_2, are associated with them. The stiffness matrix is the same for both elements. Neglecting the axial deformations we use the 2×2 stiffness matrix, considering rotational degrees of freedom only. With the fixed-end moments found from Table 10.1a, the force–displacement relations for the two elements are

$$\begin{bmatrix} M_1 \\ M_2 \end{bmatrix} = \begin{bmatrix} 80\,000 & 40\,000 \\ 40\,000 & 80\,000 \end{bmatrix} \overset{\displaystyle 0 \qquad r_1}{\begin{bmatrix} 0 \\ r_1 \end{bmatrix}}$$

$$\begin{bmatrix} M_1 \\ M_2 \end{bmatrix} = \begin{bmatrix} 80\,000 & 40\,000 \\ 40\,000 & 80\,000 \end{bmatrix} \overset{\displaystyle r_2 \qquad 0}{\begin{bmatrix} r_2 \\ 0 \end{bmatrix}} + \begin{bmatrix} 60 \\ -60 \end{bmatrix}$$

The moment–rotation relationship for the flexible bending connection is (Eq. (13.6))

$$\begin{bmatrix} M_1 \\ M_2 \end{bmatrix} = \begin{bmatrix} 10\,000 & -10\,000 \\ -10\,000 & 10\,000 \end{bmatrix} \overset{\displaystyle r_1 \qquad r_2}{\begin{bmatrix} r_1 \\ r_2 \end{bmatrix}}$$

Since moments are the same in the local and global coordinates, no transformation is required. Assembly of the above stiffness matrices and load

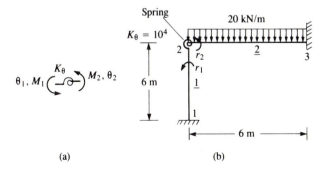

(a)   (b)

FIGURE 13.4

vector results in the following equilibrium equations:

$$\begin{matrix} r_1 & r_2 \\ \begin{bmatrix} 90\,000 & -10\,000 \\ -10\,000 & 90\,000 \end{bmatrix} \begin{bmatrix} r_1 \\ r_2 \end{bmatrix} = \begin{bmatrix} 0 \\ -60 \end{bmatrix} \end{matrix}$$

The solution of the above equations is

$$\begin{bmatrix} r_1 \\ r_2 \end{bmatrix} = -10^{-5} \begin{bmatrix} 7.5 \\ 67.5 \end{bmatrix}$$

Substitution of these values in the element force–displacement relations gives the following member-end moments in kN · m.

$$\begin{matrix} \text{Member} \rightarrow & 1 & 2 \\ \begin{bmatrix} M_1 \\ M_2 \end{bmatrix} : & \begin{bmatrix} -3.0 \\ -6.0 \end{bmatrix} & \begin{bmatrix} -6.0 \\ -87.0 \end{bmatrix} \end{matrix}$$

13.5   VARIABLE-SECTION MEMBERS

Members with variable cross-sections can be modeled by describing the variation by step changes of the moment of inertia and cross-sectional area. Each segment with constant cross-sectional area and moment of inertia may be considered as a separate element. For steep changes of cross-section this would result in a large number of elements. In such cases variable-section elements can be used, each element incorporating as many steps as required, thereby reducing the number of degrees of freedom substantially. The accuracy of the representation can be improved by increasing the number of steps without increasing the number of degrees of freedom.

Force–Displacement Relations for Variable-Section Beam Elements

The relations between the member-end forces and displacements are obtained by integration of the governing differential equation of the beam, applying the boundary conditions and invoking the equilibrium equations. The derivation follows the same steps as in Section 10.2 for beam elements with constant cross-sections. The step variations of the section properties are described by the step function as discussed in Section 9.5. Consider a variable-section beam element having n segments with different cross-sections as shown in Fig. 13.5.

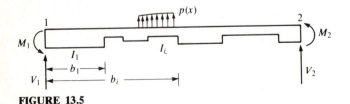

FIGURE 13.5

Using the analysis sign convention, with counterclockwise moments as positive, the bending moment at section x is

$$M = M_1 - V_1 x + m(x) \tag{13.7}$$

where $m(x)$ is the moment due to the applied loads (without their reactions), calculated at section x. The differential equation for the beam is

$$v'' = -\frac{M}{EI} \tag{13.8}$$

In view of the discussion of Section 9.5 for sections with step changes, we can express the flexibility EI as

$$\frac{1}{EI} = \frac{1}{EI_1} \left(1 + \sum_{i=1}^{n-1} r_i H_i \right) \tag{13.9a}$$

where $H_i = H(x - b_i)$ and n is the number segments.

$$\boxed{r_i = EI_1 \left[\frac{1}{EI_{i+1}} - \frac{1}{EI_i} \right]} \tag{13.9b}$$

Substitution of Eqs. (13.7) and (13.9a) into (13.8) results in

$$EI_1 v'' = -M_1 + V_1 x - m(x) - M_1 \sum_{i=1}^{n-1} r_i H_i + V_1 \sum_{i=1}^{n-1} r_i x H_i - \sum_{i=1}^{n-1} r_i m(x) H_i$$

Upon integration (Section 9.2) we get

$$EI_1 v' = -M_1 x + \tfrac{1}{2} V_1 x^2 - f_1(x) - M_1 \sum_{i=1}^{n-1} r_i(x - b_i) H_i$$

$$+ \tfrac{1}{2} V_1 \sum_{i=1}^{n-1} r_i(x^2 - b_i^2) H_i - \sum_{i=1}^{n-1} r_i f_{2i}(x) + c_1$$

$$EI_1 v = -\tfrac{1}{2} M_1 x^2 + \tfrac{1}{6} V_1 x^3 - g_1(x) - \tfrac{1}{2} M_1 \sum_{i=1}^{n-1} r_i(x - b_i)^2 H_i$$

$$+ \tfrac{1}{2} V_1 \sum_{i=1}^{n-1} r_i[\tfrac{1}{3}(x^3 - b_i^3) - b_i^2(x - b_i)] H_i - \sum_{i=1}^{n-1} r_i g_{2i}(x) + c_1 x + c_2$$

where

$$f_1(x) = \int_0^x m(x)\, dx \tag{13.10a}$$

$$f_{2i}(x) = \int_0^x m(x) H_i\, dx \tag{13.10b}$$

$$g_1(x) = \int_0^x f_1(x)\, dx \tag{13.10c}$$

$$g_{2i}(x) = \int_0^x f_{2i}(x)\, dx \tag{13.10d}$$

Applying the boundary conditions, at $x = 0$, $v' = \theta_1$, $v = v_1$, since the constants c_1 and c_2 were included in the integrations, which are the values of v' and v at $x = 0$ within a factor,

$$f_1(0) = f_{2i}(0) = g_1(0) = g_{2i}(0) = 0$$

we find $EI_1\theta_1 = c_1$, $EIv_1 = c_2$.

Also, at $x = l$, $v' = \theta_2$, $v = v_2$, upon substitution for c_1 and c_2 we get

$$EI_1\theta_2 = -M_1 l + \tfrac{1}{2}V_1 l^2 - f_1(l) - M_1 \sum_{i=1}^{n-1} r_i(l - b_i)H_i + \tfrac{1}{2}V_1 \sum_{i=1}^{n-1} r_i(l^2 - b_i^2)$$

$$- \sum_{i=1}^{n-1} r_i f_{2i}(l) + EI_1\theta_1$$

$$EIv_2 = -\tfrac{1}{2}M_1 l^2 + \tfrac{1}{6}V_1 l^3 - g_1(l) - \tfrac{1}{2}M_1 \sum_{i=1}^{n-1} r_i(l - b_i)^2$$

$$+ \tfrac{1}{2}V_1 \sum_{i=1}^{n-1} r_i[\tfrac{1}{3}(l^3 - b_i^3) - b_i^2(l - b_i)] - \sum_{i=1}^{n-1} r_i g_{2i}(l) + EI_1 l \theta_1 + EI_1 v_1$$

f_1, f_{2i}, g_1, and g_{2i} at $x = l$ are found upon integration of Eqs. (13.10) and substitution of l for x. The last two equations in matrix form become

$$\begin{bmatrix} \alpha_1 & \alpha_2 \\ \alpha_3 & \alpha_4 \end{bmatrix} \begin{bmatrix} V_1 \\ M_1 \end{bmatrix} = EI_1 \begin{bmatrix} 0 & -1 & 0 & 1 \\ -1 & -l & 1 & 0 \end{bmatrix} \begin{bmatrix} v_1 \\ \theta_1 \\ v_2 \\ \theta_2 \end{bmatrix} + \begin{bmatrix} f \\ g \end{bmatrix} \qquad (13.11)$$

with

$$\alpha_1 = \tfrac{1}{2}\left[l^2 + \sum_{i=1}^{n-1} r_i(l^2 - b_i^2) \right] \qquad (13.12a)$$

$$\alpha_2 = -\left[l + \sum_{i=1}^{n-1} r_i(l - b_i) \right] \qquad (13.12b)$$

$$\alpha_3 = \tfrac{1}{6}\left\{ l^3 + \sum_{i=1}^{n-1} r_i[(l^3 - b_i^3) - 3b_i^2(l - b_i)] \right\} \qquad (13.12c)$$

$$\alpha_4 = -\tfrac{1}{2}\left[l^2 + \sum_{i=1}^{n-1} r_i(l - b_i)^2 \right] \qquad (13.12d)$$

$$f = f_1(l) + \sum_{i=1}^{n-1} r_i f_{2i}(l) \qquad (13.13a)$$

$$g = g_1(l) + \sum_{i=1}^{n-1} r_i g_{2i}(l) \qquad (13.13b)$$

Multiplying Eq. (13.11) by the inverse of the left matrix, that is by,

$$\alpha \begin{bmatrix} \alpha_4 & -\alpha_3 \\ -\alpha_2 & \alpha_1 \end{bmatrix} \quad \text{where} \quad \alpha = \frac{1}{\alpha_1 \alpha_4 - \alpha_2 \alpha_3} \qquad (13.14)$$

we find the shear and moment at the left end of the element (which give the first two rows of the stiffness matrix) as

$$\begin{bmatrix} V_1 \\ M_1 \end{bmatrix} = EI_1 \alpha \begin{bmatrix} \alpha_2 & \alpha_2 l - \alpha_4 & -\alpha_2 & \alpha_4 \\ -\alpha_1 & \alpha_3 - \alpha_1 l & \alpha_1 & -\alpha_3 \end{bmatrix} \begin{bmatrix} v_1 \\ \theta_1 \\ v_2 \\ \theta_2 \end{bmatrix} + \alpha \begin{bmatrix} \alpha_4 f - \alpha_2 g \\ -\alpha_3 f + \alpha_1 g \end{bmatrix}$$

$$(13.15a,b)$$

Because the elements of the stiffness matrix must be symmetric, in Eqs. (13.15a,b) we can set $\alpha_2 l - \alpha_4$ equal $-\alpha_1$. With this substitution we get

$$\begin{bmatrix} V_1 \\ M_1 \end{bmatrix} = EI_1 \alpha \begin{bmatrix} \alpha_2 & -\alpha_1 & -\alpha_2 & \alpha_4 \\ -\alpha_1 & \alpha_5 & \alpha_1 & -\alpha_3 \end{bmatrix} \begin{bmatrix} v_1 \\ \theta_1 \\ v_2 \\ \theta_2 \end{bmatrix} + \begin{bmatrix} V_1^F \\ M_1^F \end{bmatrix} \qquad (13.16a,b)$$

where

$$\alpha_5 = \alpha_3 - \alpha_1 l \qquad (13.17)$$

and the fixed-end forces are

$$V_1^F = \alpha(\alpha_4 f - \alpha_2 g) \qquad (13.18a)$$

$$M_1^F = \alpha(-\alpha_3 f + \alpha_1 g) \qquad (13.18b)$$

The equilibrium equations are

$$V_1 + q + V_2 = 0$$

$$M_1 + m(l) - V_1 l + M_2 = 0$$

This gives

$$V_2 = -(V_1 + q) \qquad (13.19a)$$

$$M_2 = -M_1 - m(l) + V_1 l \qquad (13.19b)$$

where

$$q = \int_0^l p(x)\, dx \qquad (13.20)$$

is the total load on the span.

Upon substitution for V_1 and M_1, we find the last two rows of the stiffness matrix. Note that in Eqs. (13.21a,b) the terms symmetric with those in Eq. (13.16a,b) are represented by the same notations.

$$\begin{bmatrix} V_2 \\ M_2 \end{bmatrix} = EI\alpha \begin{bmatrix} -\alpha_2 & \alpha_1 & \alpha_2 & -\alpha_4 \\ \alpha_4 & -\alpha_3 & -\alpha_4 & \alpha_6 \end{bmatrix} \begin{bmatrix} v_1 \\ \theta_1 \\ v_2 \\ \theta_2 \end{bmatrix} + \begin{bmatrix} V_2^F \\ M_2^F \end{bmatrix} \qquad (13.21a,b)$$

$$\alpha_6 = \alpha_3 + \alpha_4 l \qquad (13.22)$$

$$V_2^F = -(V_1^F + q) \qquad (13.23a)$$

$$M_2^F = -M_1^F - m(l) + V_1^F l \qquad (13.23b)$$

The final forms of the stiffness matrix and fixed-end forces are given in the summary at the end of this section, along with all the terms needed to calculate their values. The fixed-end forces for uniform and concentrated loads are calculated below.

Uniform load. For a uniform load p applied over the whole span, the moment of the load at x is

$$m(x) = -\tfrac{1}{2}px^2$$

$$f_1(x) = -\tfrac{1}{2}p \int_0^x x^2 \, dx = -\tfrac{1}{6}px^3 \qquad (13.24a)$$

Integration of expressions involving step functions was discussed in Section 9.2. By Eq. (9.6)

$$f_{2i}(x) = -\tfrac{1}{2}p \int_0^x x^2 H_i \, dx = -\tfrac{1}{6}p(x^3 - b_i^3)H_i \qquad (13.24b)$$

$$g_1(x) = \int_0^x f_1(x) \, dx = -\tfrac{1}{24}px^4 \qquad (13.24c)$$

$$g_{2i}(x) = \int_0^x f_{2i}(x) \, dx = -\tfrac{1}{6}p[\tfrac{1}{4}(x^4 - b_i^4) - b_i^3(x - b_i)]H_i \qquad (13.24d)$$

and

$$f_1(l) = -\tfrac{1}{6}pl^3 \qquad (13.25a)$$

$$f_{2i}(l) = -\tfrac{1}{6}p(l^3 - b_i^3) \qquad (13.25b)$$

$$g_1(l) = -\tfrac{1}{24}pl^4 \qquad (13.25c)$$

$$g_{2i}(l) = -\tfrac{1}{6}p[\tfrac{1}{4}(l^4 - b_i^4) - b_i^3(l - b_i)] \qquad (13.25d)$$

$$q = \int_0^l p(x) \, dx = pl, \qquad m(l) = -\tfrac{1}{2}pl^2 \qquad (13.25e,f)$$

From the above values we can calculate the fixed-end forces by Eqs. (13.18) and (13.23), once α_i are found.

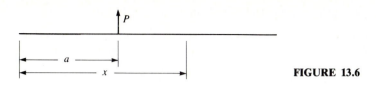

FIGURE 13.6

Concentrated load. Consider a concentrated load acting at distance a from the left end of the beam in Fig. 13.6. Referring to Section 9.3, the moment of the load at x is

$$m(x) = -P(x-a)H_a \qquad (13.26a)$$

By Eq. (9.5),

$$f_1(x) = \int_0^x m(x)\, dx = -P\int_0^x (x-a)H_a\, dx = -\tfrac{1}{2}P(x-a)^2 H_a \qquad (13.26b)$$

$$f_{2i}(x) = -P\int_0^x (x-a)H_a H_i\, dx = -P\int_0^x (x-a)H_{\delta_i}\, dx$$

$$= -P[\tfrac{1}{2}(x^2-\delta_i^2) - a(x-\delta_i)]H_{\delta_i} \qquad (13.26c)$$

$$g_1(x) = \int_0^x f_1(x)\, dx = -\tfrac{1}{2}P\int_0^x (x-a)^2 H_a\, dx = -\tfrac{1}{6}P(x-a)^3 H_a \qquad (13.26d)$$

$$g_{2i}(x) = \int_0^x f_{2i}(x)\, dx = -P\{\tfrac{1}{2}[\tfrac{1}{3}(x^3-\delta_i^3) - \delta_i^2(x-\delta_i) - a(x-\delta_i)^2]\}H_{\delta_i} \qquad (13.26e)$$

where

$$\delta_i = \max(a, b_i) \qquad (13.27a)$$

$$f_1(l) = -\tfrac{1}{2}P(l-a)^2 \qquad (13.27b)$$

$$f_{2i}(l) = -P[\tfrac{1}{2}(l^2-\delta_i^2) - a(l-\delta_i)] \qquad (13.27c)$$

$$g_1(l) = -\tfrac{1}{6}P(l-a)^3 \qquad (13.27d)$$

$$g_{2i}(l) = -\tfrac{1}{2}P[\tfrac{1}{3}(l^3-\delta_i^3) - \delta_i^2(l-\delta_i) - a(l-\delta_i)^2] \qquad (13.27e)$$

Note that the values of the expressions $f_{2i}(l)$ and $g_{2i}(l)$ depend on b_i, as they are used repeatedly in Eqs. (13.13a,b).

$$q = \int_0^l p(x)\, dx = \text{total load on the span} = P \qquad (13.28a)$$

$$m(l) = -P(l-a) \qquad (13.28b)$$

The fixed-end forces are calculated by Eqs. (13.18) and (13.23) once α_i are found. A summary of the expressions for calculating the fixed-end forces is given at the end of this section.

Rotational Stiffness and Carry-Over Factors

Rotational stiffness and carry-over factors are used in the moment distribution method. We can find their values for variable-section beams using the above results. In Eqs. (13.16b) and (13.21b) the terms relating M_1 to θ_1 and M_2 to θ_2, when the other end rotation and displacements are zero, are rotational stiffness for the two ends of the member. Thus,

$$K_1 = EI_1 \alpha \alpha_5 \tag{13.29a}$$

$$K_2 = EI_1 \alpha \alpha_6 \tag{13.29b}$$

The carry-over factors are the ratios of M_1/M_2 and M_2/M_1 for $v_1 = v_2 = 0$ when $\theta_1 = 1$, $\theta_2 = 0$ and $\theta_1 = 0$, $\theta_2 = 1$, that is,

$$C_{12} = -\frac{\alpha_3}{\alpha_5} \tag{13.30a}$$

$$C_{21} = -\frac{\alpha_3}{\alpha_6} \tag{13.30b}$$

The fixed-end moments are found from Eqs. (13.18b) and (13.23b). Note that if the above values are used, the absolute stiffness terms $4EI/l$ must also be used for members with constant cross-sections instead of stiffness factors.

Axial Stiffness

Equations (13.16) and (13.21) represent the bending stiffness and fixed-end forces for a beam element with variable sections. If axial deformations are to be included in the analysis, then the above equations must be augmented to include the axial forces. The terms of the axial stiffness matrix are derived here. A variable-section member under axial forces is shown in Fig. 13.7 with the end forces U_1 and U_2 and n segments with given lengths and cross-sectional areas.

For a distributed axial load w, by equilibrium of the free body, we find that the axial force at x is

$$U = -\left[U_1 + \int_0^x w(x)\, dx \right]$$

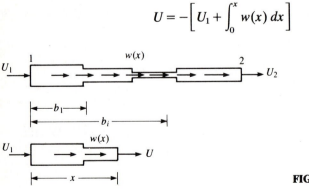

FIGURE 13.7

The differential equation for a member under axial loads was derived in Section 5.1, Eq. (5.3b) as $u' = U/EA$. Thus,

$$-u' = \frac{U_1}{EA} + \frac{Q(x)}{EA} \tag{13.31}$$

with

$$Q(x) = \int_0^x w(x)\, dx \tag{13.32a}$$

For a concentrated axial load W acting at $x = d$,

$$Q(x) = WH(x - d) \tag{13.32b}$$

where $H(x - d)$ is the unit step function defined by Eq. (9.3).

The axial flexibility EA can be expressed in a manner similar to Eq. 9.10a as follows:

$$\frac{1}{EA} = \frac{1}{EA_1}\left(1 + \sum_{i=1}^{n-1} t_i H_i\right) \tag{13.33a}$$

where

$$t_i = EA_1\left[\frac{1}{EA_{i+1}} - \frac{1}{EA_i}\right] \tag{13.33b}$$

Again, $H_i = H(x - b_i)$ is the unit step function. Upon substitution of Eq. 13.33a in Eq. 13.31 and integration, in view of Eq. (9.5), we get

$$-EA_1 u = U_1\left[x + \sum_{i=1}^{n-1} t_i(x - b_i)H_i\right] + h_1(x) + \sum_{i=1}^{n-1} t_i h_{2i}(x) + c_3$$

where we have defined

$$h_1(x) = \int_0^x Q(x)\, dx \tag{13.34a}$$

and

$$h_{2i}(x) = \int_0^x Q(x)H_i\, dx \tag{13.34b}$$

Applying the boundary conditions $u(0) = u_1$ and $u(l) = u_2$, with $H_i = 0$ for $x = 0$ and $H_i = 1$ for $x = l$, we get

$$-EA_1 u_1 = c_3$$

and

$$-EA_1 u_2 = U_1\left[l + \sum_{i=1}^{n-1} t_i(l - b_i)\right] + h_1(l) + \sum_{i=1}^{n-1} t_i h_{2i}(l) - EA_1 u_1$$

or

$$-EA_1 u_2 = U_1 \beta_1 + h - EA_1 u_1$$

where we have defined

$$\beta_1 = l + \sum_{i=1}^{n-1} t_i(l - b_i) \tag{13.35a}$$

$$h = h_1(l) + \sum_{i=1}^{n-1} t_i h_{2i}(l) \tag{13.35b}$$

Thus,

$$U_1 = \frac{EA_1}{\beta_1}(u_1 - u_2) + U_1^F \tag{13.36a}$$

$$U_1^F = -\frac{h}{\beta_1} \tag{13.36b}$$

Then by equilibrium of the element,

$$U_1 + U_2 + Q(l) = 0 \quad \text{or} \quad U_2 = -[U_1 + Q(l)],$$

which gives

$$U_2 = -\frac{EA}{\beta_1}(u_1 - u_2) + U_2^F \tag{13.37a}$$

where

$$U_2^F = -[U_1^F + Q(l)] \tag{13.37b}$$

In matrix form

$$\begin{bmatrix} U_1 \\ U_2 \end{bmatrix} = \frac{EA_1}{\beta_1} \begin{bmatrix} 1 & -1 \\ -1 & 1 \end{bmatrix} \begin{bmatrix} u_1 \\ u_2 \end{bmatrix} + \begin{bmatrix} U_1^F \\ U_2^F \end{bmatrix} \tag{13.38}$$

Summary of the Force–Displacement Relations for Variable Sections

The 6×6 stiffness matrix including axial degrees of freedom is given below.

$$\begin{bmatrix} U_1 \\ V_1 \\ M_1 \\ U_2 \\ V_2 \\ M_2 \end{bmatrix} = EI_1\alpha \begin{bmatrix} \beta & 0 & 0 & -\beta & 0 & 0 \\ 0 & \alpha_2 & -\alpha_1 & 0 & -\alpha_2 & \alpha_4 \\ 0 & -\alpha_1 & \alpha_5 & 0 & \alpha_1 & -\alpha_3 \\ -\beta & 0 & 0 & \beta & 0 & 0 \\ 0 & -\alpha_2 & \alpha_1 & 0 & -\alpha_1 & -\alpha_4 \\ 0 & \alpha_4 & -\alpha_3 & 0 & -\alpha_4 & \alpha_6 \end{bmatrix} \begin{bmatrix} u_1 \\ v_1 \\ \theta_1 \\ u_2 \\ v_2 \\ \theta_2 \end{bmatrix} + \begin{bmatrix} U_1^F \\ V_1^F \\ M_1^F \\ U_2^F \\ V_2^F \\ M_2^F \end{bmatrix} \tag{13.39}$$

where the elements of the stiffness matrix and fixed-end force vector are defined as follows.

$$\beta = \frac{A_1}{\alpha I_1 \beta_1} \qquad \beta_1 = l + \sum_{i=1}^{n-1} t_i(l - b_i) \qquad \alpha = \frac{1}{(\alpha_1 \alpha_4 - \alpha_2 \alpha_3)}$$

$$\alpha_1 = \tfrac{1}{2}\left[l^2 + \sum_{i=1}^{n-1} r_i(l^2 - b_i^2) \right] \qquad \alpha_2 = -\left[l + \sum_{i=1}^{n-1} r_i(l - b_i) \right]$$

$$\alpha_3 = \frac{1}{6}\left\{ l^3 + \sum_{i=1}^{n-1} r_i[(l^3 - b_i^3) - 3b_i^2(l - b_i)] \right\} \alpha_i$$

$$\alpha_4 = -\tfrac{1}{2}\left[l^2 + \sum_{i=1}^{n-1} r_i(l - b_i)^2 \right] \qquad \alpha_5 = \alpha_3 - \alpha_1 l \qquad \alpha_6 = \alpha_3 + \alpha_4 l$$

$$t_i = EA_1\left[\frac{1}{EA_{i+1}} - \frac{1}{EA_i} \right] \qquad r_i = EI_1\left[\frac{1}{EI_{i+1}} - \frac{1}{EI_i} \right]$$

Fixed-end forces

$$U_1^F = \frac{-h}{\beta_1}$$

$$U_2^F = -[U_1^F + Q(l)]$$

$$h = h_1(l) + \sum_{i=1}^{n-1} t_i h_{2i}(l)$$

$$Q(l) = \int_0^l w(x)\, dx = \text{total axial load on the span}$$

$$h_1(x) = \int_0^x Q(x)\, dx \qquad h_{2i}(x) = \int_0^x Q(x)H_i\, dx$$

$$V_1^F = \alpha(\alpha_4 f - \alpha_2 g) \qquad V_2^F = -(V_1^F + q)$$

$$M_1^F = \alpha(-\alpha_3 f + \alpha_1 g) \qquad M_2^F = -M_1^F - m(l) + V_1^F l$$

$$q = \int_0^l p(x)\, dx = \text{total lateral load on the span}$$

$$f = f_1(l) + \sum_{i=1}^{n-1} r_i f_{2i}(l) \qquad g = g_1(l) + \sum_{i=1}^{n-1} r_i g_{2i}(l)$$

$$f_1(x) = \int_0^x m(x)\, dx \qquad f_{2i}(x) = \int_0^x m(x)H_i\, dx$$

$$g_1(x) = \int_0^x f_1(x)\, dx \qquad g_{2i}(x) = \int_0^x f_{2i}(x)\, dx$$

$H_i = H(x - b_i)$ and b_i is the distance from the left end of the beam to the end of the ith segment. For uniform axial and lateral loads w and p, a concentrated axial load W at d, and a concentrated lateral load P at a, we have

Uniform load	**Concentrated load**

$$h_1(l) = \frac{wl^2}{2} \qquad\qquad\qquad h_1(l) = W(l - d)$$

$$h_{2i}(l) = \tfrac{1}{2}w(l^2 - b_i^2) \qquad\qquad h_{2i}(l) = W(l - \gamma_i), \ \gamma_i = \max(d, b_i)$$

$$Q(l) = wl \qquad\qquad\qquad\qquad Q(l) = W$$

$$f_1(l) = -\tfrac{1}{6}pl^3 \qquad\qquad\qquad f_1(l) = -\tfrac{1}{2}P(l - a)^2$$

$$f_{2i}(l) = -\tfrac{1}{6}p(l^3 - b_i)^3 \qquad\qquad f_{2i}(l) = -P[\tfrac{1}{2}(l^2 - \delta_i^2) - a(l - \delta_i)]$$

$$g_1(l) = -\tfrac{1}{24}pl^4 \qquad\qquad\qquad g_1(l) = -\tfrac{1}{6}P(l - a)^3$$

$$g_{2i}(l) = -\tfrac{1}{6}p[\tfrac{1}{4}(l^4 - b_i^4) - b_i^3(l - b_i)] \qquad g_{2i}(l) = -\tfrac{1}{2}P[\tfrac{1}{3}(l^3 - \delta_i^3) - \delta_i^2(l - \delta_i) - a(l - \delta_i)^2]$$

$$q = pl, \qquad m(l) = -\tfrac{1}{2}pl^2 \qquad\qquad q = P, \qquad m(l) = -P(l - a) \quad \delta_i = \max(a, b_i)$$

Rotational stiffness and carry-over factors

$$K_1 = EI_1\alpha\alpha_5 \qquad K_2 = EI_1\alpha\alpha_6$$

$$C_{12} = -\frac{\alpha_3}{\alpha_5}, \qquad C_{21} = -\frac{\alpha_3}{\alpha_6}$$

Example 13.5. Find the 6×6 stiffness matrix for the beam element with a linear hunch as shown in Fig. 13.8. The depth of the beam is 3 ft at the left end of the hunch. The depth of the uniform segment is 1 ft. The width is 0.75 ft. Also find the fixed-end forces for a 10-kip concentrated lateral gravity load acting at the mid-span. $E = 4.32 \times 10^5$ ksf. If the moment distribution method is used instead of the stiffness method, find the rotational stiffness terms and the carry-over factors for this beam.

Solution. Here we represent the hunch by two steps with equal lengths of 3 ft. The depths at mid-lengths of the steps (1.5 ft and 4.5 ft) are used in finding the cross-sectional areas and moments of inertia. Thus,

$$h_1 = 2.5 \text{ ft} \qquad h_2 = 1.5 \text{ ft} \qquad A_1 = 2.5 \times 0.75 = 1.875 \text{ ft}^2$$

$$A_2 = 1.5 \times 0.75 = 1.125 \text{ ft}^2 \qquad\qquad A_3 = 1 \times 0.75 = 0.75 \text{ ft}^2$$

$$I_1 = \frac{0.75 \times 2.5^3}{12} = 0.977 \text{ ft}^4, \qquad I_2 = \frac{0.75 \times 1.5^3}{12} = 0.211 \text{ ft}^4$$

$$I_3 = \frac{0.75 \times 1^3}{12} = 0.063 \text{ ft}^4 \qquad b_1 = 3 \text{ ft} \qquad b_2 = 6 \text{ ft}$$

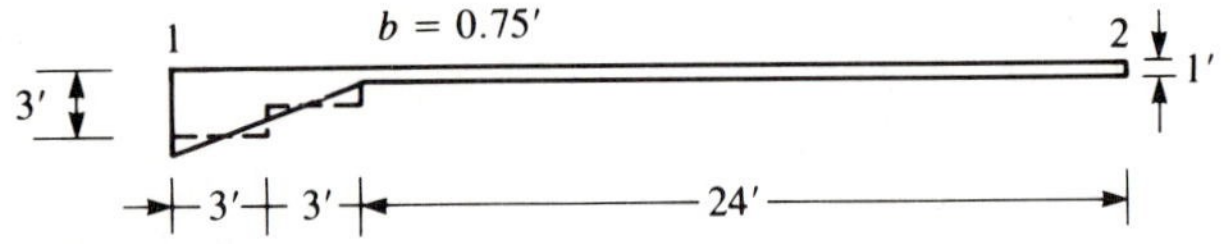

FIGURE 13.8

The force–displacement equation (Eq. (13.39)) can be found by calculating the parameters summarized above. Thus, we get

$$t_1 = 1.875\left(\frac{1}{1,125} - \frac{1}{1.875}\right) = 0.67 \qquad t_2 = 1.875\left(\frac{1}{0.75} - \frac{1}{1.125}\right) = 0.83$$

$$\beta_1 = [30 + 0.67(30 - 3) + 0.83(30 - 6)] = 68$$

and

$$r_1 = 0.977\left(\frac{1}{0.211} - \frac{1}{0.977}\right) = 3.63 \qquad r_2 = 0.977\left(\frac{1}{0.063} - \frac{1}{0.211}\right) = 10.88$$

With the above values,

$$\alpha_1 = \tfrac{1}{2}[30^2 + 3.63(30^2 - 3^2) + 10.88(30^2 - 6^2)] = 6\,767$$

$$\alpha_2 = -[30 + 3.63(30 - 3) + 10.88(30 - 6)] = -389$$

$$\alpha_3 = \tfrac{1}{6}\{30^3 + 3.63[(30^3 - 3^3) - 3 \times 3^2(30 - 3)]$$

$$\qquad + 10.88[(30^3 - 6^3) - 3 \times 6^2(30 - 6)]\} = 64\,246$$

$$\alpha_4 = -\tfrac{1}{2}[30^2 + 3.63(30 - 3)^2 + 10.88(30 - 6)^2] = -4\,907$$

$$\alpha_5 = 64\,246 - 6\,767 \times 30 = -138\,764 \qquad \alpha_6 = 64\,246 - 4\,907 \times 30 = -82\,964$$

$$\alpha = 1/[(6\,767)(-4\,907) - (-389)(64\,246)] = -1.217\,44 \times 10^{-7}$$

$$\beta = -\frac{1.875}{(-1.217\,44 \times 10^{-7} \times 0.977)68} = -2.3182 \times 10^5$$

The stiffness matrix is with $EI_1\alpha = 5.1384 \times 10^{-2}$

$$5.1384 \times 10^{-2}
\begin{bmatrix}
2.3182 \times 10^5 & 0 & 0 & -2.3182 \times 10^5 & 0 & 0 \\
0 & 389 & 6\,767 & 0 & -389 & 4\,907 \\
0 & 6\,767 & 138\,764 & 0 & -6\,767 & 64\,246 \\
-2.3102 \times 10^5 & 0 & 0 & 2.3182 \times 10^5 & 0 & 0 \\
0 & -389 & -6\,767 & 0 & 6\,767 & -4\,907 \\
0 & 4\,907 & 64\,246 & 0 & -4\,907 & 82\,964
\end{bmatrix}$$

For the concentrated load $P = -10$ kips, the fixed-end forces are found by calculating

$$\delta_i = \max(15, b_i), \qquad \delta_1 = \delta_2 = 15$$

$$f_1(l) = \tfrac{1}{2} \times 10(30 - 15)^2 = 1\,125$$

$$f_{2i}(l) = 10[\tfrac{1}{2}(30^2 - \delta_i^2) - 15(30 - \delta_i)]$$

$$g_1(l) = \tfrac{1}{6} \times 10(30 - 15)^3 = 5\,625$$

$$g_{2i}(l) = 10\{\tfrac{1}{2}[\tfrac{1}{3}(30^3 - \delta_i^3) - \delta_i^2(30 - \delta_i)] - 15(30 - \delta_i)^2]\}$$

$$q = -10 \qquad m(l) = 10(30 - 15) = 150$$

Then,

$$f = 1\,125 + (10)\{3.63[\tfrac{1}{2}(30^2 - 15^2) - 15(30 - 15)]$$
$$+ 10.88[\tfrac{1}{2}(30^2 - 15^2) - 15(30 - 15)]\} = 17\,449$$

$$g = 5\,625 + 10 \times \{3.63 \times \tfrac{1}{2}[\tfrac{1}{3}(30^3 - 15^3) - 15^2(30 - 15) - 15(30 - 15)^2]$$
$$+ 10.88 \times \tfrac{1}{2}[\tfrac{1}{3}(30^3 - 15^3) - 15^2(30 - 15) - 15(30 - 15)^2]\} = 87\,244$$

Finally, the fixed-end forces are found as

$$V_1^F = -1.217\,44 \times 10^{-7}[-4\,907 \times (17\,449) + 389(87\,244)] = 6.29 \text{ kips}$$

$$M_1^F = -1.217\,44 \times 10^{-7}[-64\,246 \times 17\,449 + 6\,767 \times 87\,244] = 64.6 \text{ kip} \cdot \text{ft}$$

$$V_2^F = -(6.29 - 10) = 3.71 \text{ kips}$$

$$M_2^F = -64.6 - 150 + 6.29 \times 30 = -25.9 \text{ kip} \cdot \text{ft}$$

For moment distribution the rotational stiffness terms are

$$K_1 = EI_1\alpha\alpha_s = 4.32 \times 10^5 \times 0.977 \times (-1.217\,44 \times 10^{-7}) \times (-138\,764) = 7130$$

$$K_2 = 4.32 \times 10^5 \times 0.977(-1.217\,44 \times 10^{-7})(-82\,964) = 4263$$

and the carry-over factors are

$$C_{12} = -\frac{64\,246}{-138\,764} = 0.46 \quad \text{and} \quad C_{21} = -\frac{64\,246}{-82\,964} = 0.77$$

M_1^F and M_2^F are the same as above.

13.6 THERMAL STRESS

The coefficient of thermal expansion, α, was discussed in Section 1.8. This is the expansion produced by a unit temperature change in a material fiber with unit length.

If the temperature changes at the top and bottom fibers of the beam (Fig. 13.9) from a reference temperature are T_u and T_l then, assuming a linear variation, the temperature change at a distance y from the neutral axis is

$$T_y = \frac{T_u + T_l}{2} + \frac{T_u - T_l}{h} y \qquad (13.40a)$$

or

$$T_y = T_m + \frac{\Delta T}{h} y \qquad (13.40b)$$

where $T_m = (T_u + T_l)/2$, $\Delta T = T_u - T_l$, with h the depth of the member.

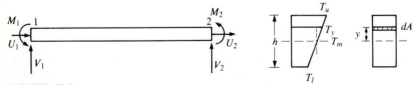

FIGURE 13.9

If the fibers are free to expand, then T_m causes an expansion of the beam axis, and the beam bends because of the temperature differential ΔT. The strain in a fiber at a distance y from the neutral axis is

$$\varepsilon = \alpha T_y = \alpha\left(T_m + \frac{\Delta T}{h}y\right) \tag{13.41a}$$

If the expansion is prevented, then a normal stress σ is developed in this fiber, where

$$\sigma = E\varepsilon = E\alpha\left(T_m + \frac{\Delta T}{h}y\right) \tag{13.41b}$$

If the ends of the beam are fixed, then fixed-end moments and axial forces are developed. The fixed-end moments are required to keep the end rotations from taking place, while the fixed-end axial forces keep the fibers from expanding. The fixed-end axial forces, U_1 and U_2, are

$$U_1^F = -U_2^F = \int_A \sigma \, dA = E\alpha \int_A \left(T_m + \frac{\Delta T}{h}y\right) dA$$

Since the first moment with respect to the neutral axis is zero,

$$\int_A y \, dA = 0 \tag{13.42}$$

we get

$$U_1^F = -U_2^F = EA\alpha T_m \tag{13.43}$$

If the top fibers of the beam are at a higher temperature, they tend to expand more than the bottom fibers, causing rotations at the ends of the beam. The fixed-end moments are equal to the moment of the internal forces and their signs are such that they render the end rotations zero by providing a compressive force near the top and a tensile force near the bottom of the beam. That is,

$$M_1^F = -M_2^F = -\int_A \sigma y \, dA = -E\alpha \int_A \left[T_m y + \frac{\Delta T}{h}y^2\right] dA$$

$$= -E\alpha \frac{\Delta T}{h} I \qquad \text{as } I = \int_A y^2 \, dA$$

Therefore,

$$M_1^F = -M_2^F = -EI\alpha \frac{\Delta T}{h} \tag{13.44}$$

Since temperature changes cause only expansion and contraction of the fibers, preventing the expansion can induce only axial forces or moments. Therefore, no fixed-end shear forces are developed because of changes in temperature. In the stiffness method, temperature effects are incorporated through the above fixed-end forces.

Example 13.6. The first span of a two-span beam (Fig. 13.10) is exposed to a uniform temperature change throughout this span. The temperature varies linearly through the depth of the beam, from 200°F at the top to 100°F at the bottom. Find the member forces for $\alpha = 6 \times 10^{-6}$ in/in/°F and $E = 29\,000$ ksi. $I = 200$ in^4, $h = 20$ in, and $l_1 = l_2 = 20$ ft.

Solution. Since the axial force developed due to the average temperature change can be calculated easily, we determine the effect of moments by considering the rotational degrees of freedom only. With $\Delta T = 200 - 100 = 100$, the fixed-end moments are, Eq. 13.44

$$\begin{bmatrix} M_1^F \\ M_2^F \end{bmatrix} = -\frac{29\,000 \times 200}{20} \times 6 \times 10^{-6} \times 100 \begin{bmatrix} 1 \\ -1 \end{bmatrix} = \begin{bmatrix} -174 \\ 174 \end{bmatrix}$$

Thus,

$$\begin{bmatrix} M_1 \\ M_2 \end{bmatrix}^1 = \frac{2EI}{l} \begin{bmatrix} 2 & 1 \\ 1 & 2 \end{bmatrix} \begin{bmatrix} \theta_1 \\ \theta_2 \end{bmatrix}^1 + \begin{bmatrix} M_1^F \\ M_2^F \end{bmatrix}^1 = \begin{bmatrix} 96\,667 & 48\,333 \\ 48\,333 & 96\,667 \end{bmatrix} \begin{bmatrix} 0 \\ r_1 \end{bmatrix} + \begin{bmatrix} -174 \\ 174 \end{bmatrix}$$

$$\begin{bmatrix} M_1 \\ M_2 \end{bmatrix}^2 = \begin{bmatrix} 96\,667 & 48\,333 \\ 48\,333 & 96\,667 \end{bmatrix} \begin{bmatrix} r_1 \\ 0 \end{bmatrix}$$

There is only one degree of freedom, r_1, rotation at 2.

$$K_s = 96\,667 + 96\,667 = 193\,334$$

$$R = -(M_2^F) = -174$$

$$K_s r_1 = R \quad \text{or} \quad 193\,334 r_1 = -174$$

$$r_1 = -0.0009 \text{ rad.}$$

Then,

$$\begin{bmatrix} M_1 \\ M_2 \end{bmatrix}^1 = \begin{bmatrix} 96\,667 & 48\,333 \\ 48\,333 & 96\,667 \end{bmatrix} \begin{bmatrix} 0 \\ -0.0009 \end{bmatrix} + \begin{bmatrix} -174 \\ 174 \end{bmatrix}$$

$$= \begin{bmatrix} -43.5 \\ -87 \end{bmatrix} + \begin{bmatrix} -174 \\ 174 \end{bmatrix} = \begin{bmatrix} -217.5 \\ 87 \end{bmatrix}$$

$$\begin{bmatrix} M_1 \\ M_2 \end{bmatrix}^2 = \begin{bmatrix} 96\,667 & 48\,333 \\ 48\,333 & 96\,667 \end{bmatrix} \begin{bmatrix} -0.0009 \\ 0 \end{bmatrix} = \begin{bmatrix} -87 \\ -43.5 \end{bmatrix}$$

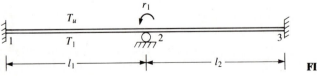

FIGURE 13.10

From the fixed-end forces calculated above, we can see that if all the spans of a continuous beam with fixed ends have the same temperature profile with no variation along the length, then no deflections and rotations are developed and the moments are the same as the fixed-end moments.

13.7 INCLINED ROLLER SUPPORTS

Inclined roller supports provide restraints in a direction other than those of the global coordinate axes. Such roller supports can be modeled by a link with a large axial stiffness, hinged to the structure at the same point as the roller is and hinged to the ground at its other end. Thus, the link extends a short distance on both sides and perpendicularly to the inclined surface. Treatment of internal hinges was discussed in Section 13.2. Alternatively, a member with a small bending resistance can be used.

13.8 INFLUENCE LINES BY THE STIFFNESS METHOD

The stiffness method can be used to find influence lines of the different effects (reaction, moment, shear, etc.) at a section of a structure. Although multiple runs for the different locations of a unit load may be used, a more efficient way is to use the Muller–Breslau principle (Section 8.3). Thus, to find the influence line of a reaction we remove the restraint against the reaction so that the reaction force undergoes a unit displacement. The deformed shape of the structure gives the influence line of that reaction. In practice this is done by applying a load, say 10 kips, in the direction of the reaction, with the restraint removed. The displacement is calculated throughout the structure. These displacements are then divided by the displacement at the support. A plot of these scaled deflections is the influence line of the reaction. Similarly, the influence line of shear (moment) can be obtained by releasing the resistance of the section against shear (moment), allowing displacements (rotations) on both sides of the section, and applying forces (moments) on both sides of the section so as to induce a positive internal shear (moment). The displacements found are then scaled so that the change in the displacement (slope) on the two sides of the section is equal to unity.

Note that for statically determinate structures, removing a resistance would add a degree of freedom, producing an unstable structure. However, in the latter case the influence lines can be found by the Muller–Breslau procedure and the ordinates can be obtained by geometry. Therefore, it is not necessary to use the stiffness method to find influence lines of statically determinate structures.

13.9 FLEXIBILITY METHOD

The flexibility method, also known as the force method, is essentially the matrix formulation of the method of consistent deformations, discussed in

Chapter 7. The flexibility method is not as commonly used as the stiffness method and does not lend itself as easily to computer solutions. The purpose of this section is to present the basic concept of the flexibility method and its application to beams, trusses and frames.

Flexibility Matrix

In the flexibility method the structure is divided into a number of elements. Here we assume that only concentrated loads and moments are acting on the structure. The elements consist of the segments with constant cross-sections between the applied loads and points of required moments. If necessary each segment can also be divided into more than one element. The basic variables in the flexibility method are a set of independent member forces. If only rotational degrees of freedom are considered, then the member-end moments are used as the basic variables. The end shear forces can be found from the moments and thus are not independent forces.

The relationship between the end forces and displacements for a beam element was derived in Chapter 10. If the in-span loads and end displacements are zero, then the end rotations can be solved for by Eqs. (10.6) and (10.7) in terms of the end moments as

$$\begin{bmatrix} \theta_1 \\ \theta_2 \end{bmatrix} = \frac{l}{6EI} \begin{bmatrix} 2 & -1 \\ -1 & 2 \end{bmatrix} \begin{bmatrix} M_1 \\ M_2 \end{bmatrix} \tag{13.45a}$$

or

$$\mathbf{s} = \mathbf{FS} \tag{13.45b}$$

where

$$\mathbf{F} = \frac{l}{6EI} \begin{bmatrix} 2 & -1 \\ -1 & 2 \end{bmatrix} \tag{13.45c}$$

is the element flexibility matrix with rotational degrees of freedom.

Solution Process

The first step in the solution of problems by the flexibility method consists of dividing the structure into elements. Each segment between two loads must be considered as an element. However, a segment may be divided into more than one element. After choosing the redundant forces, a relationship is developed between the unknown displacements and rotations and the applied and redundant forces and moments. From this relation the redundant forces and moments are found, which in turn lead to the forces and moments at the ends of each element.

For a structure that is indeterminate to the nth degree, n redundant forces must be used in the formulation. For example, the structure of Fig. 13.11 is divided into three elements, 1, 2, and 3. The internal forces are the moments at the ends of elements, M_1^1, M_2^1, M_1^2, M_2^2, M_1^3, and M_2^3. This

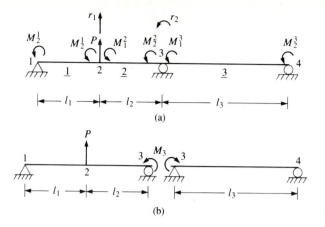

FIGURE 13.11

structure is statically indeterminate to the first degree. Thus, the moment at joint 3, M_3, is used as the redundant. The applied load is P.

If the redundant is set equal to zero, then a statically determinate (primary) structure results.

Force Transformation

The relation between the member rotations and moments for all the members can be combined together in one matrix relation. For the structure of Fig. 13.11 this is

$$
\begin{bmatrix} \theta_1^1 \\ \theta_2^1 \\ \theta_1^2 \\ \theta_2^2 \\ \theta_1^3 \\ \theta_2^3 \end{bmatrix} = \begin{bmatrix} \dfrac{l_1}{6EI_1}\begin{bmatrix} 2 & -1 \\ -1 & 2 \end{bmatrix} & & \\ & \dfrac{l_2}{6EI_2}\begin{bmatrix} 2 & -1 \\ -1 & 2 \end{bmatrix} & \\ & & \dfrac{l_3}{6EI_3}\begin{bmatrix} 2 & -1 \\ -1 & 2 \end{bmatrix} \end{bmatrix} \begin{bmatrix} M_1^1 \\ M_2^1 \\ M_1^2 \\ M_2^2 \\ M_1^3 \\ M_2^3 \end{bmatrix}
$$

(13.64a)

or

$$\mathbf{s}_c = \mathbf{F}_c \mathbf{S}_c \tag{13.46b}$$

where $\mathbf{s}_c$ and $\mathbf{S}_c$ are the vectors of composite end rotations and internal moments for all the members. $\mathbf{F}_c$ is the composite or unassembled flexibility matrix. Note that in Eq. (13.64a) the blank spaces indicate zero values for the composite flexibility matrix.

As mentioned earlier the primary structure is obtained by setting the redundant forces, in this case M_3, equal to zero. The internal moments can be related to the external and redundant forces by using the primary structure and setting the external redundant moments one at a time equal to unity while all the other applied and redundant forces are set equal to zero. For the structure of Fig. 13.11, this will give

$$
\begin{bmatrix} M_1^1 \\ M_2^1 \\ M_1^2 \\ M_2^2 \\ M_1^3 \\ M_2^3 \end{bmatrix} = \begin{bmatrix} 0 & 0 \\ -\dfrac{l_1 l_2}{l_1 + l_2} & \dfrac{l_1}{l_1 + l_2} \\ \dfrac{l_1 l_2}{l_1 + l_2} & -\dfrac{l_1}{l_1 + l_2} \\ 0 & 1 \\ 0 & -1 \\ 0 & 0 \end{bmatrix} \begin{bmatrix} P \\ M_3 \end{bmatrix}
\tag{13.47a}
$$

or

$$
\mathbf{S}_c = \mathbf{BR}
\tag{13.47b}
$$

Note that the moments M_1^1, M_2^1, etc., in the above equation are the moments at the ends of the members necessary for maintaining the end rotations. They are positive if they act in the counterclockwise direction. In other words, positive end moments cause positive (counterclockwise) rotations. $\mathbf{S}_c$ is the vector of the composite internal moments (member-end moments for all the elements) and $\mathbf{R}$ is the vector of applied and redundant forces. $\mathbf{R}^T = [\mathbf{R}_1 \, \mathbf{R}_2]^T$, with $\mathbf{R}_1$ representing the vector of applied loads and $\mathbf{R}_2$ the vector of unknown redundants. For the beam of Fig. 13.11a, $\mathbf{R}_1$ is the single load P and $\mathbf{R}_2$ is the redundant moment M_3.

We can find a relationship between the member and structure displacements similar to Eq. (10.47b) by using the principle of real work (Eq. (6.10)), which states that for an elastic system in equilibrium, the internal and external works are equal,

$$
W_i = W_e
\tag{13.48a}
$$

The internal work is the work of the internal moments through the rotations, $W_i = \frac{1}{2}\mathbf{S}_c^T \mathbf{s}_c$, and the external work is the work of the external loads and redundants through the associated displacements, $W_e = \frac{1}{2}\mathbf{R}^T \mathbf{r}$. Thus, Eq. (13.48a) becomes

$$
\tfrac{1}{2}\mathbf{S}_c^T \mathbf{s}_c = \tfrac{1}{2}\mathbf{R}^T \mathbf{r}
\tag{13.48b}
$$

where $\mathbf{R}$ and $\mathbf{r}$ are the vectors of the applied loads plus redundant forces and the vector of the associated displacements. Substitution for $\mathbf{S}_c$ from Eq. (13.47b) gives

$$
\mathbf{R}^T \mathbf{B}^T \mathbf{s}_c = \mathbf{R}^T \mathbf{r}
$$

or

$$
\mathbf{B}^T \mathbf{s}_c = \mathbf{r}
\tag{13.49}
$$

Structure Displacement–Force Relations

By Eq. (13.46b), $\mathbf{s}_c = \mathbf{F}_c \mathbf{S}_c$. Substituting for $\mathbf{S}_c$ from Eq. (13.47b), multiplying both sides of the resulting equation by $\mathbf{B}^T$, and using Eq. (13.49), we get

$$\mathbf{r} = \mathbf{B}^T \mathbf{F}_c \mathbf{B} \mathbf{R} \tag{13.50a}$$

or

$$\mathbf{r} = \mathbf{F}_s \mathbf{R} \tag{13.50b}$$

with

$$\mathbf{F}_s = \mathbf{B}^T \mathbf{F}_c \mathbf{B} \tag{13.50c}$$

Where $\mathbf{F}_s$ is the structure or assembled flexibility matrix. Equation (13.50b) enables us to find the unknown displacements from the applied loads.

Determination of Displacements and Member Moments

Equation (13.50b) can be restated, by partitioning $\mathbf{R}$ into the applied and redundant forces $\mathbf{R}_1$ and $\mathbf{R}_2$, as

$$\begin{bmatrix} \mathbf{r}_1 \\ \mathbf{r}_2 \end{bmatrix} = \begin{bmatrix} \mathbf{F}_{11} & \mathbf{F}_{12} \\ \mathbf{F}_{21} & \mathbf{F}_{22} \end{bmatrix} \begin{bmatrix} \mathbf{R}_1 \\ \mathbf{R}_2 \end{bmatrix} \tag{13.51}$$

$\mathbf{r}_1$ is the vector of unknown displacements associated with the applied loads and $\mathbf{r}_2$ is the vector of displacements associated with the redundants. These are the changes of rotation from one side of the redundant moments to the other, $\Delta\theta_i$. Owing to continuity of rotations at the redundant moments, $\mathbf{r}_2 = 0$. Thus,

$$\mathbf{r}_1 = \mathbf{F}_{11}\mathbf{R}_1 + \mathbf{F}_{12}\mathbf{R}_2 \qquad 0 = \mathbf{F}_{21}\mathbf{R}_1 + \mathbf{F}_{22}\mathbf{R}_2$$

which gives

$$\mathbf{R}_2 = -(\mathbf{F}_{22})^{-1}\mathbf{F}_{21}\mathbf{R}_1 \tag{13.52a}$$

and the vector of unknown displacements

$$\mathbf{r}_1 = [\mathbf{F}_{11} - \mathbf{F}_{12}(\mathbf{F}_{22})^{-1}\mathbf{F}_{21}]\mathbf{R}_1 \tag{13.52b}$$

Example 13.7. Find the member moments for the two-span beam of Fig. 13.11, using the flexibility method, for $P = -5$ kips. $l_1 = l_2 = l_3/2 = 10$ ft. E and I are constant.

Solution. The force transformation matrix $\mathbf{B}$ is found by setting the load P and the redundant moment M_3 equal to 1 and finding the internal moments. This gives

$$S_c = BR$$

or

$$
\begin{bmatrix} M_1^1 \\ M_2^1 \\ M_1^2 \\ M_2^2 \\ M_1^3 \\ M_2^3 \end{bmatrix} = \begin{bmatrix} 0 & 0 \\ -5 & 0.5 \\ 5 & -0.5 \\ 0 & 1 \\ 0 & -1 \\ 0 & 0 \end{bmatrix} \begin{bmatrix} P \\ M_3 \end{bmatrix}
$$

The composite stiffness matrix is found by using the element flexibility matrices.

$$
\mathbf{F}_c = \frac{1}{6EI} \begin{bmatrix} 20 & -10 & & & & \\ -10 & 20 & & & & \\ & & 20 & -10 & & \\ & & -10 & 20 & & \\ & & & & 40 & -20 \\ & & & & -20 & 40 \end{bmatrix}
$$

The structure flexibility matrix is $\mathbf{F}_s = \mathbf{B}^T \mathbf{F}_c \mathbf{B}$, or

$$
\mathbf{F}_s = \frac{1}{6EI} \begin{bmatrix} 1\,000 & -150 \\ -150 & 80 \end{bmatrix}
$$

Thus, the structure displacement-force relation, $\mathbf{r} = \mathbf{F}_s \mathbf{R}$, with $P = -5$ kips, is

$$
\begin{bmatrix} r_1 \\ r_2 \end{bmatrix} = \frac{1}{6EI} \begin{bmatrix} 1\,000 & 150 \\ 150 & 80 \end{bmatrix} \begin{bmatrix} -5 \\ M_3 \end{bmatrix}
$$

The vertical deflection at the point of application of the load is r_1:

$$
6EI(r_1) = 1\,000(-5) - 150 M_3
$$

$$
0 = -150(-5) + 80 M_3
$$

$$
M_3 = -\frac{(15 \times 5)}{8} = -9.375
$$

$$
r_1 = -\frac{359}{EI}
$$

Then, substitution of M_3 into the force transformation equation $\mathbf{S}_c = \mathbf{B}\mathbf{R}$ gives

$$
\mathbf{S}_c = \begin{bmatrix} M_1^1 \\ M_2^1 \\ M_1^2 \\ M_2^2 \\ M_1^3 \\ M_2^3 \end{bmatrix} = \begin{bmatrix} 0 \\ 20.310 \\ -20.310 \\ -9.375 \\ 9.375 \\ 0 \end{bmatrix}
$$

Flexibility Method for Trusses and Frames

The solution process described above applies to trusses and frames in a similar manner. For trusses, the two end forces are not independent and are equal and opposite by equilibrium. Thus, the element flexibility is l/EA and is obtained from the displacement–force relation Eq. 5.4c at one end,

$$u = \frac{l}{EA} U \qquad (13.53)$$

Therefore, only one end force, representing the bar force, is used in the calculations. For both trusses and frames the primary structures are obtained in a manner similar to that described for beams, by removing the redundant forces. For frames, as for beams, the member-end moments are not independent of end shear forces. Thus, it is sufficient to use the end moments in the calculations.

Example 13.8. Find the displacement of joint 1 and the bar forces for the truss of Fig. 13.12a under the applied loads. The area of cross-section of all the bars is $8\ \text{in}^2$ and $E = 29\,000$ ksi.

Solution. The bar forces are referred to by the element numbers indicated in Fig. 13.12a. Since the truss is statically indeterminate internally to the first degree, we choose the force in element 2 as the redundant force. The removal of this force then gives the primary structures, Fig. 13.12b. The bar forces in the primary structure due to the applied loads and a unit value of the redundant force are found separately. The displacements in the members are associated with the member forces through the element flexibility matrices. This relation for all the bar forces provides the composite flexibility matrix, $\mathbf{F}_c$. Thus,

$$\begin{bmatrix} u_1 \\ u_2 \\ u_3 \end{bmatrix} = \frac{1}{29\,000} \begin{bmatrix} 17.7 & 0 & 0 \\ 0 & 15 & 0 \\ 0 & 0 & 17.7 \end{bmatrix} \begin{bmatrix} U_1 \\ U_2 \\ U_3 \end{bmatrix}$$

or

$$\mathbf{s}_c = \mathbf{F}_c \mathbf{S}_c$$

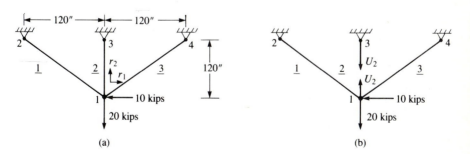

(a)   (b)

FIGURE 13.12

The force transformation relating the internal forces S_c to the external loads and redundant forces R is

$$\begin{bmatrix} U_1 \\ U_2 \\ U_3 \end{bmatrix} = \begin{bmatrix} 0.71 & -0.71 & -0.71 \\ 0 & 0 & 1 \\ -0.71 & -0.71 & -0.71 \end{bmatrix} \begin{bmatrix} R_1 \\ R_2 \\ U_2 \end{bmatrix}$$

or

$$S_c = BR$$

where R_1 and R_2 are the applied horizontal and vertical loads and U_2 is the redundant force. The columns of the above 3×3 matrix are found by setting the horizontal load, the vertical load, and the redundant force in turn equal to unity and calculating the bar forces of the primary structure.

The structure flexibility matrix by Eq. (13.50c) is

$$F_s = B^T F_c B = \frac{1}{29\,000} \begin{bmatrix} 17.85 & 0 & 0 \\ 0 & 17.85 & 17.85 \\ 0 & 17.85 & 32.85 \end{bmatrix}$$

and by Eq. (13.51)

$$\begin{bmatrix} r_1 \\ r_2 \\ \text{--} \\ u_2 \end{bmatrix} = \frac{1}{29\,000} \begin{bmatrix} 17.85 & 0 & \vdots & 0 \\ 0 & 17.85 & \vdots & 17.85 \\ \text{--} & \text{--} & \text{--} & \text{--} \\ 0 & 17.85 & \vdots & 32.85 \end{bmatrix} \begin{bmatrix} R_1 \\ R_2 \\ U_2 \end{bmatrix}$$

where assuming the directions of the X and Y coordinates to be to the right and upward, $R_1 = -10$ kips, $R_2 = -20$ kips. u_2 is the displacement in member 2 and it must be set equal to zero to insure continuity of this bar. Thus, proceeding as in Eqs. (13.51), the unknown redundant force and the joint displacements can be found by Eqs. (13.52a) and (13.52b). Therefore,

$$U_2 = -(32.85)^{-1}[10 \quad 17.85] \begin{bmatrix} -10 \\ -20 \end{bmatrix} = 10.87 \text{ kips}$$

and the horizontal and vertical displacements of joint 1 are

$$\begin{bmatrix} r_1 \\ r_2 \end{bmatrix} = \frac{1}{29\,000} \left(\begin{bmatrix} 17.85 & 0 \\ 0 & 17.85 \end{bmatrix} - \begin{bmatrix} 0 \\ 17.85 \end{bmatrix} \frac{1}{32.85} [0 \quad 17.85] \right) \begin{bmatrix} -10 \\ -20 \end{bmatrix}$$

or

$$\begin{bmatrix} r_1 \\ r_2 \end{bmatrix} = \frac{1}{29\,000} \begin{bmatrix} -178.5 \\ -163.0 \end{bmatrix} \text{ in,}$$

$$\begin{Bmatrix} U_1 \\ U_2 \\ U_3 \end{Bmatrix} = \begin{bmatrix} 0.71 & -0.71 & -0.71 \\ 0 & 0 & 1 \\ -0.71 & -0.71 & -0.71 \end{bmatrix} \begin{Bmatrix} -10 \\ -20 \\ 10.87 \end{Bmatrix} = \begin{Bmatrix} -0.62 \\ 10.87 \\ 13.58 \end{Bmatrix}$$

Example 13.9. Find the member-end moments for the frame of Fig. 13.13.

Solution. The redundant force can be chosen as the moment at any point or one of the reactions. Here the horizontal reaction of joint 4, U_4, is used as the redundant force (Fig. 13.13b). Because of the concentrated load on the beam, the

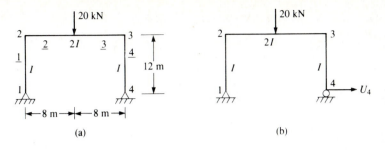

FIGURE 13.13

latter is divided into two elements. In the primary structure the moments at the ends of each element are found in the analysis sign convention. Note that the member-end moments are the moments that maintain the deflected shape of the structure. Using the element flexibility matrices, we obtain the composite displacement–force relation as

$$
\begin{bmatrix} \theta_1^1 \\ \theta_2^1 \\ \theta_1^2 \\ \theta_2^2 \\ \theta_1^3 \\ \theta_2^3 \\ \theta_1^4 \\ \theta_2^4 \end{bmatrix} = \frac{1}{6EI} \begin{bmatrix} 24 & 12 & & & & & & \\ -12 & 24 & & & & & & \\ & & 8 & -4 & & & & \\ & & -4 & 8 & & & & \\ & & & & & & & \\ & & & & -4 & 8 & & \\ & & & & & & 24 & 12 \\ & & & & & & -12 & 24 \end{bmatrix} \begin{bmatrix} M_1^1 \\ M_2^1 \\ M_1^2 \\ M_2^2 \\ M_1^3 \\ M_2^3 \\ M_1^4 \\ M_2^4 \end{bmatrix}
$$

and the force transformation matrix is

$$
\begin{bmatrix} M_1^1 \\ M_2^1 \\ M_1^2 \\ M_2^2 \\ M_1^3 \\ M_2^3 \\ M_1^4 \\ M_2^4 \end{bmatrix} = \begin{bmatrix} 0 & 0 \\ 0 & 12 \\ 0 & -12 \\ -4 & 12 \\ 4 & -12 \\ 0 & 12 \\ 0 & -12 \\ 0 & 0 \end{bmatrix} \begin{bmatrix} R_1 \\ U_4 \end{bmatrix}
$$

By Eq. (13.50c) the structure flexibility matrix is found as

$$
\mathbf{F}_s = \frac{1}{6EI} \begin{bmatrix} 256 & 1\,152 \\ 1\,152 & 13\,824 \end{bmatrix}
$$

Then, partitioning the displacement–force relations for the structure (Eq. (13.51)) and setting the horizontal displacement at point 4 (associated with the redundant force) equal to zero, Eq. (13.52a) gives

$$
0 = 1\,152(-20) - 13\,824 U_4 \qquad \text{or} \qquad U_4 = 1.67 \text{ kN}
$$

Substitution of this value and $R_1 = -20$ kN in the force transformation relation gives the member end moments as follows:

$$M_1^1 = 0 \qquad M_1^2 = 20\,\text{kN}\cdot\text{m} \qquad M_1^3 = -60\,\text{kN}\cdot\text{m} \qquad M_1^4 = 20\,\text{kN}\cdot\text{m}$$

$$M_2^1 = -20\,\text{kN}\cdot\text{m} \qquad M_2^2 = 60\,\text{kN}\cdot\text{m} \qquad M_2^3 = -20\,\text{kN}\cdot\text{m} \qquad M_2^4 = 0$$

LABORATORY EXPERIMENTS

Support Settlement

Use spring steel beam or frame models and induce support settlements. Observe that for statically determinate structures, support settlements do not induce any member deformations and thus induce no additional internal forces, while for statically indeterminate structures they do. A two-span and a one-span beam provide the simplest models.

Elastic Support

Support the end of a cantilever beam by a spring (or a rubber band for a small flexible model), apply a load there, and measure the deflection. Show that this is the same as the theoretical value.

Elastic Connections

Build a flexibly connected frame model such as that in Fig. 13.4. The elastic connection may be a coil or flexible torsional element with its axis normal to the plane of the frame. In other words, the columns and the beam will not be in the same plane but will be connected by the torsional element.

Apply a load at the middle of the beam. Measure the deflection. Compare this deflection for the same lateral load applied to similar frames, one with hinge instead of the elastic connection and the other with a rigid joint.

Flexibility Matrix

Using the beam element of Laboratory Experiments in Chapter 11, keep one end fixed and the other free. Apply a force and a moment (one at a time) and measure the displacement and rotation in each case. Compare these with the theoretical values of the flexibility matrix for this structure.

Using the frame of Fig. L12.1, apply a horizontal load and measure the horizontal and vertical displacements it induces at the free end. Scale to find the displacements for a unit horizontal load, and call these displacements F_{11} and F_{21}. Then remove the horizontal load and apply a vertical load, again measuring the horizontal and vertical displacements. Scale these to get the displacements F_{12} and F_{22} for a unit vertical load. Calculate the structure flexibility matrix associated with the horizontal and vertical displacements at the free end and compare the two results.

PROBLEMS

13.1. The right support of the supported cantilever in Fig. P13.1 settles by 1.5 in. Find the member-end forces and compare them to the case without settlement. $I = 1\,000$ in^4; $E = 3\,000$ ksi.

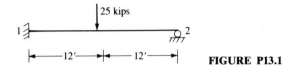

FIGURE P13.1

13.2. The middle support of the beam in Fig. P13.2 is jacked up 3 cm. Find the moment at the middle support and draw the moment diagram.

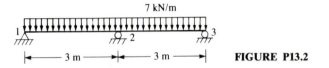

FIGURE P13.2

13.3. The two-span beam of Fig. P13.3 has a hinge at the middle of the first span. A load 10 kips is applied at the hinge. Find the displacement of the hinge and the member forces. $I = 160$ in^4; $E = 29\,000$ ksi.

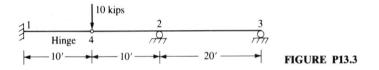

FIGURE P13.3

13.4. Find the member forces for the frame of Fig. P13.4. The two beams are continuous and the column is hinged to them. $I = 2\,000$ cm^4; $E = 200\,000$ MPa.

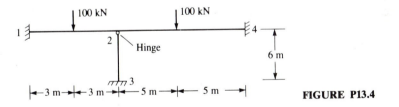

FIGURE P13.4

13.5. Find the member-end forces for the frame of Fig. P13.5 and draw the moment diagram. $I = 100$ in^4; $E = 29\,000$ ksi.

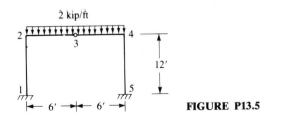

FIGURE P13.5

13.6. Find the member-end forces for the structure of Fig. P13.6. The width and depth of the horizontal members are 20 cm and 50 cm and those of the inclined members are 20 cm and 30 cm. $E = 20\,000$ MPa.

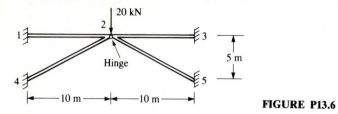

20 kN

Hinge

5 m

10 m 10 m

FIGURE P13.6

13.7. Find the member-end forces for the structure of Fig. P13.7 using the stiffness method. The moment of inertia is the same for all the members and is equal 150 in^4; $E = 29\,000$ ksi, $A = 43.4$ in^2.

Hint: Use the 6×6 element stiffness matrices with two different degrees of freedoms for the axial displacements of points 4 and 6 and two different degrees of freedom for the rotations of those points.

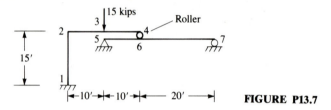

15 kips

Roller

15'

10' 10' 20'

FIGURE P13.7

13.8. The beam of Fig. P13.8 is supported by a vertical elastic support with stiffness $K = 50$ kip/in. Find the deflection and member-end forces. $I = 80$ in^4.

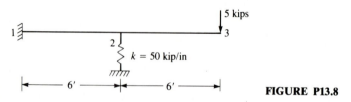

5 kips

$k = 50$ kip/in

6' 6'

FIGURE P13.8

13.9. The right end of the beam of Fig. P13.9 is supported by a vertical and a torsional spring with stiffness 100 kip/in and 200(kip · in)/rad, respectively. A uniform gravity load of 2.2 kip/ft is applied over the span. The span is 30 ft, $I = 150$ in^4, and $E = 29\,000$ ksi. Find the displacement and rotation at the right end and the moment and shear force at the left end of the beam.

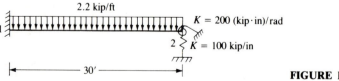

2.2 kip/ft

$K = 200$ (kip · in)/rad

$K = 100$ kip/in

30'

FIGURE P13.9

13.10. Find the member-end forces for the elastically supported structure of Fig. P13.10. $K = 50$ kN/cm for the horizontal springs and 80 kN/cm for the vertical spring. $E = 200\,000$ MPa, $A = 72$ cm^2, $I = 1\,000$ cm^4.

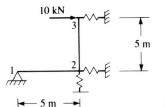

10 kN

5 m

5 m

FIGURE P13.10

13.11. A fixed-end beam is supported at its mid-span, point 5, by a simply supported beam perpendicular to it (Fig. P13.11a). The details of the supported and supporting beams are shown in Figs. P13.11b and P13.11c.
 (a) Determine the stiffness of the elastic support that the simply supported beam provides. (*Hint*: Find the mid-span deflection of this beam under a load P, calculate the stiffness and use it for a vertical string at 5.)
 (b) Calculate the member-end forces for the supported beam using the stiffness method. $I = 200$ in^4 and $E = 29\,000$ ksi for both beams.

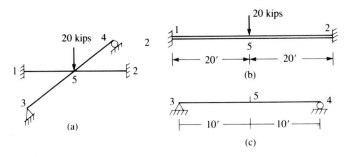

20 kips

20 kips

20'   20'

(b)

10'   10'

(a)   (c)

FIGURE P13.11

13.12. The columns of the frame in Fig. P13.12 are connected to the beam by bolted connections that provide elastic (semi-rigid) connections. The stiffness of the connections is 10^5(kip · in)/rad. Find the sway deflection of the frame and the member-end forces. The moment of inertia of each columns is 100 in^4 and that of the beam 200 in^4. $E = 29\,000$ ksi.

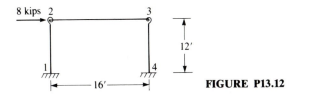

8 kips 2   3

12'

1   4

16'   FIGURE P13.12

13.13. The two cantilever beams of Fig. P13.13 are connected together at their free ends by a spring with $K = 600$ kN/cm. The moment of inertia of the beams is

2 000 cm^4 and $E = 200\,000$ MPa. Find the deflection of the free end of each cantilever for a load of 50 kN applied to the free end of the lower cantilever.

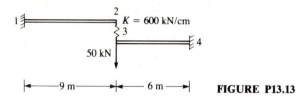

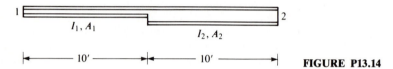

FIGURE P13.13

13.14. Compute the 6×6 stiffness matrix for the beam element of Fig. P13.14. Also calculate the fixed-end forces for a uniform load of 2.5 kip/ft and the carry-over factors, c_{12} and c_{21}. $E = 29\,000$ ksi, $I_1 = 100$ in^4, $I_2 = 200$ in^4, $A_1 = 15$ in^2, $A_2 = 30$ in^2.

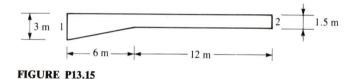

FIGURE P13.14

13.15. Determine the stiffness matrix for the reinforced-concrete beam element of Fig. P13.15. Use three segments to describe the linear hunch with I of each segment equal to I of the beam at the mid-length of the segment. The width of the beam is 50 cm. $E = 20\,000$ MPa.

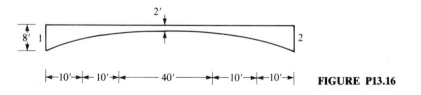

FIGURE P13.15

13.16. The symmetric prestressed concrete beam element of Fig. P13.16 has a parabolic profile and is to be modeled by using five segments with the lengths shown. The moment of inertia of each segment is equal to the moment of inertia of the beam at the mid-length of the segment. The depth of the beam at a point can be found by first finding the parabola describing the profile in a coordinate system, for example, one with the origin at the lower left end of the beam. Find the stiffness matrix of this beam element for bending. The width of the element is 10 ft. $E = 3\,000$ ksi.

FIGURE P13.16

13.17. An element of a two-dimensional frame is shown in Fig. P13.17. The

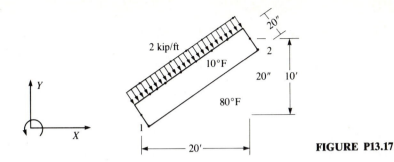

FIGURE P13.17

cross-section is 10 in wide and 20 in deep. The other dimensions are as shown. Find the equivalent joint forces (normal force, shear, and moment) in the local coordinates for the following effects. The x axis is directed from joint 1 to 2.

(a) A load 2 kip/ft normal to the beam axis as shown.

(b) A temperature gradient varying from 80°F on the bottom to 10°F on the top. $\alpha = 6.5 \times 10^{-6}$; $E = 1\,000$ ksi.

(c) Determine the transformation matrix and transform these equivalent joint forces to the global coordinates, XY.

13.18. Both spans of the beam of Fig. P13.18 are exposed to a temperature gradient $\Delta T = T_2 - T_1 = 60°F$. Find the member forces for $h = 18$ in, $\alpha = 6 \times 10^{-6}$, $I = 600$ in^4, and $E = 29\,000$ ksi. Use 3 elements as shown.

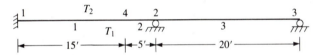

FIGURE P13.18

13.19. For the vertical member in the structure shown in Fig. P13.19 the temperature changes from 15°C on the left to 50°C on the right. Calculate the forces developed in the members because of this temperature change. The width of all the members is 60 cm and their depth is 1 m. $\alpha = 10^{-5}$(m/m)/°C; $E = 20\,000$ MPa.

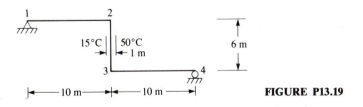

FIGURE P13.19

13.20. The deck of the slab bridge of Fig. P13.20 undergoes a temperature variation from 120°F at the top to 20°F at the bottom. Find the member-end forces due to this temperature change and the applied loads. The width of the beams and columns is 10 ft. The depth of the columns is 1 ft and that of the beams 2 ft. $\alpha = 6 \times 10^{-6}$ and $E = 3\,000$ ksi.

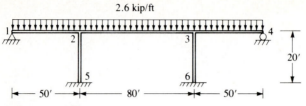

FIGURE P13.20

13.21. Find the member-end forces and the reaction of the inclined support for the frame of Fig. P13.21. $I = 120 \text{ in}^4$, $A = 10 \text{ in}^2$, and $E = 29\,000$ ksi.

> *Hint:* Use a bar oriented normal to the inclined support, with one end hinged at point 3 and the other hinged to the ground, to model the inclined support. Or take the global x axis parallel to 23.

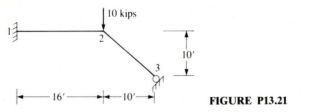

FIGURE P13.21

Find the solutions of the following beams, trusses, and frames by the flexibility method.

13.22. Find the member-end moments and draw the moment diagram. $I = 180 \text{ in}^4$, $E = 29\,000$ ksi.

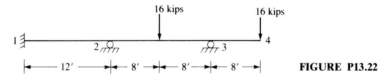

FIGURE P13.22

13.23. $I = 3\,000 \text{ cm}^4$; $E = 200\,000$ MPa.

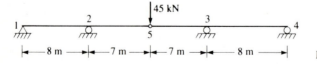

FIGURE P13.23

13.24. $A = 2 \text{ in}^2$; $E = 29\,000$ ksi.
   Hint: Element flexibility matrix $= l/EA$.

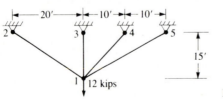

FIGURE P13.24

13.25. $A = 5$ cm^2; $E = 200\,000$ MPa.

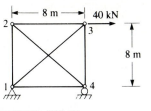

FIGURE P13.25

13.26. $I = 14\,000$ cm^4; $E = 20\,000$ MPa.

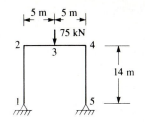

FIGURE P13.26

13.27. $I = 200$ in^4; $E = 29\,000$ ksi.

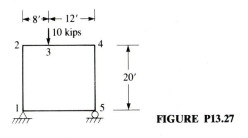

FIGURE P13.27

13.28. For the beams, $I = 2\,500$ in^4; for the columns, $I = 1\,500$ in^4. $E = 3\,000$ ksi.

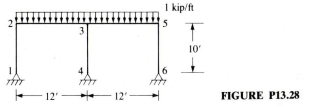

FIGURE P13.28

Hint: Find the member-end forces due to the equivalent joint forces and add the fixed-end forces.

CHAPTER
14

CABLES

A cable is a slender flexible member with zero or negligible flexural stiffness. Thus, it is only capable of developing a tensile force. It derives its capacity to resist transverse loads by undergoing significant changes in slope at the points of application of the loads. Wires, ropes, strings, and chains are examples of cables. In the following sections we discuss the forces and deformations in cables. The design sign convention (Section 4.1) is used for this purpose, with tension forces assumed as positive.

14.1 INTERNAL FORCES

Cables transmit vertical loads by undergoing sags and developing tensile forces. The horizontal component of the tensile force is called *thrust*.

The sag is usually small, about 10% of the span length. Since cables are flexible elements, their bending resistance as well as resistance against compression is practically negligible. Steel cables are usually of twisted-strand construction. The resisting area of such a cable is about two-thirds of its gross area of cross-section. The allowable stress is 1/3 of the ultimate capacity. The ultimate capacity is generally high—150 to 300 ksi. The area of cross-section and strength of cables are provided in appropriate codes for example in the AISC manual of steel construction.

For simple cables under concentrated loads, the forces and sags may be found by statics. For general loading and configurations, the differential equation developed in Section 14.2 can be integrated.

Example 14.1. Find the tension force T and sag h of a cable with span l under a single load P applied at a distance αl from the left support (Fig. 14.1). Find the location of the load that causes the maximum thrust (or tension). Compare the values of sag when the load is at the latter point and at $1/10$ of the span.

Solution. From equilibrium of the forces in the x and y directions and equilibrium of moments about support 2, we have

$$R_{1x} + R_{2x} = 0$$
$$R_{1y} + R_{2y} - P = 0$$
$$R_{1y}l - Pl(1 - \alpha) = 0$$

leading to

$$R_{1y} = P(1 - \alpha)$$
$$R_{2y} = \alpha P$$
$$R_{1x} = -R_{2x}$$

To find the horizontal force we can use an additional equation, stating that the moment capacity of the cable is zero. Thus, setting the sum of moments of the forces on segment 13 about point 3 equal to zero, we get

$$R_{1x} \times h + R_{1y} \times \alpha l = 0$$

$$R_{1x} = -\frac{R_{1y}\alpha l}{h}$$

or

$$R_{1x} = -\frac{P\alpha(1 - \alpha)l}{h}$$

Thus,

$$R_{2x} = -R_{1x} = \frac{P\alpha(1 - \alpha)l}{h}$$

We find that the thrust is the same for both segments. This is true for a cable under any number of vertical loads, owing to equilibrium of the horizontal forces for free bodies of cable segments near the vertical loads. If the trust is H,

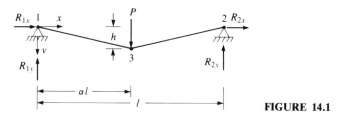

FIGURE 14.1

then by equilibrium of horizontal forces at support 1,

$$H = -R_{1x} \quad \text{or} \quad H = \frac{P\alpha(1-\alpha)l}{h}$$

To find the maximum thrust as the location of the load varies, we set the derivative of H with respect to α equal to zero,

$$\frac{d}{d\alpha}(H) = \frac{P\alpha}{h}\left(1 - \frac{\alpha}{2}\right) = 0 \quad \text{which gives} \quad \alpha = \tfrac{1}{2}$$

Therefore, when the load is at the mid-span, the maximum thrust H and tension force T occur in the cable. The tension in the cable for the load at αl is

$$T = \sqrt{R_{1x}^2 + R_{1y}^2} = \sqrt{\frac{P^2\alpha^2(1-\alpha)^2l^2}{h^2} + P^2(1-\alpha)^2} = P(1-\alpha)\sqrt{1 + \frac{\alpha^2l^2}{h^2}}$$

and the maximum tension is

$$T_{\max} = \frac{P}{2}\sqrt{1 + \frac{l^2}{4h^2}}$$

For a sag $h = 0.1l$, the values of thrust and tension for the locations of the load $\alpha = 0.1$ and $\alpha = 0.5$ are as follows

α	H	T	Difference
0.1	0.9P	1.27P	29.5%
0.5	2.5P	2.55P	2%
Difference	64%	50.2%	

From the above results we see that for common values of sag, $h \approx 0.1l$, the tension force and thrust change substantially as the load moves from $0.1l$ to $0.5l$. However, the difference between the maximum values of thrust and tension force (which occur for the load at the mid-span) is negligible.

14.2   GOVERNING EQUATION

The forces in simple cables can be found by statics. To find the cable configuration, we must consider its deformation.

Consider the cable shown in Fig. 14.2a with a distributed vertical load p per unit horizontal projection of the cable length. An infinitesimal element of the cable can be considered as a straight line if the deformations are small. In the absence of horizontal loads, equilibrium of the forces requires that the horizontal component of the cable force be constant, that is,

$$H = \text{constant}$$

Equilibrium of vertical forces gives

$$dV + p\,dx = 0$$

where V is the vertical component of the cable force at x. If θ is the angle that the tangent to the cable at x makes with the x axis, then, referring to Fig.

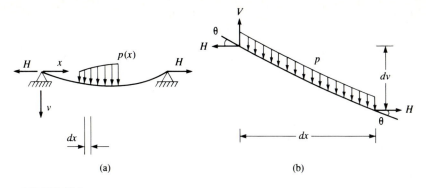

FIGURE 14.2

14.2b, the vertical and horizontal components of the cable force are related by $V = H \tan \theta$. Substitution of V in the previous equation gives

$$d(H \tan \theta) + p\, dx = 0$$

or

$$-\frac{d}{dx}(H \tan \theta) = p$$

Since H is a constant, the last equation can be written as

$$-H\frac{d}{dx}(\tan \theta) = p$$

Upon substitution for $\tan \theta$ in terms of deflection (Fig. 14.2b), that is,

$$\tan \theta = \frac{dv}{dx}$$

we get the governing equation for the cable as

$$\boxed{-Hv'' = p} \tag{14.1}$$

Equation 14.1 can be easily integrated for distributed loads. Concentrated loads can be expressed in terms of the Dirac delta function. However, it would be easier to integrate Eq. (14.1) to obtain

$$-Hv' = \int_0^x p\, dx$$

The right-hand side of the last equation is by Eq. (4.3), the negative of the shear force in an equivalent beam with the same span and load as the cable. We recall that in the design sign convention positive loads are downward while positive shear forces at the left end of a segment are upward. Thus,

$$Hv' = V \qquad (14.2)$$

Equation (14.2) is easier to use for finding the deflection of cables under concentrated loads than Eq. (14.1).

14.3   CABLE DEFLECTION

If we integrate Eqs. (14.1) or (14.2) and apply the boundary conditions, we can find an equation for the deflection in terms of the thrust. Therefore, we need an additional condition to complete the solution. This additional parameter may be the specified value of sag at a point, the cable tension (or area and allowable stress), or the total length of the cable. The thrust and then the cable deflection can be calculated as well as the cable tension.

> **Example 14.2.** Find the deflection of the cable shown in Fig. 14.3 under a uniform load p over the horizontal projection of the cable. Calculate the thrust so that the maximum sag does not exceed h. Also find the value of the thrust for $p = 5$ kip/ft, $l = 100$ ft, and $h = 8$ ft.
>
> **Solution.** Since the uniform load p is a constant, integration of Eq. (14.1) gives
>
> $$-Hv' = px + C_1$$
>
> and
>
> $$-Hv = \frac{px^2}{2} + C_1 x + C_2$$
>
> The boundary conditions, $v = 0$ at $x = 0$ and $x = l$, result in
>
> $$C_2 = 0 \qquad \text{and} \qquad C_1 = -\frac{pl}{2}$$
>
> Thus,
>
> $$Hv = \frac{p}{2} x(l - x)$$
>
> H can be calculated by specifying the maximum sag, h. Because of symmetry of the load and geometry, the maximum sag occurs at the mid-span,

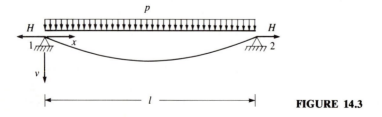

FIGURE 14.3

$x = l/2$, resulting in

$$H = \frac{pl^2}{8h}$$

and

$$v = \frac{4hx}{l^2}(l - x)$$

Note that the deflection of the cable at any point is dependent on the maximum sag. The maximum sag is also related to the internal forces. For the given numerical values, the thrust is

$$H = \frac{5 \times 100^2}{8 \times 8} = 781.25 \text{ kips}$$

The deflection at any point, in feet, is

$$v = \frac{4 \times 8x}{100^2}(100 - x) = 0.0032x(100 - x)$$

From the above results we observe that a cable under a uniform load over its projection has a parabolic deflected shape.

Singularity Method for Cables under Concentrated Loads

Since the shape of a cable changes under the applied loads, the solution is nonlinear and thus cannot be obtained by superposition. The singularity method is suitable for the solution of cables under concentrated loads. Integration of singularity functions and representation of concentrated loads were discussed in Sections 9.2 and 9.3. Here we use those results to find deflection of cables under concentrated loads.

Consider for example the cable of Fig. 14.4 under a concentrated load P at a distance a. The vertical reaction of the cable at support 1 is

$$R_1 = \frac{P(l - a)}{l}$$

and sum of the vertical forces at a distance x can be expressed as

$$V = R_1 - PH_a$$

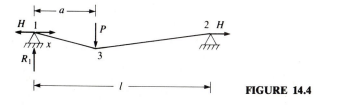

FIGURE 14.4

where $H_a = H(x - a)$ is the unit step function, defined by Eq. (9.3). H_a is zero for $x < a$ and 1 for $x \geq a$. Substitution of V in Eq. (14.2) and integration (Eq. (9.5)) yields

$$Hv = R_1 x - P(x - a)H_a + C_1$$

Since the supports are unyielding, at $x = 0$, $v = 0$, which gives $C_1 = 0$, and thus,

$$Hv = \frac{P(l - a)}{l} x - P(x - a)H_a$$

If the maximum sag at a is h, then

$$Hh = P(l - a)\frac{a}{l} \quad \text{or} \quad H = P(l - a)\frac{a}{lh}$$

and

$$v = \frac{xh}{a} - \frac{x - a}{l - a}\frac{lh}{a} H_a$$

This represents a bilinear diagram, with $v = h$ at $x = a$ and $v = 0$ at $x = 0$ and l. The maximum deflection occurs when the load is at the mid-span.

Example 14.3. A cable with a total length of 51 m is fixed at its ends spanning a distance of 50 m (Fig. 14.5). The effective area of cross-section of the cable is 0.785 cm² and its allowable stress is 1 000 MPa. What is the maximum magnitude of the concentrated load that can be applied at a distance of 30 m from the left-hand end?.

Solution. With the total length of 51 m and the load at 30 m from the left, the sag h can be calculated from geometry of the deflected cable.

$$l_1 + l_2 = 51$$

or

$$\sqrt{30^2 + h^2} + \sqrt{20^2 + h^2} = 51$$

Squaring both sides, we get

$$30^2 + h^2 + 20^2 + h^2 + 2\sqrt{(30^2 + h^2)(20^2 + h^2)} = 51^2$$

or

$$\sqrt{(30^2 + h^2)(20^2 + h^2)} = 650.5 - h^2$$

Squaring both sides again and solving for h, we find $h = 4.93$ m.

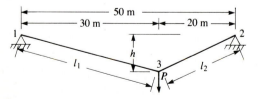

FIGURE 14.5

Then, from the previous results,

$$H = P(l - a)\frac{a}{lh}$$

with $x = 30$, $h = 4.93$ m we get

$$H = P(50 - 30)\frac{30}{50 \times 4.93} = 2.43P$$

The slopes of the cable on the left and right of the load are

$$\alpha_1 = \tan^{-1}(4.93/30) = 9.33° \quad \text{and} \quad \alpha_2 = \tan^{-1}(4.93/20) = 13.85°$$

Since the horizontal component of the tension force is constant, the largest force occurs in the segment with the largest slope, where the vertical component is the largest. Thus, the tension in the right segment of the cable in Fig. 14.5 is larger than that in the left segment. This force is

$$T = \frac{H}{\cos \alpha_2} = \frac{H}{0.971} = \frac{2.43P}{0.971} = 2.50P$$

The allowable tension force in the cable is

$$T_a = \sigma_a A = \frac{1\,000 \times 10^6 \times 0.785}{10\,000} = 78\,500\,\text{N} = 78.50\,\text{kN}$$

Setting the actual tension force equal to the allowable tension force we get

$$2.50P = 78.50\,\text{kN} \quad \text{or} \quad P = 31.40\,\text{kN}$$

Example 14.4. Find the equation for the deflection of a cable under n concentrated loads P_i acting at distances a_i (Fig. 14.6).

Solution. The left reaction is

$$R_1 = \sum_{i=1}^{n} \frac{P_i}{l}(l - a_i)$$

and shear at x is

$$V = R_1 - \sum_{i=1}^{n} P_i H_i$$

where $H_i = H(x - a_i)$ is the unit step function.

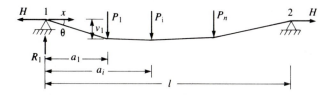

FIGURE 14.6

The differential equation of the cable (Eq. (14.2)) integrates to

$$Hv = R_1 x - \sum_{i=1}^{n} P_i(x - a_i)H_i + C_1$$

The constant C_1 is found from the condition of zero deflection at the left end. At $x = 0$, $H_i = 0$ and $C_1 = 0$. Substitution of R_1 and C_1 in the last equation gives

$$Hv = \sum_{i=1}^{n} P_i \left[(l - a_i)\frac{x}{l} - (x - a_i)H_i \right] \tag{14.3}$$

If the sag at a is h, then

$$H = \sum_{i=1}^{n} \frac{P_i}{h} \left[(l - a_i)\frac{a}{l} - (a - a_i)H(a - a_i) \right] \tag{14.4}$$

If the maximum sag, h_m, is specified we can find its location as follows. Since H is a constant, the maximum sag will occur where the right-hand side of Eq. (14.3) is a maximum, Since one needs to calculate the sag at all the load points, the right-hand side of Eq. (14.3) is first calculated for the different a_i or until the succeeding value becomes smaller than the previous one. A more efficient alternative for finding the location of maximum sag is to calculate successive values of the derivative of the right hand side of Eq. (14.3) i.e. $\sum_i P_i[(l - a_i)/l - Hi]$. As soon as the value of the latter changes sign the previous value of i gives a_i the location of the maximum sag. Having found the location of the maximum defection, a_m, in this manner, we can set $a = a_m$ and $h = h_m$ in Eq. (14.4) and find the maximum thrust H. Then the values of the right-hand side of Eq. (14.3), which were calculated before, can be scaled down by H to give the actual values of sag throughout the cable. The tension force is then found by

$$T = \frac{H}{\cos \theta} = H\sqrt{1 + [(h_{i+1} - h_i)/(a_{i+1} - a_i)]^2} \tag{14.5}$$

where a_i and h_i are the distance from the left-hand end to load P_i and the sag at load P_i. If the two ends of the cable are at the same level, then h_i is the same as v_i. The required area of the cable is

$$A = \frac{T}{\sigma_a}$$

In the last equation, σ_a is the allowable stress in the cable.

Example 14.5. The cable structure of Fig. 14.7 is supported by a tower. The base of the tower is hinged in order to avoid large bending moments there. The sag at point 2 is not to exceed 5 ft below the left support. Find the sag of the cable at points 2 and 3, the tension force in the different segments of the cable, and the vertical force in the tower.

Solution. Considering the free body of the tower (Fig. 14.7b), the cable forces at 4 have horizontal components H_{45} and H_{56} (Fig. 14.7c). Since the base of the tower is hinged, the equation of condition requires that sum of the moments be zero at the base, leading to $H_{45} = H_{56}$. Thus, the cable thrust is constant in all the segments. The slope of the chord 14 is 7/60, which results in a maximum sag at 2

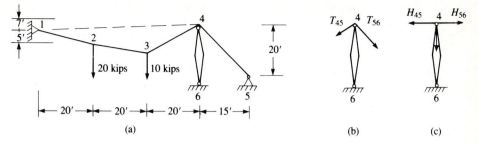

FIGURE 14.7

from the chord equal to $5 + (7/60) \times 20 = 7.33$ ft. By Eq. (14.3) at $x = 20$ ft,

$$7.33H = 20\left[(60 - 20) \times \frac{20}{60}\right] + 10\left[(60 - 40) \times \frac{20}{60}\right]$$

which gives $H = 45.48$ kips. With this value of thrust, the sag at point 3, measured from the chord 14, can be found by Eq. (14.3), as

$$v_3 = \frac{20}{45.48}\left[(60 - 20) \times \frac{40}{60}\right] + \frac{10}{45.48}\left[(60 - 40) \times \frac{40}{60}\right] = 14.66 \text{ ft}$$

The sag at point 3, from the horizontal axis passing through point 1, is

$$h_3 = 14.66 - (7/60) \times 40 = 9.99 \text{ ft}$$

Using the displacements measured from the horizontal axis at point 1, the tension forces in the different segments are

$$T_{12} = 45.48\sqrt{1 + [5/20]^2} = 46.89 \text{ kips}$$

$$T_{23} = 45.48\sqrt{1 + [(9.99 - 5)/(40 - 20)]^2} = 46.87 \text{ kips}$$

$$T_{34} = 45.48\sqrt{1 + [(9.99 + 4.66)/(60 - 40)]^2} = 59.68 \text{ kips}$$

$$T_{45} = 45.48\sqrt{1 + [20/15]^2} = 75.80 \text{ kips}$$

The vertical force in the tower can be found from the vertical components of the forces in the cable on both sides of point 4:

$$V_4 = 59.68 \times \frac{16.99}{\sqrt{16.99^2 + 20^2}} + 75.80 \times \frac{20}{\sqrt{15^2 + 20^2}} = 99.28 \text{ kips}$$

Catenary

A catenary is the shape of a cable that is hanging from two points and is under its own weight (Fig. 14.8). The solution of this problem is similar to the ones already discussed. However, the algebra is more cumbersome.

The differential equation (14.1) is for general vertical loading and thus is applicable to this case. However, in that equation p is the load per unit length of the horizontal projection of the cable. The weight of the cable over the horizontal length dx is $p\,dx = w\,ds$, where w is the weight of a unit length of

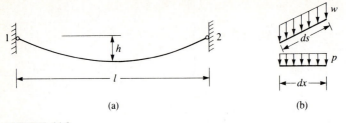

(a)   (b)

FIGURE 14.8

the cable and ds is the length of the cable with projection dx. Thus, the load per unit horizontal projection is $p = w(ds/dx)$. Since

$$ds = \sqrt{dx^2 + dv^2} = \sqrt{1 + v'^2}\, dx$$

we get

$$p = w\sqrt{1 + v'^2}$$

Therefore, the governing equation becomes

$$-Hv'' = p = w\sqrt{1 + v'^2}$$

or

$$\frac{v''}{\sqrt{1 + v'^2}} = -\frac{w}{H}$$

If the slope of the cable at x is θ, then

$$v' = \frac{dv}{dx} = \tan\theta$$

$$v'' = \frac{dv'}{dx} = \frac{dv'}{d\theta}\frac{d\theta}{dx} = (1 + \tan^2\theta)\frac{d\theta}{dx}$$

and

$$\frac{v''}{\sqrt{1 + v'^2}} = \frac{1 + \tan^2\theta}{\sqrt{1 + \tan^2\theta}}\frac{d\theta}{dx} = \sqrt{1 + \tan^2\theta}\,\frac{d\theta}{dx} = \frac{1}{\cos\theta}\frac{d\theta}{dx}$$

leading to

$$\frac{1}{\cos\theta}\frac{d\theta}{dx} = -\frac{w}{H} \quad \text{or} \quad \frac{d\theta}{\cos\theta} = -\frac{w}{H}dx$$

With the aid of tables of integrals, or upon expressing $\cos\theta$ in terms of $\tan(\theta/2)$, the above equation can be integrated as

$$\log\tan\left(\frac{\theta}{2} + \frac{\pi}{4}\right) = -\frac{w}{H}x + C_1$$

By symmetry, at $x = l/2$, $\theta = 0$, which gives $C_1 = wl/2H$. Thus,

$$\log\tan\left(\frac{\theta}{2} + \frac{\pi}{4}\right) = \frac{w}{H}\left(\frac{l}{2} - x\right)$$

or

$$\tan\left(\frac{\theta}{2}+\frac{\pi}{4}\right) = \exp\left[\frac{w}{H}\left(\frac{l}{2}-x\right)\right]$$

Since

$$\tan\left(\frac{\theta}{2}+\frac{\pi}{4}\right) = \frac{1+\tan(\theta/2)}{1-\tan(\theta/2)}$$

we have

$$\tan\left(\frac{\theta}{2}+\frac{\pi}{4}\right) - \cot\left(\frac{\theta}{2}+\frac{\pi}{4}\right) = 2\tan\theta$$

or upon substitution for $\tan(\theta/2 + \pi/4)$ and $\tan\theta$ we get

$$\exp\left[\frac{w}{H}\left(\frac{l}{2}-x\right)\right] - \exp\left(-\left[\frac{w}{H}\left(\frac{l}{2}-x\right)\right]\right) = 2v'$$

which gives

$$v = \frac{H}{2w}\left(\exp\left[\frac{w}{H}\left(\frac{l}{2}-x\right)\right] - \exp\left\{-\left[\frac{w}{H}\left(\frac{l}{2}-x\right)\right]\right\}\right) + C_2$$

Because at $x = l/2$, $v = h$, we get $C_2 = h$. Therefore,

$$v = \frac{H}{w}\cosh\left[\frac{w}{H}\left(\frac{l}{2}-x\right)\right] + h$$

14.4 STIFFNESS METHOD FOR CABLE STRUCTURES

As we saw in the previous sections the internal forces in a cable are not only dependent on the applied loads but also on the deflection. In fact (see pretensioned cables in Laboratory Experiments for this chapter) cables become stiffer with deflections induced by the loads. Thus, the stiffness of a cable is not constant and the force-displacement relationship is not linear. Therefore, the solution of cable structures with the stiffness method must be carried out by one of the methods for non-linear structures; that is, by an incremental, iterative or a combination of these two methods. The force-displacement relation for a cable segment is similar to that of a truss element, but the stiffness is set equal to zero whenever the force in the element becomes compressive. The force-displacement relation is written in terms of the increment in the tension force and increment in the elongation. The latter is in turn expressed in terms of the end displacements. If the weight of the cable is not negligible then the catenary shape of the cable must be considered. For initially straight cables under loads the solution is started by adding springs (Section 13.3) to the lateral displacement degrees of freedom (see R. D. Cooke, *Concepts and Applications of Finite Element Analysis*, 2nd ed. Wiley, 1981).

LABORATORY EXPERIMENTS

Load Transfer

Loads are transferred in flexible cables by changes in the geometry of the cable. The forces in a cable can be found by equilibrium.

Apply tension to a flexible cable by a weight hanging from one end of the cable that goes over a pulley. Add a lateral (vertical) load to the cable at the mid-span. Record the load magnitude and geometry. Increase the load in steps and record the load and geometry again. Plot the load deflection curve. A cable attached to two points may also be used for this experiment.

Apply a non-symmetric load. Observe and record the geometry and location and magnitude of the load. Calculate the horizontal projection of the cable force and observe that it is constant throughout the cable. Compare the calculated cable force to the tension force applied to the end of the cable if the pulley and end load is used. From the measured geometry and load, calculate the tension force in the cable. Observe where its maximum value occurs. Calculate the horizontal projection of the cable force. Note that this is again constant throughout the cable (under vertical loads only). Compare the calculated cable force with the pretension applied to the end of the cable. Change the pretension force to twice its value and repeat the above steps.

FIGURE L14.1

Internal Forces

Place a cable with a 6 in to 8 in initial sag between two points (Fig. L14.1). Attach a load cell to the cable near one of the supports. Note that a spring with known stiffness can also be placed in line. The extension of the spring will then reveal the internal force. Record the geometry and internal force for a number of symmetrical and unsymmetrical loads and for different values of sag. For the same loading condition, what happens to the internal force when the sag increases?

Pretensioned Cables

To avoid significant changes of geometry during loading, pretensioned cables are used. For a cable suspended from two points, apply a series of uniformly spaced loads (say 6 at 10 lb each). Note the change in geometry and internal force. These loads cause a tension in the cable. Next apply loads as in the previous experiment and compare the results. Explain the effect of pretension on the internal forces and geometry.

Cable Systems

Hang two cables with their supports about 16 inches from one another (one maybe on top of the other or next to it). Insert a load cell at one end of a cable

FIGURE L14.2

(the uppermost support) and zero the strain indicator. Using a number of turnbuckles (4 to 6) pull the cables together to produce a pretension (of 20 to 30 lb) (Fig. L14.2). Load the system as in the previous experiments (apply loads at turnbuckle anchorages). Record the changes in force and geometry. Increase the prestress (to about double the initial value) and repeat the experiment. Compare the two results and compare with those in the above experiments.

PROBLEMS

14.1. Find the reactions of the cable of Fig. P14.1, the tension force in all the segments, and the sag at the locations of the loads.

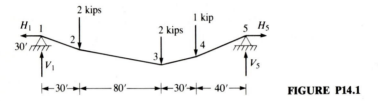

FIGURE P14.1

14.2. Calculate the sag of the loaded points on the cable in Fig. P14.2 from the horizontal axis passing through point 1. The total length of the cable is 200 ft.

    Hint: Calculate the sag from the chord 1–5, then add the distance between the chord and the horizontal axis.

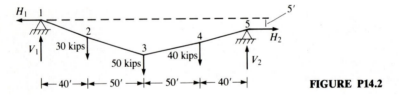

FIGURE P14.2

14.3. Find the tension force in each segment of the cable of Fig. P14.3. See the hint in Problem 14.2.

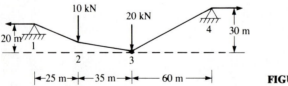

FIGURE P14.3

14.4. Find the maximum tension force in a cable under a uniform load of 40 kN/m over the 60 m span of the cable (Fig. P14.4). The maximum sag is 5 m.

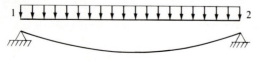

FIGURE P14.4

14.5. Find the tension forces in the segments of the cable shown in Fig. P14.5.

    Hint: The equation of condition for the hinge at the base of the tower indicates that no net horizontal force exists at the top of the tower. Therefore, the horizontal components of the cable forces on the two sides of the tower are equal.

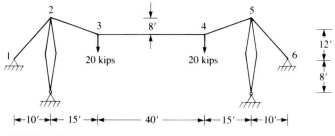

FIGURE P14.5

14.6. Find the sag at the different points of the cable of Fig. P14.6 and the tension force in the different segments.

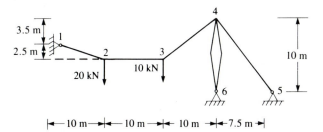

FIGURE P14.6

CHAPTER
15

ARCHES

Arches are curved structures that carry their loads primarily by developing axial compression. That is why arches have been widely used in the past for masonry and concrete construction. The choice of the arch shape depends on the dominant loading conditions and ease of construction. Circular arches are easier to build. However, parabolic arches are more economical for uniform loads.

In the following sections we discuss the forces and deformations of statically determinate and indeterminate arches. The design sign convention (Section 4.1) is used, except for the stiffness method in Section 15.3, where the analysis sign convention (Section 10.1) is used.

15.1 STATICALLY DETERMINATE ARCHES

Before determining the axial force, moment and shear in an arch, we must usually calculate the support reactions.

The arches shown in Figs. 15.1a,b,c are statically determinate because each one has three unknown reactions that can be determined by the equations of statics. The structures in Fig. 15.1a, and especially 15.1c, develop large bending moments. Because of this, the structure of Fig. 15.1c is sometimes

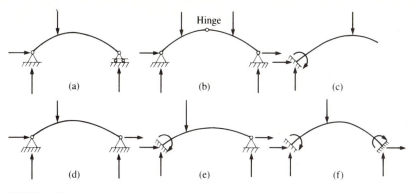

FIGURE 15.1

categorized as a curved beam rather than an arch. However, the three-hinged arch of Fig. 15.1b is primarily under axial compression and is therefore commonly used in practice.

The structures of Figs. 15.1d,e,f are statically indeterminate. Statically indeterminate arches are more economical to build if the foundation allows it. If there is a possibility of support rotation and displacement because of the foundation material, then a statically determinate arch would provide a safer alternative.

Normal Force, Shear, and Moment

Once the shape of the arch is chosen, the equation of the arch expressing y in terms of x can be determined. The support reactions are then calculated and the normal force, shear, and moment can be calculated at any point in the arch. Normal force, shear, and moment diagrams are obtained by calculating the values at different points or from their equations. The values are plotted along and normal to the axis of the arch.

Referring to Fig. 15.2, in the design sign convention a positive normal force induces tension in the arch, a positive shear effects an upward displacement of the left end of a segment and a positive moment causes

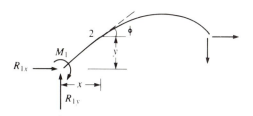

FIGURE 15.2

compression in the top fibers of the arch. The normal force N, shear V, and moment M, at section x, Fig. 15.2, are

$$N = -R_{1y} \sin \phi - R_{1x} \cos \phi$$

$$V = R_{1y} \cos \phi - R_{1x} \sin \phi$$

$$M = M_1 + R_{1y}x - R_{1x}y$$

Since the slope of the curve is $\tan \phi = y'$ with

$$\sin \phi = \frac{\tan \phi}{\sqrt{1 + \tan^2 \phi}} \quad \text{and} \quad \cos \phi = \frac{1}{\sqrt{1 + \tan^2 \phi}}$$

we have,

$$\sin \phi = \frac{y'}{\sqrt{1 + y'^2}} \quad \text{and} \quad \cos \phi = \frac{1}{\sqrt{1 + y'^2}}$$

If any loads are applied between the left support and the section under consideration, they must be included in the above equations. Note that for circular arches it is easier to use polar coordinates. For other arches, given the equation of the arch and the specified loads, the normal force, shear, and moment diagrams can be determined. As in the case of cables, the horizontal component of the force is called the *thrust* of the arch.

Example 15.1. Draw the normal force, shear, and moment diagrams for the three-hinged semi-circular arch of Fig. 15.3a under its own weight. The weight of the arch is 2 kip/ft of the centerline. The centerline of the arch has a radius of 50 ft.

Solution. The four reactions can be found by the equations of statics, and the equation of condition for the internal hinge. Here we use polar coordinates with

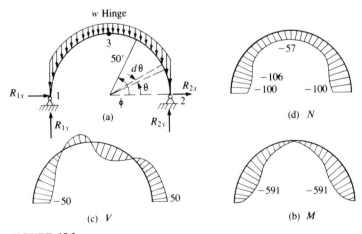

FIGURE 15.3

angle ϕ varying (in the counterclockwise sense) from zero to π to find the internal forces. Thus, we first calculate the vertical reaction at support 2. By symmetry,

$$R_{2y} = w\pi\frac{a}{2} = 2 \times \pi \times \frac{50}{2} = 157 \text{ kips}$$

The equation of condition is obtained by setting the moment of the forces on segment 23 with respect to the hinge at 3 equal to zero. Since the moment and shear at ϕ due to the load are needed later, we calculate their values, which will also be used in calculating the reactions. To accomplish this end, we choose the variable θ, which varies from zero to ϕ. The normal force at ϕ due to the load is

$$N_p = \int_0^\phi wa\, d\theta \cos\phi = wa \cos\phi \int_0^\phi d\theta = wa\phi \cos\phi$$

The shear at ϕ due to the load is

$$V_p = \int_0^\phi wa\, d\theta \sin\phi = wa\phi \sin\phi$$

and the moment at ϕ due to the load is

$$M_p = -\int_0^\phi wa\, d\theta\, a(\cos\theta - \cos\phi) = -wa^2\int_0^\phi \cos\theta\, d\theta - wa^2 \cos\phi\int_0^\phi d\theta$$

$$= -wa^2(\sin\phi - \phi\cos\phi)$$

Thus, the moment of the forces on the right half of the arch calculated at the crown hinge ($\phi = \pi/2$) gives

$$R_{2y} \times 50 + R_{2x} \times 50 - 2 \times 50^2 = 0$$

which gives

$$R_{2x} = -157 + 100 = -57 \text{ kips.}$$

With the above values, the normal force in the arch at section ϕ is

$$N = -R_{2y} \cos\phi + R_{2x} \sin\phi + wa\phi \cos\phi$$

$$= 100 \cos\phi(\phi - 1.57) - 57 \sin\phi$$

The shear is

$$V = -R_{2y} \sin\phi - R_{2x} \cos\phi + wa\phi \sin\phi$$

$$= 100(\phi - 1.57) \sin\phi + 57 \cos\phi$$

and the moment at ϕ is

$$M = R_{2y}a(1 - \cos\phi) + R_{2x}a \sin\phi - wa^2(\sin\phi - \phi\cos\phi)$$

$$= 157 \times 50(1 - \cos\phi) - 57 \times 50 \sin\phi - 2 \times 50^2(\sin\phi - \phi\cos\phi)$$

$$= 7\,850 - 7\,850 \sin\phi - 5\,000(1.57 - \phi) \cos\phi$$

To verify the above equations we can consider the equilibrium of half of the arch, and find that by symmetry shear at the crown is equal to zero and that the normal force at the crown is equal to the horizontal reaction at the support. Figures 15.3b,c,d show the plots of N, V, and M.

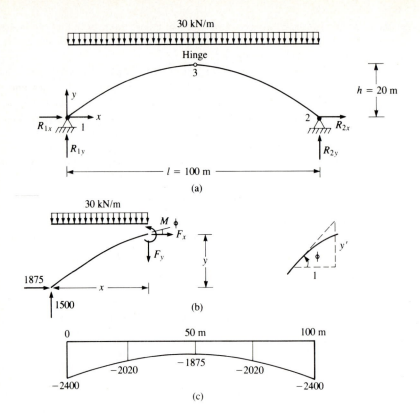

FIGURE 15.4

Example 15.2. The three-hinged parabolic arch of Fig. 15.4a, has a span of 100 m with the two supports at the same level. The two supports are hinged and a third hinge is located at the crown, with a rise of 20 m. Find the normal force, shear, and moment diagrams in the arch for a uniform load of 30 kN/m acting over the horizontal projection of the arch.

Solution. The origin of the coordinates is taken at the left support. To find the equation of the arch, we assume a parabola with unknown parameters, which will be determined from the geometry of the arch. Let

$$y = a_0 + a_1 x + a_2 x^2$$

Since $y = 0$ at $x = 0$ and l, and $y = h$ at $x = l/2$, we find

$$a_0 = 0 \qquad a_1 = \frac{4h}{l} \qquad a_2 = -\frac{4h}{l^2}$$

Substituting a_0, a_1, and a_2 into the first equation gives

$$y = \frac{4hx}{l}\left(1 - \frac{x}{l}\right)$$

For the given values ($l = 100$ m, $h = 20$ m) we get

$$y = 0.8x - 0.008x^2$$

Since this is a statically determinate structure, the four unknown reactions at supports 1 and 2 can be found from the three equations of statics and an equation of condition. The equation of condition is found by setting the sum of moments of the left half of the arch, with respect to the hinge at 3, equal to zero. Thus,

$$\sum F_y = 0 \qquad R_{1y} + R_{2y} - 30 \times 100 = 0$$

$$\sum M_2 = 0 \qquad R_{1y} \times 100 - 30 \times 100 \times \frac{100}{2} = 0$$

$$\sum F_x = 0 \qquad R_{1x} + R_{2x} = 0$$

The moment of the forces on part 13 with respect to 3 gives the equation of condition,

$$\sum M_3 = 0 \qquad R_{1y} \times 50 - R_{1x} \times 20 - 30 \times \frac{50^2}{2} = 0$$

resulting in

$$R_{1x} = 1\,875 \text{ kN} \qquad R_{1y} = 1\,500 \text{ kN}$$

$$R_{2x} = -1\,875 \text{ kN} \qquad R_{2y} = 1\,500 \text{ kN}$$

To find the normal force, shear, and moment at an arbitrary section, we must first find the orientation of the section. Because of symmetry we need to consider only half of the arch. The slope of the arch is

$$y' = \tan \phi = 0.8 - 0.016x$$

with

$$\cos \phi = \frac{1}{\sqrt{1 + \tan^2 \phi}} = \frac{1}{\sqrt{1.64 - 0.025\,6x + 0.000\,256x^2}}$$

$$\sin \phi = \frac{\tan \phi}{\sqrt{1 + \tan^2 \phi}} = \frac{0.8 - 0.016x}{\sqrt{1.64 - 0.025\,6x + 0.000\,256x^2}}$$

Calculating F_x, F_y, and M at the section, by equilibrium of a segment (Fig. 15.4b), we get

$$F_x + 1\,875 = 0 \qquad F_x = -1\,875$$

$$-F_y + 1\,500 - 30x = 0 \qquad F_y = 1\,500 - 30x$$

$$M - 1\,500x + 30\frac{x^2}{2} + 1\,875y = 0$$

$$M = 1\,500x - 15x^2 - 1\,875(0.8x - 0.008x^2) = 0$$

The normal force and shear can be found from the horizontal and vertical force components F_x and F_y:

$$N = F_x \cos \phi - F_y \sin \phi$$

$$V = F_x \sin \phi + F_y \cos \phi$$

or

$$N = \frac{-1\,875}{\sqrt{1.64 - 0.025\,6x + 0.000\,256x^2}} - \frac{(1\,500 - 30x)(0.8 - 0.016x)}{\sqrt{1.64 - 0.025\,6x + 0.000\,256x^2}}$$

$$= \frac{-3\,075 + 48x - 0.48x^2}{\sqrt{1.64 - 0.025\,6x + 0.000\,256x^2}}$$

$$V = \frac{-1\,875(0.8 - 0.016x)}{\sqrt{1.64 - 0.025\,6x + 0.000\,256x^2}} + \frac{1\,500 - 30x}{\sqrt{1.64 - 0.025\,6x + 0.000\,256x^2}} = 0$$

These results show that for a parabolic arch under a uniform load over the span, the shear and moment are zero throughout the arch and the load is carried by compressive normal forces. Note that the axial compression varies as a parabola throughout the arch with only 22% difference between the maximum and minimum values (Fig. 15.4c).

15.2 STATICALLY INDETERMINATE ARCHES

In statically indeterminate arches the number of reactions is larger than the number of equations of statics plus the equations of condition. The additional equations needed for finding the redundant reactions are determined by considering the deformation of the arch. the method of consistent deformations (Section 7.1) may be used for the analysis of statically indeterminate arches. For more complex arches the stiffness method (Section 15.3) can be used.

Example 15.3. Find the reactions of the two-hinged circular arch under a uniform load 25 kN/m over the horizontal projection (Fig. 15.5). The radius of the arch is 120 m and its span is 100 m.

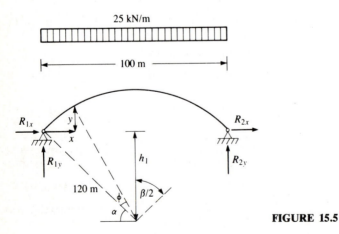

FIGURE 15.5

Solution. Taking R_{2x}, the horizontal reaction of the arch at support 2, as the redundant force, we find this force by setting the horizontal displacement of point 2 equal to zero. To find u_2, the horizontal displacement at the right support, we use Castigliano's second theorem (Eq. (6.22)). Thus,

$$\frac{\partial U}{\partial R_{2x}} = u_2 = 0$$

since

$$U = \int_0^\beta \frac{M^2 a}{2EI} \, d\phi$$

or

$$u_2 = \frac{a}{EI} \int_0^\beta M \frac{\partial M}{\partial R_{2x}} \, d\phi = 0 \qquad \text{or} \qquad \int_0^\beta M \frac{\partial M}{\partial R_{2x}} \, d\phi = 0$$

β is the angle of the arch. Since $\beta/2 = \sin^{-1}(50/120) = 0.43$ rad, $\beta = 0.86$ rad.

The moment in the arch will be expressed in terms of the reaction R_{2x}. The equilibrium equations are

$$R_{1x} + R_{2x} = 0$$

$$R_{1y} + R_{2y} - 25 \times 100 = 0$$

$$R_{1y} \times 100 - 25 \times \frac{100^2}{2} = 0$$

leading to

$$R_{1y} = 1\,250 \text{ kN}$$

$$R_{2y} = 5\,000 - R_{1y} = 5\,000 - 2\,500 = 1\,250 \text{ kN}$$

$$R_{1x} = -R_{2x}$$

with

$$h_1 = \sqrt{120^2 - 100^2} = 109.09 \text{ m}$$

$$\alpha = \sin^{-1}(h_1/a) = \sin^{-1}(109.09/120) = 1.14 \text{ rad.}$$

$$x = 50 - a\cos(\alpha + \phi) = 50 - 120\cos(1.14 + \phi)$$

$$y = a\sin(\alpha + \phi) - 109.09 = 120\sin(1.14 + \phi) - 109.09$$

Thus,

$$M = R_{1y}x - R_{1x}y - \frac{25x^2}{2}$$

$$= 1\,250[50 - \cos(1.14 + \phi)] + R_{2x}[120\sin(1.14 + \phi) - 109.09]$$
$$- 12.5[100 - 120\cos(1.14 + \phi)]^2$$

and

$$\frac{\partial M}{\partial R_{2x}} = [120\sin(1.14 + \phi) - 109.09]$$

From

$$\int_0^{0.86} M \frac{\partial M}{\partial R_{2x}} \, d\phi = 0$$

we have

$$7\,726 + 2.716R_{2x} = 0 \quad \text{or} \quad R_{2x} = -2\,845 \text{ kips}$$

15.3 STIFFNESS METHOD FOR ARCHES

The solution of statically indeterminate arches, especially those with several segments with different curvatures or in combination with other members such as beams, would be rather tedious with classical methods. The stiffness method, on the other hand, provides a straightforward solution method for such complex problems. With the aid of a computer the solution can be carried out efficiently.

Stiffness Matrix for Circular Arches

We derive the stiffness matrix of a circular arch segment by finding and inverting the flexibility matrix. To find the flexibility matrix we need to calculate the end displacements in terms of the end forces. Castigliano's second theorem can be used for this purpose. An arch segment is shown in Fig. 15.6. Points 1 and 2 are chosen such that the arch angle β is positive (counterclockwise) as we move from 1 to 2. The forces acting in the positive directions in the analysis sign convention are shown in Fig. 15.6. Positive moments are counterclockwise. The bending moment at a point on the arch at angle ϕ is

$$M = M_1 + U_1 a(1 - \cos\phi) - V_1 a \sin\phi$$

The strain energy of the arch is

$$U = \frac{1}{2EI}\int_0^\beta M^2 a\,d\phi$$

By Castigliano's theorem the displacements at the first end are

$$u_1 = \frac{a}{EI}\int_0^\beta M\frac{\partial M}{\partial U_1}\,d\phi \qquad (15.1a)$$

$$v_1 = \frac{a}{EI}\int_0^\beta M\frac{\partial M}{\partial V_1}\,d\phi \qquad (15.1b)$$

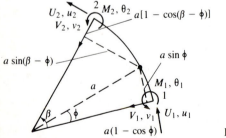

FIGURE 15.6

$$\theta_1 = \frac{a}{EI} \int_0^\beta M \frac{\partial M}{\partial M_1} d\phi \tag{15.1c}$$

Upon substituting for M and carrying out the integrals, we can find the displacements, which in matrix form are

$$\begin{bmatrix} u_1 \\ v_1 \\ \theta_1 \end{bmatrix} = \begin{bmatrix} f_{11} & f_{12} & f_{13} \\ f_{21} & f_{22} & f_{23} \\ f_{31} & f_{32} & f_{33} \end{bmatrix} \begin{bmatrix} U_1 \\ V_1 \\ M_1 \end{bmatrix} \tag{15.2a}$$

or

$$\mathbf{s}_1 = \mathbf{F}_{11}\mathbf{S}_1 \tag{15.2b}$$

where

$$f_{11} = \frac{a^2}{EI} (\tfrac{3}{2}\beta - 2 \sin \beta + \tfrac{1}{4} \sin 2\beta) \tag{15.2c}$$

$$f_{12} = f_{21} = \frac{a^2}{EI} (-\tfrac{1}{2} + \cos \beta - \tfrac{1}{2} \cos^2 \beta) \tag{15.2d}$$

$$f_{13} = f_{31} = \frac{a^2}{EI} (\beta - \sin \beta) \tag{15.2e}$$

$$f_{22} = \frac{a^3}{EI} (\tfrac{1}{2}\beta - \tfrac{1}{4} \sin 2\beta) \tag{15.2f}$$

$$f_{23} = f_{32} = \frac{a^2}{EI} (-1 + \cos \beta) \tag{15.2g}$$

$$f_{33} = \frac{a\beta}{EI} \tag{15.2h}$$

If we invert Eq. (15.2a) we find the force–displacement relationship, and the stiffness matrix. However, this is only for degrees of freedom at the left end of the arch. For degrees of freedom at both ends of the arch, the force–displacement relation would have the following form

$$\begin{bmatrix} U_1 \\ V_1 \\ M_1 \\ U_2 \\ V_2 \\ M_2 \end{bmatrix} = \begin{bmatrix} \mathbf{K}_{11} & \vdots & \mathbf{K}_{12} \\ \cdots & \vdots & \cdots \\ \mathbf{K}_{21} & \vdots & \mathbf{K}_{22} \end{bmatrix} \begin{bmatrix} u_1 \\ v_1 \\ \theta_1 \\ u_2 \\ v_2 \\ \theta_2 \end{bmatrix} \tag{15.3a}$$

$$\mathbf{S} = \mathbf{K}\mathbf{s} \tag{15.3b}$$

where the submatrix

$$\mathbf{K}_{11} = \mathbf{F}_{11}^{-1} \tag{15.4a}$$

Thus,

$$S_1 = K_{11}s_1 \tag{15.4b}$$

The submatrix K_{22} can be found in a similar manner. If we calculate the moment at a section with angle ϕ, from the forces at the second end we have

$$M = -M_2 - U_2a[1 - \cos(\beta - \phi)] - V_2a\sin(\beta - \phi)$$

Using Castigliano's theorem to find displacements u_2, v_2, θ_2, that is, applying Eqs. (15.1) while replacing subscript 1 by 2, we get expressions similar to those of Eq. (15.2a):

$$\begin{bmatrix} u_2 \\ v_2 \\ \theta_2 \end{bmatrix} = \begin{bmatrix} f_{11} & -f_{12} & f_{13} \\ -f_{12} & f_{22} & -f_{23} \\ f_{31} & -f_{23} & f_{33} \end{bmatrix} \begin{bmatrix} U_2 \\ V_2 \\ M_2 \end{bmatrix} \tag{15.5a}$$

or

$$s_2 = F_{22}S_2 \tag{15.5b}$$

As we can see, except for some sign changes, the flexibility matrices, F_{11} and F_{22} have the same terms. Inverting Eq. (15.5b) we find

$$K_{22} = F_{22}^{-1} \tag{15.6}$$

Since the stiffness matrix must be symmetric, K_{12} in Eq. (15.3a) must be the transpose of K_{21}. To find K_{21}, consider the equilibrium of the forces in Fig. 15.7 in the tangent and normal directions at 2. We have

$$U_1 \cos \beta + V_1 \sin \beta + U_2 = 0$$

$$-U_1 \sin \beta + V_1 \cos \beta + V_2 = 0$$

$$M_1 + U_1a(1 - \cos \beta) - V_1a \sin \beta + M_2 = 0$$

Casting the above equations in matrix form, we get

$$\begin{bmatrix} U_2 \\ V_2 \\ M_2 \end{bmatrix} = \begin{bmatrix} -\cos \beta & -\sin \beta & 0 \\ \sin \beta & -\cos \beta & 0 \\ -a(1 - \cos \beta) & a \sin \beta & -1 \end{bmatrix} \begin{bmatrix} U_1 \\ V_1 \\ M_1 \end{bmatrix} \tag{15.7a}$$

or

$$S_2 = QS_1 \tag{15.7b}$$

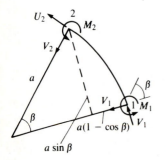

FIGURE 15.7

Substitution from Eq. (15.4b) into (15.7b) gives

$$\mathbf{S}_2 = \mathbf{QK}_{11}\mathbf{s}_1 \tag{15.8a}$$

or

$$\begin{bmatrix} U_2 \\ V_2 \\ M_2 \end{bmatrix} = \mathbf{QK}_{11} \begin{bmatrix} u_1 \\ v_1 \\ \theta_1 \end{bmatrix} \tag{15.8b}$$

Comparing this relation with Eq. (15.3a), we see that,

$$\mathbf{K}_{21} = \mathbf{QK}_{11} \tag{15.9}$$

Therefore, given the radius a, angle β, and EI, the stiffness matrix for the arch can be computed by Eqs. (15.2), (15.4a), (15.5), (15.6), (15.7), and (15.9).

Transformation Matrix

Consider the arch segment of Fig. 15.8a, with an arch angle β and the first end at an angle α, measured from the y axis in the positive sense (counterclockwise). Referring to Fig. 15.8a, the forces in the local coordinates can be expressed in terms of the forces in the global coordinates as follows

$$U_1 = \bar{U}_1 \cos \alpha + \bar{V}_1 \sin \alpha$$

$$V_1 = -\bar{U}_1 \sin \alpha + \bar{V}_1 \cos \alpha$$

$$M_1 = \bar{M}_1$$

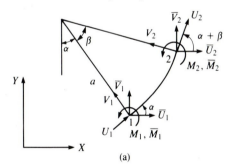

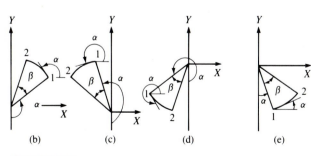

FIGURE 15.8

and

$$U_2 = \bar{U}_2 \cos(\alpha + \beta) + \bar{V}_2 \sin(\alpha + \beta)$$
$$V_2 = -\bar{U}_1 \sin(\alpha + \beta) + \bar{V}_2 \cos(\alpha + \beta)$$
$$M_2 = \bar{M}_2$$

or

$$
\begin{bmatrix} U_1 \\ V_1 \\ M_1 \\ U_2 \\ V_2 \\ M_2 \end{bmatrix} =
\begin{bmatrix}
\cos\alpha & \sin\alpha & 0 & 0 & 0 & 0 \\
-\sin\alpha & \cos\alpha & 0 & 0 & 0 & 0 \\
0 & 0 & 1 & 0 & 0 & 0 \\
0 & 0 & 0 & \cos(\alpha+\beta) & \sin(\alpha+\beta) & 0 \\
0 & 0 & 0 & -\sin(\alpha+\beta) & \cos(\alpha+\beta) & 0 \\
0 & 0 & 0 & 0 & 0 & 1
\end{bmatrix}
\begin{bmatrix} \bar{U}_1 \\ \bar{V}_1 \\ \bar{M}_1 \\ \bar{U}_2 \\ \bar{V}_2 \\ \bar{M}_2 \end{bmatrix}
$$

$$(15.10a)$$

or

$$\mathbf{S} = \mathbf{T}\bar{\mathbf{S}} \qquad (15.10b)$$

The transformation matrix $\mathbf{T}$ is an orthogonal matrix, that is, $\mathbf{T}^{-1} = \mathbf{T}^T$. Thus

$$\bar{\mathbf{S}} = \mathbf{T}^T\mathbf{S} \qquad (15.10c)$$

A similar expression can be written for displacements:

$$\mathbf{s} = \mathbf{T}\bar{\mathbf{s}} \qquad (15.11a)$$

$$\bar{\mathbf{s}} = \mathbf{T}^T\mathbf{s} \qquad (15.11b)$$

Figs. 15.8b to 15.8e show the angle α for different orientations of arch elements.

Stiffness Matrix in the Global Coordinates

Substitution of Eq. (15.11a) into (15.3b) gives

$$\mathbf{S} = \mathbf{K}\mathbf{T}\bar{\mathbf{s}} = \mathbf{B}\bar{\mathbf{s}} \qquad (15.12)$$

Given the displacements in the global coordinates, Eq. (15.12) allows the determination of the forces in the local coordinates (normal force, shear, and moments).

Multiplying both sides of Eq. (15.12) by $\mathbf{T}^T$ we get

$$\mathbf{T}^T\mathbf{S} = \mathbf{T}^T\mathbf{K}\mathbf{T}\bar{\mathbf{s}}$$

In view of Eq. (15.10c) we have

$$\bar{\mathbf{S}} = \bar{\mathbf{K}}\bar{\mathbf{s}} \qquad (15.13a)$$

where we have defined the global stiffness matrix

$$\bar{\mathbf{K}} = \mathbf{T}^T\mathbf{K}\mathbf{T} \qquad (15.13b)$$

The assembly of the structure stiffness matrix and the solution process for arches is similar to that of continuous beams (Section 11.2) and frames (Sections 12.2 to 12.4). If there are internal hinges, then two different degrees of freedom can be used for the rotations at the two sides of the hinges, similar to the procedure of Section 13.2.

The force–displacement relations developed in this section did not include the loads over the elements, but only joint loads. Such loads can easily be included in the derivation of the force–displacement relations, in a manner similar to that in Section 12.4, and expressions can be obtained for fixed-end forces of arch elements for common loading conditions.

Example 15.4. Find the member-end forces for the fixed-end circular arch of Fig. 15.9 with a radius 100 ft, arch angle 120°, under a concentrated load of 15 kips acting at the crown of the arch. The moment of inertia of the arch cross-section is $I = 46\,656$ in^4.

Solution. The arch is divided into two elements. The degrees of freedom are shown in Fig. 15.9. The arch angle β is the same for both elements. The start angle α is 120° for the first element, on the right, and 180° for the second element. The stiffness matrix for the two elements in the local coordinates is

$$\begin{bmatrix} U_1 \\ V_1 \\ M_1 \\ U_2 \\ V_2 \\ M_2 \end{bmatrix} = \begin{bmatrix} 400 & 200 & 40\,400 & -300 & 200 & -45\,600 \\ 200 & 100 & 29\,300 & -200 & 100 & -20\,300 \\ 40\,400 & 29\,300 & 9\,487\,400 & -45\,600 & 20\,300 & -3\,263\,900 \\ -300 & -200 & -45\,600 & 400 & -200 & 40\,400 \\ 200 & 100 & 20\,300 & -200 & 100 & -29\,300 \\ -45\,600 & -20\,300 & -3\,263\,900 & 40\,400 & -29\,300 & 9\,487\,400 \end{bmatrix} \begin{bmatrix} u_1 \\ v_1 \\ \theta_1 \\ u_2 \\ v_2 \\ \theta_2 \end{bmatrix}$$

The transformation matrices for the two elements are

$$\mathbf{T}_1 = \begin{bmatrix} -0.5 & 0.866 & 0 & 0 & 0 & 0 \\ -0.866 & -0.5 & 0 & 0 & 0 & 0 \\ 0 & 0 & 1.0 & 0 & 0 & 0 \\ 0 & 0 & 0 & -1.0 & 0 & 0 \\ 0 & 0 & 0 & 0 & -1.0 & 0 \\ 0 & 0 & 0 & 0 & 0 & 1.0 \end{bmatrix}$$

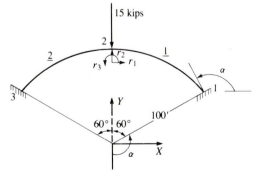

FIGURE 15.9

$$
\mathbf{T}_2 = \begin{bmatrix}
-1.0 & 0 & 0 & 0 & 0 & 0 \\
0 & -1.0 & 0 & 0 & 0 & 0 \\
0 & 0 & 1.0 & 0 & 0 & 0 \\
0 & 0 & 0 & -0.5 & -0.866 & 0 \\
0 & 0 & 0 & 0.866 & -0.5 & 0 \\
0 & 0 & 0 & 0 & 0 & 1.0
\end{bmatrix}
$$

The force–displacement relations in the global coordinates are $\bar{\mathbf{S}} = \bar{\mathbf{K}}\bar{\mathbf{s}}$, with $\bar{\mathbf{K}} = \mathbf{T}^T \mathbf{K}\mathbf{T}$. This results in the force–displacement relations in the global coordinates as follows.

$$
\begin{bmatrix} \bar{U}_1 \\ \bar{V}_1 \\ \bar{M}_1 \\ \bar{U}_2 \\ \bar{V}_2 \\ \bar{M}_2 \end{bmatrix}^1 =
\begin{bmatrix}
400 & -200 & -45\,600 & -400 & 200 & 40\,400 \\
-200 & 100 & 20\,300 & 200 & -100 & -29\,300 \\
-45\,600 & 20\,300 & 9\,487\,400 & 45\,600 & -20\,300 & -3\,263\,900 \\
-400 & 200 & 45\,600 & 400 & -200 & -40\,400 \\
200 & -100 & -20\,300 & -200 & 100 & 29\,300 \\
-40\,400 & -29\,300 & -3\,263\,900 & -40\,400 & 29\,300 & 9\,487\,400
\end{bmatrix}
\begin{bmatrix} 0 \\ 0 \\ 0 \\ r_1 \\ r_2 \\ r_3 \end{bmatrix}
$$

(columns labelled: 0, 0, 0, r_1, r_2, r_3)

$$
\begin{bmatrix} \bar{U}_1 \\ \bar{V}_1 \\ \bar{M}_1 \\ \bar{U}_2 \\ \bar{V}_2 \\ \bar{M}_2 \end{bmatrix}^2 =
\begin{bmatrix}
400 & 200 & -40\,400 & -400 & -200 & 45\,600 \\
200 & 100 & -29\,300 & -200 & -100 & 20\,300 \\
-40\,400 & -29\,300 & 9\,487\,400 & 40\,400 & 29\,300 & -3\,263\,900 \\
-400 & -200 & 40\,400 & 400 & 200 & -45\,600 \\
-200 & -100 & 29\,300 & 200 & 100 & -20\,300 \\
45\,600 & 20\,300 & -3\,263\,900 & -45\,600 & -20\,300 & 9\,487\,400
\end{bmatrix}
\begin{bmatrix} r_1 \\ r_2 \\ r_3 \\ 0 \\ 0 \\ 0 \end{bmatrix}
$$

(columns labelled: r_1, r_2, r_3, 0, 0, 0)

The degrees of freedom associated with the columns of the global stiffness matrices are indicated above the matrices. The structure stiffness matrix is found by assembling the global stiffness matrices in the same way as discussed for continuous beams (Section 11.2). This gives the following equilibrium equations:

$$
\begin{bmatrix}
800 & 0 & -80\,800 \\
0 & 200 & 0 \\
-80\,800 & 0 & 18\,974\,800
\end{bmatrix}
\begin{bmatrix} r_1 \\ r_2 \\ r_3 \end{bmatrix} =
\begin{bmatrix} 0 \\ -10 \\ 0 \end{bmatrix}
$$

(columns labelled: r_1, r_2, r_3)

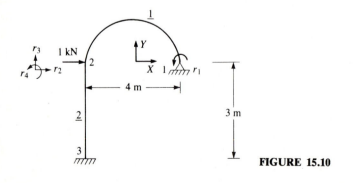

FIGURE 15.10

The solution of the above equation gives the displacements,

$$
\begin{bmatrix} r_1 \\ r_2 \\ r_3 \end{bmatrix} = \begin{bmatrix} 0 \\ -0.075 \\ 0 \end{bmatrix}
$$

Thus, the member displacement vectors in the global coordinates are fund and calculated by Eq. (15.12), as

$$
\text{Element} \rightarrow \qquad 1 \qquad\qquad 2
$$

$$
\mathbf{S} = \begin{bmatrix} U_1 \\ V_1 \\ M_1 \\ U_2 \\ V_2 \\ M_2 \end{bmatrix} : \begin{bmatrix} 8 \\ 5 \\ 828 \\ -8 \\ -5 \\ -1\,193 \end{bmatrix} \begin{bmatrix} 8 \\ 5 \\ 1\,193 \\ -8 \\ 5 \\ -828 \end{bmatrix}
$$

Example 15.5. The structure of Fig. 15.10 is constructed of steel pipes with inner and outer diameters of the cross-section 9 cm and 10 cm, respectively. The radius of the centerline of the curved part is 2 m and the height of the post to the beginning of the curved part is 3 m. Find the member-end forces for a horizontal force of 1 kN acting at joint 2. $E = 200\,000$ MPa.

Solution. Both elements have the same cross-section. The moment of inertia of the section is

$$
I = \frac{\pi(d_0^4 - d_i^4)}{64} = \frac{\pi(10^4 - 9^4)}{64} = 169 \text{ cm}^4
$$

The area of the cross-section used in calculating the stiffness of element 2 is $A = \pi(10^2 - 9^2)/4 = 14.9$ cm^2. The force–displacement relations in the local coordinates are for the arch element, by the equations derived in this section,

$$
\begin{bmatrix} U_1 \\ V_1 \\ M_1 \\ U_2 \\ V_2 \\ M_2 \end{bmatrix}^1 = \begin{bmatrix} 88 & 0 & -17\,188 & 85 & 0 & 27\,288 \\ 0 & 453 & 57\,766 & 0 & 453 & 57\,766 \\ -17\,188 & 57\,766 & 2\,511\,645 & -17\,188 & 57\,766 & 5\,636\,151 \\ 85 & 0 & -17\,188 & 85 & 0 & 17\,188 \\ 0 & 453 & 57\,766 & 0 & 453 & 57\,766 \\ -17\,188 & -57\,766 & -5\,636\,151 & -17\,188 & -57\,766 & 12\,511\,645 \end{bmatrix} \begin{bmatrix} u_1 \\ v_1 \\ \theta_1 \\ u_2 \\ v_2 \\ \theta_2 \end{bmatrix}^1
$$

and for the beam element, by Eq. (12.7a),

$$
\begin{bmatrix} U_1 \\ V_1 \\ M_1 \\ U_2 \\ V_2 \\ M_2 \end{bmatrix}^2 = \begin{bmatrix} 994\,837 & 0 & 0 & -994\,837 & 0 & 0 \\ 0 & 480 & 72\,000 & 0 & -480 & 72\,000 \\ 0 & 72\,000 & 14\,400\,000 & 0 & -72\,000 & 7\,200\,000 \\ -994\,837 & 0 & 0 & 994\,837 & 0 & 0 \\ 0 & -480 & -72\,000 & 0 & 480 & -72\,000 \\ 0 & 72\,000 & 7\,200\,000 & 0 & -72\,000 & 14\,400\,000 \end{bmatrix} \begin{bmatrix} u_1 \\ v_1 \\ \theta_1 \\ u_2 \\ v_2 \\ \theta_2 \end{bmatrix}^2
$$

The transformation matrices, by Eqs. (15.10a) and (12.8b) are

$$\mathbf{T}_1 = \begin{bmatrix} 0 & 1 & 0 & 0 & 0 & 0 \\ -1 & 0 & 0 & 0 & 0 & 0 \\ 0 & 0 & 1 & 0 & 0 & 0 \\ 0 & 0 & 0 & 0 & -1 & 0 \\ 0 & 0 & 0 & 1 & 0 & 0 \\ 0 & 0 & 0 & 0 & 0 & 1 \end{bmatrix}$$

$$\mathbf{T}_2 = \begin{bmatrix} 0 & -1 & 0 & 0 & 0 & 0 \\ 1 & 0 & 0 & 0 & 0 & 0 \\ 0 & 0 & 1 & 0 & 0 & 0 \\ 0 & 0 & 0 & 0 & -1 & 0 \\ 0 & 0 & 0 & 1 & 0 & 0 \\ 0 & 0 & 0 & 0 & 0 & 1 \end{bmatrix}$$

Then the force–displacement relations in the global coordinates are

$$\begin{bmatrix} \bar{U}_1 \\ \bar{V}_1 \\ \bar{M}_1 \\ \bar{U}_2 \\ \bar{V}_2 \\ \bar{M}_2 \end{bmatrix}^1 = \begin{bmatrix} 453 & 0 & -57\,766 & -453 & 0 & 57\,766 \\ 0 & 85 & -17\,188 & 0 & -85 & -17\,188 \\ -57\,766 & -17\,188 & 12\,511\,645 & 57\,766 & 17\,188 & -5\,636\,151 \\ -453 & 0 & 57\,766 & 453 & 0 & -57\,766 \\ 0 & -85 & 17\,188 & 0 & 85 & 17\,188 \\ 57\,766 & -17\,188 & -5\,636\,151 & -57\,766 & 17\,188 & 12\,511\,645 \end{bmatrix} \begin{bmatrix} 0 \\ 0 \\ r_1 \\ r_2 \\ r_3 \\ r_4 \end{bmatrix}$$

with column labels $0\quad 0\quad r_1\quad r_2\quad r_3\quad r_4$

$$\begin{bmatrix} \bar{U}_1 \\ \bar{V}_1 \\ \bar{M}_1 \\ \bar{U}_2 \\ \bar{V}_2 \\ \bar{M}_2 \end{bmatrix}^2 = \begin{bmatrix} 480 & 0 & 72\,000 & -480 & 0 & 72\,000 \\ 0 & 994\,837 & 0 & 0 & -994\,837 & 0 \\ 72\,000 & 0 & 14\,400\,000 & -72\,000 & 0 & 7\,200\,000 \\ -480 & 0 & -72\,000 & 480 & 0 & -72\,000 \\ 0 & -994\,837 & 0 & 0 & 994\,837 & 0 \\ 72\,000 & 0 & 7\,200\,000 & -72\,000 & 0 & 14\,400\,000 \end{bmatrix} \begin{bmatrix} r_2 \\ r_3 \\ r_4 \\ 0 \\ 0 \\ 0 \end{bmatrix}$$

with column labels $r_2\quad r_3\quad r_4\quad 0\quad 0\quad 0$

The structure stiffness matrix is fouund by assembling the above stiffness matrices in the global coordinates. The only nonzero component of the load vector is a 1000-N force acting in the second degree of freedom. Thus, the equilibrium equations are

$$\begin{bmatrix} 1.251\,164\,5 & 0.005\,776\,6 & 0.001\,718\,8 & -0.563\,615\,1 \\ 0.005\,776\,6 & 0.000\,093\,3 & -0.000\,000\,0 & 0.001\,423\,3 \\ 0.001\,718\,8 & -0.000\,000\,0 & 0.099\,492\,3 & 0.001\,718\,8 \\ -0.563\,615\,1 & 0.001\,423\,3 & 0.001\,718\,8 & 2.691\,164\,5 \end{bmatrix} \begin{bmatrix} r_1 \\ r_2 \\ r_3 \\ r_4 \end{bmatrix} = \begin{bmatrix} 0 \\ 1\,000 \\ 0 \\ 0 \end{bmatrix}$$

with column labels $r_1\quad r_2\quad r_3\quad r_4$

The solution of the equilibrium equtions gives the displacements,

$$\begin{bmatrix} r_1 \\ r_2 \\ r_3 \\ r_4 \end{bmatrix} = \begin{bmatrix} -0.008\,92 \\ 1.665\,32 \\ 0.000\,20 \\ -0.002\,75 \end{bmatrix}$$

The element displacement vectors are indicated in the global force–displacement relations. Using the latter vectors with the values found here, and applying Eq. (15.12), we find the member-end forces in the local coordinates as

$$
\begin{array}{ccc}
\text{Element} \rightarrow & 1 & 2 \\
\mathbf{S} = \begin{bmatrix} U_1 \\ V_1 \\ M_1 \\ U_2 \\ V_2 \\ M_2 \end{bmatrix} : & \begin{bmatrix} 200 \\ 398 \\ 0 \\ 200 \\ 398 \\ -80\,291 \end{bmatrix} & \begin{bmatrix} -200 \\ 601 \\ 80\,291 \\ 200 \\ -601 \\ 100\,097 \end{bmatrix}
\end{array}
$$

LABORATORY EXPERIMENTS

Arch Effect

Use a segment of spring steel longer than the distance between the wooden blocks and place it between the blocks to form an arch (Fig. L15.1a). Apply a load to the arch (by hanging the weight through the hole with a string) and find the displacement at the load.

Remove the steel segment and use it as a beam with the blocks as supports (Fig. L15.1b). Apply the same load as before and at the same distance. Compare the deflections of the beam and the arch.

Thrust

Using the set up of Fig. L15.1 with one end on a roller or allowed to move, with both ends at a level, attach the roller end to a string that is connected to a can after passing over a pulley (Fig. L15.2). Place sand or water in the can so that it balances a load applied at the mid-span (the roller will be close to an end block but not touching it). Calculate the thrust of the circular arch under

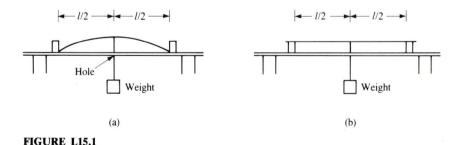

(a)   (b)

FIGURE L15.1

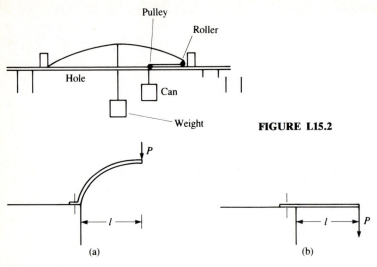

FIGURE L15.2

FIGURE L15.3

the applied load and compare it to the measured value. Do this experiment for a straight strip forced between the blocks as an arch and one formed into an arch with a span equal to the distance between the blocks prior to putting it in place.

Circular Arch

Use a steel strip about 1/16 in thick and 1 in wide and form it into a quarter of a circular arch. Bend one end to be used for fixing the arch (Fig. L15.3). Apply a vertical load and measure the horizontal and vertical deflections of the free end. Use a segment of the same material as a cantilever beam with a span equal to the radius of the arch. Compare the deflections of the two systems.

PROBLEMS

15.1. Derive the equations for the normal force, shear and moment of the circular arch with a radius of 5 m and angle 60° under a vertical force 10 kN (Fig. P15.1). Draw the normal force, shear and moment diagrams.

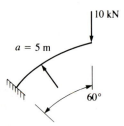

FIGURE P15.1

15.2. Draw the normal force, shear, and moment diagrams for the three-hinged parabolic arch of Fig. P15.2 under its weight of 2 kip/ft. The hinge is at mid-span.

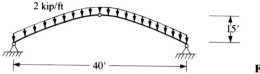

2 kip/ft

15'

40'

FIGURE P15.2

15.3. Find the equations for the normal force, shear, and moment of the circular three-hinged arch of Fig. P15.3 under a uniform load of 30 kN/m over its horizontal projection. The radius of the centerline of the arch is 30 m and the angles on the left and right of the middle hinge are 90° and 45°.

30 kN/m

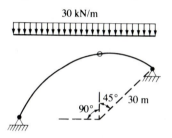

45° 30 m

90°

FIGURE P15.3

15.4. Find the redundant reaction of the two-hinged parabolic arch of Fig. P15.4 under a uniformly distributed load of 3 kip/ft over the horizontal projection. $I = 6\,000\text{ in}^4$; $E = 3\,100\text{ ksi}$.

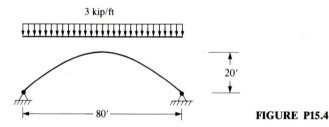

3 kip/ft

20'

80'

FIGURE P15.4

15.5. Calculate the redundant reactions of the circular arch of Fig. P15.5 under a concentrated load of 20 kN applied at the crown of the arch. The radius of the mid-thickness of the arch is 60 m and its angle 45°. $I = 9\,000\text{ cm}^4$ and $E = 200\,000$ MPa.

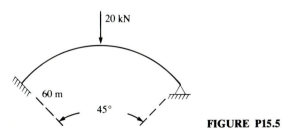

20 kN

60 m

45°

FIGURE P15.5

15.6. Find the displacements and rotation at the right-hand end and the shear and moment at the left-hand end of the circular arch of Fig. P15.6 using the stiffness method. The arch radius is 10 ft, the arch angle is 90°, $I = 200 \text{ in}^4$ and $E = 29\,000$ ksi.

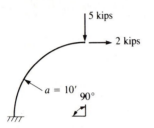

FIGURE P15.6

15.7. Find the deflection and the moment at the crown of the circular fixed-ended arch of Fig. P15.7. The span length is 100 ft and the moment of inertia of the cross-section 2.25 ft⁴. The rise of the arch is equal to one-fourth of its span. A load of 20 kips is applied at the crown of the arch and $E = 29\,000$ ksi.

Compare these results to the moment and deflection of a fixed-ended beam with the same span and moment of inertia.

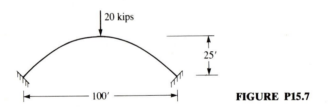

FIGURE P15.7

15.8. Find the forces at points 1 and 2 of the ring shown in Fig. P15.8. $I = 1/768 \text{ in}^3$; $a = 4$ in; $E = 29\,000$ ksi.

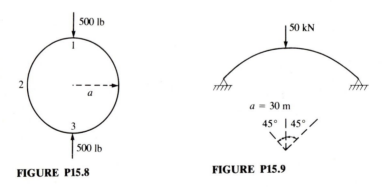

FIGURE P15.8   FIGURE P15.9

15.9. Find the forces in the circular arch of Fig. P15.9. $I = 1/24 \text{ m}^4$; $E = 20\,000$ MPa.

15.10. Find the member forces in the circular arch of Fig. P15.10. $I = 3\,000 \text{ in}^3$ and $E = 3\,100$ ksi.

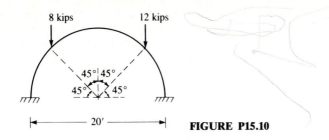

FIGURE P15.10

15.11. Find the forces at points 2 and 3 of the circular hook of Fig. P15.11. The hook has a circular section with $\frac{1}{2}$-in diameter.

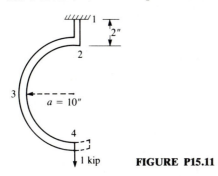

FIGURE P15.11

15.12. Calculate the member forces for the structure of Fig. P15.12. $I = 100 \text{ in}^4$ and $E = 29\,000$ ksi. The radius of the semi-circular members is 6 ft. $A = 35.48 \text{ in}^2$.

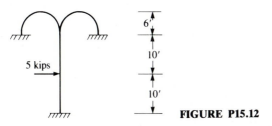

FIGURE P15.12

15.13. Find the member-end forces for the symmetric structure of Fig. P15.13 under the applied loads. The radius of the circular segment is 8.48 m and its angle 90°. $I = 0.05 \text{ m}^4$; $E = 20\,000$ MPa, $A = 793 \text{ cm}^2$.

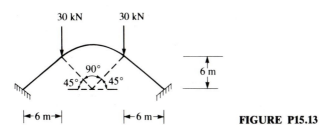

FIGURE P15.13

15.14. Find the member-end forces for the structure of Fig. P15.14. The radius of the smaller arches is 50 ft and their angles are 90°. The radius of the larger arch is 100 ft and its angle also 90°. The larger arch is supported by the smaller arches at an angle of 22.5° as shown in the figure. For the small arches $I = 300$ in⁴ and for the larger one $I = 400$ in⁴. $E = 29\,000$ ksi.

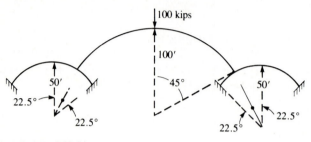

FIGURE P15.14

15.15. Find the stiffness and transformation matrices for a parabolic arch (Fig. P15.15) segment with span l, rise h, modulus of elasticity E, moment of inertia of the cross-section I, and start angle γ.

FIGURE P15.15

CHAPTER
16

BEAM–
COLUMNS
AND BUCKLING

The study of beam-columns and stability has become more important in structural engineering because of the continuing trend to use thin-walled members and lighter sections. When the moment caused by the axial force is not negligible, as compared with the moment of the lateral loads, the solution cannot be obtained by superposition of the axial and lateral load effects because of the nonlinear nature of the problem. Thus, the solution must be obtained by the simultaneous application of the lateral and axial loads. A large axial load, beyond the capacity of a slender member, can cause buckling. As will be seen, the buckling load is independent of the type and magnitude of the lateral loads.

In practice columns are under both axial load and bending moment induced by the lateral loads. Moments may be also due to the continuity of the members, for example, due to rigid connections in frames, or because of the eccentricity of the axial loads. Design codes specify a minimum eccentricity even when the loads are applied in a concentric manner.

The design sign convention is used for the solution of beam–columns, except for the stiffness method (Section 16.6), where the analysis sign convention is used. However, with either sign convention axial compression is assumed positive for beam–columns.

16.1 GOVERNING EQUATION

As discussed in Chapter 4, the moment and shear in beams are usually calculated ignoring the deflection of the beam, that is, with reference to the geometry of the beam before the application of the loads. Thus, for the beam of Fig. 16.1a, with a load P applied at the mid-span, the moment at x, on the left of the load, is $\frac{1}{2}Px$. The distance x is measured on the axis of the beam before deflection has taken place. This is a realistic approximation because the distance after deflection has taken place is very nearly the same as that before deflection.

On the other hand, for the beam of Fig. 16.1b, which is under a compressive axial force N, if we ignore the deformation of the beam the moment would be zero throughout the beam. If N is sufficiently large its moment, Nv, may not be negligible, and may even cause buckling of the beam. This is because even for small v the moment $M = Nv$ will be significant for large values of N. Therefore, for a beam–column, in addition to the moment due to the lateral loads, the moment due to the axial load must be included. The effect of this moment is called the P–Δ effect, where P and Δ are alternate notations for the axial force N and lateral deflection v.

The moment in a beam–column may be expressed in the general form

$$M = Nv + f(x) \tag{16.1}$$

where $f(x)$ is the moment in the beam due to lateral loads (and their reactions), while Nv is the moment due to the axial force. For example, for a uniform load p (Fig. 16.2),

$$f(x) = \frac{pl}{2}x - \frac{px^2}{2}$$

The differential equation (Eq. (5.16)), $-EIv'' = M$, becomes

$$-EIv'' = Nv + f(x)$$

or

$$EIv'' + Nv = -f(x)$$

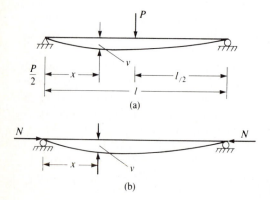

(a)

(b)

FIGURE 16.1

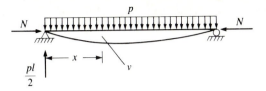

FIGURE 16.2

or

$$v'' + \frac{N}{EI} v = -\frac{f(x)}{EI}$$

or

$$v'' + \alpha^2 v = -\frac{f(x)}{EI} \tag{16.2a}$$

with

$$\alpha^2 = \frac{N}{EI} \tag{16.2b}$$

The differential equation $(16.2a)$ is the governing equation of beam–columns. For a constant axial force it is a linear differential equation with constant coefficients. The general solution of this equation is the sum of the complementary and particular solutions. The complementary solution is the solution of the homogeneous equation, the differential equation with the right-hand side equal to zero. The particular solution is any solution satisfying Eq. $(16.2a)$. The particular solution can be found by trial or by variation of parameters. As will be seen in Section 16.2, the buckling load for beam–columns is independent of the lateral loads. Therefore, in calculating the buckling load of a column, the moments due to the lateral loads are set equal to zero.

Complementary Solution—Buckling Load

The complementary solution is the solution of

$$v'' + \alpha^2 v = 0 \tag{16.3}$$

This is the governing differential equation for a simply supported beam under axial load N and no lateral loads. The solution of this equation gives the buckling load of the beam. This linear differential equation with constant coefficients has the general solution

$$v = e^{mx} \tag{16.4}$$

where m is a parameter to be determined. The total complementary solution is the sum of the independent solutions multiplied by arbitrary constants. The number of independent solutions is equal to the order of the differential equation (the highest derivative of v). Substitution of v from Eq. (16.4) in Eq. (16.3) gives

$$m^2 e^{mx} + \alpha^2 e^{mx} = 0 \qquad \text{or} \qquad e^{mx}(m^2 + \alpha^2) = 0$$

In the above equation, either $e^{mx} = 0$ or $m^2 + \alpha^2 = 0$. If e^{mx} is zero, then by Eq. (16.4) the column keeps its original straight shape, without buckling. This is the so-called trivial solution. To find the solution of the buckled beam we must set $m^2 + \alpha^2 = 0$. This gives $m^2 = -\alpha^2$, or $m = \pm\alpha\sqrt{-1} = \pm\alpha i$. The two independent solutions are found by substituting the values of m in Eq. (16.4). These are $e^{\alpha i x}$ and $e^{-\alpha i x}$. Thus, the general solution is

$$v = C_1 e^{\alpha i x} + C_2 e^{-\alpha i x}$$

Using Euler's relations, $e^{\pm\alpha i x} = \cos \alpha x \pm i \sin \alpha x$, we get

$$v = C_1 \cos \alpha x + iC_1 \sin \alpha x + C_2 \cos \alpha x - iC_2 \sin \alpha x$$

or

$$v = (C_1 + C_2) \cos \alpha x + i(C_1 - C_2) \sin \alpha x$$

If instead of constants C_1 and C_2 we use two new constants, $A = C_1 + C_2$ and $B = i(C_1 - C_2)$, we get

$$v = A \cos \alpha x + B \sin \alpha x \qquad (16.5)$$

Equation (16.5) is the solution of the differential equation (16.3). It should be noted that although C_1 and C_2 are complex numbers, A and B are real.

To find the constants A and B we apply the boundary conditions $v = 0$ at $x = 0$ and $x = l$, to get

$$A = 0 \qquad \text{and} \qquad B \sin \alpha l = 0$$

Again $B = 0$ leads to the trivial solution and thus for a nontrivial solution we must have $\sin \alpha l = 0$. Referring to the trigonometric circle (Fig. 16.3), this results in $\alpha l = n\pi$, with $n = 1, 2, \ldots$

Since by definition (Eq. (16.2b)), $\alpha^2 = N/EI$, and by the previous relation, $\alpha^2 l^2 = n^2 \pi^2$, we get

$$\frac{N}{EI} l^2 = n^2 \pi^2 \qquad \text{which results in} \qquad N = \frac{n^2 \pi^2 EI}{l^2}$$

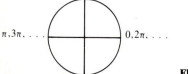

$\pi, 3\pi, \ldots$ $0, 2\pi, \ldots$

FIGURE 16.3

We are interested in the critical load, that is, the smallest load causing buckling. This is, for the smallest value of n ($n = 1$). Thus,

$$N_{cr} = N_e = \frac{\pi^2 EI}{l^2} \tag{16.6a}$$

This is the critical or the Euler buckling load for a column with hinged ends. Substituting for I in terms of the radius of gyration $r = \sqrt{I/A}$ (where $I = r^2 A$) gives

$$N_e = \frac{\pi^2 EA}{(l/r)^2}$$

or

$$\sigma_e = \frac{N_e}{A} = \frac{\pi^2 E}{(l/r)^2} \tag{16.6b}$$

σ_e is the *critical* or *Euler stress* and l/r is called the *slenderness ratio*.

Example 16.1. Find the buckling load of a column (Fig. 16.4) with one end fixed and the other free. Calculate this load for $l = 20$ ft, $E = 29\,000$ ksi, $I = 200$ in^4 and $A = 25$ in^2.

Solution. Denoting the fixed-end moment by M_0, we have $M = M_0 + Nv$. The left reaction is zero by equilibrium of the vertical forces. The differential equation $-EIv'' = M$, gives

$$EIv'' + Nv = -M_0$$

or

$$v'' + \alpha^2 v = -\frac{M_0}{EI}$$

where α^2 was defined by Eq. (16.2b).

As before, the solution of the above linear differential equation is the sum of the complementary solution (Eq. (16.5)) plus a particular solution satisfying the differential equation. The particular solution can be found by inspection. Trying a constant D for the particular solution and substituting it the differential equation we get

$$\alpha^2 D = -\frac{M_0}{EI} \quad \text{or} \quad D = -\frac{M_0}{EI\alpha^2}$$

The total solution of the problem becomes

$$v = A \cos \alpha x + B \sin \alpha x - \frac{M_0}{EI\alpha^2}$$

FIGURE 16.4

$\pi/2, 5\pi/2, \ldots$

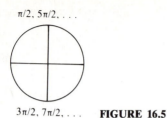

$3\pi/2, 7\pi/2, \ldots$ **FIGURE 16.5**

The boundary conditions are $v = v' = 0$ at $x = 0$, and $M = -EIv'' = 0$ at $x = l$. These conditions result in three equations from which the three unknowns A, B, and M_0 can be found. The three equations are

$$v(0) = 0 = A - \frac{M_0}{EI\alpha^2} \quad \text{or} \quad A = \frac{M_0}{EI\alpha^2}$$

$$v'(0) = 0 = B$$

and

$$-EIv''(l) = 0 = A\alpha^2 \cos \alpha l + B\alpha^2 \sin \alpha l$$

The last equation, upon substitution for A and B from the previous ones, gives

$$A\alpha^2 \cos \alpha l = \frac{M_0}{EI} \cos \alpha l = 0$$

Again, if $M_0 = 0$ then $A = 0$ and we get the trivial solution. The nontrivial solution is found for $\cos \alpha l = 0$, which leads (Fig. 16.5) to

$$\alpha l = \frac{n+1}{2} \pi \quad n = 0, 2, \ldots$$

The critical load is found for the smallest n, $n = 0$. Then, by Eq. (16.2b),

$$\alpha^2 = \frac{N_{cr}}{EI} = \frac{\pi^2}{4l^2}$$

$$N_{cr} = \frac{\pi^2 EI}{4l^2} = \frac{\pi^2 EAr^2}{4l^2}$$

or

$$\frac{N_{cr}}{A} = \sigma_{cr} = \frac{\pi^2 E}{(2l/r)^2}$$

16.2 COLUMN BUCKLING LOAD

The buckling loads found for simply supported and fixed-free columns were found in the above examples. The buckling load for other columns can be obtained in a similar manner. The buckling stress can be written in the general form

$$\frac{N_{cr}}{A} = \sigma_{cr} = \frac{\pi^2 E}{(Kl/r)^2} \tag{16.7}$$

TABLE 16.1
Effective length factor K

Column					
$K \rightarrow$	$\frac{1}{2}$	0.7	1	2	1

where K is called the *effective length factor,* and Kl the *effective length.* The values of K for the different end conditions are given in Table 16.1. It should be noted that the solution of the second column in Table 16.1 can only be obtained by numerical or approximate methods. If the unbraced lengths are different in the two planes, then Kl/r must be calculated in the weak and strong plane (using I_y and I_x) and the smaller Kl/r must be used in Eq. (16.7).

16.3 DEFLECTION OF BEAM-COLUMNS

In addition to axial loads, columns are generally under lateral loads. These may be due to the applied lateral loads or to a minimum eccentricity specified by the design codes. Thus, compression members are rarely under axial load alone, but carry axial load and bending. In other words, they generally act as beam–columns.

In the case of ties that carry an axial tensile force and bending, the solution is similar to that of beam-columns. However, for such ties buckling is not a problem.

The solution of beam–columns is obtained by solving Eq. (16.2a) to find the deflection of the member and its internal forces. The relationships between the different forces are derived below, from which the differential equation of the beam–column is obtained in terms of the lateral loads.

Consider an infinitesimal element dx of the beam–column (Fig. 16.6a). This element, with the forces acting on it, is shown in Fig. 16.6b.

In writing the equilibrium equations for this element we make the usual assumption that the slope of the member is small, that is, $\tan \theta = \theta = dv/dx = v'$. The slope as well as the axial and the distributed loads are assumed to be constant over the infinitesimal element. Equilibrium of the vertical forces gives

$$p \, dx - V + V + dV = 0 \quad \text{or} \quad \frac{dV}{dx} = -p \quad (16.8)$$

Setting sum of the moments at the right end of the element (Fig. 16.6b) equal to zero gives

$$M + V \, dx + N \, dv - \frac{p(dx)^2}{2} - (M + dM) = 0$$

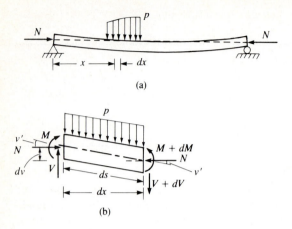

(a)

(b)

FIGURE 16.6

Ignoring the higher-order term $p(dx)^2/2$, we get

$$V = \frac{dM}{dx} - N\frac{dv}{dx} \tag{16.9}$$

Differentiating Eq. (16.9) and substituting from Eq. (16.8), we get

$$-p = \frac{d^2M}{dx^2} - \frac{d}{dx}\left(N\frac{dv}{dx}\right)$$

Since, by Eq. (5.12),

$$EI\frac{d^2v}{dx^2} = -M$$

we find

$$-p = -\frac{d^2}{dx^2}\left(EI\frac{d^2v}{dx^2}\right) - \frac{d}{dx}\left(N\frac{dv}{dx}\right)$$

Denoting the derivative with respect to x by a prime, the differential equation of beam-columns, in terms of the loads, becomes

$$(EIv'')'' + (Nv')' = p \tag{16.10}$$

When EI and N are constants

$$EIv^{iv} + Nv'' = p \tag{16.11}$$

or

$$v^{iv} + \alpha^2 v'' = \frac{p}{EI} \tag{16.12}$$

with α^2 defined by Eq. 16.2b

Equation (6.12) can be used in the solution of problems. However, it is generally easier to use Eq. (16.2a), because it is a lower-order differential equation. For members with variable sections or variable axial loads, Eq. (16.10) must be used. For members with step changes in the axial load or I, Eq. (16.12) can be applied to each segment with matching conditions at the segment boundaries.

Example 16.2. Find the deflection of the simply supported beam–column of Fig. 16.7 under a uniform lateral load p and a compressive axial load N.

Solution. The moment at x is

$$M = \frac{pl}{2}x - \frac{px^2}{2} + Nv$$

By Eq. (16.9) the shear force is

$$V = \frac{dM}{dx} - N\frac{dv}{dx} = \frac{pl}{2} - px$$

The last equation indicates that the axial force does not contribute to the shear force. Therefore, we can find the shear in a beam–column in the same manner as shear in a beam. However, the moment is affected by the axial force. The differential equation,

$$-EIv'' = M = \frac{pl}{2}x - \frac{px^2}{2} + Nv$$

leads to

$$v'' + \alpha^2 v = -\frac{plx}{2EI} + \frac{px^2}{2EI}$$

The complementary solution is given by Eq. (16.5). The particular solution can be found by inspection. Since the right-hand side is a polynomial, we can try a polynomial of the same order with unknown coefficients such as

$$v_p = C + Dx + Fx^2$$

Substituting v_p into the differential equation gives

$$2F + \alpha^2 C + \alpha^2 Dx + \alpha^2 Fx^2 = -\frac{plx}{2EI} + \frac{px^2}{2EI}$$

Since this equation must be satisfied for every value of x, it must be an identity.

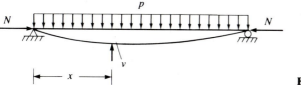

FIGURE 16.7

That is, the constants and the coefficients of x and x^2 must be equal on both sides i.e.,

$$2F + \alpha^2 C = 0 \qquad \alpha^2 D = -\frac{pl}{2EI} \qquad \alpha^2 F = \frac{p}{2EI}$$

These give

$$D = -\frac{pl}{2\alpha^2 EI} \qquad F = \frac{p}{2\alpha^2 EI}$$

and

$$C = -\frac{2F}{\alpha^2} = -\frac{p}{\alpha^4 EI}$$

Therefore,

$$v_p = -\frac{p}{\alpha^4 EI} - \frac{plx}{2\alpha^2 EI} + \frac{px^2}{2\alpha^2 EI}$$

and

$$v = A \cos \alpha x + B \sin \alpha x - \frac{p}{\alpha^4 EI} - \frac{plx}{2\alpha^2 EI} + \frac{px^2}{2\alpha^2 EI}$$

The boundary conditions are $v = 0$ at $x = 0, l$, giving

$$0 = A - \frac{p}{\alpha^4 EI} \qquad A = \frac{p}{\alpha^4 EI}$$

$$0 = A \cos \alpha l + B \sin \alpha l - \frac{p}{\alpha^4 EI}$$

$$B = -\frac{p}{\alpha^4 EI} \cot \alpha l + \frac{p}{\alpha^4 EI \sin \alpha l}$$

Thus,

$$v = \frac{p}{\alpha^4 EI} \cos \alpha x - \frac{p}{\alpha^4 EI} \cot \alpha l \sin \alpha x + \frac{p}{\alpha^4 EI \sin \alpha l} \sin \alpha x$$

$$- \frac{p}{\alpha^4 EI} - \frac{plx}{2\alpha^2 EI} + \frac{px^2}{2\alpha^2 EI}$$

The deflection at the mid-span is

$$v_c = \frac{p}{\alpha^4 EI} \cos \frac{\alpha l}{2} - \frac{p}{\alpha^4 EI} \cot \alpha l \sin \frac{\alpha l}{2} + \frac{p}{\alpha^4 EI \sin \alpha l} \sin \frac{\alpha l}{2}$$

$$- \frac{p}{\alpha^4 EI} - \frac{pl^2}{8\alpha^2 EI}$$

Letting $\alpha l/2 = \beta$, then $\alpha^2 = 4\beta^2/l^2$, and substituting for $\cos 2\beta = 2 \cos^2 \beta - 1$ and simplifying we get

$$v_c = \frac{pl^4}{16\beta^4 EI} \cos \beta - \frac{pl^4}{32\beta^4 EI} \frac{\cos 2\beta}{\cos \beta} + \frac{pl^4}{32\beta^4 EI \cos \beta}$$

$$- \frac{pl^4}{16\beta^4 EI} - \frac{pl^4}{16\beta^2 EI} + \frac{pl^4}{32\beta^2 EI}$$

The above expression can be cast, in terms of the mid-span deflection of a simply supported beam under a uniform load, as

$$v_c = \frac{5pl^4}{384EI} \frac{12(2 \sec \beta - 2 - \beta^2)}{5\beta^4}$$

When the value of the axial load is close to the buckling load, the deflection becomes very large, that is, $v_c \to \infty$. For the deflection to become large, the denominator of the above equation must approach zero. However, when $\beta = 0$ both the denominator and numerator are zero. Furthermore, $\beta = 0$ gives the trivial solution $N = 0$, which is not of interest. On the other hand, if $\cos \beta = 0$, then $\sec \beta = \infty$ (as $\beta = 0$) and $v_c \to \infty$. The smallest angle with zero cosine is $\pi/2$. Thus,

$$\beta = \frac{\alpha l}{2} = \frac{\pi}{2} \quad \text{and} \quad \alpha^2 l^2 = \frac{N}{EIl^2} = \pi^2$$

and the buckling (Euler) load is $N_e = \pi^2 EI/l^2$.

This is the same result as Eq. (16.6a), which was found by considering the axial load only. The same value would be found for any other lateral load. Therefore, the buckling load is independent of lateral loads and if the purpose of the analysis is to determine the buckling load, the lateral loads can be ignored.

Initially Bent Columns

Consider the simply supported column of Fig. 16.8. Assume that the column is initially bent. The initial shape may be approximated by

$$v_0 = a \sin(\pi x/l)$$

where a is the initial displacement of the mid-length from the axis of the column. In this case the moment is

$$M = N(v + v_0) \quad \text{or} \quad M = Nv + Na \sin \frac{\pi x}{l}$$

Note that v is the deflection from the initially deformed shape and not from the horizontal axis. The differential equation for the column becomes

$$v'' + \alpha^2 v = -\frac{Na}{EI} \sin \frac{\pi x}{l} = -\alpha^2 a \sin \frac{\pi x}{l}$$

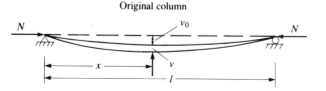

FIGURE 16.8

with

$$\alpha^2 = \frac{N}{EI}$$

The particular solution can be found by trying $v_p = C \sin(\pi x/l)$, which gives, upon substitution in the differential equation,

$$C = -\frac{a}{1 - (\pi^2/\alpha^2 l^2)} = -\frac{a}{1 - (N_e/N)}$$

where the Euler buckling load N_e is defined by Eq. (16.6a).

The total solution from the initially deformed shape is the sum of the complementary (Eq. (16.5)) and particular solutions,

$$v = A \cos \alpha x + B \sin \alpha x - \frac{a}{1 - (N_e/N)} \sin \frac{\pi x}{l}$$

The boundary conditions are $v = 0$ at $x = 0, l$. The first boundary condition gives $A = 0$. The second condition requires that either $B = 0$ or $\alpha l = \pi$. The latter leads to Euler's buckling load. However for N smaller than the buckling load, $A = B = 0$ and

$$v = -\frac{a}{1 - (N_e/N)} \sin \frac{\pi x}{l}$$

The deflection from the horizontal position is

$$v_t = v_0 + v = a \sin \frac{\pi x}{l} - \frac{a}{1 - N_e/N} \sin \frac{\pi x}{l} = \frac{a}{1 - (N/N_e)} \sin \frac{\pi x}{l} \quad (16.13)$$

The maximum deflection at $x = l/2$ is $a/[1 - (N/N_e)]$.

The variation of the coefficient for the maximum deflection, $1/[1 - (N/N_e)]$ with N/N_e is shown in Fig. 16.9. From this figure we can see that deflection is affected by axial forces smaller than N_e. However, the factor increases rapidly as N approaches N_e. Therefore, the effect of the initial imperfection is to amplify the deflection.

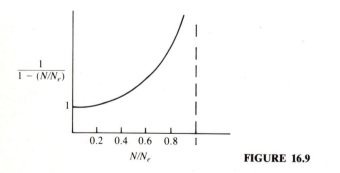

FIGURE 16.9

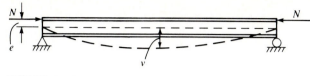

FIGURE 16.10

Eccentric Axial Loads

Consider the simply supported column of Fig. 16.10 with an axial load applied at eccentricity e.

In this case, $M = N(v + e)$ and the differential equation $EIv'' = -M$ becomes

$$v'' + \alpha^2 v = -e\alpha^2$$

where α was defined by Eq. (16.2b). This last differential equation has the general solution

$$v = A \cos \alpha x + B \sin \alpha x - e$$

Application of the boundary conditions, $v = 0$ at $x = 0, l$, gives

$$A = e \qquad B = \frac{1 - \cos \alpha l}{\sin \alpha l} e$$

Thus,

$$v = \left[\cos \alpha x + \frac{1 - \cos \alpha l}{\sin \alpha l} \sin \alpha x - 1 \right] e$$

The maximum deflection at $x = l/2$ is, after simplification,

$$v_c = \left(\sec \frac{\alpha l}{2} - 1 \right) e = \left(\sec \frac{\pi}{2} \left[\frac{N}{N_e} \right]^{1/2} - 1 \right) e \qquad (16.14)$$

The plot of this equation (Fig. 16.11) is similar to that of the column with initial deflection. That is, the load eccentricity amplifies the deflection.

Design codes usually require a minimum eccentricity. Otherwise, they specify a safety factor to account for eccentric column loads.

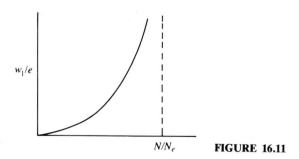

w_1/e

N/N_e **FIGURE 16.11**

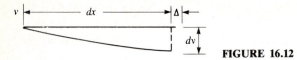

FIGURE 16.12

16.4 ENERGY OF DEFORMATION OF BEAM–COLUMNS

The energy of deformation of a beam–column is due to the bending deformation and the work done by the axial force. The elastic energy for bending of a beam was given by Eq. (6.7). That is,

$$\tfrac{1}{2} \int_0^l EI(v'')^2 \, dx$$

We assume that the axial load does not change its magnitude and direction. Because of lateral deformations, the length of the column changes from its original value. Therefore, the axial force N performs work that is equal to the product of the change in the column length and the force. After deformation, an infinitesimal element dx (Fig. 16.12) with end displacements v and $v + dv$ would have a length $\sqrt{dx^2 + dv^2}$. If the displacement is zero then its length will be dx. Therefore, the change of length Δ of the infinitesimal element due to displacement is (Fig. 16.12)

$$\Delta = \sqrt{dx^2 + dv^2} - dx = dx\sqrt{1 + v'^2} - dx \approx dx(1 + v'^2/2) - dx = (v'^2/2)\, dx$$

The work of the axial force is thus

$$-N\Delta = -\frac{N}{2} \int_0^l v'^2 \, dx$$

The negative sign in the above expression is in view of the convention (Section 6.1) that the elastic energy of a system is considered positive, while the work of the external load that opposes this internal work is considered negative. The total potential energy of a beam–column element is

$$\Pi = \tfrac{1}{2} \int_0^l (EIv''^2 - Nv'^2 - 2pv) \, dx \tag{16.15}$$

where p is the distributed lateral load.

16.5 BUCKLING LOADS BY ENERGY METHODS

Approximate buckling loads can be found by energy methods such as the Rayleigh–Ritz method discussed in Section 6.7. As was seen in that section, the deformed shape of the structure is approximately described by the sum of a set of functions satisfying the displacement boundary conditions multiplied by a number of unknown parameters. The unknown parameters are found by minimizing the energy of the system, which is equivalent to satisfying the equilibrium equations. Unlike deflection problems, the number of terms does not improve the accuracy of buckling load, but gives the buckling loads for the higher buckling mode shapes. The latter will provide the buckling loads should the buckling modes with lower buckling loads be prevented. If a term of the approximating function describes the exact mode shape, then the exact buckling load for that mode shape is obtained. Otherwise, an approximate solution is obtained whose accuracy depends on the nature of the approximating function and the number of boundary conditions that it satisfies. Since buckled shapes are often trigonometric functions, the latter functions, which are also easier to use, are usually considered. Thus, the deflected shape is described as

$$v = \sum_{j=1}^{n} a_j \phi_j \qquad (16.16)$$

where ϕ_j are the approximating functions satisfying the displacement boundary conditions and a_j are unknown parameters. Substitution of the above displacement function in the expression for the potential energy of the system (Eq. (16.15)) and setting the variation of the energy equal to zero leads to

$$\delta \Pi = \sum_{j=1}^{n} \frac{\partial \Pi}{\partial a_j} \delta a_j = 0$$

Since δa_j are arbitrary variations, this results in n equations,

$$\frac{\partial \Pi}{\partial a_j} = 0$$

In view of Eq. (16.15), with the lateral load p set equal to zero, the above equation becomes

$$\frac{\partial}{\partial a_j} \left(\int_0^l EI v''^2 \, dx - \int_0^l N v'^2 \, dx \right) = 0$$

$$\int_0^l EI v'' \frac{\partial v''}{\partial a_j} \, dx - \int_0^l N v' \frac{\partial v'}{\partial a_j} \, dx = 0$$

which gives

$$N = \left(\int_0^l EI v'' \frac{\partial v''}{\partial a_j} \, dx \right) \Big/ \left(\int_0^l v' \frac{\partial v'}{\partial a_j} \, dx \right) \qquad (16.17)$$

If n terms are used for the approximating function, then Eq. (16.17) yields n values of buckling loads for the different buckling modes. The procedure described here is especially useful for finding buckling loads of variable-section columns.

Example 16.3. Find the approximate buckling load for the column of Fig. 16.13.

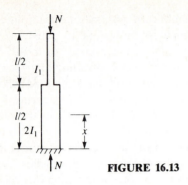

FIGURE 16.13

The moments of inertia are $2I_1$ and I_1 for the lower and upper halves of the column. The column is fixed at its base and free at the top.

Solution. We can assume the function

$$v = a_1\left(1 - \cos\frac{\pi x}{2l}\right)$$

for the lowest buckling mode, which satisfies the displacement and rotation boundary condition at the base and the displacement boundary condition at the top. It also satisfies the condition of zero moment at the top (as $EIv'' = 0$ there), but not zero shear ($EIv''' \neq 0$ at the top). Substitution of the above v into Eq. (16.17) with

$$v' = a_1\frac{\pi}{2l}\sin\frac{\pi x}{2l} \qquad \text{and} \qquad v'' = a_1\frac{\pi^2}{4l^2}\cos\frac{\pi x}{2l}$$

$$\int_0^l EIv''\frac{\partial v''}{\partial a_1}\,dx = \int_0^{l/2} 2EI_1a_1\frac{\pi^4}{16l^4}\cos^2\frac{\pi x}{2l}\,dx + \int_{l/2}^l EI_1a_1\frac{\pi^4}{16l^4}\cos^2\frac{\pi x}{2l}$$

$$= EI_1a_1\frac{\pi^4}{16l^4}\frac{2l + 3\pi l}{4\pi}$$

and

$$\int_0^l v'\frac{\partial v'}{\partial a_1}\,dx = \int_0^l a_1\frac{\pi^2}{4l^2}\sin^2\frac{\pi x}{2l}\,dx = \frac{\pi^2}{4l^2}a_1\frac{l}{4}$$

gives $N = 4.487EI_1/l^2$.

16.6 FINITE-ELEMENT FORMULATION OF BEAM–COLUMNS

The finite-element method provides a general technique for the solution of beam–columns and stability of columns and frames. The finite-element technique is similar to the stiffness method discussed in Chapters 11, 12, and 13, for beams, trusses, and frames. The solution involves the division of the structure into a number of elements, determination of the stiffness matrices for the elements, assembly of the elements to find the structure stiffness matrix and equilibrium equations, and the solution of the equilibrium equations for the beam–column to find the displacements and member forces or the buckling load.

The main difference between the stiffness and finite-element methods is the way the stiffness matrix is derived. In the stiffness method the stiffness matrix is derived in terms of the exact displacement equation, for example, by solving the governing differential equation for the element. The finite-element provides an alternative procedure for deriving the stiffness matrix when it is not possible to solve the differential equation directly, as for plate and shell problems, or when such a solution is cumbersome and not amenable to easy numerical solution, as for stability problems. In such cases, assumed displacement functions are used, which include a number of unknown parameters. The unknown parameters are usually taken as the nodal displacements and rotations. Minimization of the potential energy of the structure or the virtual work principle leads to the stiffness matrix and other pertinent matrices as well as the equilibrium equations. Other steps in the solution are the same as for the stiffness method.

Stiffness Matrix for Beam–Columns

The stiffness matrix for a beam element was given in Section 11.3 (Eq. (11.10a)). This was derived in Section 10.2 by integrating the differential equation of the beam and deriving the relationship between the member-end forces and displacements. For beam–columns the stiffness matrix can be derived in a similar manner by solving the differential equation and finding the member-end force–displacement relations. However, in this case the axial load parameter N is implicit in the terms of the stiffness matrix. An incremental stiffness method may be used by increasing N. Buckling is indicated by the change of sign in the determinant of the structure stiffness matrix. Here we use the finite-element approach, which allows for general solutions in a straightforward manner. The stiffness matrix is found by an assumed approximate displacement function rather than the exact displacement function. Since the accuracy can be increased by using a larger number of elements, this will not impose a serious restriction on the solution procedure. The stiffness matrix is derived from the potential energy of the element.

Shape Functions

In the finite-element method, as in the stiffness method, the nodal displacements and rotations are used as the basic variables of the problem. To find the stiffness matrix using the energy of the element, we need the expression for the defletion of the beam in terms of the nodal displacements. Since there are four boundary conditions, we assume a third-degree polynomial with four unknown parameters a_i as

$$v = a_0 + a_1 x + a_2 x^2 + a_3 x^3$$

To express the unknown parameters a_i in terms of the end displacements and rotations $[v_1 \;\; \theta_1 \;\; v_2 \;\; \theta_2]$, we apply the boundary conditions to v and v' at $x = 0, l$. Thus, at $x = 0$, $v = v_1$, and $v' = \theta_1$, and at $x = l$, $v = v_2$, $v' = \theta_2$, resulting in

$$v_1 = a_0$$

$$\theta_1 = a_1$$

$$v_2 = a_0 + a_1 l + a_2 l^2 + a_3 l^3$$

$$\theta_2 = a_1 + 2a_2 l + 3a_3 l^2$$

Solving for a_i in terms of the end displacements and substituting into the equation for v gives

$$v = \boldsymbol{\phi} \mathbf{s} \tag{16.18a}$$

with the row vector

$$\boldsymbol{\phi} = \langle \phi_1 \;\; \phi_2 \;\; \phi_3 \;\; \phi_4 \rangle \tag{16.18b}$$

and

$$\mathbf{s}^T = \langle v_1 \;\; \theta_1 \;\; v_2 \;\; \theta_2 \rangle^T \tag{16.18c}$$

where $\boldsymbol{\phi}$ is the vector of shape functions and $\mathbf{s}$ the nodal displacement vector,

$$\phi_1 = 1 - \frac{3x^2}{l^2} + \frac{2x^3}{l^3} \tag{16.19a}$$

$$\phi_2 = x - \frac{2x^2}{l} + \frac{x^3}{l^2} \tag{16.19b}$$

$$\phi_3 = \frac{3x^2}{l^2} - \frac{2x^3}{l^3} \tag{16.19c}$$

$$\phi_4 = -\frac{x^2}{l} + \frac{x^3}{l^2} \tag{16.19d}$$

Differentiation of Eq. (16.18a) with respect to x gives $v' = \boldsymbol{\phi}' \mathbf{s}$ and $v'' = \boldsymbol{\phi}'' \mathbf{s}$. To express the expressions in matrix notation we note that v''^2 and v'^2 are scalars (1×1 matrices) and thus $v''^2 = (v'')^T v''$ and $v'^2 = (v')^T v'$. The potential energy for the element is

$$\Pi = \tfrac{1}{2} \int_0^l [EI(v'')^T v'' - N(v')^T v' - 2v^T p] \, dx$$

$$= \tfrac{1}{2} \int_0^l [EI\mathbf{s}^T(\boldsymbol{\phi}'')^T \boldsymbol{\phi}'' \mathbf{s} - N\mathbf{s}^T(\boldsymbol{\phi}')^T \boldsymbol{\phi}' \mathbf{s} - 2\mathbf{s}^T \boldsymbol{\phi}^T p] \, dx$$

or

$$\Pi = \tfrac{1}{2}\mathbf{s}^T \left[\int_0^l [EI(\boldsymbol{\phi}'')^T \boldsymbol{\phi}'' - N(\boldsymbol{\phi}')^T \boldsymbol{\phi}'] \, dx \right] \mathbf{s} - \mathbf{s}^T \int_0^l \boldsymbol{\phi}^T p \, dx$$

or

$$\Pi = \tfrac{1}{2}\mathbf{s}^T \mathbf{K}\mathbf{s} - \mathbf{s}^T \mathbf{P}$$

Note that this is similar to the energy of a spring. In the latter case the only degree of freedom is the axial deformation v and the internal force is $S = Kv$. Thus, for a spring

$$\Pi = \tfrac{1}{2}\mathbf{S}^T \mathbf{K}\mathbf{S} = \tfrac{1}{2}Kv^2$$

For the beam–column, however, we have

$$\mathbf{K} = \mathbf{K}_E + \mathbf{K}_G \qquad\qquad (16.20a)$$

where

$$\mathbf{K}_E = \int_0^l EI(\boldsymbol{\phi}'')^T \boldsymbol{\phi}'' \, dx \qquad\qquad (16.20b)$$

is the elastic stiffness matrix (which is the same as the stiffness matrix for a beam element found in Section 11.3).

$$\mathbf{K}_G = -N \int_0^l (\boldsymbol{\phi}')^T \boldsymbol{\phi}' \, dx \qquad\qquad (16.20c)$$

is the geometric stiffness matrix, and

$$\mathbf{P} = \int_0^l \boldsymbol{\phi}^T p \, dx \qquad\qquad (16.20d)$$

is the member load vector, which is the negative of the fixed-end forces. Substituting from Eq. (16.18b) and carrying out the algebra, we find the complete matrices. Inspection would indicate that each term of the matrices can be expressed independently as

$$(K_E)_{ij} = \int_0^l EI\phi_i''\phi_j'' \, dx \qquad\qquad (16.20e)$$

and

$$(K_G)_{ij} = -N \int_0^l \phi_i'\phi_j' \, dx \qquad\qquad (16.20f)$$

$$P_i = \int_0^l \phi_i p \, dx \qquad\qquad (16.20g)$$

In this form each term can be calculated separately. For example,

$$(K_G)_{23} = -N \int_0^l \phi_2'\phi_3' \, dx = -N \int_0^l \left(1 - \frac{4x}{l} + \frac{3x^2}{l^2}\right)\left(\frac{6x}{l^2} - \frac{6x^2}{l^3}\right) dx$$

$$= -\frac{N}{10}$$

The complete stiffness matrices are

$$\mathbf{K}_E = \frac{EI}{l^3} \begin{bmatrix} 12 & 6l & -12 & 6l \\ & 4l^2 & -6l & 2l^2 \\ & & 12 & -6l \\ & & & 4l^2 \end{bmatrix} \tag{16.21a}$$

$$\mathbf{K}_G = -N \begin{bmatrix} \dfrac{6}{5l} & \dfrac{1}{10} & \dfrac{-6}{5l} & \dfrac{1}{10} \\[2ex] & \dfrac{2l}{15} & \dfrac{-1}{10} & \dfrac{-l}{30} \\[2ex] & & \dfrac{6}{5l} & \dfrac{-1}{10} \\[2ex] & & & \dfrac{2l}{15} \end{bmatrix} \tag{16.21b}$$

Note that for beams under concentrated and uniform loads the above shape functions provide the exact displaced shapes. Therefore, Eq. (16.20g) can be used as an alternative to Eqs. (10.8), (10.9), (10.12), and (10.13) for finding the fixed-end forces for beam elements.

Equilibrium Equations

The total potential energy of the structure is the sum of the energies for all the elements. This can be expressed in terms of the structure stiffness matrix and load vector. The assembly of the structure stiffness matrix and load vector for beam–column elements is similar to that of the direct stiffness method for other elements such as beams (Section 11.2). The elastic and geometric stiffness matrices for the structure can be assembled separately from the elastic and geometric stiffness matrices for the elements after transformation to the global coordinates. This would lead to an expression for the total potential energy of the system as

$$\Pi_S = \tfrac{1}{2}\mathbf{r}^T \mathbf{K}_S \mathbf{r} - \mathbf{r}^T \mathbf{R} \tag{16.22}$$

Where $\mathbf{r}$ is the structure displacement vector, which includes all the degrees of freedom. $\mathbf{K}_S$ is the structure stiffness matrix, the sum of the elastic and geometric stiffness matrices for the structure, and $\mathbf{R}$ is the structure load vector. By the principle of stationary potential energy, variation of Π_S must be zero. This leads to the equilibrium equations

$$\mathbf{K}_S \mathbf{r} = \mathbf{R} \tag{16.23a}$$

or

$$(\mathbf{K}_{ES} + \mathbf{K}_{GS})\mathbf{r} = \mathbf{R} \tag{16.23b}$$

where $\mathbf{K}_{ES}$ and $\mathbf{K}_{GS}$ are the elastic and geometric stiffness matrices for the structure. For buckling problems, the unknown axial load parameter N can be factored out of $\mathbf{K}_{GS}$, that is, we can write $\mathbf{K}_{GS} = -N\mathbf{G}_S$, and the lateral load vector $\mathbf{R}$ is zero. Thus, in this case, Eq. (16.23b) can be written as

$$\mathbf{K}_{ES}\mathbf{r} = N\mathbf{G}_S\mathbf{r} \qquad (16.23c)$$

This is an eigenvalue problem, with the eigenvalue as the axial load parameter N. Efficient procedures are available for the solution of such problems. In this section a simple procedure due to Stodola is presented for finding the buckling loads.

Solution Procedure

The steps in the solution of beam–column and stability problems are similar to those of stiffness method for other problems. The difference is that beam-columns involve geometric as well as elastic stiffness matrices. In addition the determination of the buckling load involves the solution of an eigenvalue problem instead of that of a system of equations. The steps in the solution of beam–column and buckling problems are as follows.

1. Number the elements, joints and the structure degrees of freedom.
2. Find the elastic and geometric stiffness matrices for each element by Eqs. (16.21). The element stiffness matrix is the sum of the elastic and geometric stiffness matrices. Indicate the number of structure degrees of freedom above and to the side of each term of the stiffness matrices.
3. Set up the $n \times n$ elastic and geometric structure stiffness matrices. Add the terms of the element stiffness matrices to the appropriate locations in the structure stiffness matrices; see Section 11.2.
4a. For the solution of beam–columns, find the load vector (Section 11.2) described under the steps in the direct stiffness method. Each term of the load vector is the sum of the applied joint loads plus the equivalent joint load (negative of sum of the fixed-end forces). These lead to the equilibrium equations as $\mathbf{K}_S\mathbf{r} = \mathbf{R}$. Where $\mathbf{K}_S$ is the structure stiffness matrix (sum of the elastic and geometric stiffness matrices for the structure), $\mathbf{r}$ is the displacement vector and $\mathbf{R}$ the load vector. Solve for the displacements and calculate the member-end forces for the beam-column elements.
4b. To find the buckling load, express the axial load for the different elements in terms of the parameter N, the lowest axial load in the members. Thus, a member carrying twice as much axial load as the lowest value would involve $2N$ in the geometric stiffness matrix, from which N can be factored out. Because N is an unknown value, the elastic and geometric stiffness matrices cannot be combined. Thus, they are assembled separately, with the geometric stiffness matrix expressed in terms of the unknown N. Since lateral loads do not affect the buckling load $\mathbf{R} = 0$ in Eq. (16.23a) and for

non-trivial solution we must have $\det(\mathbf{K}_S) = 0$. Thus, the equilibrium equations are reduced to a standard eigenvalue problem (Eq. (16.23c)). $\mathbf{K}_{ES}$ and $\mathbf{G}_S$ are assembled from the elastic and geometric stiffness matrices for the element in the usual way. The solution of the latter eigenvalue problem gives the buckling load parameter N. The Stodola method discussed below provides a simple but efficient way of finding the buckling load.

Stodola Method

The eigenvalue problem involved in finding the buckling load (Eq. (16.23c)) can be written as

$$(1/N)\mathbf{r} = \mathbf{Qr} \qquad (16.24a)$$

where the stability matrix

$$\mathbf{Q} = \mathbf{K}_{ES}^{-1}\mathbf{G}_S \qquad (16.24b)$$

Assume that we have the correct displacement vector (buckling mode shape) $\mathbf{r}$. We can substitute this in Eq. (16.24a), multiply the right-hand side, and factor out a coefficient from the resulting vector so that on the right-hand side we have a factor α times the displacement vector on the left. Since on the left we have $(1/N)$ times the displacement vector, we can conclude that $1/N = \alpha$. However, we do not know the correct displacement vector at this stage. We can assume a vector $\mathbf{r}_0$. It is easier to use a normalized vector, with one of the terms equal to 1. For example, all the elements of $\mathbf{r}_0$ can be taken as 1. Substituting into the right-hand side of Eq. (16.24a) and carrying out the multiplication, we find a new vector. We can normalize this vector so that one of the terms (say the largest) becomes 1. Then we use this normalized vector as the new trial vector. If we repeat this process, after a number of iterations all the terms of the trial vector and the normalized vector will be the same. Thus, the normalizing factor on the right-hand side must be $1/N$. The mechanics of this procedure are shown in Examples 16.5 and 16.6. In these examples we are finding the buckling load, that is, the lowest load that can cause buckling of the system. Other possible mode shapes and the associated eigenvalues or buckling loads can be found by using the Stodola method in conjunction with the sweeping matrix procedure. This procedure is discussed in most texts on structural dynamics (e.g. *Dynamics of Structures* by R. W. Clough and J. Penzien, McGraw-Hill, 1975).

Example 16.4. Find the mid-span displacement and the member forces for the fixed-end beam-column of Fig. 16.14. The axial force is $20\,000$ kN, and a concentrated load of 5 kN is applied at the mid-span. $E = 2 \times 10^9$ kN/m^2 and $I = 2 \times 10^{-3}$ m^4.

Solution. Using two elements for the beam–column the only degrees of freedom are the deflection and rotation of the mid-span. In this case the deformation is

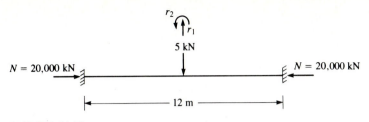

FIGURE 16.14

symmetric, that is, the mid-span rotation is zero. Although this information can be used to reduce the number of degrees of freedom, this is not done here. The elastic and geometric stiffness matrices for the two elements and the associated degrees of freedom are as follows.

$$
\mathbf{K}_{E1} = 100
\begin{matrix}
0 & 0 & r_1 & r_2 \\
\begin{bmatrix}
0.2222 & 0.6667 & -0.2222 & 0.6667 \\
0.6667 & 2.6667 & -0.6667 & 1.3333 \\
-0.2222 & -0.6667 & 0.2222 & -0.6667 \\
0.6667 & 1.3333 & -0.6667 & 2.6667
\end{bmatrix}
\begin{matrix} 0 \\ 0 \\ r_1 \\ r_2 \end{matrix}
\end{matrix}
$$

$$
\mathbf{K}_{G1} =
\begin{matrix}
0 & 0 & r_1 & r_2 \\
\begin{bmatrix}
-4 & -2 & 4 & -2 \\
-2 & -16 & 2 & 4 \\
4 & 2 & -4 & 2 \\
-2 & 4 & 2 & -16
\end{bmatrix}
\begin{matrix} 0 \\ 0 \\ r_1 \\ r_2 \end{matrix}
\end{matrix}
$$

$$
\mathbf{K}_{E2} = 100
\begin{matrix}
r_1 & r_2 & 0 & 0 \\
\begin{bmatrix}
0.2222 & 0.6667 & -0.2222 & 0.6667 \\
0.6667 & 2.6667 & -0.6667 & 1.3333 \\
-0.2222 & -0.6667 & 0.2222 & -0.6667 \\
0.6667 & 1.3333 & -0.6667 & 2.6667
\end{bmatrix}
\begin{matrix} r_1 \\ r_2 \\ 0 \\ 0 \end{matrix}
\end{matrix}
$$

$$
\mathbf{K}_{G2} =
\begin{matrix}
r_1 & r_2 & 0 & 0 \\
\begin{bmatrix}
-4 & -2 & 4 & -2 \\
-2 & -16 & 2 & 4 \\
4 & 2 & -4 & 2 \\
-2 & 4 & 2 & -16
\end{bmatrix}
\begin{matrix} r_1 \\ r_2 \\ 0 \\ 0 \end{matrix}
\end{matrix}
$$

The 2×2 structure stiffness matrix is $\mathbf{K}_S = \mathbf{K}_E + \mathbf{K}_G$, where $\mathbf{K}_E$ and $\mathbf{K}_G$ are the assembled elastic and geometric stiffness matrices for the structure, and the equilibrium equations, $\mathbf{K}_S \mathbf{r} = \mathbf{R}$, are as follows

$$
\begin{matrix}
r_1 & r_2 \\
\begin{bmatrix}
36.44 & 0 \\
0 & 501.33
\end{bmatrix}
\begin{bmatrix} r_1 \\ r_2 \end{bmatrix}
=
\begin{bmatrix} -5 \\ 0 \end{bmatrix}
\end{matrix}
$$

The equivalent joint shear forces are zero and the applied joint force vector is as indicated on the right of the above equations. The solution of the equilibrium equations gives the joint displacements as

$$\begin{bmatrix} r_1 \\ r_2 \end{bmatrix} = \begin{bmatrix} -0.1372 \\ 0 \end{bmatrix}$$

The member-end displacement vectors are found by identifying the structure degrees of freedom with the corresponding element degrees of freedom. This gives

$$\mathbf{s}_1 = \begin{bmatrix} 0 \\ 0 \\ -0.1372 \\ 0 \end{bmatrix} \qquad \mathbf{s}_2 = \begin{bmatrix} -0.1372 \\ 0 \\ 0 \\ 0 \end{bmatrix}$$

The member-end forces are $\mathbf{S}_i = \mathbf{K}_i \mathbf{s}_i$. With $\mathbf{K}_i$ the sum of elastic and geometric stiffness matrices for each element. This leads to

$$\mathbf{S}_1 = \begin{bmatrix} V_1 \\ M_1 \\ V_2 \\ M_2 \end{bmatrix}^1 = \begin{bmatrix} 25 \\ 8.87 \\ -25 \\ -8.87 \end{bmatrix} \qquad \mathbf{S}_2 = \begin{bmatrix} V_1 \\ M_1 \\ V_2 \\ M_2 \end{bmatrix}^2 = \begin{bmatrix} -25 \\ -8.87 \\ 25 \\ -8.87 \end{bmatrix}$$

Note that in the absence of the axial force N the shear forces will be the same, but the magnitudes of moments would be 7.5 instead of 8.87.

Example 16.5. Find the buckling load of the column of Fig. 16.15a with hinged ends. Calculate the value of N using one and two elements and compare the results with the Euler buckling load $N_e = \pi^2 EI/l^2$. The span length is 20 ft; $E = 29\,000$ ksi; $I = 100$ in^4.

Solution. The degrees of freedom with one element are shown in Fig. 16.15b. The elastic and geometric stiffness matrices for this element are as follows.

$$\mathbf{K}_E = \begin{array}{c c c c c} & 0 & r_1 & 0 & r_2 \\ \begin{bmatrix} 2.52 & 302 & -2.52 & 302 \\ 302 & 48\,333 & -302 & 24\,167 \\ -2.52 & -302 & 2.52 & -302 \\ 302 & 24\,167 & -302 & 48\,333 \end{bmatrix} & \begin{matrix} 0 \\ r_1 \\ 0 \\ r_2 \end{matrix} \end{array}$$

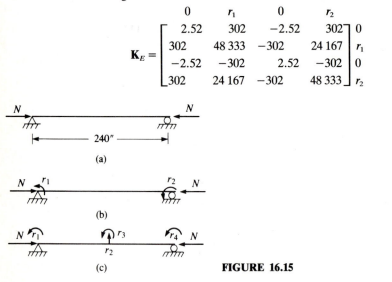

FIGURE 16.15

$$
\mathbf{K}_G = -N
\begin{array}{cccc}
0 & r_1 & 0 & r_2 \\
\end{array}
\begin{bmatrix}
0.005 & 0.1 & -0.005 & 0.1 \\
0.1 & 32.0 & -0.1 & -8.0 \\
-0.005 & -0.1 & 0.005 & -0.1 \\
0.1 & -8.0 & -0.1 & 32.0
\end{bmatrix}
\begin{array}{c}
0 \\ r_1 \\ 0 \\ r_2
\end{array}
$$

The 2×2 structure elastic and geometric stiffness matrices and the equilibrium equations are

$$
\left(
\begin{bmatrix}
48\,333 & 24\,167 \\
24\,167 & 48\,333
\end{bmatrix}
- N
\begin{bmatrix}
32 & -8 \\
-8 & 32
\end{bmatrix}
\right)
\begin{bmatrix}
r_1 \\ r_2
\end{bmatrix}
=
\begin{bmatrix}
0 \\ 0
\end{bmatrix}
$$

The determinant of $\mathbf{K}_S$ (the sum of the two matrices in parenthesis) set equal to zero gives

$$
N^2 - 3\,625N + 1\,825\,104 = 0.
$$

The lowest root of this equation gives the buckling load and is $N = 604.17$ kips.

Using two equal-length elements (Fig. 16.15c), the element stiffness matrices with the structure degrees of freedom are as follows.

$$
\mathbf{K}_{E1} =
\begin{array}{cccc}
0 & r_1 & r_2 & r_3 \\
\end{array}
\begin{bmatrix}
20 & 1\,208 & -20 & 1\,208 \\
1\,208 & 96\,666 & -1\,208 & 48\,333 \\
-20 & -1\,208 & 20 & -1\,208 \\
1\,208 & 48\,333 & -1\,208 & 96\,666
\end{bmatrix}
\begin{array}{c}
0 \\ r_1 \\ r_2 \\ r_3
\end{array}
$$

$$
\mathbf{K}_{G1} = -N
\begin{array}{cccc}
0 & r_1 & r_2 & r_3 \\
\end{array}
\begin{bmatrix}
0.01 & 0.10 & -0.01 & 0.10 \\
0.10 & 16.00 & -0.10 & -4.00 \\
0.01 & -0.10 & 0.01 & -0.10 \\
0.10 & -4.00 & -0.10 & 16.00
\end{bmatrix}
\begin{array}{c}
0 \\ r_1 \\ r_2 \\ r_3
\end{array}
$$

$$
\mathbf{K}_{E2} =
\begin{array}{cccc}
r_2 & r_3 & 0 & r_4 \\
\end{array}
\begin{bmatrix}
20 & 1\,208 & -20 & 1\,208 \\
1\,208 & 96\,666 & -1\,208 & 48\,333 \\
-20 & -1\,208 & 20 & -1\,208 \\
1\,208 & 48\,333 & -1\,208 & 96\,666
\end{bmatrix}
\begin{array}{c}
r_2 \\ r_3 \\ 0 \\ r_4
\end{array}
$$

$$
\mathbf{K}_{G2} = -N
\begin{array}{cccc}
r_2 & r_3 & 0 & r_4 \\
\end{array}
\begin{bmatrix}
0.01 & 0.10 & -0.01 & 0.10 \\
0.10 & 16.00 & -0.10 & -4.00 \\
0.01 & -0.10 & 0.01 & -0.10 \\
0.10 & -4.00 & -0.10 & 16.00
\end{bmatrix}
\begin{array}{c}
r_2 \\ r_3 \\ 0 \\ r_4
\end{array}
$$

The structure stiffness matrix and the equilibrium equations are given below. For stability solution, the load vector is zero,

$$\left(\begin{bmatrix} 96\,666 & -1\,208 & 48\,333 & 0 \\ -1\,208 & 40 & 0 & 1\,208 \\ 48\,333 & 0 & 1\,933\,322 & \\ 0 & 1\,208 & 4\,8333 & 96\,666 \end{bmatrix} - N \begin{bmatrix} 16 & -0.1 & -4 & 0.1 \\ -0.1 & 0.02 & 0 & 0.1 \\ -4 & 0 & 32 & -4 \\ 0 & 0.1 & -4 & 16 \end{bmatrix}\right) \begin{bmatrix} r_1 \\ r_2 \\ r_3 \\ r_4 \end{bmatrix} = \begin{bmatrix} 0 \\ 0 \\ 0 \\ 0 \end{bmatrix}$$

which leads to the eigenvalue problem

$$\begin{bmatrix} 96\,666 & -1\,208 & 48\,333 & 0 \\ -1\,208 & 40 & 0 & 1\,208 \\ 48\,333 & 0 & 1\,933\,322 & \\ 0 & 1\,208 & 48\,333 & 96\,666 \end{bmatrix} \begin{bmatrix} r_1 \\ r_2 \\ r_3 \\ r_4 \end{bmatrix} = N \begin{bmatrix} 16 & -0.1 & -4 & 0.1 \\ -0.1 & 0.02 & 0 & 0.1 \\ -4 & 0 & 32 & -4 \\ 0 & 0.1 & -4 & 16 \end{bmatrix} \begin{bmatrix} r_1 \\ r_2 \\ r_3 \\ r_4 \end{bmatrix}$$

Dividing both sides by N and pre-multiplying by the inverse of the matrix on the left, we get

$$\frac{1}{N} \begin{bmatrix} r_1 \\ r_2 \\ r_3 \\ r_4 \end{bmatrix} = \frac{1}{1\,000} \begin{bmatrix} 0.2960 & 0.0213 & 0.0019 & -0.1215 \\ 10.1901 & 1.7843 & 2.5482 & -10.0627 \\ -0.0064 & 0 & 0.0186 & -0.0064 \\ -0.1242 & -0.0213 & -0.0825 & 0.2945 \end{bmatrix} \begin{bmatrix} r_1 \\ r_2 \\ r_3 \\ r_4 \end{bmatrix}$$

Here for simplicity we have factored out 1000.

As shown in the following table, we start out with v_0 as a vector of ones, multiply this vector by the stability matrix Q (on the right hand side) and normalize the resulting vector ($1000v_1$) by dividing by its second (the largest) term. We repeat this process until convergence is reached.

1 000Q				v_0	$1\,000v_1$	$\bar{v}_1$	$1\,000v_2$	$\bar{v}_2$	$1000v_3$
0.2960	0.0213	0.0019	−0.1215	1	0.1977	0.0443	0.0326	0.0156	0.0272
10.1901	1.7843	2.5482	−10.0627	1	4.4599	1.000	2.0891	1.0000	2.0510
−0.0064	0	0.0186	−0.0064	1	0.0058	0.0013	−0.0004	−0.0002	−0.0000
−0.1242	−0.0213	−0.0825	0.2945	1	0.0665	0.0149	−0.0225	−0.0108	−0.0264

Thus, after three iterations, we have $1/N = 1/(2.051/1000)$, or

$$N = 487.567 \text{ kips}$$

The exact solution is

$$N_e = \frac{\pi^2 EI}{l^2} = \frac{\pi^2 \times 29\,000 \times 100}{240^2} = 496.9 \text{ kips}$$

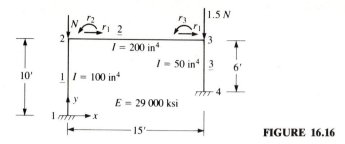

FIGURE 16.16

Note that the solution can be improved by increasing the number of iterations in the Stodola method. Furthermore, in the solution of buckling problems the accuracy of the solution is substantially increased by using more than one element. For the present problem the solution with one element differs from the exact value by 22%, while with two elements the difference is 1.9%. Note also that when the deformed structure is symmetric, the symmetry may be used to reduce the number of degrees of freedom.

Example 16.6. Find the buckling load for the frame of Fig. 16.16. The compressive axial load on member 34 is twice as large as that on 12.

Solution. Since axial deformations are neglected, the same degree of freedom is used for the horizontal displacements of joints 2 and 3.

$$
\mathbf{K}_{E_1} =
\begin{array}{cccc}
r_1 & r_2 & 0 & 0 \\
\end{array}
\begin{bmatrix}
20 & 1\,208 & -20 & 1\,208 \\
1\,208 & 96\,667 & -1\,208 & 48\,333 \\
-20 & -1\,208 & 20 & -1\,208 \\
1\,208 & 48\,333 & -1\,208 & 96\,667
\end{bmatrix}
\begin{array}{c}
r_1 \\ r_2 \\ 0 \\ 0
\end{array}
$$

$$
\mathbf{K}_{G_1} = -N
\begin{array}{cccc}
r_1 & r_2 & 0 & 0 \\
\end{array}
\begin{bmatrix}
0.01 & 0.10 & -0.01 & 0.10 \\
0.10 & 16.00 & -0.10 & -4.00 \\
-0.01 & -0.10 & 0.01 & -0.10 \\
0.10 & -4.00 & -0.10 & 16.00
\end{bmatrix}
\begin{array}{c}
r_1 \\ r_2 \\ 0 \\ 0
\end{array}
$$

$$
\mathbf{K}_{E_2} =
\begin{array}{cccc}
0 & r_2 & 0 & r_3 \\
\end{array}
\begin{bmatrix}
10 & 1\,070 & -10 & 1\,070 \\
1\,070 & 128\,890 & -1\,070 & 64\,440 \\
-10 & -1\,070 & 10 & 1\,070 \\
1\,070 & 64\,440 & -1\,070 & 128\,890
\end{bmatrix}
\begin{array}{c}
0 \\ r_2 \\ 0 \\ r_3
\end{array}
$$

$$
\mathbf{K}_{E_3} =
\begin{array}{cccc}
r_1 & r_3 & 0 & 0 \\
\end{array}
\begin{bmatrix}
47 & 1\,678 & -47 & 1\,678 \\
1\,678 & 80\,556 & -1\,678 & 40\,278 \\
-47 & -1\,678 & 47 & -1\,678 \\
1\,678 & 40\,278 & -1\,678 & 80\,556
\end{bmatrix}
\begin{array}{c}
r_1 \\ r_3 \\ 0 \\ 0
\end{array}
$$

$$\mathbf{K}_{G_3} = -N \begin{array}{c} \\ \end{array} \overset{\begin{array}{cccc} r_1 & r_3 & 0 & 0 \end{array}}{\begin{bmatrix} 0.025 & 0.15 & -0.025 & 0.15 \\ 0.150 & 14.40 & -0.150 & -3.60 \\ -0.025 & -0.15 & 0.025 & -0.15 \\ 0.150 & -3.60 & -0.150 & 14.40 \end{bmatrix}} \begin{array}{c} r_1 \\ r_3 \\ 0 \\ 0 \end{array}$$

Note that the directions of the local coordinates for the elements of this square frame are shown downward and to the right. That is, the first joints for the columns are the higher joints, while for the beam the first joint is the one on the left. Thus, as described in Section 12.3, the shear forces in the local and global coordinates are the same and no transformation of coordinates is necessary. Since element 2 is not under any axial load, no geometric stiffness matrix exists for that element.

The equilibrium equation, Eq. (16.23c), $\mathbf{K}_{ES}\mathbf{r} = N\mathbf{G}_{ES}\mathbf{r}$,

$$10^5 \begin{bmatrix} 0.0007 & 0.0121 & 0.0168 \\ 0.0121 & 2.2556 & 0.6444 \\ 0.0168 & 0.6444 & 2.0944 \end{bmatrix} \begin{bmatrix} r_1 \\ r_2 \\ r_3 \end{bmatrix} = N \begin{bmatrix} 0.0350 & 0.1000 & 0.1500 \\ 0.1000 & 16.0000 & 0 \\ 0.1500 & 0 & 14.4000 \end{bmatrix} \begin{bmatrix} r_1 \\ r_2 \\ r_3 \end{bmatrix}$$

gives $(1/N)\,\mathbf{r} = \mathbf{Q}\mathbf{r}$, with the stability matrix, $\mathbf{Q} = \mathbf{K}_{ES}^{-1}\mathbf{G}_{ES}$, as

$$\mathbf{Q} = 10^{-3} \begin{bmatrix} 0.6594 & 0.9059 & 0.9713 \\ -0.0020 & 0.0747 & -0.0248 \\ -0.0040 & -0.0303 & 0.0686 \end{bmatrix}$$

By the Stodola method,

1 000Q			$v_0\ 1\,000v_1$	$\bar{v}_1$	$1\,000v_2$	$\bar{v}_2$	$1\,000v_3$	$\bar{v}_3$	$1\,000v_4$
0.6594	0.9059	0.9713	1 2.5366	1.0000	0.6897	1.0000	0.6532	1.0000	0.6504
−0.0020	0.0747	−0.0248	1 0.0480	0.0189	−0.0009	−0.0013	−0.0019	−0.0029	−0.0020
−0.0040	−0.0303	0.0686	1 0.0344	0.0136	−0.0036	−0.0052	−0.0043	−0.0066	−0.0043

The buckling load is $N = 1/(0.6504/1000) = 1\,537$ kips.

The deformed shape of the buckled frame can be drawn using the mode shape found, with $r_1 = -1$, $r_2 = 0.0029$ and $r_3 = 0.0066$. That is, with the beam deflecting to the left and joints 2 and 3 undergoing counterclockwise rotations.

LABORATORY EXPERIMENTS

Beam–Column Interaction Diagrams

Using square section wood or reinforced-concrete specimens of about 1 in × 1 in × 12 in (Fig. L16.1a), find the ultimate compressive axial load capacity for different eccentricities. Test the specimens for increasing values of eccentricity, say from 0 to 2.5 in. Measure the load at failure in a universal testing machine. Plot the interaction diagram of moment versus load, Fig. L16.1b.

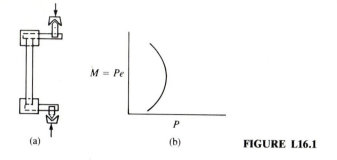

$$M = Pe$$

(a) (b) **FIGURE L16.1**

Column Stability

Buckling loads are more difficult to measure than other loads. This is because of the sudden nature of buckling and the effects of crookedness and eccentricity of the load. Using a strip or a $\frac{1}{4}$-in diameter round aluminum or steel bar (see Fig. L16.2):

(*a*) Measure the buckling load for hinged boundary conditions.

(*b*) Repeat part (*a*) for different l/r values where l is the column length and r is the radius of gyration, $r = \sqrt{I/A}$.

FIGURE L16.2

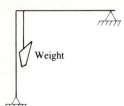

FIGURE L16.3

(c) Kink the bar so that it has an initial displacement v_0 at the mid-point. Verify that under axial load N the total mid-length displacement from the axis through the end points of the bar is

$$v = \frac{1}{1-(N/N_e)}\, v_0$$

where $N_e = \pi^2 EI/l^2$ is the Euler buckling load.

Frame Buckling

Use a frame made of spring steel or similar material, with the end of the column hinged and the end of the beam hinged or on a roller. Load the frame near the beam–column connection with sand poured in a can (Fig. L16.3). Find the buckling load. Calculate the buckling load by the stiffness method and compare the results.

Lateral Buckling of Cantilever Beams

Although not discussed in the text, lateral buckling is a critical failure mode in thin-walled structures. This experiment should provide some insight into the problem.

Use a strip of steel, say $\frac{1}{16}$ in × 3 in × 24 in, and fix one end to form a cantilever beam with depth 3″.

(a) Using the ultimate stress for the steel used, calculate the ultimate load P that the cantilever beam can carry at its free end in the direction of the 3 in dimension (Fig. L16.4).

(b) Apply a gradually increasing end load (may use weights or sand in a pail) at points near the top, mid-depth and near the bottom (along the depth) of the beam. Each time determine the load-carrying capacity (the ultimate load that the beam can carry).

(c) Calculate the critical lateral buckling load using the following formula: (from Timoshenko and Gere, *Theory of Elastic Stability*, McGraw-Hill, New York, 1961):

$$P_{cr} = \frac{4.013}{l^2} \sqrt{EIGJ} \left(1 - \frac{a}{l}\sqrt{\frac{EI}{GJ}}\right)$$

FIGURE L16.4

Compare the results to the measured values in part (*b*) and those point of the load positive when above the neutral axis of the beam, *I* is the moment of inertia of the cross-section, *J* is the torsion constant (see Table 17.1), *l* is the span length, and *E* and *G* are the modulus of elasticity and shear modulus.

PROBLEMS

16.1. Find the buckling load of a fixed-end column (Fig. P16.1) from the solution of the differential equation.

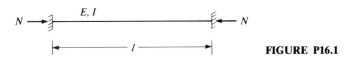

FIGURE P16.1

16.2. Determine the buckling load of the column of Fig. P16.2 from the governing differential equation.

FIGURE P16.2

Hint: Solve the resulting equation either numerically or graphically for *αl*.

16.3. Find the deflection of the fixed-end column with an initial deflection $v_0 = a(1 - \cos 2\pi x/l)$.

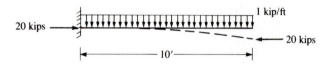

FIGURE P16.3

16.4. Determine the percentage increase in the deflection of the free end and moment at the fixed-end for the beam–column of Fig. P16.4 due to the axial force. $I = 200 \text{ in}^4$; $E = 29\,000$ ksi.

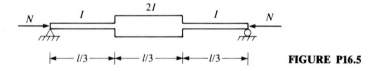

FIGURE P16.4

16.5. Use the Ritz method and assumed deflected shape $v = a \sin \pi x/l$ to find the buckling load of the simply supported column in Fig. P16.5.

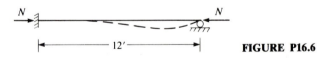

FIGURE P16.5

Find the buckling load of the following structures by the finite-element method.

16.6. Use two or more elements for the supported column of Fig. P16.6 with $I = 150 \text{ in}^4$, $E = 29\,000$ ksi.

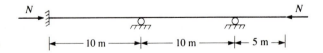

FIGURE P16.6

16.7. $I = 1\,000 \text{ cm}^4$; $E = 200\,000$ MPa.

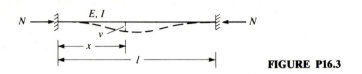

FIGURE P16.7

16.8. $I_1 = 150$ in^4; $I_2 = 200$ in^4; $E = 29\,000$ ksi.

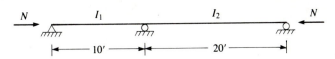

FIGURE P16.8

16.9. $I = 200$ in^4; $E = 29\,000$ ksi.

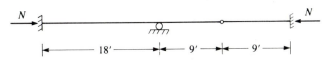

FIGURE P16.9

16.10. $E = 3\,000$ ksi.

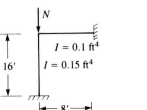

FIGURE P16.10

16.11. $E = 200\,000$ MPa.

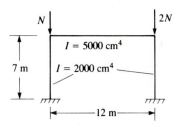

FIGURE P16.11

16.12. Equilateral frame with the length of the sides $l = 100$ in, $I = 100$ in^4, $E = 29\,000$ ksi, $A = 12.16$ in^2.

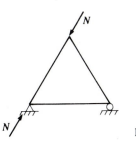

FIGURE P16.12

16.13. Find the stiffness matrix of a constant cross-section truss element using the energy of deformation (Eq. (6.6)) and the displacement function $u = a_0 + a_1 x$. The area of the cross-section is A, the length is l, and modulus of elasticity is E.

16.14. Using the energy method find the stiffness matrix for a truss element with a cross-section linearly varying from A_1 to A_2 (Fig. P16.14). The width of the section is b and the heights are d_1 and d_2. The length of the element is l and the modulus of elasticity is E.

Assume the same displacement function as for Problem 16.13.

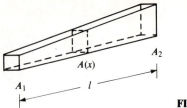

FIGURE P16.14

16.15. Find the stiffness matrix for a beam element on a distributed elastic foundation with foundation modulus k. The length of the element is l, its moment of inertia is I, and its modulus of elasticity is E. Use the same shape functions as those of beam elements (Eqs. (16.19)). The energy of deformation for such a member is

$$\Pi = \tfrac{1}{2} \int_0^l (EIv''^2 + kv^2) \, dx.$$

16.16 Find the stiffness matrix for a beam element including the shear deformations (see also Chapter 6 and problem 11.15).

16.17 Determine the load vector for a beam element under a uniformly distributed load over its span, using Eq. (16.20g).

16.18 The consistent mass matrix for dynamic analysis of beams and frames uses the beam shape functions (Eqs. (16.19)). This matrix for an element with mass m per unit length can be shown to be

$$M_{ij} = \int_0^l m\varphi_i\varphi_j \, dx.$$

Calculate the elements of the mass matrix in terms of m and l.

CHAPTER
17

MEMBERS
UNDER
TORSION

The problem of torsion has become more significant in structural engineering in recent years because of the advances in the material research and analysis techniques. The codes have become more sophisticated and less conservative, allowing more slender structural members. Thin-walled members are used for many structural components. In such members, torsion may be a critical problem. Three-dimensional frames, grid structures, and curved beams also involve torsion. Spandrel beams, placed at the periphery of concrete buildings, are another example where significant torques are involved. Many machine parts are subjected to torsion. Unlike structural components these members have circular cross-sections. Some common cross-sections for structural members are I, rectangular shape, and box sections.

Although two different sign conventions can be used for the applied and reactive torques in the design sign convention, we can also use the same sign for both. In the analysis sign convention all positive torques, whether applied or reactive must act in the positive direction of a right-handed coordinate system. The analysis convention is used in the stiffness method for the solution of problems involving torsion.

17.1 TORQUE DIAGRAMS

The torque at an arbitrary section of a member can be calculated in a manner similar to that for shear and moment. Thus, a portion of the member is separated as a free body and the torque at a section is found by considering equilibrium of the free body. This would indicate that the torque at a section can be found as the sum of the torques acting on the left or right of that section. A diagram depicting the torque at different sections of a member is called the *torque diagram*.

Only one equilibrium equation, equilibrium of the torques about the axis of the member, is available for finding the unknown torques in a member. Thus, a member with unknown torques at the two ends is statically indeterminate. Therefore, we must consider the twisting deformations in the solution of this problem.

Example 17.1. Find the reaction T_0 of the member shown in Fig. 17.1a under the applied torques, and draw the torque diagram.

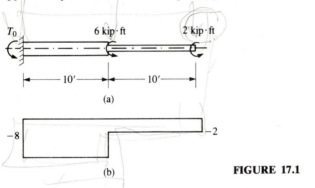

(a)

(b)

FIGURE 17.1

Solution. This is a statically determinate structure since the fixed-end torque can be found by the equilibrium equation alone. Setting the sum of the torques acting on the member equal to zero, we get

$$T_0 + 6 + 2 = 0 \quad \text{or} \quad T_0 = -8 \text{ kip} \cdot \text{ft}$$

It is seen that the fixed-end torque is independent of the variation of the cross-section. This is true for all statically determinate structures as was for moment and shear. Calculating the torque at each section, by summing the torques on the left of that section, and plotting them, we obtain the torque diagram of Fig. 17.1b.

17.2 RELATION BETWEEN TORQUE AND ANGLE OF TWIST

The angles of twist are needed not only for calculating the deformation of members under torsion but also for solving statically indeterminate problems involving torsion.

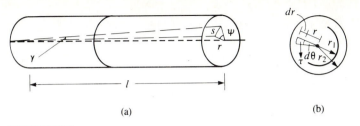

FIGURE 17.2

Experimental results show that when members with circular cross-sections twist, their cross-sections remain plane. For members other than solid and hollow cylinders, the cross-sections do not remain plane but warp into a curved surface, causing differential axial displacements at different points in the section.

For a member with a constant torque along its length, the rate of twist remains constant. This is called *uniform torsion*. If the torque varies along the length, then the rate of twist is variable and we have a nonuniform torsion problem. Nonuniform torsion also prevails when warping of the cross-section is restricted, such as at fixed ends. Here we consider uniform torsion only.

Consider the member of Fig. 17.2a with a circular cross-section, subjected to a constant torque. If the length of the member is l then the angle of twist ψ at the end of the member is equal to the rate of twist ψ' times the length, that is, $\psi = \psi'l$. The stress induced by torsion is shear stress. The change of angle of a longitudinal fiber at radius r is

$$\gamma = \frac{s}{l} = \frac{r\psi}{l} \tag{17.1}$$

By Hooke's law, the relationship between the shear stress τ and shear strain γ is

$$\tau = G\gamma \quad \text{or} \quad \tau = \frac{Gr\psi}{l} \tag{17.2}$$

where $G = E/[2(1+v)]$ is the shear modulus, E is the modulus of elasticity, and v is Poisson's ratio. v is equal 0.3 for steel and about 0.2 for concrete.

The internal torque T at a section of a hollow cylinder, Fig. 17.2b, is

$$T = \int_{r_1}^{r_2} \int_0^{2\pi} \tau \cdot r \, d\theta \, dr \cdot r = \int_{r_1}^{r_2} \tau r^2 \, dr \int_0^{2\pi} d\theta = 2\pi \int_{r_1}^{r_2} \tau r^2 \, dr$$

Upon substitution for τ, we get

$$T = \frac{2\pi G\psi}{l} \int_{r_1}^{r_2} r^3 \, dr = \frac{G\psi}{l} \frac{\pi(r_2^4 - r_1^4)}{2}$$

or

$$T = \frac{GJ}{l}\psi = GJ\psi' \tag{17.3}$$

where $J = \frac{1}{2}\pi(r_2^4 - r_1^4)$ is the polar moment of inertia of the circular section. GJ/l is called the torsional stiffness.

Equation (17.3) holds true for sections other than circular if the warping deformation is not restrained. However, the torsion constant J in this case is different from the polar moment of inertia. Torsion constants for different sections are derived in Section 17.5 and the values for common cross-sections are listed in Table 17.1.

Statically indeterminate problems can be solved by using Eq. (17.3).

Example 17.2. Find the fixed-end torques T_1 and T_2 for the members in Fig. 17.3. The ends are fixed against rotation about the x axis. A torque T_c is applied at a distance l_1 from the left-hand end. The torsional constants are J_1 and J_2 for the segments with lengths l_1 and l_2. Find T_1 and T_2 in terms of J_1 and J_2 and T_c and calculate their values for $T_c = 5$ kN · m, $l_1 = l_2 = 12$ m, $J_1 = 2J_2 = 0.03$ m^4. Also find T_1 and T_2 when $J_1 = J_2 = 0.03$ m^4.

Solution. If the fixed-end torques are T_1 and T_2, then the torques in segments l_1 and l_2 are constant and equal to T_1 and T_2, respectively. The angle of twist at any point can be calculated by Eq. (17.3). Since the angle of twist ψ_c at l_1 is the same whether it is calculated from the left or the right segment,

$$\psi_c = \frac{T_1 l_1}{GJ_1} = \frac{T_2 l_2}{GJ_2} \quad \text{or} \quad T_2 = \frac{J_2 l_1}{J_1 l_2} T_1$$

By the equilibrium equation, the sum of the torques acting on the structure must be zero, that is,

$$T_1 + T_2 + T_c = 0 \quad \text{or} \quad T_1 + T_2 = -T_c$$

Substituting for T_2 in terms of T_1, in the equilibrium equation we get

$$T_1 + \frac{J_2 l_1}{J_1 l_2} T_1 = -T_c$$

or

$$T_1\left[1 + \frac{J_2 l_1}{J_1 l_2}\right] = -T_c$$

which gives

$$T_1 = -\frac{T_c}{[1 + (J_2 l_1/J_1 l_2)]}$$

and

$$T_2 = -\frac{[J_2 l_1/J_1 l_2]}{[1 + (J_2 l_1/J_1 l_2)]} T_c$$

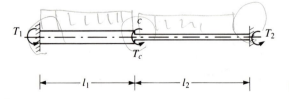

FIGURE 17.3

For the given values and $J_1 = 2J_2$ we get

$$T_1 = -\frac{5}{1 + [(0.015 \times 12)/(0.03 \times 12)]} = -3.333 \text{ kN} \cdot \text{m}$$

$$T_1 = -\frac{(0.015 \times 12/0.0^3 \times 12)}{1 + [(0.015 \times 12/0.03 \times 12)]} = -1.667 \text{ kN} \cdot \text{m}$$

For constant cross-sections, $J_1 = J_2$,

$$T_1 = -\frac{T_c}{(1 + l_1/l_2)} = -\frac{l_2}{l} T_c$$

$$T_2 = -\frac{l_1}{l} T_c$$

where $l = l_1 + l_2$.

Note that in the latter case each support reaction is proportional to the distance of the torque T_c from the other support. This is similar to the reaction of a beam under a concentrated load. With the given values,

$$T_1 = T_2 = -\tfrac{12}{24} \times 5 = -2.5 \text{ kN} \cdot \text{m}$$

17.3 SHEAR STRESS DUE TO TORSION

For circular sections we can find the shear stress due to torsion at an arbitrary point of the cross-section located at a distance r from the origin.

By Eqs. (17.2) and (17.3), $\tau = Gr\psi/l$ and $\psi = Tl/GJ$. Therefore, we can find

$$\tau = \frac{Gr}{l} \times \frac{l}{GJ} T$$

or

$$\tau = \frac{Tr}{J} \tag{17.4}$$

The maximum stress occurs at the outer surface of a cylinder.

For noncircular sections the stress is proportional to the torque and depends on the location of the point in the section. It is also inversely related to J. For a solid rectangular section, with the larger and smaller dimensions a and b, respectively, the maximum shear stress[1] due to torsion is

$$\tau_{\text{max}} = \frac{3a + 1.8b}{8a^2b^2} T \tag{17.5}$$

---

[1] R. J. Roark, *Formulas for Stress and Strain*, 3rd ed., McGraw-Hill, New York, 1954.

17.4 ENERGY OF TORSION

In view of Eqs. (6.4) and (17.2) the elastic energy of torsion is

$$U_t = \tfrac{1}{2} \int_{\text{vol}} \tau \gamma \, d(\text{vol}) = \tfrac{1}{2} \int \frac{\tau^2}{G} \, d(\text{vol}) \qquad (17.6a)$$

For circular sections, using Eq. (17.4), we have,

$$U_t = \tfrac{1}{2} \int_A \int_l \frac{T^2 r^2}{GJ^2} \, r \, d\theta \, dr \, dx = \int_l \frac{T^2}{GJ^2} J \, dx$$

or by Eq. (17.3)

$$U_t = \tfrac{1}{2} \int_l \frac{T^2}{GJ} \, dx = \tfrac{1}{2} \int_l GJ \psi'^2 \, dx \qquad (17.6b)$$

Equation (17.6b) is valid for all sections. However, for noncircular sections J is not the polar moment of inertia.

Energy methods can be used for solving problems involving torsion plus other effects. Such a problem was presented in Example 6.10.

17.5 TORSION CONSTANTS

As we saw in Section 17.2, torsion constants are needed for members subjected to torsion. For circular sections the torsion constant is the polar moment of inertia of the cross-section.

Determination of torsion constants for noncircular sections is significantly more complex with or without warping restraints. If warping is not restrained, then the torsion problem is called the St. Venant torsion. A general theory exists[2] that allows the calculation of torsion constants by assuming the warping deformation to be a function of the section coordinates and proportional to the rate of twist. For restrained warping, the existing theory applies to thin-walled closed and open sections only. Warping deformations are less critical for solid sections than thin-walled members.

Many structural shapes, such as wide-flange I-sections (W-sections), box sections, etc., are composed of thin-walled members. The procedures for determining torsion constants for open and closed sections will be discussed below. These include both circular and noncircular cross-sections that are commonly used.

---

[2] S. P. Timoshenko and J. N. Goodier, *Theory of Elasticity*, 2nd ed., McGraw-Hill, New York, 1951.

Thick Circular Sections

The torsion constant for an annular circular section with the inner and outer radii r_1 and r_2 was found in Section 17.2. This was

$$J = \frac{\pi(r_2^4 - r_1^4)}{2} \tag{17.7}$$

In this case J is the polar moment of inertia. For a solid circular section, $r_1 = 0$, $r_2 = a$, and

$$J = \frac{\pi a^4}{2} \tag{17.8}$$

Thin-Walled Closed Sections

Consider a constant-cross-section member with its thickness t the smallest dimension of the section (Fig. 17.4a). The thickness may vary along the circumference. The shear force per unit length of the circumference, the shear flow, is

$$q = \tau t \tag{17.9}$$

For an infinitesimal element with dimensions ds along the circumference and dx along the length, the forces are as indicated in Fig. 17.4b.

Equilibrium of the forces on the infinitesimal element in the x and s directions gives

$$\left(q + \frac{\partial q}{\partial s} ds\right) dx - q\, dx = 0 \qquad \text{or} \qquad \frac{\partial q}{\partial s} = 0$$

and

$$\left(q + \frac{\partial q}{\partial x} dx\right) ds - q\, ds = 0 \qquad \text{or} \qquad \frac{\partial q}{\partial x} = 0$$

It is easy to prove that if a function of two variables has zero derivatives

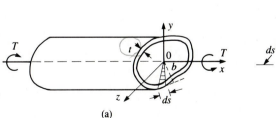

(a) (b)

FIGURE 17.4

with respect to both variables then it must be a constant. Therefore, the shear flow is constant along the length and the circumference, that is $q = $ constant.

The torque induced by the shear flow over ds, about the x axis, is $bq\,ds$, where b is the normal distance from the origin to the tangent at the mid-point of ds. $b\,ds$ is equal to twice the cross-hatched area in Fig. 17.4a, since it is the product of the base and the height of a triangular section. The total torque resisted by the section is

$$T = \int bq\,ds = q\int b\,ds$$

or

$$T = q(2A_0) \tag{17.10}$$

where A_0 is the area enclosed by the mid-thickness circumference of the section. Since for thin-walled sections the thickness is small, the area enclosed by the inner circumference may be used instead.

The shear stress at a point can be found from Eqs. (17.9) and (17.10),

$$\tau = \frac{q}{t} = \frac{T}{2A_0 t} \tag{17.11}$$

For uniform torsion that we are considering here, a torque applied to the end of a member produces a rotation ψ, with a constant rate of twist. The torque T is constant over the length of the member. We can find the relation between the applied torque T and the angle of twist ψ by setting the internal and external work of the system equal (Eq. (6.10)). In this case the internal work is the internal energy of the member and is equal to the energy per unit length times the length of the member. The external work is the work of the torque through the angle of rotation. Thus, using Eq. (17.6a) the internal energy per unit length is,

$$U_t = \frac{1}{2}\int \frac{\tau^2}{G}t\,ds = \frac{1}{2}\int \frac{T^2}{4A_0^2 t^2 G}t\,ds = \frac{T^2}{8A_0^2 G}\int \frac{ds}{t}$$

and the external work per unit length, with $\psi' = \psi/l$, is

$$W_e = \tfrac{1}{2}T\psi'$$

Thus,

$$\frac{T^2}{8A_0^2 G}\int \frac{ds}{t} = \tfrac{1}{2}T\psi'$$

which gives

$$T = \frac{4A_0^2 G}{\displaystyle\int ds/t}\,\psi' \tag{17.12}$$

Comparison of Eq. (17.12) with (17.3) gives

$$J = \frac{4A_0^2}{\displaystyle\int ds/t} \qquad\qquad (17.13a)$$

If the thickness t is constant, then

$$J = \frac{4A_0^2 t}{\Gamma} \qquad\qquad (17.13b)$$

where Γ is the mid-thickness circumference of the section.

Solid Noncircular Sections

The torsion constant J for noncircular sections is more difficult to determine, because of warping of the cross-section.

For rectangular sections, with a and b the larger and the smaller dimensions, respectively, this results in (Ref. footnote 1)

$$J = \frac{1}{3}\left[1 - 0.63\frac{b}{a}\left(1 - \frac{b^4}{12a^4}\right)\right]ab^3 \qquad\qquad (17.14)$$

Thin-Walled Open Sections

For a thin rectangular section where b/a is small (say < 0.1), the value in the brackets in Eq. (17.14) is approximately equal to 1, and thus

$$J = \tfrac{1}{3}ab^3 \qquad\qquad (17.15a)$$

For compound thin-walled open sections the torsion constant can be found by calculating the value of J for each branch of the section and summing up the results. Thus, for an open section with n branches in the section circumference

$$J = \tfrac{1}{3}\sum_{i=1}^{n} a_i b_i^3 \qquad\qquad (17.15b)$$

For wide-flange I-sections (W shapes), the Manual of Steel Construction (AISC) gives the torsion as well as warping constants. The above results for the torsion constant J are summarized in Table 17.1.

TABLE 17.1
Torsion constants J

Section	$\psi = Tl/GJ$ Torsion constant J	Maximum shear stress
Solid circular section	 $\frac{1}{2}\pi a^4$	$\frac{2T}{\pi a^3}$ on boundary
Hollow circular section	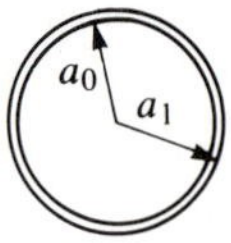 $\frac{1}{2}\pi(a_1^4 - a_0^4)$	$\frac{2a_1 T}{\pi(a_1^4 - a_0^4)}$
Thin-walled closed section 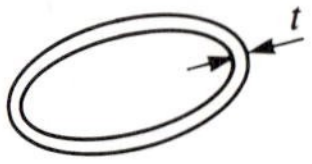	$4A_0^2 t/\Gamma$	$\frac{T}{2A_0 t}$
Rectangular (b = smaller dimension)	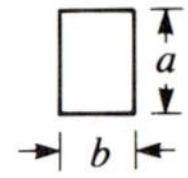 $\frac{ab^3}{3}\left[1 - 0.63\frac{b}{a}\left(1 - \frac{b^4}{12a^4}\right)\right]$	$\frac{3a + 1.8b}{8a^2 b^2}T$
Thin-walled open section	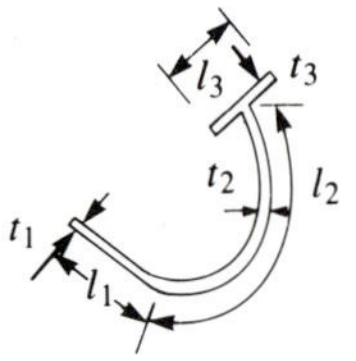 $\frac{1}{3}\sum_{i=1}^{n} l_i t_i^3$	$\frac{3\Gamma + 1.8t}{\Gamma^2 t^2}T$

A_0 = area enclosed by the median line.　　ψ = angle of twist
Γ = length of median line.　　　　　　T = torque.
G = shear modulus.

Example 17.4. Find the torsion constants for a circular thin-walled section with a slit along its length and a closed section with the same thickness and radius (Fig. 17.5a,b). Find the ratio of the angles of twist for the two members having the same length and under the same torque. The mid-thickness radius of both sections is a and the thickness is $0.1a$.

Solution. From Table 17.1, the torsion constant for the open section is

$$J_1 = \tfrac{1}{3}\Gamma t^3 = \tfrac{1}{3}(2\pi a)(0.1a)^3 = 0.0021a^4$$

and for the closed section is

$$J_2 = \frac{4A_0^2 t}{\Gamma} = \frac{4(\pi a^2)^2(0.1a)}{2\pi a} = 0.1\pi a^4 = 0.314a^4$$

The ratio of the angles of twist is

$$\frac{\psi_1}{\psi_2} = \frac{Tl/GJ_1}{Tl/GJ_2} = \frac{J_2}{J_1} = \frac{0.314a^4}{0.002\,1a^4} = 150$$

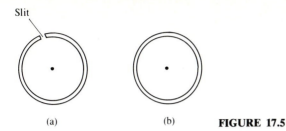

Slit

(a)   (b)   **FIGURE 17.5**

As we see, the angle of twist for the open section is 150 times larger than that of the closed section; that is, the closed section is much stiffer.

17.6 STIFFNESS METHOD FOR TORSIONAL MEMBERS

The solution of members under torsion can be obtained by the stiffness method. For straight members under torsion and bending, the two effects can be included in the derivation of the stiffness matrix. Since torsional deformations are independent of bending, the two effects can be calculated separately and then combined. The torsional stiffness is derived in this section, while the bending stiffness was derived in Section 11.3 (Eq. (11.10a)). However, if there is interaction between torsion and bending, as in bow-girders (Section 17.8), the effects must be included in the derivation.

Force–Displacement Relations for Torsional Members

Consider the constant-cross-section element of Fig. 17.6 with length l and torsion constant J. The member is under end torques T_1 and T_2, which cause the end rotations ψ_1 and ψ_2.

The torque at the right end is T_2 and the relative angle of twist at that end is $\psi_2 - \psi_1$. By Eq. (17.2) we can write

$$T_2 = \frac{GJ}{l}(\psi_2 - \psi_1)$$

Also, by equilibrium, $T_1 + T_2 = 0$, giving,

$$T_1 = \frac{GJ}{l}(\psi_1 - \psi_2)$$

Writing the above expressions in matrix form we get the torque–twist

T_1, ψ_1 —————————————————— T_2, ψ_2

FIGURE 17.6

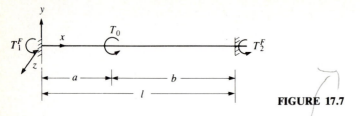

FIGURE 17.7

angle relationship. If in-span torques are also present, they exhibit themselves in the form of fixed-end torques. The general expression is thus

$$\begin{bmatrix} T_1 \\ T_2 \end{bmatrix} = \frac{GJ}{l}\begin{bmatrix} 1 & -1 \\ -1 & 1 \end{bmatrix}\begin{bmatrix} \psi_1 \\ \psi_2 \end{bmatrix} + \begin{bmatrix} T_1^F \\ T_2^F \end{bmatrix} \tag{17.16a}$$

where

$$\mathbf{K}_T = \frac{GJ}{l}\begin{bmatrix} 1 & -1 \\ -1 & 1 \end{bmatrix} \tag{16.16b}$$

is the torsional stiffness matrix of the member.

Fixed-End Torques

For a member under a concentrated applied torque T_0, the fixed-end torques were found in Example 17.2. If the cross-section is constant (Fig. 17.7), then the fixed-end torques are proportional to the distance to the other end of the member. This is similar to fixed-end shear. Thus,

$$T_1^F = -\frac{bT_0}{l} \tag{17.17a}$$

$$T_2^F = -\frac{aT_0}{l} \tag{17.17b}$$

Example 17.5. The structure of Fig. 17.8 is under the concentrated torques as shown. Find the rotations and torques at 8-ft intervals. $G = 12\,500$ ksi; $J = 100\,\text{in}^4$ for the end segments and 50 in⁴ for the middle segment.

Solution. The torque–twist angle relation for element 1 is

$$\begin{bmatrix} T_1 \\ T_2 \end{bmatrix}^1 = \frac{12\,500 \times 100}{8 \times 12}\begin{bmatrix} 1 & -1 \\ -1 & 1 \end{bmatrix}\begin{bmatrix} \psi_1 \\ \psi_2 \end{bmatrix} = \begin{bmatrix} 13\,020 & -13\,020 \\ -13\,020 & 13\,020 \end{bmatrix}\begin{bmatrix} 0 \\ r_1 \end{bmatrix}$$

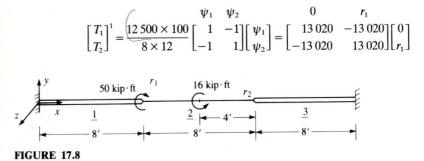

FIGURE 17.8

For element 2 with the fixed-end torques calculated by Eqs. (17.17)

$$\begin{bmatrix} T_1 \\ T_2 \end{bmatrix}^2 = \begin{matrix} r_1 & r_2 \\ \begin{bmatrix} 6\,510 & -6\,510 \\ -6\,510 & 6\,510 \end{bmatrix} \end{matrix} \begin{bmatrix} r_1 \\ r_2 \end{bmatrix} + \begin{bmatrix} -8 \\ -8 \end{bmatrix}$$

and for element 3

$$\begin{bmatrix} T_1 \\ T_2 \end{bmatrix}^3 = \begin{matrix} r_2 & 0 \\ \begin{bmatrix} 13\,020 & -13\,020 \\ -13\,020 & 13\,020 \end{bmatrix} \end{matrix} \begin{bmatrix} r_2 \\ 0 \end{bmatrix}$$

The assembly of the element stiffness matrices and fixed-end torques into the structure stiffness matrix and load vector is similar to that for other structures (Section 11.2). This results in the equilibrium equations,

$$\begin{matrix} r_1 & r_2 \\ \begin{bmatrix} 19\,530 & -6\,510 \\ -6\,510 & 19\,530 \end{bmatrix} \end{matrix} \begin{bmatrix} r_1 \\ r_2 \end{bmatrix} = \begin{bmatrix} -50 + 8 \\ 8 \end{bmatrix}$$

The solution of the above equations gives the member-end twist angles as

$$\begin{bmatrix} r_1 \\ r_2 \end{bmatrix} = \begin{bmatrix} -0.0011 \\ -0.0033 \end{bmatrix}$$

The member-end torques are found upon substitution of the above twist angles into the torque–twist angle relations for each element. This results in the following member-end torques:

Element →	1	2	3
$\begin{bmatrix} T_1 \\ T_2 \end{bmatrix}$:	$\begin{bmatrix} 14.32 \\ -14.32 \end{bmatrix}$	$\begin{bmatrix} 20.64 \\ -36.64 \end{bmatrix}$	$\begin{bmatrix} -42.96 \\ 42.96 \end{bmatrix}$

17.7 GRID SYSTEMS

Grids are systems composed of rigidly connected beams that undergo bending and torsional deformations. Girder bridge decks and floor systems with longitudinal and transverse beams are examples of grid systems. Although three-dimensional beam elements (Section 18.3) can be used to model grid elements, here we discuss a grid element with a smaller number of degrees of freedom than that of three-dimensional beam elements.

Stiffness Matrix for a Grid Element

A grid element with its end forces and deformations is shown in Fig. 17.9. Since the bending and torsional deformations are independent, we can combine the force–displacement relations for bending and torsion, found in Sections 11.3 and 17.6, to obtain the force–displacement relationship for grid

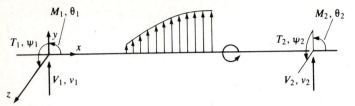

FIGURE 17.9

members. This results in

$$
\begin{bmatrix} V_1 \\ T_1 \\ M_1 \\ V_2 \\ T_2 \\ M_2 \end{bmatrix} = \begin{bmatrix} \dfrac{12EI}{l^3} & 0 & \dfrac{6EI}{l^2} & -\dfrac{12EI}{l^3} & 0 & \dfrac{6EI}{l^2} \\ 0 & \dfrac{GJ}{l} & 0 & 0 & -\dfrac{GJ}{l} & 0 \\ \dfrac{6EI}{l^2} & 0 & \dfrac{4EI}{l} & -\dfrac{6EI}{l^2} & 0 & \dfrac{2EI}{l} \\ -\dfrac{12EI}{l^3} & 0 & -\dfrac{6EI}{l^2} & \dfrac{12EI}{l^3} & 0 & -\dfrac{6EI}{l^2} \\ 0 & -\dfrac{GJ}{l} & 0 & 0 & \dfrac{GJ}{l} & 0 \\ \dfrac{6EI}{l^2} & 0 & \dfrac{2EI}{l} & -\dfrac{6EI}{l^2} & 0 & \dfrac{4EI}{l} \end{bmatrix} \begin{bmatrix} v_1 \\ \psi_1 \\ \theta_1 \\ v_2 \\ \psi_2 \\ \theta_2 \end{bmatrix} + \begin{bmatrix} V_1^F \\ T_1^F \\ M_1^F \\ V_2^F \\ T_2^F \\ M_2^F \end{bmatrix}
$$

$$(17.18a)$$

or
$$\mathbf{S} = \mathbf{Ks} + \mathbf{S}^F \tag{17.18b}$$

Note that in deriving the above relations a right-handed coordinate system is used with the y axis directed upward (Fig. 17.9). This convention must be followed in the solution of problems.

Transformation of Coordinates

Although grid elements are located in the same plane, they usually have arbitrary orientations in that plane. If the plane of the grid is the XY plane, with deflection taking place in the direction of the Y axis (Fig. 17.10), we can find the transformation matrix relating the forces in the local and global coordinates.

Referring to Fig. 17.10, the end forces in the local coordinates can be expressed in terms of their counterparts in the global coordinates. Thus,

$$V_1 = \bar{V}_1$$
$$T_1 = \bar{M}_{1X} \cos \alpha - \bar{M}_{1Z} \sin \alpha$$
$$M_1 = \bar{M}_{1X} \sin \alpha + \bar{M}_{1Z} \cos \alpha$$

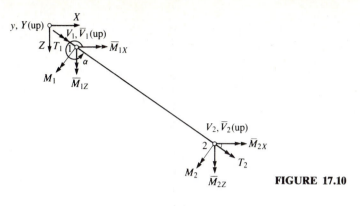

FIGURE 17.10

with

$$\cos \alpha = \frac{X_2 - X_1}{l} \quad \text{and} \quad \sin \alpha = \frac{-(Z_2 - Z_1)}{l}$$

We can write the above expressions in matrix form as

$$\begin{bmatrix} V_1 \\ T_1 \\ M_1 \end{bmatrix} = \begin{bmatrix} 1 & 0 & 0 \\ 0 & \cos \alpha & -\sin \alpha \\ 0 & \sin \alpha & \cos \alpha \end{bmatrix} \begin{bmatrix} \bar{V}_1 \\ \bar{M}_{1X} \\ \bar{M}_{1Z} \end{bmatrix} \tag{17.19a}$$

or

$$\mathbf{S}_1 = \mathbf{C}\bar{\mathbf{S}}_1 \tag{17.19b}$$

The relation for forces at end 2 is similar, as well as for displacements. Thus, for the forces at both ends, we have

$$\begin{bmatrix} V_1 \\ T_1 \\ M_1 \\ V_2 \\ T_2 \\ M_2 \end{bmatrix} = \begin{bmatrix} \mathbf{C} & \mathbf{0} \\ \mathbf{0} & \mathbf{C} \end{bmatrix} \begin{bmatrix} \bar{V}_1 \\ \bar{M}_{1X} \\ \bar{M}_{1Z} \\ \bar{V}_2 \\ \bar{M}_{2X} \\ \bar{M}_{2Z} \end{bmatrix} \tag{17.20a}$$

or

$$\mathbf{S} = \mathbf{T}\bar{\mathbf{S}} \tag{17.20b}$$

where $\mathbf{T}$ is the transformation matrix.

Note that in the above transformation matrix the y axes in both local (member) and global (structure) coordinates is in the upward direction. This convention must be followed in solving problems.

$\mathbf{T}$ is an orthogonal matrix and it is easy to verify that, $\mathbf{T}^{-1} = \mathbf{T}^T$. Thus,

$$\bar{\mathbf{S}} = \mathbf{T}^T \mathbf{S} \tag{17.20c}$$

Similarly, the relationships between the displacements in the local and global coordinates are

$$s = T\bar{s} \qquad (17.21a)$$

and

$$\bar{s} = T^T s \qquad (17.21b)$$

The member forces can be expressed in terms of the global joint displacements by substituting Eq. (17.21a) into (17.18b), resulting in

$$S = KT\bar{s} + S^F \qquad (17.22)$$

Premultiplying both sides of Eq. (17.22) by T^T and using Eq. (17.20c), we get

$$\bar{S} = \bar{K}\bar{s} + \bar{S}^F \qquad (17.23a)$$

where we have defined the global stiffness matrix as

$$\bar{K} = T^T K T \qquad (17.23b)$$

$$\bar{S}^F = T^T S^F \qquad (17.23c)$$

The force–displacement relation (Eq. (17.23a)) can be used in assembling the structure stiffness matrix and load vector. The solution procedure is otherwise similar to that of other structures using the stiffness method.

Example 17.6. Find the member-end forces for the two-bar grid shown in Fig. 17.11. Both span lengths are 10 ft with uniform gravity loads of 2 kip/ft acting over them. In addition, a concentrated load of 8 kips is acting downward at joint 2. For both members $I = 100$ in⁴; $J = 22.42$; $E = 29\,000$ ksi; $G = 11\,150$ ksi.

Solution. The degrees of freedom are the vertical displacement of joint 2 and its rotations about the X and Z axes. The force–displacement relations in the local coordinates are the same for both elements. By Eq. (17.18a) these are

$$
\begin{bmatrix} V_1 \\ T_1 \\ M_1 \\ V_2 \\ T_2 \\ M_2 \end{bmatrix}^1
=
\begin{bmatrix}
20 & 0 & 1\,208 & -20 & 0 & 1\,208 \\
0 & 2\,083 & 0 & 0 & -2\,083 & 0 \\
1\,208 & 0 & 96\,667 & -1\,208 & 0 & 48\,333 \\
-20 & 0 & -1\,208 & 20 & 0 & -1\,208 \\
0 & -2\,083 & 0 & 0 & 2\,083 & 0 \\
1\,208 & 0 & 48\,333 & -1\,208 & 0 & 96\,667
\end{bmatrix}
\begin{bmatrix} v_1 \\ \psi_1 \\ \theta_1 \\ v_2 \\ \psi_2 \\ \theta_2 \end{bmatrix}^1
+
\begin{bmatrix} 10 \\ 0 \\ 200 \\ 10 \\ 0 \\ -200 \end{bmatrix}
$$

$$
\begin{bmatrix} V_1 \\ T_1 \\ M_1 \\ V_2 \\ T_2 \\ M_2 \end{bmatrix}^2
=
\begin{bmatrix}
20 & 0 & 1\,208 & -20 & 0 & 1\,208 \\
0 & 2\,083 & 0 & 0 & -2\,083 & 0 \\
1\,208 & 0 & 96\,667 & -1\,208 & 0 & 48\,333 \\
-20 & 0 & -1\,208 & 20 & 0 & -1\,208 \\
0 & -2\,083 & 0 & 0 & 2\,083 & 0 \\
1\,208 & 0 & 48\,333 & -1\,208 & 0 & 96\,667
\end{bmatrix}
\begin{bmatrix} v_1 \\ \psi_1 \\ \theta_1 \\ v_2 \\ \psi_2 \\ \theta_2 \end{bmatrix}^2
+
\begin{bmatrix} 10 \\ 0 \\ 200 \\ 10 \\ 0 \\ -200 \end{bmatrix}
$$

The transformation matrices for the two elements are (Eqs. (17.20*a,b*) and (17.19*a*))

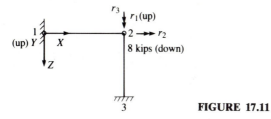

FIGURE 17.11

$$
\mathbf{T}_1 = \begin{bmatrix} 1 & 0 & 0 & 0 & 0 & 0 \\ 0 & 1 & 0 & 0 & 0 & 0 \\ 0 & 0 & 1 & 0 & 0 & 0 \\ 0 & 0 & 0 & 1 & 0 & 0 \\ 0 & 0 & 0 & 0 & 1 & 0 \\ 0 & 0 & 0 & 0 & 0 & 1 \end{bmatrix}
\qquad
\mathbf{T}_2 = \begin{bmatrix} 1 & 0 & 0 & 0 & 0 & 0 \\ 0 & 0 & 1 & 0 & 0 & 0 \\ 0 & -1 & 0 & 0 & 0 & 0 \\ 0 & 0 & 0 & 1 & 0 & 0 \\ 0 & 0 & 0 & 0 & 0 & 1 \\ 0 & 0 & 0 & 0 & -1 & 0 \end{bmatrix}
$$

As mentioned before, both the local and global *y* axes are upward. Upon transformation (Eq. (17.23*a*)) the force–displacement relations for the two elements in the global coordinates and the associated degrees of freedom are

$$
\begin{bmatrix} \bar{V}_1 \\ \bar{T}_1 \\ \bar{M}_1 \\ \bar{V}_2 \\ \bar{T}_2 \\ \bar{M}_2 \end{bmatrix}^1
=
\begin{bmatrix}
20 & 0 & 1\,208 & -20 & 0 & 1\,208 \\
0 & 2\,083 & 0 & 0 & -2\,083 & 0 \\
1\,208 & 0 & 96\,667 & -1\,208 & 0 & 48\,333 \\
-20 & 0 & -1\,208 & 20 & 0 & -1\,208 \\
0 & -2\,083 & 0 & 0 & 2\,083 & 0 \\
1\,208 & 0 & 48\,333 & -1\,208 & 0 & 96\,667
\end{bmatrix}
\begin{bmatrix} 0 \\ 0 \\ 0 \\ r_1 \\ r_2 \\ r_3 \end{bmatrix}
+
\begin{bmatrix} 10 \\ 0 \\ 200 \\ 10 \\ 0 \\ -200 \end{bmatrix}
$$

with column labels: 0, 0, 0, r_1, r_2, r_3

$$
\begin{bmatrix} \bar{V}_1 \\ \bar{T}_1 \\ \bar{M}_1 \\ \bar{V}_2 \\ \bar{T}_2 \\ \bar{M}_2 \end{bmatrix}^2
=
\begin{bmatrix}
20 & -1\,208 & 0 & -20 & -1\,208 & 0 \\
-1\,208 & 96\,667 & 0 & 1\,208 & 48\,333 & 0 \\
0 & 0 & 2\,083 & 0 & 0 & -2\,083 \\
-20 & 1\,208 & 0 & 20 & 1\,208 & 0 \\
-1\,208 & 48\,333 & 0 & 1\,208 & 96\,667 & 0 \\
0 & 0 & -2\,083 & 0 & 0 & 2\,083
\end{bmatrix}
\begin{bmatrix} r_1 \\ r_2 \\ r_3 \\ 0 \\ 0 \\ 0 \end{bmatrix}
+
\begin{bmatrix} 10 \\ -200 \\ 0 \\ 10 \\ 100 \\ 0 \end{bmatrix}
$$

with column labels: r_1, r_2, r_3, 0, 0, 0

After assembling the structure stiffness matrix and the load vector, we get the following equilibrium equations:

$$
\begin{bmatrix}
40 & -1\,208 & -1\,208 \\
-1\,208 & 98\,750 & 0 \\
-1\,208 & 0 & 98\,750
\end{bmatrix}
\begin{bmatrix} r_1 \\ r_2 \\ r_3 \end{bmatrix}
=
\begin{bmatrix} -28 \\ 200 \\ 200 \end{bmatrix}
$$

with column labels: r_1, r_2, r_3

The solution of the above equations yields the displacement vector:

$$\begin{bmatrix} r_1 \\ r_2 \\ r_3 \end{bmatrix} = \begin{bmatrix} -2.158\,0 \\ -0.024\,4 \\ -0.024\,4 \end{bmatrix}$$

With these values, the element displacement vectors in the global coordinates will be as indicated in the global force–displacement relations. Using the latter values, the member forces in the local coordinates are found by Eq. (17.22) as follows:

$$\text{Member} \rightarrow \qquad 1 \qquad\qquad 2$$

$$\mathbf{S} = \begin{bmatrix} V_1 \\ T_1 \\ M_1 \\ V_2 \\ T_2 \\ M_2 \end{bmatrix} : \begin{bmatrix} 240 \\ 508 \\ 1\,629 \\ -40 \\ -508 \\ 508 \end{bmatrix} \begin{bmatrix} -40 \\ -508 \\ -508 \\ 240 \\ 508 \\ -1\,629 \end{bmatrix}$$

17.8 BOW-GIRDERS

Bow-girders are curved beams that, unlike arches, deform out of their planes. In such beams, torsion is a significant parameter. Although three-dimensional beam elements can be used to model bow-girders by a series of straight segments, it is easier to model these structures by curved elements. In this section we derive the stiffness matrix for a circular bow-girder element.

Stiffness Matrix for Circular Bow-Girders

An element of a circular bow-girder in the horizontal plane is shown in Fig. 17.12. Again the vertical axis in the local coordinates, y, is the same as that in the global coordinates, Y.

We follow the same procedure used for arches (Section 15.3) to derive

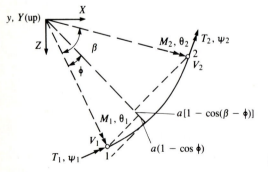

FIGURE 17.12

the stiffness matrix by Castigliano's theorem. Since bow-girders are under bending and torsion, their elastic energy is

$$U_S = \frac{a}{2GJ} \int_0^\beta T^2 \, d\phi + \frac{a}{2EI} \int_0^\beta M^2 \, d\phi \qquad (17.24)$$

Here a is the radius of the centerline of the girder, M and T are the moment and torque at an arbitrary angle ϕ, and β is the angle of the bow-girder. E, I, and J are the modulus of elasticity, moment of inertia, and torsion constant for the cross-section.

Calculated from the left-hand end, the torque and moment with respect to the tangent and normal axes at a point with angle ϕ are

$$T = V_1 a(1 - \cos \phi) + T_1 \cos \phi - M_1 \sin \phi \qquad (17.25a)$$

and

$$M = V_1 a \sin \phi + T_1 \sin \phi + M_1 \cos \phi \qquad (17.25b)$$

By Castigliano's second theorem (Eq. (6.22))

$$v_1 = \frac{\partial U_S}{\partial V_1} = \frac{a}{GJ} \int_0^\beta T \frac{\partial T}{\partial V_1} \, d\phi + \frac{a}{EI} \int_0^\beta M \frac{\partial M}{\partial V_1} \, d\phi \qquad (17.26a)$$

$$\psi_1 = \frac{\partial U_S}{\partial T_1} = \frac{a}{GJ} \int_0^\beta T \frac{\partial T}{\partial T_1} \, d\phi + \frac{a}{EI} \int_0^\beta M \frac{\partial M}{\partial T_1} \, d\phi \qquad (17.26b)$$

$$\theta_1 = \frac{\partial U_S}{\partial M_1} = \frac{a}{GJ} \int_0^\beta T \frac{\partial T}{\partial M_1} \, d\phi + \frac{a}{EI} \int_0^\beta M \frac{\partial M}{\partial M_1} \, d\phi \qquad (17.26c)$$

Substitution of Eqs. (17.25) into Eqs. (17.26) and carrying out the algebra, which may easily be done by a symbolic computer program (Section 19.1), gives the following relation, in terms of the symmetric flexibility matrix F:

$$\begin{bmatrix} v_1 \\ \psi_1 \\ \theta_1 \end{bmatrix} = \begin{bmatrix} F_{11} & F_{12} & F_{13} \\ F_{12} & F_{22} & F_{23} \\ F_{13} & F_{23} & F_{33} \end{bmatrix} \begin{bmatrix} V_1 \\ T_1 \\ M_1 \end{bmatrix} \qquad (17.27a)$$

or

$$\mathbf{s}_1 = \mathbf{F}_{11}\mathbf{S}_1 \qquad (17.27b)$$

with

$$F_{11} = \frac{a^3}{4EI} (6\Gamma\beta + 2\beta - \sin 2\beta - 8\Gamma \sin \beta + \Gamma \sin 2\beta) \qquad (17.28a)$$

$$F_{12} = \frac{a^2}{4EI} (-2\Gamma\beta - 2\beta + \sin 2\beta + 4\Gamma \sin \beta - \Gamma \sin 2\beta) \qquad (17.28b)$$

$$F_{13} = \frac{a^2}{4EI} (-1 + \cos 2\beta - 3\Gamma + 4\Gamma \cos \beta - \Gamma \cos 2\beta) \qquad (17.28c)$$

$$F_{22} = \frac{a}{4EI}(2\Gamma\beta + 2\beta - \sin 2\beta + \Gamma \sin 2\beta) \qquad (17.28d)$$

$$F_{23} = \frac{a}{4EI}(1 - \cos 2\beta - \Gamma + \Gamma \cos 2\beta) \qquad (17.28e)$$

$$F_{33} = \frac{a}{4EI}(2\Gamma\beta + 2\beta + \sin 2\beta - \Gamma \sin 2\beta) \qquad (17.28f)$$

$$\Gamma = \frac{EI}{GJ} \qquad (17.28g)$$

Similarly, if we express the torque and moment at ϕ in terms of the values at the right end, we get

$$T = -V_2 a[1 - \cos(\beta - \phi)] - T_2 \cos(\beta - \phi) - M_2 \sin(\beta - \phi) \quad (17.29a)$$
$$M = -V_2 a \sin(\beta - \phi) + T_2 \sin(\beta - \phi) - M_2 \cos(\beta - \phi) \qquad (17.29b)$$

Then Eqs. (17.26) give the right end displacements upon replacing the subscript 1 by 2, leading to

$$\begin{bmatrix} v_2 \\ \psi_2 \\ \theta_2 \end{bmatrix} = \begin{bmatrix} F_{11} & F_{12} & -F_{13} \\ F_{12} & F_{22} & -F_{23} \\ -F_{13} & -F_{23} & F_{33} \end{bmatrix} \begin{bmatrix} V_2 \\ T_2 \\ M_2 \end{bmatrix} \qquad (17.30a)$$

or

$$\mathbf{s}_2 = \mathbf{F}_{22}\mathbf{S}_2 \qquad (17.30b)$$

where F_{ij} were defined by Eqs. (17.28). The force–displacement relations are

$$\begin{bmatrix} V_1 \\ T_1 \\ M_1 \\ V_2 \\ T_2 \\ M_2 \end{bmatrix} = \begin{bmatrix} \mathbf{K}_{11} & \mathbf{K}_{12} \\ \mathbf{K}_{21} & \mathbf{K}_{22} \end{bmatrix} \begin{bmatrix} v_1 \\ \psi_1 \\ \theta_1 \\ v_2 \\ \psi_2 \\ \theta_2 \end{bmatrix} \qquad (17.31a)$$

or

$$\begin{bmatrix} \mathbf{S}_1 \\ \mathbf{S}_2 \end{bmatrix} = \begin{bmatrix} \mathbf{K}_{11} & \mathbf{K}_{12} \\ \mathbf{K}_{21} & \mathbf{K}_{22} \end{bmatrix} \begin{bmatrix} \mathbf{s}_1 \\ \mathbf{s}_2 \end{bmatrix} \qquad (17.31b)$$

with

$$\mathbf{K}_{11} = \mathbf{F}_{11}^{-1} \qquad (17.31c)$$

and

$$\mathbf{K}_{22} = \mathbf{F}_{22}^{-1} \qquad (17.31d)$$

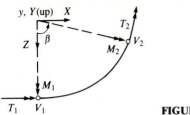

FIGURE 17.13

$\mathbf{K}_{12}$ is the transpose of $\mathbf{K}_{21}$. To find $\mathbf{K}_{21}$ we can use the equilibrium equations, $\Sigma F_y = 0$, $\Sigma T_1 = 0$, $\Sigma M_1 = 0$. Referring to Figs. 17.12 and 17.13,

$$V_1 + V_2 = 0$$

$$T_1 + V_2 a(1 - \cos \beta) + M_2 \sin \beta + T_2 \cos \beta = 0$$

$$M_1 + V_2 a \sin \beta + M_2 \cos \beta - T_2 \sin \beta = 0$$

$$\begin{bmatrix} V_1 \\ T_1 \\ M_1 \end{bmatrix} = \begin{bmatrix} -1 & 0 & 0 \\ -a(1 - \cos \beta) & -\cos \beta & -\sin \beta \\ -a \sin \beta & \sin \beta & -\cos \beta \end{bmatrix} \begin{bmatrix} V_2 \\ T_2 \\ M_2 \end{bmatrix} \qquad (17.32)$$

Inversion of the above equation gives

$$\begin{bmatrix} V_2 \\ T_2 \\ M_2 \end{bmatrix} = \begin{bmatrix} -1 & 0 & 0 \\ -a(1 - \cos \beta) & -\cos \beta & \sin \beta \\ a \sin \beta & -\sin \beta & -\cos \beta \end{bmatrix} \begin{bmatrix} V_1 \\ T_1 \\ M_1 \end{bmatrix} \qquad (17.33a)$$

or

$$\mathbf{S}_2 = \mathbf{Q}\mathbf{S}_1 \qquad (17.33b)$$

By Eqs. (17.27b) and (17.31c),

$$\mathbf{S}_2 = \mathbf{Q}\mathbf{K}_{11}\mathbf{s}_1$$

and we see from the latter equation and Eq. (17.31b) that

$$\mathbf{K}_{21} = \mathbf{Q}\mathbf{K}_{11} \qquad (17.34)$$

Thus, the complete stiffness matrix is obtained.

Transformation of Coordinates

Consider a segment of a circular bow-girder, with the angle α at the beginning of the segment, with respect to the Z axis of a right-handed coordinate system, and the angle β of the bow girder (Fig. 17.14).

Expressing the forces in the local coordinates in terms of their com-

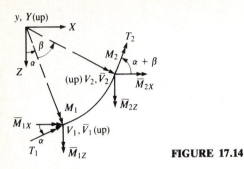

FIGURE 17.14

ponents in the global coordinates, we get

$$V_1 = \bar{V}_1$$

$$T_1 = \bar{M}_{1X} \cos \alpha - \bar{M}_{1Z} \sin \alpha$$

$$M_1 = \bar{M}_{1X} \sin \alpha + \bar{M}_{1Z} \cos \alpha$$

$$V_2 = \bar{V}_2$$

$$T_2 = \bar{M}_{2X} \cos(\alpha + \beta) - \bar{M}_{2Z} \sin(\alpha + \beta)$$

$$M_2 = \bar{M}_{2X} \sin(\alpha + \beta) + \bar{M}_{2Z} \cos(\alpha + \beta)$$

or

$$
\begin{bmatrix} V_1 \\ T_1 \\ M_1 \\ V_2 \\ T_2 \\ M_2 \end{bmatrix} =
\begin{bmatrix}
1 & 0 & 0 & 0 & 0 & 0 \\
0 & \cos \alpha & -\sin \alpha & 0 & 0 & 0 \\
0 & \sin \alpha & \cos \alpha & 0 & 0 & 0 \\
0 & 0 & 0 & 1 & 0 & 0 \\
0 & 0 & 0 & 0 & \cos(\alpha + \beta) & -\sin(\alpha + \beta) \\
0 & 0 & 0 & 0 & \sin(\alpha + \beta) & \cos(\alpha + \beta)
\end{bmatrix}
\begin{bmatrix} \bar{V}_1 \\ \bar{M}_{1X} \\ \bar{M}_{1Z} \\ \bar{V}_2 \\ \bar{M}_{2X} \\ \bar{M}_{2Z} \end{bmatrix}
$$

$$(17.35a)$$

or

$$\mathbf{S} = \mathbf{T}\bar{\mathbf{S}} \qquad\qquad (17.35b)$$

The force–displacement relations in the global coordinates can be found in a similar manner to Eqs. (17.20c) to (17.23c). The solution procedure is also similar to that of grids and other structures.

Example 17.7. The structure of Fig. 17.15 is composed of a bow-girder and a grid element. The length of the grid element is 10 m. The angle of the bow-girder is 90° and its radius is 5 m. A gravity load of 75 kN is acting at the intersection of the two elements. For both members $E = 200\,000$ MPa and $I = 2\,500$ cm⁴, $J = 900$ cm⁴, and $G = 80\,000$ MPa. Find the displacement at the intersection and the member-end forces.

Solution. There are three degrees of freedom at joint 2—the vertical displacement, rotation about the X axis, and rotation about the Z axis. The first and

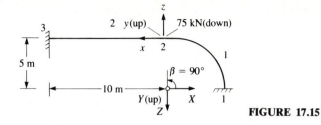

FIGURE 17.15

second joints of the bow-girder element are chosen such that the angle β is counterclockwise. The force–displacement relations for the bow-girder in its local coordinates are calculated by Eq. (17.31a) as

$$
\begin{bmatrix} V_1 \\ T_1 \\ M_1 \\ V_2 \\ T_2 \\ M_2 \end{bmatrix}^1 =
\begin{bmatrix}
91.3 & 112.4 & 344.3 & -91.3 & -112.4 & 344.3 \\
112.4 & 345.6 & 522.4 & -112.4 & -39.6 & 216.4 \\
344.3 & 522.4 & 1\,505.0 & -344.3 & -216.4 & 1\,198.9 \\
-91.3 & -112.4 & -344.3 & 91.3 & 112.4 & -344.3 \\
-112.4 & -39.6 & -216.4 & 112.4 & 345.6 & -522.4 \\
344.3 & 216.4 & 1\,198.9 & -344.3 & -522.4 & 1\,505.0
\end{bmatrix}
\begin{bmatrix} v_1 \\ \psi_1 \\ \theta_1 \\ v_2 \\ \psi_2 \\ \theta_2 \end{bmatrix}^1
$$

Member 2 is a grid element whose force–displacement relation in the local coordinates is, by eq. (17.18a),

$$
\begin{bmatrix} V_1 \\ T_1 \\ M_1 \\ V_2 \\ T_2 \\ M_2 \end{bmatrix}^2 =
\begin{bmatrix}
60 & 0 & 300 & -60 & 0 & 300 \\
0 & 72 & 0 & 0 & -72 & 0 \\
300 & 0 & 2\,000 & -300 & 0 & 1\,000 \\
-60 & 0 & -300 & 60 & 0 & -300 \\
0 & -72 & 0 & 0 & 72 & 0 \\
300 & 0 & 1\,000 & -300 & 0 & 2\,000
\end{bmatrix}
\begin{bmatrix} v_1 \\ \psi_1 \\ \theta_1 \\ v_2 \\ \psi_2 \\ \theta_2 \end{bmatrix}^2
$$

The transformation matrices for the two elements are as follows (Eqs. (17.35a) and (17.20a)). For the first element $\alpha = 90°$ and $\alpha + \beta = 180°$.

$$
\mathbf{T}_1 =
\begin{bmatrix}
1 & 0 & 0 & 0 & 0 & 0 \\
0 & 0 & -1 & 0 & 0 & 0 \\
0 & 1 & 0 & 0 & 0 & 0 \\
0 & 0 & 0 & 1 & 0 & 0 \\
0 & 0 & 0 & 0 & -1 & 0 \\
0 & 0 & 0 & 0 & 0 & -1
\end{bmatrix}
$$

$$
\mathbf{T}_2 =
\begin{bmatrix}
1 & 0 & 0 & 0 & 0 & 0 \\
0 & -1 & 0 & 0 & 0 & 0 \\
0 & 0 & -1 & 0 & 0 & 0 \\
0 & 0 & 0 & 1 & 0 & 0 \\
0 & 0 & 0 & 0 & -1 & 0 \\
0 & 0 & 0 & 0 & 0 & -1
\end{bmatrix}
$$

The force–displacement relations in the global coordinates are

$$\begin{bmatrix} \bar{V}_1 \\ \bar{M}_{1X} \\ \bar{M}_{1Z} \\ \bar{V}_2 \\ \bar{M}_{2X} \\ \bar{M}_{2Z} \end{bmatrix}^1 = \begin{matrix} 0 & 0 & 0 & r_1 & r_2 & r_3 \\ \begin{bmatrix} 91.3 & 344.3 & -112.4 & -91.3 & 112.4 & -344.3 \\ 344.3 & 1\,505.2 & -522.4 & -344.3 & 216.4 & -1\,198.9 \\ -112.4 & -522.4 & 345.6 & 112.4 & -39.6 & 216.4 \\ -91.3 & -344.3 & 112.4 & 91.3 & -112.4 & 344.3 \\ 112.4 & 216.4 & -39.6 & -112.4 & 345.6 & -522.4 \\ -344.3 & -1\,198.9 & 216.4 & 344.3 & -522.4 & 1\,505.0 \end{bmatrix} \end{matrix} \begin{bmatrix} 0 \\ 0 \\ 0 \\ r_1 \\ r_2 \\ r_3 \end{bmatrix}$$

$$\begin{bmatrix} \bar{V}_1 \\ \bar{M}_{1X} \\ \bar{M}_{1Z} \\ \bar{V}_2 \\ \bar{M}_{2X} \\ \bar{M}_{2Z} \end{bmatrix}^2 = \begin{matrix} r_1 & r_2 & r_3 & 0 & 0 & 0 \\ \begin{bmatrix} 60 & 0 & -300 & -60 & 0 & -300 \\ 0 & 72 & 0 & 0 & -72 & 0 \\ -300 & 0 & 2\,000 & 300 & 0 & 1\,000 \\ -60 & 0 & 300 & 60 & 0 & 300 \\ 0 & -72 & 0 & 0 & 72 & 0 \\ -300 & 0 & 1\,000 & 300 & 0 & 2\,000 \end{bmatrix} \end{matrix} \begin{bmatrix} r_1 \\ r_2 \\ r_3 \\ 0 \\ 0 \\ 0 \end{bmatrix}$$

The equilibrium equations are

$$\begin{bmatrix} 151.3 & -112.4 & 044.3 \\ -112.4 & 417.6 & -522.4 \\ 44.3 & -522.4 & 3\,505.0 \end{bmatrix} \begin{bmatrix} r_1 \\ r_2 \\ r_3 \end{bmatrix} = \begin{bmatrix} -75 \\ 0 \\ 0 \end{bmatrix}$$

whose solution gives the displacements

$$\begin{bmatrix} r_1 \\ r_2 \\ r_3 \end{bmatrix} = \begin{bmatrix} -0.6365 \\ -0.1982 \\ -0.0215 \end{bmatrix}$$

With these values, the member-end forces in the local coordinates can be found as $\mathbf{S} = \mathbf{KT\bar{s}}$, or

$$\begin{matrix} \text{Member} \rightarrow & 1 & 2 \\ \begin{bmatrix} V_1 \\ T_1 \\ M_1 \\ V_2 \\ T_2 \\ M_2 \end{bmatrix} : & \begin{bmatrix} 43.26 \\ 68.35 \\ 202.03 \\ -43.26 \\ -14.27 \\ 147.94 \end{bmatrix} & \begin{bmatrix} -31.73 \\ 14.27 \\ -147.94 \\ 31.73 \\ -14.27 \\ -169.45 \end{bmatrix} \end{matrix}$$

LABORATORY EXPERIMENTS

Open and Closed Sections

Use tubular square or round sections with about 1/2 in dimensions of the cross-section and 2 ft length and provide end supports so that one end is fixed and the other can twist without otherwise displacing. Apply a torque (via a

FIGURE L17.1

loaded lever) to the end that is free to twist (Fig. L17.1). Compare the twisting angle for the same torque for tubular sections with and without a slit along the length. Observe the distortion of the end cross-sections.

Compare the measured values of the twisting angle for the given torque with the theoretical value. Note that J is the polar moment of inertia for the tubular section and $J = \frac{1}{3}bt^3$ for the open section (the one with slit), where b is the perimeter of the section and t is its thickness.

Structures Under Torsion

Build models of structures in Figs. P17.2, P17.6, P17.7, and P17.10 and load them with symmetrical and unsymmetrical loads. Observe the effect of the torques on the structures and compare the results to the analytically found values.

PROBLEMS

Draw the torque diagrams for the following structures.

17.1.

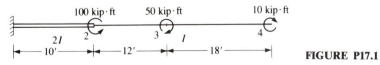

FIGURE P17.1

17.2.

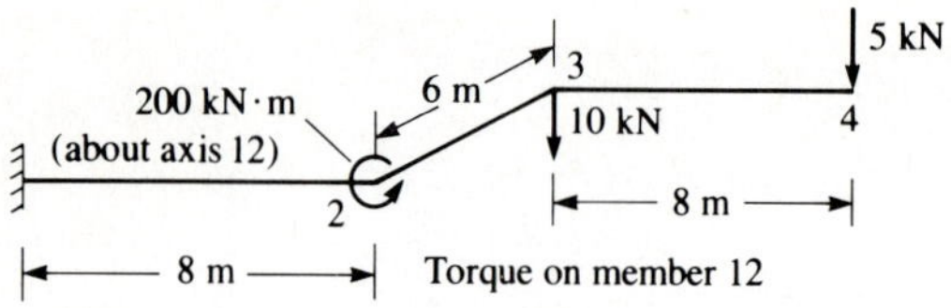

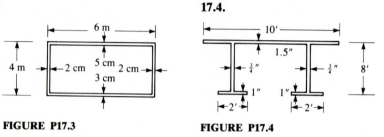

FIGURE P17.2

Find the torsion constants for the following sections.

17.3.   **17.4.**

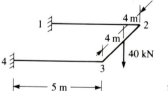

FIGURE P17.3   FIGURE P17.4

Find the joint displacements and the member-end forces for the following structures by the stiffness method.

17.5. Solid circular sections $D = 5$ in and 2.5 in under the torques shown, $E = 29\,000$ ksi, $G = 11\,200$ ksi.

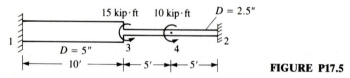

FIGURE P17.5

17.5. Solid circular sections with $D = 5$ in and 2.5 in under the torques shown, $E = 29\,000$ ksi, $G = 11\,200$ ksi.

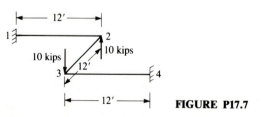

FIGURE P17.6

17.7. $I = 200$ in^4, $J = 100$ in^4 for member 23; $I = 150$ in^4, $J = 90$ in^4 for members 12, 34. $E = 29\,000$ ksi; $G = 11\,200$ ksi.

FIGURE P17.7

17.8. $I = 0.2$ ft^4, $J = 0.1$ ft^4 for member 12; $I = 0.3$ ft^4 $J = 0.25$ ft^4 for members 35, 68. $E = 3\,100$ ksi; $v = 0.2$.

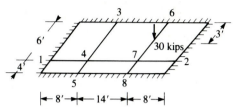

FIGURE P17.8

17.9. $I = 0.04$ m^4, $J = 0.02$ m^4 $E = 20\,000$ MPa, $v = 0.2$ for all members.

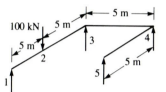

FIGURE P17.9

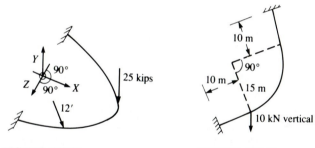

FIGURE P17.10

FIGURE P17.11

17.10. $I = 170$ in^4, $J = 80$ in^4, $E = 29\,000$ ksi, $v = 0.3$.

17.11. $I = 10\,000$ cm^4, $J = 3\,000$ cm^4, $E = 200\,000$ MPa, $v = 0.3$.

17.12. $I = 200$ in^4, $J = 100$ in^4, $E = 29\,000$ ksi, $G = 11\,200$ ksi.
Points 1, 2 and 3 are supports while point 4 is the location of the 20 kip vertical load.

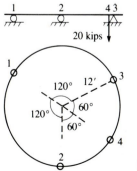

FIGURE P17.12

CHAPTER
18

THREE-DIMENSIONAL TRUSSES AND FRAMES

Unlike two-dimensional structures, where all the members and forces are in the same plane, the members and forces in three-dimensional structures are not in the same plane but in a three-dimensional space. In this chapter we discuss three-dimensional trusses and frames. As mentioned in Chapter 3 for two-dimensional trusses, truss members may not necessarily be connected by hinges, but may be welded or bolted. However, whenever the members primarily develop axial forces, the structure can be analyzed as a truss. This is the case for structures with slender members with the loads applied at the joints. Statically determinate three-dimensional trusses and frames may be analyzed by statics. Simple statically indeterminate trusses and frames may be analyzed by the method of consistent deformations in conjunction with energy methods for finding the deflections. Such a three-dimensional structure was analyzed in Chapter 6 (Example 6.10).

For complex and statically indeterminate trusses and frames the stiffness method, discussed in Sections 18.2 and 18.3, provides a more suitable solution technique.

18.1 STATICS OF THREE-DIMENSIONAL TRUSSES AND FRAMES

In this section the equations of statics and the methods based on application of these equations are discussed for three-dimensional structures. These include

504

the equilibrium equations, determination of reaction components, internal forces in frames, and the method of joints for trusses.

Force Components

The forces applied to three-dimensional structures act in a three-dimensional space. A force in space can be specified by its magnitude, direction, and line of action, or by its components along three mutually perpendicular coordinate axes.

As shown in Fig. 18.1, the line of action of an arbitrary force makes three planes with the X, Y, and Z axes. The angles θ_X, θ_Y, and θ_Z in these planes between the force P and its components P_X, P_Y, and P_Z are called the *direction angles*. Since for trusses the forces act along the axes of the members, the direction angles are the angles of the axes of the members with the coordinate axes. The cosines of the direction angles are called *direction cosines* and are used in calculating the force components. The direction cosines can be found in terms of the coordinates of the joints, the end points of the members. For a member that is connected to joints 1 and 2, the direction cosines in terms of the coordinates of these joints are

$$\cos \theta_X = \frac{X_2 - X_1}{l} = \frac{l_X}{l} \tag{18.1a}$$

$$\cos \theta_Y = \frac{Y_2 - Y_1}{l} = \frac{l_Y}{l} \tag{18.1b}$$

$$\cos \theta_Z = \frac{Z_2 - Z_1}{l} = \frac{l_Z}{l} \tag{18.1c}$$

and the length of the member

$$l = [(X_2 - X_1)^2 + (Y_2 - Y_1)^2 + (Z_2 - Z_1)^2]^{1/2} \tag{18.1d}$$

Therefore, the direction cosines are the ratios of the projected lengths and the length of the member.

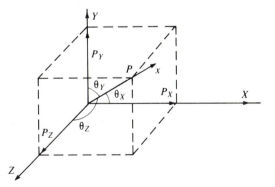

FIGURE 18.1

The components of an axial force in terms of the direction cosines are

$$P_X = P \cos \theta_X = P \frac{l_X}{l} \qquad (18.2a)$$

$$P_Y = P \cos \theta_Y = P \frac{l_Y}{l} \qquad (18.2b)$$

$$P_Z = P \cos \theta_Z = P \frac{l_Z}{l} \qquad (18.2c)$$

and

$$P = [P_X^2 + P_Y^2 + P_Z^2]^{1/2}$$

From the above equations it is easy to see that the force components are proportional to the projected lengths, that is,

$$\frac{P_X}{l_X} = \frac{P_Y}{l_Y} = \frac{P_Z}{l_Z} = \frac{P}{l} \qquad (18.3)$$

Thus, if we know one component of the force, the other two can be found by geometry. Similarly, the moments acting about an arbitrary axis can be decomposed into three components about three mutually orthogonal axes. These components will have the same form as the forces in Eqs. (18.2).

For three-dimensional structures the equilibrium equations consist of equilibrium of forces in three mutually perpendicular directions, plus equilibrium of moments about three mutually perpendicular axes. If the coordinate axes are X, Y, and Z we must have

$$\sum F_X = 0 \qquad \sum F_Y = 0 \qquad \sum F_Z = 0 \qquad (18.4a,b,c)$$

and

$$\sum M_X = 0 \qquad \sum M_Y = 0 \qquad \sum M_Z = 0 \qquad (18.4d,e,f)$$

These equations may be used to find the reactions as well as the bar forces in statically determinate trusses and the axial force, shear, moment, and torque in three-dimensional beams and frames. Three-dimensional frame members may experience bending about two axes as well as torsion.

It should be noted that instead of using the force components the forces can be expressed in vector form. This is done in Section 18.3 for finding expressions for direction cosines. The equilibrium equation would then be $\sum \mathbf{F} = 0$ and $\sum \mathbf{M} = 0$ where the last equation includes the moment of forces $\mathbf{r} \times \mathbf{F}$. In the latter cross product $\mathbf{r}$ is the position vector for the force vector $\mathbf{F}$.

Support Conditions and Reactions

For three-dimensional trusses and frames, sufficient support restraints must be provided in order to produce a stable structure. Restraints may be against

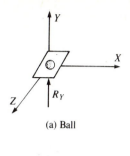

(a) Ball

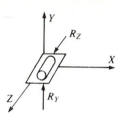

(b) Roller

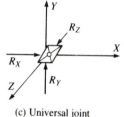

(c) Universal joint

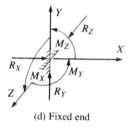

(d) Fixed end

FIGURE 18.2

displacements in any of the three mutually perpendicular directions or against rotation about any of those axes. The reactions due to support restraints are determined by applying the equations of statics and, if applicable, equations of condition. As discussed in Chapter 2, equations of condition are set up for portions of the structure connected together with hinges or other devices that transfer only some force or moment components. Support restraints may be provided by balls, rollers, universal joints, or fixed ends, as described below.

Balls. The support provided by a ball allows displacements in two directions while developing a reaction in the third direction.

For example, the ball in Fig. 18.2a can move in the X and Z directions but no displacement is possible in the Y direction. Thus, there is a reaction R_Y in the Y direction. Similarly, balls may provide restraints in the X and Z directions.

Rollers. A roller allows displacement in one direction only. Figure 18.2b shows a roller that can displace in the X direction. The reactions R_Y and R_Z are developed in the Y and Z directions.

Universal joints. A universal joint (Fig. 18.2c) does not allow displacement in any direction but allows rotations about three axes. It therefore induces three reactions R_X, R_Y, and R_Z.

Fixed ends. Fixed ends allow no displacement or rotation in the direction of or about an axis (Fig. 18.2d). Thus, an end may be fixed against bending moments but not against torque or against moment in one plane.

Determinacy

Equations of statics are used to find the unknown reactions. In addition, if the structure consists of parts connected together by hinges or rollers, then additional equations (equations of condition) can be written as discussed in Chapter 2. For example, if two parts of a truss are connected together by a universal joint, then the sum of the moments of the forces acting on one side of this joint calculated at the joint must be zero about the three axes. This would lead to three equations of condition. If the number of equations of condition, c, plus 6 (the number of equations of statics) is less than the number of unknown reaction components, r, then the structure is statically indeterminate. The solution of such structures requires the consideration of displacements. Indeterminacy in this case is due to reactions (external forces). Therefore, a structure that is indeterminate because of redundant reactions is called *externally indeterminate*. As discussed in Section 2.4, for a stable structure the supports must be properly designed. In other words, a system may satisfy the above conditions but still be unstable, because of its support conditions. In addition, a truss may also be indeterminate because the bar forces cannot be determined by the equations of statics alone. In this case the structure is internally indeterminate.

To establish internal determinacy we can refer to the method of joints discussed below. In this method, a joint with the bars connecting to it is separated as a free body. Since all the forces pass through the same point (the joint), the moment equilibrium equations are identically satisfied. Thus, there are three force equations allowing only three unknown forces or reactions to be determined at each joint. If there are j joints we have $3j$ equations. For internal determinacy this must be equal to the number of bars plus the number of unknown reactions. On the basis of the above discussions the following conditions can be established for external and internal determinacy of three dimensional structures.

External determinacy

If	$r < 6 + c$	unstable	(18.5a)
If	$r = 6 + c$	statically determinate	(18.5b)
If	$r > 6 + c$	statically indeterminate externally	(18.5c)

r is the number of reactions and c is the number of equations of condition.

Internal determinacy for trusses

If	$b + r < 3j$	unstable	(18.5d)
If	$b + r = 3j$	statically determinate	(18.5e)
If	$b + r > 3j$	statically indeterminate internally	(18.5f)

b is the number of bars and j is the number of truss joints.

The above conditions are necessary but not sufficient. In other words, a truss may satisfy those conditions and still be unstable. If in doubt, a simple reduced-scale model will help determine whether the system is stable.

Method of Joints

As for two-dimensional trusses, in the method of joints a joint with the members connecting to it is isolated and the unknown forces are found by the application of the equilibrium equations. Since all the members converge at the joints, the moment equations are identically satisfied and only the three force equations are available. Thus, only three unknown bar forces per joint can be determined. The solution process must thus start at a joint with only three unknown forces and proceed to other joints.

To find the bar forces, we separate each joint with its bars and apply the equilibrium equations. The unknown forces are shown by their components. These components can always be indicated in the positive directions; that is, assuming tensile bar forces, with joint forces pulling on the joints. Negative values of forces would then indicate compression in the member. Since it may be confusing to draw all three components of the forces in the same sketch, it would be easier to show all the bars attached at the joint but indicate the force components in one plane at a time. Thus, even the bars that are not in the plane of the figure, but have force components parallel to the plane, will be indicated in the sketch. In this way, two sketches may be sufficient for showing all the force components, say xz and xy planes. The calculations are best performed in a table, where the member lengths and their projections along the three coordinate axes are recorded first. The force components are recorded as they are calculated, and other components are found by Eq. (18.4). Such a table is shown in Example 18.1, Table 18.1.

An alternative for applying the joint method to three-dimensional trusses is to express the forces in vector form, in terms of unit base vectors. The force components in the directions of the coordinates are then found by the dot product of the force vector with the unit base vectors. The calculation in terms of the force components is easier to perform and this procedure is used here.

Example 18.1. Find the reactions and the bar forces for the truss of Fig. 18.3a. The truss is in the form of a tripod, with an equilateral triangle at the base. The length of the members at the base is 20 ft. The loads consist of a -40-kip load in the Z direction and a 20-kip load in the X-direction at joint 4. The perspective view is shown in Fig. 18.3a, and Fig. 18.3b shows the vertical and horizontal views. The supports 1 and 2 are rollers allowing displacements in the X direction, while the roller at support 3 allows displacement in the Y direction.

Solution. The origin of the coordinates is taken at the centroid of the base triangle. The three rollers provide six reactions. By Eq. (18.5b) we can see that the structure is statically determinate externally. Furthermore, since there are 6 bars and 4 joints, by Eq. (18.5e) the truss is internally determinate. Therefore, the reactions can be found by the equations of statics and the bar forces can be

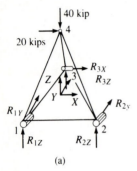

(a)

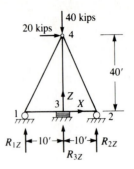

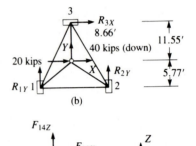

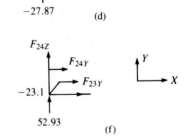

(b)

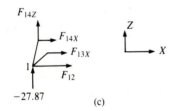

(c)

(d)

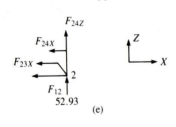

(e)

(f)

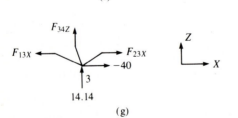

(g)

FIGURE 18.3

determined by the method of joints. Applying the equations of statics we get,

$$\sum F_X = 0 \qquad R_{3X} + 20 = 0$$

$$\sum F_Y = 0 \qquad R_{1Y} + R_{2Y} = 0$$

$$\sum F_Z = 0 \qquad R_{1Z} + R_{2Z} + R_{3Z} - 40 = 0$$

and

$$\sum M_X = 0, \text{ about line } 1\text{--}2,$$

$$R_{3Z} \times 17.32 - 40 \times 5.77 = 0$$

$\sum M_Y = 0$, about a line parallel to Y through joint 1,

$$-R_{2Z} \times 20 - R_{3Z} \times 10 + 20 \times 40 + 40 \times 10 = 0$$

$\sum M_Z = 0$, about a vertical line through joint 4,

$$-R_{1Y} \times 10 + R_{2Y} \times 10 - R_{3X} \times 11.55 = 0$$

These yield the six reaction components as follows:

$$R_{3X} = -20 \text{ kips} \qquad R_{1Y} = 23.10 \text{ kips} \qquad R_{2Y} = -23.1 \text{ kips}$$

$$R_{1Z} = -27.07 \text{ kips} \qquad R_{2Z} = 52.93 \text{ kips} \qquad R_{3Z} = 14.14 \text{ kips}$$

To calculate the bar forces, we first find the projections of the member lengths in the three coordinate directions. These are shown in Table 18.1. Starting with joint 1 (Figs. 18.3c,d) equilibrium of forces gives

$$\sum F_X = 0 \qquad F_{12} + F_{13X} + F_{14X} = 0$$

$$\sum F_Z = 0 \qquad -27.07 + F_{14Z} = 0$$

$$\sum F_Y = 0 \qquad 23.1 + F_{13Y} + F_{14Y} = 0$$

Expressing the components in terms of the forces, the above equations

TABLE 18.1

Member	Length Projections				Force components			
	l_X	l_Y	l_Z	Length l	F_X	F_Y	F_Z	Force, F
12	20	0	0	20	−44.46	0	0	−44.46
13	10	17.32	0	20	−28.83	−49.94	0	−57.67
14	101	5.77	40	41.63	46.78	26.99	0	194.75
23	10	17.32	0	20	−4.68	−8.10	0	−9.35
24	10	5.77	40	41.63	−13.25	−7.64	52.98	−55.14
34	0	11.55	40	41.63	0	−4.09	−14.14	−14.73

become

$$F_{12} + \frac{l_{13X}}{l_{13}} F_{13} + \frac{l_{14X}}{l_{14}} F_{14} = 0$$

$$-27.07 + \frac{l_{14Z}}{l_{14}} F_{14} = 0$$

$$23.1 + \frac{l_{13Y}}{l_{13}} F_{13} + \frac{l_{14Y}}{l_{14}} F_{14} = 0$$

or, with the lengths and their projection from Table 18.1,

$$F_{12} + 0.5F_{13} + 0.24F_{14} = 0$$

$$-27.07 + 0.139F_{14} = 0$$

$$23.1 + 0.87F_{13} + 0.139F_{14} = 0$$

This leads to $F_{12} = -44.46$ kips, $F_{13} = -4.56$ kips, and $F_{14} = 194.75$ kips.

These forces are recorded in the table and their components are found in proportion to the projections of the length of the member. Next we consider joint 2 (Figs. 18.3e,f). From Fig. 18.3e,

$$\sum F_X = 0 \qquad -F_{12} - F_{23X} - F_{24X} = 0$$

$$\sum F_Z = 0 \qquad 52.93 + F_{24Z} = 0$$

or

$$-F_{12} - \frac{l_{23X}}{l_{23}} F_{23} - \frac{l_{24X}}{l_{24}} F_{24} = 0$$

$$52.93 + \frac{l_{24Z}}{l_{24}} F_{24} = 0$$

Substituting for F_{12} (which has already been found), and for the lengths, we get

$$44.46 + 0.5F_{23} + 0.24F_{24} = 0$$

$$52.93 + 0.96F_{24} = 0$$

These two equations give $F_{23} = -9.35$ kips and $F_{24} = -55.14$ kips. Finally, equilibrium of joint 3 in the Z direction (Fig. 18.3g) gives the last unknown bar force F_{34},

$$14.14 + F_{34Z} = 0 \qquad \text{or} \qquad 14.14 + \frac{l_{34Z}}{l_{34}} F_{34} = 0.$$

$$14.14 + 0.96F_{34} = 0$$

or

$$F_{34} = -14.73 \text{ kips.}$$

Normal Force, Shear, Moment, and Torque Diagrams

To find the internal force and moment diagrams, we consider a free body of the structure, indicate the internal forces in the positive directions and find their values by establishing the equilibrium of the free body. Since bending can take place in two planes, shear and moment diagrams may have to be drawn in two planes. The diagrams for both planes may be combined in one sketch or drawn separately. In addition, local coordinates x, y, z may be used for each member to establish where positive values should be drawn. Alternatively, positive values may be drawn on the tension side of the member.

Example 18.2. Find the reactions of the structure in Fig. 18.4a. The left end is fixed against displacement and rotation in all three directions. Also draw the normal force diagram, shear and moment diagrams in two planes, and torque diagram for the structure.

Solution. There are six reactions at the left end, point 1. Therefore, the six equations of statics are sufficient for finding the reactions:

$$\sum F_X = 0 \qquad\qquad R_X + 50 = 0 \qquad R_X = -50 \text{ kN}$$

$$\sum F_Y = 0 \qquad\qquad R_Y - 20 = 0 \qquad R_Y = 20 \text{ kN}$$

$$\sum F_Z = 0 \qquad R_Z - 15 \times 5 - 100 = 0 \qquad R_Z = 175 \text{ kN}$$

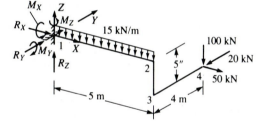

(a)

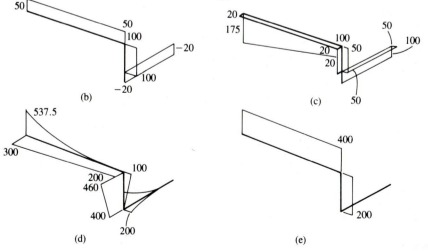

(b)

(c)

(d)

(e)

FIGURE 18.4

The moment equations, with the sum of moments about X, Y, and Z axes, calculated at the left end, give

$$M_X - 100 \times 4 - 200 \times 3 = 0 \qquad\qquad M_X = 460 \text{ kN} \cdot \text{m}$$

$$M_Y + 15 \times (5^2/2) + 100 \times 5 - 50 \times 3 = 0 \qquad M_Y = 537.5 \text{ kN} \cdot \text{m}$$

$$M_Z - 20 \times 5 + 50 \times 4 = 0 \qquad\qquad M_Z = -100 \text{ kN} \cdot \text{m}$$

The normal force, shear, and moment are found by considering free body diagrams and calculating the sum of these actions on one side of the section. Figure 18.4*b* shows the normal force diagram. Figures 18.4*c,d* show the shear and moment diagrams in both *xy* and *xz* planes. The diagrams are plotted in the plane of bending, with the positive values on the tension side. Figure 18.4*e* shows the torque diagram.

18.2   THREE-DIMENSIONAL TRUSSES—STIFFNESS METHOD

Analysis of three-dimensional trusses by the stiffness method is similar to application of this method to other structures already discussed. The first step in the process involves the determination of the element stiffness matrix in the local coordinates and transformation of the coordinates to obtain the element stiffness matrix and force-displacement relation in the global coordinates.

Stiffness Matrix for Three-Dimensional Truss Elements

The bar forces and displacements in the local (element) coordinates are shown in Fig. 18.5. The stiffness matrix is obtained by finding the relationship between the end forces and displacements. This is the same as for elements of plane trusses derived in Section 12.1. The force–displacement equation given by Eq. (12.1*a*) is

$$\begin{bmatrix} U_1 \\ U_2 \end{bmatrix} = \frac{EA}{l} \begin{bmatrix} 1 & -1 \\ -1 & 1 \end{bmatrix} \begin{bmatrix} u_1 \\ u_2 \end{bmatrix} \tag{18.6}$$

or

$$\mathbf{S} = \mathbf{Ks} \tag{18.6b}$$

with

$$\mathbf{K} = \frac{EA}{l} \begin{bmatrix} 1 & -1 \\ -1 & 1 \end{bmatrix} \tag{18.6c}$$

U_1, u_1 U_2, u_2

1   2   **FIGURE 18.5**

Transformation Matrix

The force U_1 has three components $\bar{U}_1$, $\bar{V}_1$, and $\bar{W}_1$ in the global coordinates as shown in Fig. 18.6. Thus,

$$U_1 = \bar{U}_1 \cos \theta_X + \bar{V}_1 \cos \theta_Y + \bar{W}_1 \cos \theta_Z \tag{18.7}$$

where θ_X, θ_Y, and θ_Z are the angles between the bar axis x and the X, Y, and Z axes. The cosines of these angles are the direction cosines of the x axis with respect to the global coordinate axes. With

$$c_1 = \cos \theta_X \qquad c_2 = \cos \theta_Y \qquad c_3 = \cos \theta_Z \tag{18.8a,b,c}$$

the forces at both ends of the bar in terms of their global components can be written in matrix form as

$$
\begin{bmatrix} U_1 \\ U_2 \end{bmatrix} = \begin{bmatrix} c_1 & c_2 & c_3 & 0 & 0 & 0 \\ 0 & 0 & 0 & c_1 & c_2 & c_3 \end{bmatrix} \begin{bmatrix} \bar{U}_1 \\ \bar{V}_1 \\ \bar{W}_1 \\ \bar{U}_2 \\ \bar{V}_2 \\ \bar{W}_2 \end{bmatrix} \tag{18.9a}
$$

or

$$\mathbf{S} = \mathbf{T}\bar{\mathbf{S}} \tag{18.9b}$$

The procedure for calculating the direction cosines c_i is discussed at the end of this section. Conversely, the forces $\bar{\mathbf{S}}$ in the global coordinates can be expressed in terms of the values in the local coordinates $\mathbf{S}$ as

$$\bar{U}_1 = U_1 \cos \theta_X \tag{18.10a}$$

$$\bar{V}_1 = U_1 \cos \theta_Y \tag{18.10b}$$

$$\bar{W}_1 = U_1 \cos \theta_Z \tag{18.10c}$$

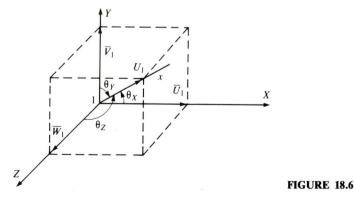

FIGURE 18.6

We can write, with the short notation of c_i = cosine of the angles,

$$
\begin{bmatrix} \bar{U}_1 \\ \bar{V}_1 \\ \bar{W}_1 \\ \bar{U}_2 \\ \bar{V}_2 \\ \bar{W}_2 \end{bmatrix} = \begin{bmatrix} c_1 & 0 \\ c_2 & 0 \\ c_3 & 0 \\ 0 & c_1 \\ 0 & c_2 \\ 0 & c_3 \end{bmatrix} \begin{bmatrix} U_1 \\ U_2 \end{bmatrix} \tag{18.11a}
$$

or

$$
\bar{\mathbf{S}} = \mathbf{T}^T \mathbf{S} \tag{18.11b}
$$

Joint 1 can have three displacements—u_1, v_1, and w_1 in the X, Y, and Z directions. The end displacements in the global coordinates can be related to those in the local coordinates in a manner similar to that discussed for forces. Thus,

$$
\begin{bmatrix} \bar{u}_1 \\ \bar{v}_1 \\ \bar{w}_1 \\ \bar{u}_2 \\ \bar{v}_2 \\ \bar{w}_2 \end{bmatrix} \begin{bmatrix} u_1 \\ u_2 \end{bmatrix} = \begin{bmatrix} c_1 & c_2 & c_3 & 0 & 0 & 0 \\ 0 & 0 & 0 & c_1 & c_2 & c_3 \end{bmatrix} \tag{18.12a}
$$

or

$$
\mathbf{s} = \mathbf{T}\bar{\mathbf{s}} \tag{18.12b}
$$

and

$$
\bar{\mathbf{s}} = \mathbf{T}^T \mathbf{s} \tag{18.13}
$$

Now consider Eq. (18.6b). Substitution for $\mathbf{s}$ from Eq. (18.12b) gives

$$
\mathbf{S} = \mathbf{KT}\bar{\mathbf{s}} \tag{18.14a}
$$

or

$$
\mathbf{S} = \mathbf{B}\bar{\mathbf{s}} \tag{18.14b}
$$

with

$$
\mathbf{B} = \mathbf{KT} \tag{18.14c}
$$

or

$$
\mathbf{B} = \frac{EA}{l} \begin{bmatrix} c_1 & c_2 & c_3 & -c_1 & -c_2 & -c_3 \\ -c_1 & -c_2 & -c_3 & c_1 & c_2 & c_3 \end{bmatrix} \tag{18.14d}
$$

Thus, if we have the displacement vector for an element in the global coordinates, Eq. (18.14b) can be used to find the forces in the local coordinates. Multiplying both sides of Eq. (18.14a) by $\mathbf{T}^T$ we get

$$
\mathbf{T}^T \mathbf{S} = \mathbf{T}^T \mathbf{KT}\bar{\mathbf{s}}
$$

In view of Eq. (18.11b), this becomes

$$\bar{S} = T^T K T \bar{s}$$

or

$$\bar{S} = \bar{K} \bar{s} \tag{18.15a}$$

$$\bar{K} = T^T K T \tag{18.15b}$$

Or, after carrying out the algebra, with

$$C = \begin{bmatrix} c_1^2 & c_1 c_2 & c_1 c_3 \\ c_1 c_2 & c_2^2 & c_2 c_3 \\ c_1 c_3 & c_2 c_3 & c_3^2 \end{bmatrix} \tag{18.15c}$$

the 6×6 stiffness matrix in the global coordinates is

$$\bar{K} = \frac{EA}{l} \begin{bmatrix} C & -C \\ -C & C \end{bmatrix} \tag{18.15d}$$

Direction Cosines

The direction cosines, that is, the cosines of the angles that the member axis makes with the global coordinate axes, can be evaluated as follows. The joints 1 and 2 of member 12 (Fig. 18.7) have the coordinates (X_1, Y_1, Z_1) and (X_2, Y_2, Z_2) in the global X, Y, Z coordinate system. Here joint 1 is shown at the origin of the coordinates because it simplifies the figure. The length of member 12 is

$$l = \sqrt{(X_2 - X_1)^2 + (Y_2 - Y_1)^2 + (Z_2 - Z_1)^2} \tag{18.16}$$

and the direction cosines are

$$c_1 = \frac{X_2 - X_1}{l} \qquad c_2 = \frac{Y_2 - Y_1}{l} \qquad c_3 = \frac{Z_2 - Z_1}{l} \tag{18.17a,b,c}$$

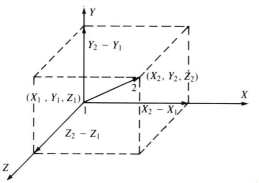

FIGURE 18.7

Solution Process

The steps in the solution process are as follows.

1. Number the elements, joints and the degrees of freedom in the global coordinates.
2. Calculate the global stiffness matrices for all the elements and indicate the pertinent degrees of freedom for each term.
3. Assemble the global stiffness matrices into the structure stiffness matrix.
4. Establish the load vector.
5. Solve the equilibrium equations to get the joint displacements.
6. Find the bar forces in the local coordinates (axial forces).

Example 18.3. The three-dimensional truss of Fig. 18.8a is in the form of a tripod with an additional vertical bar. The horizontal and vertical projections of the truss are shown in Fig. 18.8b. The loads consist of a horizontal load of 20 kips and a vertical load of 50 kips, as shown. The coordinates of the joints in feet are (X, Y, Z) equal to $(0, 0, 36)$ for joint 1, $(-10.39, 6, 0)$ for joint 2, $(10.39, 6, 0)$ for joint 3 and $(0, -12, 0)$ for joint 4. Find the member forces. $A = 10\,in^2$ and $E = 29\,000$ ksi for all members.

Solution. The member lengths and direction cosines c_1, c_2, and c_3 are found by

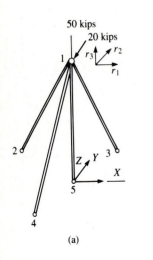

(a)

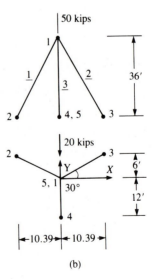

(b)

FIGURE 18.8

Eqs. (18.16) and (18.17). For the four elements, these are

Element→	1	2	3	4
l	455.37	455.37	455.37	432 (in)
c_1	−0.27	0.27	0.00	0
c_2	0.16	0.16	−0.32	0
c_3	−0.95	−0.95	−0.95	−1

The global stiffness matrices calculated by Eq. (18.15d) are

$$
\mathbf{K}_1 =
\begin{array}{c}
\begin{array}{cccccc} r_1 & r_2 & r_3 & 0 & 0 & 0 \end{array} \\
\left[
\begin{array}{rrrrrr}
47.76 & -27.58 & 165.46 & -47.76 & 27.58 & -165.46 \\
-27.58 & 15.92 & -95.53 & 27.58 & -15.92 & 95.53 \\
165.46 & -95.53 & 573.16 & -165.46 & 95.53 & -573.16 \\
-47.76 & 27.58 & -165.46 & 47.76 & -27.58 & 165.46 \\
27.58 & -15.92 & 95.53 & -27.58 & 15.92 & -95.53 \\
-165.46 & 95.53 & -573.16 & 165.46 & -95.53 & 573.16
\end{array}
\right]
\begin{array}{c} r_1 \\ r_2 \\ r_3 \\ 0 \\ 0 \\ 0 \end{array}
\end{array}
$$

$$
\mathbf{K}_2 =
\begin{array}{c}
\begin{array}{cccccc} r_1 & r_2 & r_3 & 0 & 0 & 0 \end{array} \\
\left[
\begin{array}{rrrrrr}
47.76 & 27.58 & -165.46 & -47.76 & -27.58 & 165.46 \\
27.58 & 15.92 & -95.53 & -27.58 & -15.92 & 95.53 \\
-165.46 & -95.53 & 573.16 & 165.46 & 95.53 & -573.16 \\
-47.76 & -27.58 & 165.46 & 47.76 & 27.58 & -165.46 \\
-27.58 & -15.92 & 95.53 & 27.58 & 15.92 & -95.53 \\
165.46 & 95.53 & -573.16 & -165.46 & -95.53 & 573.16
\end{array}
\right]
\begin{array}{c} r_1 \\ r_2 \\ r_3 \\ 0 \\ 0 \\ 0 \end{array}
\end{array}
$$

$$
\mathbf{K}_3 =
\begin{array}{c}
\begin{array}{cccccc} r_1 & r_2 & r_3 & 0 & 0 & 0 \end{array} \\
\left[
\begin{array}{rrrrrr}
0 & 0 & 0 & 0 & 0 & 0 \\
0 & 63.68 & 191.05 & 0 & -63.68 & -191.05 \\
0 & 191.05 & 573.16 & 0 & -191.05 & -573.16 \\
0 & 0 & 0 & 0 & 0 & 0 \\
0 & -63.68 & -191.05 & 0 & 63.68 & 191.05 \\
0 & -191.05 & -573.16 & 0 & 191.05 & 573.16
\end{array}
\right]
\begin{array}{c} r_1 \\ r_2 \\ r_3 \\ 0 \\ 0 \\ 0 \end{array}
\end{array}
$$

$$
\mathbf{K}_4 =
\begin{array}{c}
\begin{array}{cccccc} r_1 & r_2 & r_3 & 0 & 0 & 0 \end{array} \\
\left[
\begin{array}{rrrrrr}
0 & 0 & 0 & 0 & 0 & 0 \\
0 & 0 & 0 & 0 & 0 & 0 \\
0 & 0 & 671.29 & 0 & 0 & -671.29 \\
0 & 0 & 0 & 0 & 0 & 0 \\
0 & 0 & 0 & 0 & 0 & 0 \\
0 & 0 & -671.29 & 0 & 0 & -671.29
\end{array}
\right]
\begin{array}{c} r_1 \\ r_2 \\ r_3 \\ 0 \\ 0 \\ 0 \end{array}
\end{array}
$$

Note that joint 1 is considered the first joint for all the bars. The structure stiffness matrix and load vector are shown below.

$$
\begin{array}{ccc}
r_1 & r_2 & r_3
\end{array}
$$
$$
\begin{bmatrix}
95.53 & 0.00 & 0.00 \\
0.00 & 95.53 & 0.00 \\
0.00 & 0.00 & 2\,391.00
\end{bmatrix}
\begin{bmatrix}
r_1 \\ r_2 \\ r_3
\end{bmatrix}
\begin{array}{c} \mathbf{R} \\ \begin{bmatrix} 0 \\ -20 \\ -50 \end{bmatrix} \end{array}
$$

The solution of these equations gives the displacements of joint 1 as

$$
\begin{bmatrix}
r_1 \\ r_2 \\ r_3
\end{bmatrix}
=
\begin{bmatrix}
0.00 \\ -0.21 \\ -0.021
\end{bmatrix}
$$

The displacement vectors in the global coordinates, $\bar{\mathbf{s}}$, for the three elements are

$$
\text{Element} \rightarrow \qquad 1 \qquad 2 \qquad 3
$$
$$
\bar{\mathbf{s}} =
\begin{bmatrix}
\bar{u}_1 \\ \bar{v}_1 \\ \bar{w}_1 \\ \bar{u}_2 \\ \bar{v}_2 \\ \bar{w}_2
\end{bmatrix}
:
\begin{bmatrix}
0.00 \\ -0.21 \\ -0.021 \\ 0.00 \\ 0.00 \\ 0.00
\end{bmatrix}
\begin{bmatrix}
0.00 \\ -0.21 \\ -0.021 \\ 0.00 \\ 0.00 \\ 0.00
\end{bmatrix}
\begin{bmatrix}
0.00 \\ -0.21 \\ -0.021 \\ 0.00 \\ 0.00 \\ 0.00
\end{bmatrix}
$$

Matrices $\mathbf{B}_i = \mathbf{K}_i \mathbf{T}_i$ for the three elements are

$$
\mathbf{B}_1 =
\begin{bmatrix}
-174.41 & 100.69 & -604.17 & 174.41 & -100.69 & 604.17 \\
174.41 & -100.69 & 604.17 & -174.41 & 100.69 & -604.17
\end{bmatrix}
$$

$$
\mathbf{B}_2 =
\begin{bmatrix}
174.41 & 100.69 & -604.17 & -174.41 & -100.69 & 604.17 \\
-174.41 & -100.69 & 604.17 & 174.41 & 100.69 & -604.17
\end{bmatrix}
$$

$$
\mathbf{B}_3 =
\begin{bmatrix}
0 & -201.39 & -604.17 & 0 & 201.39 & 604.17 \\
0 & 201.39 & 604.17 & 0 & -201.39 & -604.17
\end{bmatrix}
$$

$$
\mathbf{B}_4 =
\begin{bmatrix}
0 & 0 & -671.30 & 0 & 0 & 671.30 \\
0 & 0 & 671.30 & 0 & 0 & -671.30
\end{bmatrix}
$$

The forces in the local coordinates are calculated by $\mathbf{S} = \mathbf{B}\bar{\mathbf{s}}$ as

$$
\text{Element} \rightarrow \qquad 1 \qquad 2 \qquad 3 \qquad 4
$$
$$
\mathbf{S} =
\begin{bmatrix}
U_1 \\ U_2
\end{bmatrix}
:
\begin{bmatrix}
-8.35 \\ 8.35
\end{bmatrix}
\begin{bmatrix}
-8.35 \\ 8.35
\end{bmatrix}
\begin{bmatrix}
54.77 \\ -54.77
\end{bmatrix}
\begin{bmatrix}
14.09 \\ -14.09
\end{bmatrix}
$$

18.3   THREE-DIMENSIONAL FRAMES— STIFFNESS METHOD

In a three-dimensional frame, the members may be oriented in any direction and subjected to shear and moment in two different planes. In addition, some members may be subjected to torque induced by bending of other members.

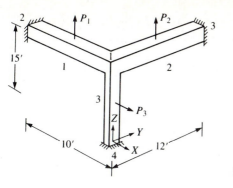

FIGURE 18.9

For example, in Fig. 18.9 member 1 is under the force P_1, causing it to deform in the vertical plane. Because of bending in member 1, joint 1 incurs a rotation. This rotation causes a twist at joint 1 in member 2. Furthermore, the bending of members 1 and 2 causes bending of member 3 in two different planes.

In this section we derive the stiffness and transformation matrices for the general case of three-dimensional beam or frame elements.

Force–Displacement Relations for Three-Dimensional Beam Elements

From the above discussion we can see that a general beam element, to be used in the analysis of three-dimensional beams and frames, must include forces in three directions and moments about three axes. Such a beam element is shown in Fig. 18.10. Here for clarity the forces are shown at end 1 only, and the moments at end 2. The double arrows indicate moments about the axis in the positive sense.

The stiffness matrix for a three-dimensional beam element is found by writing the relationship between the end forces **S** and displacements **s**, where

$$\mathbf{S}^T = [U_1 \quad V_1 \quad W_1 \quad T_1 \quad M_{y1} \quad M_{z1} \quad U_2 \quad V_2 \quad W_2 \quad T_2 \quad M_{y2} \quad M_{z2}]^T \quad (18.18)$$

$$\mathbf{s}^T = [u_1 \quad v_1 \quad w_1 \quad \psi_1 \quad \theta_{y1} \quad \theta_{z1} \quad u_2 \quad v_2 \quad w_2 \quad \psi_2 \quad \theta_{y2} \quad \theta_{z2}]^T \quad (18.19)$$

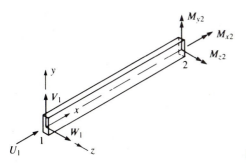

FIGURE 18.10

Thus, the stiffness matrix is a 12×12 matrix. In deriving the stiffness matrix for a two-dimensional beam element, we saw that axial deformations are independent of bending. Bending of a beam in one plane is also independent of bending in another plane. Similarly, torsion is independent of other deformations and can be treated separately. The terms of the stiffness matrix for axial deformation and bending for a three-dimensional beam element are similar to those for two-dimensional beams. The bending moments are indicated by subscripts referring to the axis about which they act.

The force–displacement relation for a three-dimensional beam element, found by combining the stiffness terms for axial deformation, bending about two axes, and torsion is as follows. Note that v_1 and V_1 are displacement and shear in the direction of the local y coordinate, while M_{y1} and θ_{y1} are the moment and rotation about the y axis. The moment of inertia associated with M_{y1} is I_y. A $(\cdot)$ indicates a zero value.

$$
\begin{bmatrix}
U_1 \\ V_1 \\ W_1 \\ T_1 \\ M_{y1} \\ M_{z1} \\ U_2 \\ V_2 \\ W_2 \\ T_2 \\ M_{y2} \\ M_{z2}
\end{bmatrix}
=
\begin{bmatrix}
\dfrac{EA}{l} & & & & & & & & & & & \\
\cdot & \dfrac{12EI_z}{l^3} & & & & \text{symmetrical} & & & & & & \\
\cdot & \cdot & \dfrac{12EI_y}{l^3} & & & & & & & & & \\
\cdot & \cdot & \cdot & \dfrac{GJ}{l} & & & & & & & & \\
\cdot & \cdot & -\dfrac{6EI_y}{l^2} & \cdot & \dfrac{4EI_y}{l} & & & & & & & \\
\cdot & \dfrac{6EI_z}{l^2} & \cdot & \cdot & \cdot & \dfrac{4EI_z}{l} & & & & & & \\
-\dfrac{EA}{l} & \cdot & \cdot & \cdot & \cdot & \cdot & \dfrac{EA}{l} & & & & & \\
\cdot & -\dfrac{12EI_z}{l^3} & \cdot & \cdot & \cdot & -\dfrac{6EI_z}{l^2} & \cdot & \dfrac{12EI_z}{l^3} & & & & \\
\cdot & \cdot & -\dfrac{12EI_y}{l^3} & \cdot & \dfrac{6EI_y}{l^2} & \cdot & \cdot & \cdot & \dfrac{12EI_y}{l^3} & & & \\
\cdot & \cdot & \cdot & -\dfrac{GJ}{l} & \cdot & \cdot & \cdot & \cdot & \cdot & \dfrac{GJ}{l} & & \\
\cdot & \cdot & -\dfrac{6EI_y}{l^2} & \cdot & \dfrac{2EI_y}{l} & \cdot & \cdot & \cdot & \dfrac{6EI_y}{l^2} & \cdot & \dfrac{4EI_y}{l} & \\
\cdot & \dfrac{6EI_z}{l^2} & \cdot & \cdot & \cdot & \dfrac{2EI_z}{l} & \cdot & -\dfrac{6EI_z}{l^2} & \cdot & \cdot & \cdot & \dfrac{4EI_z}{l}
\end{bmatrix}
\begin{bmatrix}
u_1 \\ v_1 \\ w_1 \\ \psi_1 \\ \theta_{y1} \\ \theta_{z1} \\ u_2 \\ v_2 \\ w_2 \\ \psi_2 \\ \theta_{y2} \\ \theta_{z2}
\end{bmatrix}
+
\begin{bmatrix}
U_1^F \\ V_1^F \\ W_1^F \\ T_1^F \\ M_{y1}^F \\ M_{z1}^F \\ U_2^F \\ V_2^F \\ W_2^F \\ T_2^F \\ M_{y2}^F \\ M_{z2}^F
\end{bmatrix}
$$

$$(18.20a)$$

or

$$\mathbf{S} = \mathbf{Ks} + \mathbf{S}^F \qquad (18.20b)$$

where $\mathbf{S}^F$ is the vector of fixed-end forces due to in-span loads, and $\mathbf{K}$ is the 12×12 stiffness matrix.

Transformation Matrix for Three-Dimensional Frame Elements

Consider the forces at end 1 of a three-dimensional beam element, Fig. 18.11a. These are U_1, V_1, and W_1 in the local coordinates of the beam with the x axis directed from point 1 to 2, and the y and z axes normal to x forming a right handed coordinate system (Sec. 10.1). In the global coordinates, the end forces are $\bar{U}_1$, $\bar{V}_1$, $\bar{W}_1$.

For clarity, each force component in the local coordinates of the member is shown in a separate figure, along with its components in the global coordinates in Figs. 18.11b,c, and d. We can write the following three equations, which express the force components in the local coordinates in terms of the components in the global coordinates.

$$U_1 = \bar{U}_1 \cos \theta_{xX} + \bar{V}_1 \cos \theta_{xY} + \bar{W}_1 \cos \theta_{xZ}$$
$$V_1 = \bar{U}_1 \cos \theta_{yX} + \bar{V}_1 \cos \theta_{yY} + \bar{W}_1 \cos \theta_{yZ} \qquad (18.21)$$
$$W_1 = \bar{U}_1 \cos \theta_{zX} + \bar{V}_1 \cos \theta_{zY} + \bar{W}_1 \cos \theta_{zZ}$$

Defining

$$
\begin{array}{lll}
C_{11} = \cos \theta_{xX} & C_{12} = \cos \theta_{xY} & C_{13} = \cos \theta_{xZ} \\
C_{21} = \cos \theta_{yX} & C_{22} = \cos \theta_{yY} & C_{23} = \cos \theta_{yZ} \\
C_{31} = \cos \theta_{zX} & C_{32} = \cos \theta_{yY} & C_{33} = \cos \theta_{yZ}
\end{array}
\qquad (18.22)
$$

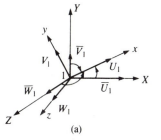

(a)

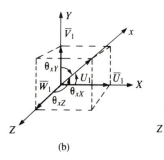 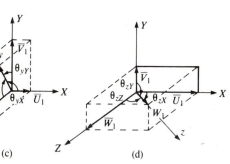

(b) (c) (d)

FIGURE 18.11

we get

$$\mathbf{F}_1 = \mathbf{C}\bar{\mathbf{F}}_1 \quad \text{and} \quad \mathbf{F}_2 = \mathbf{C}\bar{\mathbf{F}}_2 \tag{18.23}$$

$\mathbf{F}_1$ and $\bar{\mathbf{F}}_1$ are the force vectors at end 1 in the local and global coordinates with

$$\mathbf{C} = \begin{bmatrix} C_{11} & C_{12} & C_{13} \\ C_{21} & C_{22} & C_{23} \\ C_{31} & C_{32} & C_{33} \end{bmatrix} \tag{18.24}$$

$$\mathbf{F}_1 = \begin{bmatrix} U_1 \\ V_1 \\ W_1 \end{bmatrix} \quad \bar{\mathbf{F}}_1 = \begin{bmatrix} \bar{U}_1 \\ \bar{V}_1 \\ \bar{W}_1 \end{bmatrix} \quad \mathbf{F}_2 = \begin{bmatrix} U_2 \\ V_2 \\ W_2 \end{bmatrix} \quad \bar{\mathbf{F}}_2 = \begin{bmatrix} \bar{U}_2 \\ \bar{V}_2 \\ \bar{W}_2 \end{bmatrix}$$

Similar relations can be written for the moments about the three axes:

$$\mathbf{M}_1 = \mathbf{C}\bar{\mathbf{M}}_1 \quad \text{and} \quad \mathbf{M}_2 = \mathbf{C}\bar{\mathbf{M}}_2 \tag{18.25}$$

where

$$\mathbf{M}_1 = \begin{bmatrix} T_1 \\ M_{y1} \\ M_{z1} \end{bmatrix} \quad \bar{\mathbf{M}}_1 = \begin{bmatrix} \bar{M}_{X1} \\ \bar{M}_{Y1} \\ \bar{M}_{Z1} \end{bmatrix} \quad \mathbf{M}_2 = \begin{bmatrix} T_2 \\ M_{y2} \\ M_{z2} \end{bmatrix} \quad \bar{\mathbf{M}}_2 = \begin{bmatrix} \bar{M}_{X2} \\ \bar{M}_{Y2} \\ \bar{M}_{Z2} \end{bmatrix}$$

For both ends of the element, the forces in the local coordinates in terms of those in the global coordinates are

$$\begin{bmatrix} U_1 \\ V_1 \\ W_1 \\ T_1 \\ M_{y1} \\ M_{z1} \\ U_2 \\ V_2 \\ W_2 \\ T_2 \\ M_{y2} \\ M_{z2} \end{bmatrix} = \begin{bmatrix} \mathbf{C} & & & \\ & \mathbf{C} & & \\ & & \mathbf{C} & \\ & & & \mathbf{C} \end{bmatrix} \begin{bmatrix} \bar{U}_1 \\ \bar{V}_1 \\ \bar{W}_1 \\ \bar{M}_{X1} \\ \bar{M}_{Y1} \\ \bar{M}_{Z1} \\ \bar{U}_2 \\ \bar{V}_2 \\ \bar{W}_2 \\ \bar{M}_{X2} \\ \bar{M}_{Y2} \\ \bar{M}_{Z2} \end{bmatrix} \tag{18.26a}$$

or

$$\mathbf{S} = \mathbf{TS} \tag{18.26b}$$

As for other elements, the transformation matrix $\mathbf{T}$ is an orthogonal matrix, that is, its inverse is equal to its transpose. Thus, we can write

$$\bar{\mathbf{S}} = \mathbf{T}^T \mathbf{S} \tag{18.27}$$

Similar expressions can be found for displacements:

$$\mathbf{s} = \mathbf{T}\bar{\mathbf{s}} \tag{18.28}$$

$$\bar{\mathbf{s}} = \mathbf{T}^T \mathbf{s} \tag{18.29}$$

where

$$\mathbf{s}^T = [u_1 \quad v_1 \quad w_1 \quad \psi_1 \quad \theta_{y1} \quad \theta_{z1} \quad u_2 \quad v_2 \quad w_2 \quad \psi_2 \quad \theta_{y2} \quad \theta_{z2}]^T \tag{18.30a}$$

$$\bar{\mathbf{s}}^T = [\bar{u}_1 \quad \bar{v}_1 \quad \bar{w}_1 \quad \bar{\theta}_{X1} \quad \bar{\theta}_{Y1} \quad \bar{\theta}_{Z1} \quad \bar{u}_2 \quad \bar{v}_2 \quad \bar{w}_2 \quad \bar{\theta}_{X2} \quad \bar{\theta}_{Y2} \quad \bar{\theta}_{Z2}]^T$$

$$\tag{18.30b}$$

Substitution of Eq. (18.28) into (18.20b) gives the member-end forces in the local coordinates in terms of the member-end displacements in the global coordinates:

$$\mathbf{S} = \mathbf{KT}\bar{\mathbf{s}} + \mathbf{S}^F \tag{18.31a}$$

or

$$\mathbf{S} = \mathbf{B}\bar{\mathbf{s}} + \mathbf{S}^F \tag{18.31b}$$

with

$$\mathbf{B} = \mathbf{KT} \tag{18.31c}$$

Premultiplying both sides of Eq. (18.31a) by $\mathbf{T}^T$ and invoking Eq. (18.27), we get

$$\bar{\mathbf{S}} = \bar{\mathbf{K}}\bar{\mathbf{s}} + \bar{\mathbf{S}}^F \tag{18.32a}$$

with the element stiffness matrix and fixed end forces in the global coordinates as

$$\bar{\mathbf{K}} = \mathbf{T}^T \mathbf{KT} \tag{18.32b}$$

and

$$\bar{\mathbf{S}} = \mathbf{T}^T \mathbf{S}^F \tag{18.32c}$$

Calculation of Direction Cosines

The direction cosines can be calculated in different fashions depending on whether hand calculations are used or the procedure is implemented in a computer program. In both cases, however, the x axis is taken along the axis of the member.

In calculating the direction cosines by hand, the choice of the local y and z coordinate axes is arbitrary as long as the orthogonal coordinates form a right-handed system.

Example 18.4. As an example, let us consider the frame of Fig. 18.12 which is a schematic sketch of the frame in Fig. 18.9. The local coordinates x, y, z for each

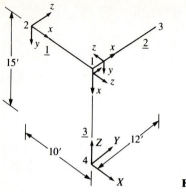

FIGURE 18.12

member are shown in the figure along with the global coordinate axes X, Y, Z. Note that the direction angles are measured from the global coordinates in the positive direction of the right-handed coordinate system. In this example, the local y axes are taken in the strong planes of the members, Fig. 18.9. For hand calculation, this is not necessary. However, as will be seen in Example 18.5, this is the convention used for procedures that are implemented in computer programs and makes it easier to identify the orientation of the section.

Solution. For the members of the frame in Fig. 18.12, the angles between each local coordinate axis and the global coordinate axes are shown in Table 18.2. Note that the angles start from the global coordinate axis and end at the local coordinate axis in the positive sense of the global coordinate system. The transformation matrix **C** is found by calculating the cosines of these angles.

The direction cosines can be found from the above direction angles,

TABLE 18.2

	Direction angles (degrees)		
Member →	**1**	**2**	**3**
θ_{xX}	0	90	90
θ_{xY}	270	0	270
θ_{xZ}	90	270	180
θ_{yX}	90	900	90
θ_{yY}	270	270	0
θ_{yZ}	180	180	270
θ_{zX}	90	180	0
θ_{zY}	0	90	270
θ_{zZ}	270	270	90

resulting in the following transformation matrices for the three elements.

$$
\begin{array}{cccc}
\text{Element} \rightarrow & 1 & 2 & 3 \\
\mathbf{C} \quad : &
\begin{bmatrix} 1 & 0 & 0 \\ 0 & 0 & -1 \\ 0 & 1 & 0 \end{bmatrix} &
\begin{bmatrix} 0 & 1 & 0 \\ 0 & 0 & -1 \\ -1 & 0 & 0 \end{bmatrix} &
\begin{bmatrix} 0 & 0 & -1 \\ 0 & 1 & 0 \\ 1 & 0 & 0 \end{bmatrix}
\end{array}
$$

Expressions for Direction Cosines

In calculating the direction cosine in an automated way, for example in computer programs, it is convenient to use an additional point to define the xy plane and establish the y and z axes of the member in the local coordinate system. This point is generally chosen in the strong plane of the member. That is, in the plane formed by the axis of the member and the longer dimension of the cross-section. It could be a joint of the structure or a new point used merely to define the local coordinate system. In the latter case, the point does not have to be part of the structure and no degree of freedom will be assigned to it.

For member 12 (Fig. 18.13), a third point, 3, is chosen in its strong plane, xy. Member 12 can be represented by a vector $\bar{r}_1$ in the direction of the x axis directed from joint 1 to 2. The strong axis of the beam is established by vector $\bar{r}$ connecting points 1 and 3. The weak plane of the beam is xz, where z is in the direction of a vector $\bar{r}_3$ normal to $\bar{r}_1$ and $\bar{r}$, and the y axis is in the direction of a vector $\bar{r}_2$, normal to $\bar{r}_1$ and $\bar{r}_3$. The vectors $\bar{r}_1$, $\bar{r}_2$, and $\bar{r}_3$ can be determined from the coordinates of points 1, 2, and 3. Once these vectors are found, the unit vectors in the local coordinates can be calculated by dividing the above vectors by their lengths. The direction cosines in the transformation matrix can then be found from dot products of the unit vectors in the local and global coordinates.

In Fig. 18.13, joint 1 is shown at the origin of the global coordinates for clarity. However, the derivations are done in terms of the coordinates of point

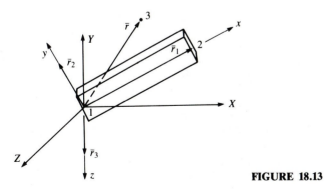

FIGURE 18.13

1. In the global X, Y, Z coordinates, vector $\bar{r}_1$ can be expressed as

$$\bar{r}_1 = \alpha_1\bar{i} + \alpha_2\bar{j} + \alpha_3\bar{k} \qquad (18.33a)$$

where

$$\alpha_1 = X_2 - X_1 \qquad \alpha_2 = Y_2 - Y_1 \qquad \alpha_3 = Z_2 - Z_1 \qquad (18.33b)$$

and $\bar{i}$, $\bar{j}$, $\bar{k}$ are the unit vectors of the global coordinate system. The vector $\bar{r}$ joining points 1 and 3 (the auxiliary point) can be written in a similar fashion,

$$\bar{r} = \beta_1\bar{i} + \beta_2\bar{j} + \beta_3\bar{k} \qquad (18.34a)$$

with

$$\beta_1 = X_3 - X_1 \qquad \beta_2 = Y_3 - Y_1 \qquad \beta_3 = Z_3 - Z_1 \qquad (18.34b)$$

Since the vectors $\bar{r}_1$ and $\bar{r}$ are in the strong plane of the beam, a vector normal to these vectors will be in the weak plane of the beam. This vector will be in the direction of the local z axis of the beam. The latter vector $\bar{r}_3$ can be found by a cross product as $\bar{r}_3 = \bar{r}_1 \times \bar{r}$, leading to

$$\bar{r}_3 = \gamma_1\bar{i} + \gamma_2\bar{j} + \gamma_3\bar{k} \qquad (18.35a)$$

with

$$\gamma_1 = \alpha_2\beta_3 - \alpha_3\beta_2 \qquad \gamma_2 = \alpha_3\beta_1 - \alpha_1\beta_3 \qquad \gamma_3 = \alpha_1\beta_2 - \alpha_2\beta_1 \qquad (18.35b)$$

The vector $\bar{r}_2$ in the strong plane normal to $\bar{r}_1$ and $\bar{r}_3$ is

$$\bar{r}_2 = \delta_1\bar{i} + \delta_2\bar{j} + \delta_3\bar{k} \qquad (18.36a)$$

with

$$\delta_1 = \gamma_2\alpha_3 - \gamma_3\alpha_2 \qquad \delta_2 = \gamma_3\alpha_1 - \gamma_1\alpha_3 \qquad \delta_3 = \gamma_1\alpha_2 - \gamma_2\alpha_1 \qquad (18.36b)$$

If d_i is the length of the vector $\bar{r}_i$, $d_i = |\bar{r}_i| = \bar{r}_i \cdot \bar{r}_i$. The lengths of the three vectors $\bar{r}_1$, $\bar{r}_2$, and $\bar{r}_3$ in the x, y, and z directions of the local coordinates are

$$d_1 = (\alpha_1^2 + \alpha_2^2 + \alpha_3^2)^{1/2} \qquad (18.37a)$$

$$d_2 = (\delta_1^2 + \delta_2^2 + \delta_3^2)^{1/2} \qquad (18.37b)$$

$$d_3 = (\gamma_1^2 + \gamma_2^2 + \gamma_3^2)^{1/2} \qquad (18.37c)$$

The unit vectors in the local x, y, z coordinates, i, j, and k are found by dividing the vectors $\bar{r}_i$ by their lengths d_i. Thus,

$$i = \frac{\alpha_1}{d_1}\bar{i} + \frac{\alpha_2}{d_1}\bar{j} + \frac{\alpha_3}{d_1}\bar{k}$$

$$j = \frac{\delta_1}{d_2}\bar{i} + \frac{\delta_2}{d_2}\bar{j} + \frac{\delta_3}{d_2}\bar{k}$$

$$k = \frac{\gamma_1}{d_3}\bar{i} + \frac{\gamma_2}{d_3}\bar{j} + \frac{\gamma_3}{d_3}\bar{k}$$

Finally, the direction cosines can be calculated by finding the dot products of the local and global unit vectors. For example, the direction cosines of the local x axis with respect to the global axes X, Y, and Z are

$$C_{11} = \cos \theta_{xX} = \mathbf{i} \cdot \mathbf{i} = \frac{\alpha_1}{d_1}$$

$$C_{12} = \cos \theta_{xY} = \mathbf{i} \cdot \mathbf{j} = \frac{\alpha_2}{d_1}$$

$$C_{13} = \cos \theta_{xZ} = \mathbf{i} \cdot \bar{\mathbf{k}} = \frac{\alpha_3}{d_1}$$

These values give the orientation of the beam axis and are the same as c_1, c_2 and c_3 given by Eq. (18.17) for three-dimensional truss elements. The other direction cosines in the transformation matrix can be found in a similar manner. They are, however, different from those found for three-dimensional trusses. This is because the orientations of the axes of the cross-section are critical in the analysis of beams and frames but immaterial for truss members. The transformation matrix $\mathbf{C}$ for three-dimensional frames is thus as follows.

$$\mathbf{C} = \begin{bmatrix} \dfrac{\alpha_1}{d_1} & \dfrac{\alpha_2}{d_1} & \dfrac{\alpha_3}{d_1} \\[2ex] \dfrac{\delta_1}{d_2} & \dfrac{\delta_2}{d_2} & \dfrac{\delta_3}{d_2} \\[2ex] \dfrac{\gamma_1}{d_3} & \dfrac{\gamma_2}{d_3} & \dfrac{\gamma_3}{d_3} \end{bmatrix} \qquad (18.38)$$

with points 1 and 2 the end points of the beam axis, and 3 a point in the strong plane of the beam, not on the axis 12.

$\alpha_1 = X_2 - X_1$	$\alpha_2 = Y_2 - Y_1$	$\alpha_3 = Z_2 - Z_1$
$\beta_1 = X_3 - X_1$	$\beta_2 = Y_3 - Y_1$	$\beta_3 = Z_3 - Z_1$
$\gamma_1 = \alpha_2\beta_3 - \alpha_3\beta_2$	$\gamma_2 = \alpha_3\beta_1 - \alpha_1\beta_3$	$\gamma_3 = \alpha_1\beta_2 - \alpha_2\beta_1$
$\delta_1 = \gamma_2\alpha_3 - \gamma_3\alpha_2$	$\delta_2 = \gamma_3\alpha_1 - \gamma_1\alpha_3$	$\delta_3 = \gamma_1\alpha_2 - \gamma_2\alpha_1$
$d_1 = (\alpha_1^2 + \alpha_2^2 + \alpha_3^2)^{1/2}$	$d_2 = (\delta_1^2 + \delta_2^2 + \delta_3^2)^{1/2}$	$d_3 = (\gamma_1^2 + \gamma_2^2 + \gamma_3^2)^{1/2}$

Thus, the terms of the transformation matrix $\mathbf{C}$ are calculated after finding α_i, β_i, γ_i, δ_i, and d_i.

Example 18.5. Find the direction cosines for the frame members in Fig. 18.9. Assume that the strong planes of the beams are the vertical planes and that of the column the YZ plane. The sketch of the structure with the local axes and global coordinates is shown in Fig. 18.12.

TABLE 18.3

Member→	1	2	3	Member→	1	2	3
1st joint	2	1	1	α_1	10	0	0
2nd joint	1	3	4	α_2	0	12	0
3rd point	4	4	3	α_3	0	0	-15
X_1	-10	0	0	β_1	10	0	0
X_2	0	0	0	β_2	0	0	12
X_3	0	0	0	β_3	-15	-15	0
Y_1	0	0	0	γ_1	0	-180	180
Y_2	0	12	0	γ_2	150	0	0
Y_3	0	0	15	γ_3	0	0	0
Z_1	15	15	15	δ_1	0	0	0
Z_2	15	15	0	δ_2	0	0	2 160
Z_3	0	0	15	δ_3	-1 500	-2 160	0
				d_1	10	12	15
				d_2	1 500	2 160	2 160
				d_3	150	180	180

Solution. The strong axes of members 1 and 2 are in the vertical planes of these elements. We use joint 4 as the third point for members 1 and 2. For member 3 the third point will be joint 3. The calculation of the direction cosines using Eqs. (18.38) is done in Table 18.3.

The transformation matrices are

$$\text{Element}\rightarrow \quad 1 \qquad\qquad 2 \qquad\qquad 3$$

$$\mathbf{C} \quad : \quad \begin{bmatrix} 1 & 0 & 0 \\ 0 & 0 & -1 \\ 0 & 1 & 0 \end{bmatrix} \begin{bmatrix} 0 & 1 & 0 \\ 0 & 0 & -1 \\ -1 & 0 & 0 \end{bmatrix} \begin{bmatrix} 0 & 0 & -1 \\ 0 & 1 & 0 \\ 1 & 0 & 0 \end{bmatrix}$$

As expected, the same values are obtained for the direction cosines as in Example 18.4.

Equilibrium Equations for Three-Dimensional Frames

The equilibrium equations can be written at each joint by setting the sum of the forces and moments in the global coordinates in each direction equal to zero, and then substituting from the force–displacement equations (Eq. (18.15a)). Thus, we obtain a number of equations equal to the number of degrees of freedom of the joints.

The equilibrium equations for three-dimensional frames are found in the same manner as for other structures with the direct stiffness method. Using the force–displacement relations in the global coordinates, each term of the structure stiffness matrix can be obtained by adding the terms from the global stiffness matrix of each element contributing to that degree of freedom. Similarly, each term of the load vector is equal to the sum of the applied forces in that degree of freedom plus the equivalent joint force in that degree of freedom (that is, the negative of sum of the fixed-end forces associated with that degree of freedom.

Solution Process for Three-Dimensional Frames

From the above discussion we can see that the solution process is similar to that for two-dimensional frames, discussed in Section 12.4. The steps in the solution consist of the following

1. Establish a global coordinate system for the structure, number the elements, joints (including points that would identify the strong planes of the members), and the degrees of freedom and indicate the structure degrees of freedom on the sketch.
2. Calculate the force–displacement relations in the local coordinates (Eq. (18.20a)) for each element. The fixed-end shear forces and moments may be obtained from Tables 10.1a and 10.1b in Chapter 10. The fixed-end torques for a member under a concentrated torque can be found by Eqs. (17.17).
3. A point in the strong plane of the element is needed to establish the direction cosines. The values of the latter can be found either directly by Eqs. (18.22) or more commonly by computing the values in Eqs. (18.38). The transformation matrices are then found by Eq. (18.26a). Transform the force–displacement relations to the global coordinate system (Eqs. (18.32)).
4. Set up the equilibrium equations by assembling the global element stiffness matrices and load vectors into the structure stiffness matrix and structure load vector. The components of the load vector are sum of the applied joint loads and the equivalent joint forces (the negatives of the fixed-end forces for each degree of freedom for the in-span loads).
5. Solve the equilibrium equations to find the joint displacements in the global coordinates.
6. Using the member-end displacement vectors in the global coordinates find the member-end forces by Eq. (18.31b).

> **Example 18.6.** Find the member forces for the three-dimensional frame of Fig. 18.9 with P_1 and P_2 gravity loads of 20 kips and 30 kips, respectively, applied at the mid-spans, and $P_3 = 0$. The frame is reinforced concrete with the cross-sections of the beams 8 in × 12 in having their larger dimension in the vertical plane. The column is 8 in × 8 in. $E = 3100$ ksi and Poisson's ratio $v = 0.2$.
>
> **Solution.** If we use the same coordinate systems as in Fig. 18.12, the transformation matrices for the three elements are as found in Examples 18.4 or 18.5. The applied loads act in the positive directions of the local y axes.
>   For elements 1 and 2, the cross-sectional area is $A = 96\,\text{in}^2$ and the

moments of inertia are $I_y = 12 \times (8^3/12) = 512 \text{ in}^4$, and $I_z = 8 \times (12^3/12) = 1152 \text{ in}^4$. For element 3, $A = 64 \text{ in}^2$, $I_y = I_z = 8^4/12 = 341 \text{ in}^4$. The torsion constants can be found by using the formula for rectangular sections in Table 17.1, that is,

$$J = \frac{ab^3}{3}\left[1 - 0.63\frac{b}{a}\left(1 - \frac{b^4}{12a^4}\right)\right]$$

Thus, for members 1 and 2,

$$J = \frac{12 \times 8^3}{3}\left[1 - 0.63 \times \frac{8}{12}\left(1 - \frac{8^4}{12 \times 12^4}\right)\right] = 1\,202 \text{ in}^4$$

and for member 3,

$$J = \frac{8 \times 8^3}{3}\left[1 - 0.63 \times \frac{8}{8}\left(1 - \frac{8^4}{12 \times 8^4}\right)\right] = 577 \text{ in}^4$$

The stiffness matrices and load vectors for the force–displacement relations for the three elements in the local coordinates are

$$\mathbf{K}_1$$

2 480	0	0	0	0	0	−2 480	0	0	0	0	0
0	20	0	0	0	1 490	0	−20	0	0	0	1 490
0	0	10	0	−660	0	0	0	−10	0	−660	0
0	0	0	12 940	0	0	0	0	0	−12 940	0	0
0	0	−660	0	52 910	0	0	0	660	0	26 450	0
0	−1 490	0	0	0	119 040	0	−1 490	0	0	0	59 520
−2 480	0	0	0	0	0	2 480	0	0	0	0	0
0	−20	0	0	0	1 490	0	20	0	0	0	−1 490
0	0	−10	0	660	0	0	0	10	0	660	0
0	0	0	−12 940	0	0	0	0	0	12 940	0	0
0	0	−660	0	26 450	0	0	0	660	0	52 910	0
0	1 490	0	0	0	59 520	0	−1 490	0	0	0	119 040

$$\mathbf{S}_1^F$$

0
−10
0
0
0
−300
0
−10
0
0
0
300

$$\mathbf{K}_2$$

2 067	0	0	0	0	0	−2 067	0	0	0	0	0
0	14	0	0	0	1 033	0	−14	0	0	0	1 033
0	0	6	0	−459	0	0	0	−6	0	−459	0
0	0	0	10 782	0	0	0	0	0	−10 782	0	0
0	0	−459	0	44 089	0	0	0	459	0	22 044	0
0	−1 033	0	0	0	99 200	0	−1 033	0	0	0	49 600
−2 067	0	0	0	0	0	2 067	0	0	0	0	0
0	−14	0	0	0	1 033	0	14	0	0	0	−1 033
0	0	−6	0	459	0	0	0	6	0	459	0
0	0	0	−10 782	0	0	0	0	0	10 782	0	0
0	0	−459	0	22 044	0	0	0	459	0	44 089	0
0	1 033	0	0	0	49 600	0	−1 033	0	0	0	99 200

$$\mathbf{S}_2^F$$

0
−15
0
0
0
−540
0
−15
0
0
0
540

$$\mathbf{K}_3$$

1 102	0	0	0	0	0	−1 102	0	0	0	0	0
0	2	0	0	0	196	0	−2	0	0	0	196
0	0	2	0	−196	0	0	0	−2	0	−196	0
0	0	0	4 141	0	−0	0	0	0	−4 141	0	0
0	0	−196	0	23 491	0	0	0	196	0	11 746	0
0	−196	0	0	0	23 491	0	−196	0	0	0	11 746
−1 102	0	0	0	0	0	1 102	0	0	0	0	0
0	−2	0	0	0	196	0	2	0	0	0	−196
0	0	−2	0	196	0	0	0	2	0	196	0
0	0	0	−4 141	0	0	0	0	0	4 141	0	0
0	0	−196	0	11 746	0	0	0	196	0	23 491	0
0	196	0	0	0	11 746	0	−196	0	0	0	23 491

Using the transformation matrices found in Examples 18.4 or 18.5, we can find the stiffness matrices and load vectors in the element force–displacement relations in the global coordinates by Eqs. (18.32) as

$$\bar{\mathbf{K}}_1$$

0	0	0	0	0	0	r_1	r_2	r_3	r_4	r_5	r_6	$\bar{\mathbf{S}}_1^F$	
2 480	0	0	0	0	0	−2 480	0	0	0	0	0	0	0
0	10	0	0	0	660	0	−10	0	0	0	660	0	0
0	0	20	0	−1 490	0	0	0	−20	0	−1 490	0	10	0
0	0	0	12 940	0	0	0	0	0	−12 940	0	0	0	0
0	0	−1 490	0	119 040	0	0	0	1 490	0	59 520	0	−300	0
0	660	0	0	0	52 910	0	−660	0	0	0	26 450	0	0
−2 480	0	0	0	0	0	2 480	0	0	0	0	0	0	r_1
0	−10	0	0	0	−660	0	10	0	0	0	−660	0	r_2
0	0	−20	0	1 490	0	0	0	20	0	1 490	0	10	r_3
0	0	0	−12 940	0	0	0	0	0	12 940	0	0	0	r_4
0	0	−1 490	0	59 520	0	0	0	1 490	0	119 040	0	300	r_5
0	660	0	0	0	26 450	0	−660	0	0	0	52 910	0	r_6

$$\bar{\mathbf{K}}_2$$

r_1	r_2	r_3	r_4	r_5	r_6	0	0	0	0	0	0	$\bar{\mathbf{S}}_2^F$	
6	0	0	0	0	−459	−6	0	0	0	0	−459	0	r_1
0	2 067	0	0	0	0	0	−2 067	0	0	0	0	0	r_2
0	0	14	1 033	0	0	0	0	−14	1 033	0	0	15	r_3
0	0	1 033	99 200	0	0	0	0	−1 033	49 600	0	0	540	r_4
0	0	0	0	10 782	0	0	0	0	0	−10 782	0	0	r_5
−459	0	0	0	0	44 089	459	0	0	0	0	22 044	0	r_6
−6	0	0	0	0	459	6	0	0	0	0	459	0	0
0	−2 067	0	0	0	0	0	2 067	0	0	0	0	0	0
0	0	−14	−1 033	0	0	0	0	14	−1 033	0	0	15	0
0	0	1 033	49 600	0	0	0	0	−1 033	99 200	0	0	−540	0
0	0	0	0	−10 782	0	0	0	0	0	10 782	0	0	0
−459	0	0	0	0	22 044	459	0	0	0	0	44 089	0	0

$$\tilde{K}_3$$

r_1	r_2	r_3	r_4	r_5	r_6	0	0	0	0	0	0	
2	0	0	0	−196	0	−2	0	0	0	−196	0	r_1
0	2	0	196	0	0	0	−2	0	196	0	0	r_2
0	0	1 102	0	0	0	0	0	−1 102	0	0	0	r_3
0	196	0	23 491	0	0	0	−196	0	11 746	0	0	r_4
−196	0	0	0	23 491	0	196	0	0	0	11 746	0	r_5
0	0	0	0	0	4 147	0	0	0	0	0	−4 147	r_6
−2	0	0	0	196	0	2	0	0	0	196	0	0
0	−2	0	−196	0	0	0	2	0	−196	0	0	0
0	0	−1 102	0	0	0	0	0	1 102	0	0	0	0
0	196	0	11 746	0	0	0	−196	0	23 491	0	0	0
−196	0	0	0	11 746	0	196	0	0	0	23 491	0	0
0	0	0	0	0	−4 147	0	0	0	0	0	4 147	0

The equilibrium equations in terms of the structure stiffness matrix and load vector are

$$
\begin{bmatrix}
2\,490 & 0 & 0 & 0 & -200 & -460 \\
0 & 2\,080 & 0 & 200 & 0 & -660 \\
0 & 0 & 1\,140 & 1\,030 & 1\,490 & 0 \\
0 & 200 & 1\,030 & 135\,630 & 0 & 0 \\
-200 & 0 & 1\,490 & 0 & 153\,310 & 0 \\
-460 & -660 & 0 & 0 & 0 & 101\,140
\end{bmatrix}
\begin{bmatrix}
r_1 \\ r_2 \\ r_3 \\ r_4 \\ r_5 \\ r_6
\end{bmatrix}
=
\begin{bmatrix}
0 \\ 0 \\ -25 \\ -540 \\ -300 \\ 0
\end{bmatrix}
$$

The solution of the equilibrium equations gives

$$
\begin{bmatrix}
r_1 \\ r_2 \\ r_3 \\ r_4 \\ r_5 \\ r_6
\end{bmatrix}
=
\begin{bmatrix}
-0.0001 \\ 0.0004 \\ -0.0161 \\ -0.0039 \\ -0.0018 \\ 0.0000
\end{bmatrix}
$$

which results in the following element displacement vectors in the global coordinates.

$$
\begin{array}{cccc}
\text{Element} \rightarrow & 1 & 2 & 3 \\
\bar{s} \quad : &
\begin{bmatrix}
0 \\
0 \\
0 \\
0 \\
0 \\
0 \\
-0.0001 \\
0.0004 \\
-0.0161 \\
-0.0039 \\
-0.0018 \\
0.0000
\end{bmatrix}
&
\begin{bmatrix}
-0.0001 \\
0.0004 \\
-0.0161 \\
-0.0039 \\
-0.0018 \\
0.0000 \\
0 \\
0 \\
0 \\
0 \\
0 \\
0
\end{bmatrix}
&
\begin{bmatrix}
-0.0001 \\
0.0004 \\
-0.0161 \\
-0.0039 \\
-0.0018 \\
0.0000 \\
0 \\
0 \\
0 \\
0 \\
0 \\
0
\end{bmatrix}
\end{array}
$$

Equation (18.31a) then gives the member-end force vectors as follows

$$
\begin{array}{cccc}
\text{Element} \rightarrow & 1 & 2 & 3 \\
S \quad : &
\begin{bmatrix}
0.35 \\
-13.08 \\
0 \\
49.94 \\
-0.19 \\
-431.10 \\
-0.35 \\
-6.92 \\
0 \\
-49.94 \\
-0.15 \\
61.70
\end{bmatrix}
&
\begin{bmatrix}
0.75 \\
-10.78 \\
0 \\
-19.42 \\
0.14 \\
-140.53 \\
-0.75 \\
-19.22 \\
0 \\
19.42 \\
-0.10 \\
748.03
\end{bmatrix}
&
\begin{bmatrix}
17.70 \\
-0.75 \\
0.35 \\
0 \\
-42.28 \\
-90.60 \\
-17.70 \\
0.75 \\
-0.35 \\
-0 \\
-21.13 \\
-45.26
\end{bmatrix}
\end{array}
$$

LABORATORY EXPERIMENTS

Suspended Roof

Using closed circular or square rings made of steel strip or circular rod (Fig. L18.1) plus cable segments, wire strands, or rods, build a suspended structure, with one or more joints outside the periphery. Load one or more joints. Draw the load–displacement curve. Compare with the analytical values.

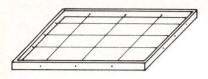

FIGURE L18.1

Three-Dimensional Structures

Build models of three-dimensional structures such as those of Figs. 18.3a, 18.9, P18.1, P18.2, P18.5, P18.9 or P18.11. Load and observe the deformation behavior, measure forces, moments, and displacements, and compare the results to those obtained analytically.

PROBLEMS

18.1. Find the reactions of the truss of Fig. P18.1. It consists of three members, hinged together at point 4 to form a tripod. The bars are hinged to the base by universal joints.

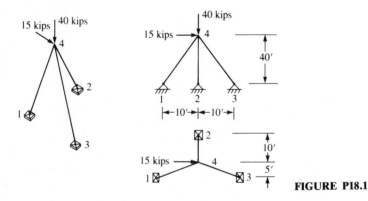

FIGURE P18.1

18.2. Draw the normal-force, shear, moment, and torque diagrams for the structure of Fig. P18.2.

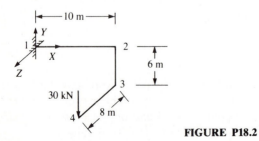

FIGURE P18.2

18.3. Find the bar forces for the three-dimensional truss of Fig. P18.3 by the method of joints.

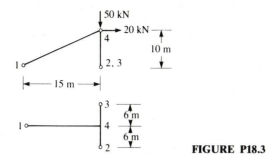

FIGURE P18.3

Find the solution of the following three-dimensional trusses and frames by the stiffness method.

18.4. The bar 12 shown in Fig. P18.4 is an element of a three-dimensional truss. Joint 1 is at the origin of the coordinates and joint 2 at $X = Y = Z = 100$ in. End 1 is fixed and end 2 undergoes a displacement 0.1 in in the Y direction and 0.2 in in the Z direction. Find the forces developed at the ends of the member. The cross-sectional area is 20 in^2 and $E = 29\,000$ ksi.

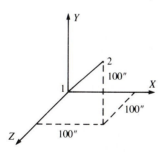

FIGURE P18.4

18.5. $A = 10$ in^2; $E = 29\,000$ ksi.

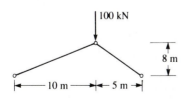

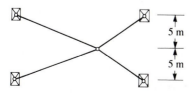

FIGURE P18.5

18.6. $A = 10 \text{ in}^2$; $E = 29\,000$ ksi.

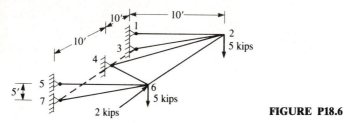

FIGURE P18.6

18.7. $A = 15 \text{ in}^2$ for the inclined members and 12 in^2 for the horizontal members. $E = 29\,000$ ksi.

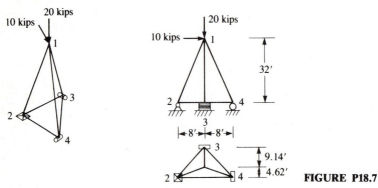

FIGURE P18.7

18.8. $A = 40 \text{ cm}^2$ for all members. $E = 200\,000$ MPa.

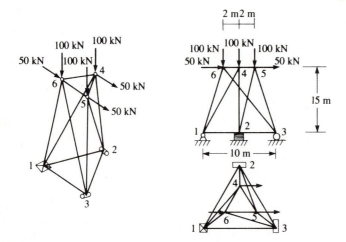

FIGURE P18.8

18.9. The top and base are both square shapes. $A = 20 \text{ in}^2$ for the diagonal bars, 10 in^2 for the top bars, and 15 in^2 for the bottom bars. $E = 29\,000$ ksi.

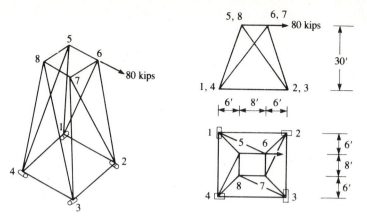

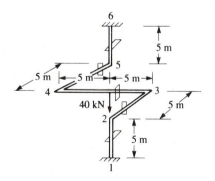

FIGURE P18.9

18.10. If the member of Fig. P18.4 is a three-dimensional beam element with end 1 fixed and end 2 undergoing the same displacements as those of the truss member in Problem 18.4 without any rotations, calculate the member-end forces. The cross-section is rectangular, 10 in × 20 in, with the longer dimension in the vertical plane. $E = 3\,100$ ksi, $v = 0.2$.

18.11. For the three-dimensional frame members in Fig. P18.11, $A = 120\,\text{cm}^2$ and the orientations of the cross-sections are as shown. The moment of inertia of the cross-sections about the strong and weak axes are $8\,000$ and $6\,000\,\text{cm}^4$, respectively. The torsion constant is $J = 5\,000\,\text{cm}^4$; $E = 200\,000$ MPa; $G = 80\,000$ MPa $(80 \times 10^6\,\text{kN/m}^2)$. A 40 kN vertical load is applied at the mid-length of member 34 as shown.

FIGURE P18.11

18.12. For the three-dimensional frame members of Fig. P18.12, $E = 3\,100$ ksi; $v = 0.2$. The sections are all 10 in × 20 in, and oriented as shown in the figure. The load consists of a uniform gravity load of 3 kip/ft acting on members 12, 23, 24 and 35.

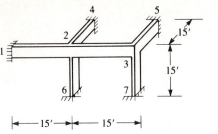

FIGURE P18.12

18.13. The two supporting members of the three-dimensional frame in Fig. P18.13 are 6 in × 12 in, with the longer dimension in the vertical plane. The other member is 8 in × 36 in, with the smaller dimension in the vertical plane. A uniform gravity load of 5 kip/ft is acting on the latter member. Find the member forces. $E = 3\,100$ ksi; $v = 0.2$.

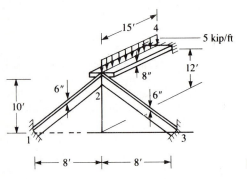

FIGURE P18.13

18.14. Calculate the bar forces for the three-dimensional frame of Fig. P18.14. The second element is horizontal and a gravity load of 10 kips is acting in the middle of this element. It is located at a height of 10 ft. The strong plane of the first member, 12, makes a 45° angle with the vertical axis, for the other two members the strong plane is in the vertical direction. All three members have a 8 in × 14 in section and $E = 3\,000$ ksi, $v = 0.2$.

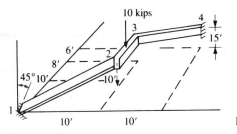

FIGURE P18.14

CHAPTER
19

APPLICATION OF COMPUTERS IN STRUCTURAL ANALYSIS

19.1 COMPUTER SOFTWARE

Computers have many applications in structural engineering. Three general categories of programs and their applications are discussed here. The first category are commercial programs, developed for business applications, that are useful to structural engineers. The second category are commercial or otherwise available programs for analysis and design of structures. The third category are programs developed for specific tasks. Examples of the first category are word processors, spreadsheets, drafting packages, and symbolic programs. General-purpose finite-element programs belong to the second group. The third category are programs developed by the user. In view of the many available programs, the third category usually consists of small, special-purpose, programs. In the following sections we briefly discuss these classes of programs as they apply to structural engineering problems.

Word Processors

Word processors are used for writing and editing reports, letters, etc. They may also be used for entering and editing program statements. Word processors offer many features, including moving, deleting, and formatting texts. A dictionary may be available for detecting and correcting spelling errors. Some word processors also provide a thesaurus of synonyms. Others allow for scientific and mathematical notations and simple drawings. Examples of word processors are Microsoft Word (Microsoft Corp., 16011 NE 36th Way, Redmond, WA 98073, U.S.A.) and Wordperfect (Wordperfect Corp., Orem, UT 84057, U.S.A.). Pc-Write, a shareware program, (Quicksoft, 219 First N. Seatle, WA 98109, U.S.A.) is an example of a basic word processor. For writing texts with mathematical formulas, Chi Writer (Horstman Software Design Corp., Ann Arbor, MI 48103, U.S.A.) would be suitable. Some word processors allow simple sketches made of horizontal and vertical lines. Others (e.g. Microsoft Word and Wordperfect) can import sketches drawn with the aid of a drafting package, or even images reproduced from photographs by a scanner.

Spreadsheets

Spreadsheets are used to perform many calculations. They allow formatting, moving, copying, etc. Their strength lies in their ability to enter a formula for a set of arguments and repeat the calculation for other arguments using the copy command. Producing a graph from a set of data is one of the many useful functions readily performed by spreadsheets. Some spreadsheets, for example Lotus (Lotus Development Corp., 5 Cambridge Parkway, Cambridge, MA 02142, U.S.A.), and the shareware spreadsheet Aseasy (Trius Inc., 15 Atkinson Street, MA 01905, U.S.A.) provide functions for matrix operations, a useful tool for solving problems by the stiffness method. Most stiffness method examples and assigned problems in this book may be solved totally or partially by using such spreadsheets. These would allow the user to avoid tedious matrix operations and concentrate on the solution process. Some spreadsheets provide macros that allow automation of many tasks. An automated solution procedure has been devised by the author using Lotus macros for the stiffness solutions of trusses and frames, using the steps outlined in Chapter 12. They seem to provide a tool for learning the stiffness method, since they show animation of the steps in the solution process.

Drafting Programs

Drafting packages can be useful in producing a variety of graphs and sketches, layouts of buildings, and design details. Some allow computation of geometric and material quantities. There are drafting packages available that combine interactive graphics and analysis programs into computer-aided design (CAD) packages. Autocad and its smaller version, Autosketch (Autodesk Corp., 2320 Marinship Way, Sausalito, CA 94965, U.S.A.) and Cadkey (Cadkey Inc., 440 Oakland Street, Manchester CT 06040, U.S.A.) are examples of drafting packages.

Symbolic Programs

Symbolic programs are different from other computer programs because they are primarily used to carry out closed-form solutions of problems. These programs can also be used for numerical solutions or a combination of numerical and symbolic manipulations. They can handle algebraic expressions, integrals, derivatives, matrix operations, solution of parametric equations, and differential equations, and can perform many other tasks symbolically. They can be useful in derivations of procedures that would otherwise involve a substantial amount of algebra. For example, derivation of stiffness matrices for use in stiffness method and finite-element programs. Examples of symbolic systems for microcomputers are MuMath (distributed by Microsoft Corp.) and Macsyma (for 386 microcomputers, Symbolics Inc., 11 Cambridge Center, Cambridge MA 02142, U.S.A.).

Programming Languages

If spreadsheets are not appropriate for performing an analytical task and other suitable programs are not available, then it may be necessary to develop a computer program for the purpose. Programming can be done in a number of languages. The common language for engineering tasks in the past has been FORTRAN. At present other possibilities exist depending on the requirements of the program. For example, if the aim of the program is merely numerical calculations, FORTRAN may still be the best tool. Many FORTRAN compilers are available; some examples are Microsoft's (see word processors) with codeview which simplifies the debugging, and WATFOR-77 (WATCOM Group Inc., 415 Phillip St., Waterloo, Ont. Canada N2L 3X2). For smaller but more interactive programs, other languages such as BASIC or PASCAL may be preferable. BASIC has become the standard language for microcomputers and is suitable for many engineering problems. It is easy to use and allows development of interactive programs and graphic output. When compiled (for example with Microsoft's Quick Basic compiler) it is comparable in speed to FORTRAN. A more powerful language for interactive programming is C, which is closer to machine language, C provides a variety of concise commands and is used for most professionally developed programs as well as for development of compilers and system programs. FORTRAN subroutines may be called from routines written in C and other languages.

Another system for performing numerical calculations of all matrix and mathematical operations in a compact form is Matlab (Matrix Laboratory, The Mathworks, Inc., 20N. Main St. Sherborn, MA 01770 U.S.A.). This is an efficient system and like the symbolic programs can be used interactively or by program functions. The high level language provided and the large number of available functions make the programing in this system very easy. Matrices can be input as a string of numbers, without requiring dimensioning. They can be included in operations by using the name of the matrix. Parts of a matrix can also be easily manipulated.

A large number of mathematical and statistical functions are provided by the IMSL FORTRAN package for both mainframes and microcomputers (IMSL, 2500 City West Blvd., Houston, TX 77042 U.S.A.).

19.2 PROGRAMMING THE STIFFNESS METHOD

This stiffness method is now the most common procedure used for the analysis of structures. The programs for different types of elements vary only slightly by requiring different stiffness matrices and a few other details. It is also not difficult to develop a program that can incorporate a variety of elements and solution procedures. Programming the stiffness method, if done with some planning and with a knowledge of the subtle aspects of bookkeeping, should not present any particular difficulty. In the following paragraphs, the main facets of programming this method that require special attention are discussed. The steps are given in a form that can easily be converted to any programming language. The trivial aspects of translating the expressions into program statements and transferring the variables from one routine to another are omitted.

Application of the stiffness method involves the element stiffness matrix, the transformation matrix, and development of the element stiffness matrix in the global coordinate system. The element stiffness matrices in the global coordinate system are assembled into the structure stiffness matrix and the load vector is established. This results in a set of equilibrium equations. The solution of this system of simultaneous linear equations yields the joint displacements and rotations. The member-end displacement vectors are obtained from the joint displacements, which in turn lead to the member-end forces.

When developing programs, modular programming eases the development and debugging of the routines. Each module should perform a specific task. A stiffness program can be divided into four parts: the main routine, the input module, the solution and the output module. In the following discussions a beam element with six degrees of freedom is used to develop a program for two-dimensional frames. The procedure can be used with some modification for other structures using the stiffness or finite-element method.

The Main Routine

The main routine represents the skeleton of the computer program. It should be clear and concise, consisting of the main lines of the solution process. It should resemble the skeletal flow diagram of the program. This routine calls other routines to perform specific tasks and should not be cluttered by details. The main routine would include the following phases.

1. Starting statements.
2. Call the input routine to read the input data. The degrees of freedom may also be calculated in the input routine.
3. Call the solution routines to carry out the solution of the problem. The following operations are needed to complete phase 3.
   a. Calculate the stiffness matrices in the local and global coordinates and assemble the global stiffness matrices into the structure stiffness matrix. In this manner each global element stiffness matrix is included into the structure stiffness matrix as soon as it is calculated. The local element stiffness matrices will be needed at the end of the solution process for finding the member forces, and can be recalculated at that time. Thus, the same two 6×6 matrices can be used to store the local and global element stiffness matrices. For microcomputers this procedure may be preferable to the alternative that keeps all the local stiffess matrices in the memory or to the one that writes them to a disk file, because of the memory and speed limitations of such computers. Although the latter alternative may become necessary if very large problems are to be handled.
   b. Call the routine to assemble the load vector.
   c. Call the routine to solve the system of linear simultaneous equations. This results in the joint displacements in the global coordinates.
   d. Call the routine to calculate the member force vectors in the local (member) coordinates.
4. Call the routine to produce on screen, or hard copy, numerical or graphical output.

It should be noted that in the solution process discussed above the displacements are found in the global coordinates while the forces are calculated in the local coordinates. This is appropriate because the forces are needed in the local coordinates for constructing shear and moment diagrams, while the displacements are required in the global coordinates. The details of each phase are given in the following paragraphs.

Starting statements. The starting statements depend on the language and the computer used. They may include dimensioning of the arrays, common statements, initializing the parameters and arrays, setting up the monitor screen mode and color for microcomputers, etc.

Input Data

The input data consist of the geometry of the structure, section, and material properties and loads. The first part of the input data provides some general information. The input data should be printed out as soon as they are read in, in order to help locate possible errors in the input data.

The arrangement of the input data depends on the design of the computer program. The latter is in turn dependent on the computer hardware. Thus, if the program is for large problems using a fast machine, such as a mainframe computer, then care must be taken that the input preparation is not overwhelming. Commerical programs may have input modules with interactive data-generation capabilities or mesh-generation routines. Even for large problems, where a large number of joints and members are used, the number of different sections, materials and loads are relatively small. Thus, the section and material properties and loads should be input separately from the elements and joints.

The first portion of the input module would require input of the following common data.

Common data

1. Title
2. Number of joints, NJ, number of elements, NE, number of different sections, NS, number of different materials, NM, number of loaded joints, NLJ.

The number of sections, materials and that of loaded joints are provided to avoid input of section and material properties for each element and load input for each joint. For most practical problems only a few different sections and materials are used and a small number of joints are loaded.

The parameters needed to calculate the stiffness and transformation matrices include the member lengths, cross-sectional areas, moments of inertia, and moduli of elasticity. The transformation matrix requires the orientation of the member as well. The best way to calculate the elements of the transformation matrix is to provide the coordinates of the joints in the global coordinate system. The member lengths as well as sines and cosines of the angles can be found using these coordinates. Information is also needed on the fixity conditions of the joints. A code can be used to indicate whether the joint is fixed or free in a given degree of freedom. For example, a number 1 may indicate that the joint is restrained while a 0 may mean that a joint is free to displace in that degree of freedom. The joint data can thus be input in the following form.

Joint data. For $J = 1$ to NJ (total number of joints)

Input, JN [joint no], X(JN), Y(JN), and F(JN, K), with $K = 1, 2, 3$ where X(JN) and Y(JN) are the X and Y coordinates of the joint, F(JN, K) indicates the fixity condition of the joint in X or Y directions and against rotation. A value of 1 would indicate fixed, while a zero value means that the joint is free to displace in that degree of freedom. $K = 1, 2, 3$ indicates the degrees of freedom of the joint (3 for frame elements) in the X and Y directions and the rotation degree of freedom. For example, $F(5, 1) = 1$ implies that joint number

5 is fixed in the X direction. The array F is used later in the program to identify the degrees of freedom for assembling the structure stiffness matrix and the load vector. The degree of freedom associated with each joint can be determined, using the following statements, after the information for all the joints is input. Determination of the degrees of freedom for the joints and the total number of degrees of freedom is done in the following loops. The brackets indicate the limits of the loops.

For J = 1 to NJ (total no. of joints)

    For K = 1 to 3 (number of dof per joint)

    If F (J, K) = 0 then Q = Q + 1, F(J, K) = Q

    else F(J, K) = 0

The degrees of freedom associated with each joint may be optionally printed out to help in the trouble shooting.

Element data. The element data consist of connectivity, that is, joint numbers and section numbers associated with each element. The element data must be provided for all the elements as follows.

    For N = 1 to NE (total no. of elements)

    Input M, NI(M), NJ(M), (if NS > 1)SN(M), (If NM > 1)MN(M)

if NS = 1, SN(M) = 1

if NM = 1, MN(M) = 1

where NI(M), NJ(M), SN(M) and MN(M) are the first and second joints, section number and material number for element M.

Section data. The section data are required for all the different sections used in the structure.

    For N = 1 to NS (total no. of sections)

    Input SN, AA(MM), IS(MM)

where SN is the section number, AA and IS are the cross-sectional area and moment of inertia of the section.

Material Data

    For M = 1 to NM (total no of materials)

    Input MN, EE(MN)

where MN is the material number and EE(MN) modulus of elasticity for that material.

LOAD DATA

    For J = 1 to NLJ (total number of loaded joints)

    Input JL(J)

    JN = JL(J)

    Input PP(JN, 1), PP(JN, 2), PP(JN, 3) (loads for 3 dof per joint)

⎡ For N = 1 to NLE (total number of loaded elements)
| Input LE(N)
| LL = LE(N)
⎣ Input P(LL) (distributed lateral load)

JL(J) is the number of the loaded joint and LE(N) the number of loaded element. Here we are providing three load components for each loaded joint. If some of the components at a given loaded joint are not present, then a zero value is entered.

SOLUTION ROUTINES

The solution portion of the program is divided into four parts, as indicated in the discussion of the main program.

Assembly of the structure stiffness matrix and load vector. The local stiffness matrix for a frame element with six degrees of freedom is given by Eq. (12.7a). The global stiffness matrix can be found either directly by Eq. (12.13d) or by carrying out the operation indicated by Eq. (12.13b). The former process is more efficient.

The structure stiffness matrix is a $Q \times Q$ matrix, where Q is the total number of degrees of freedom calculated in the input module. The assembly process consists of examining the stiffness matrix of each element in the global coordinates, identifying the joints associated with the element, and determining the degrees of freedom associated with each joint. The terms of this stiffness matrix are then added to the location of the structure stiffness matrix associated with those degrees of freedom. This can be done in the following manner.

⎡ For N = 1 to NE (total no. of elements)
|
| K1 = IS(N)   (Retrive the section properties for the
| A = AA(K1): I = IS(K1)   element)
| M1 = MN(N)   (Identify the first and second joints of the
| E = EE(M1)   element and their coordinates)
| N1 = NI(N)
| N2 = NJ(N)
| X1 = X(N1): X2 = X(N2): Y1 = Y(N1): Y2 = Y(N2)
|
| Call the routine to calculate the global stiffness matrix, KG for the element
|
| ⎡ For K = 1 to 3   (Identify the joint degrees of freedom
| | L(K) = F(N1, K)   and put them in the locater matrix L)
| ⎣ L(K + 3) = F(N2, K)
| ⎡ For I = 1 to 6   (Sweep the global stiffness matrix and
| | I1 = L(I)   add the stiffness terms with nonzero
| | ⎡ For J = 1 to 6   degrees of freedom to the pertinent
| | | J1 = L(J)   locations in the structure stiff matrix).
⎣   ⎣   ⎣ If I1 ≠ 0 and J1 ≠ 0 then KS(I1, J1) = KS(I1, J1) + KG(I, J)

Load vector. The structure load vector can be assembled in a manner similar to the structure stiffness matrix. The loaded element numbers were stored in array LE and the loaded joint numbers were stored in array JL, in the input module, and the loads in PP(J1, K), with $K = 1, 2, 3$ and P(LL). J1 is the number of loaded joint; K indicates the x, y or, θ directions and LL the loaded element number. We recover the loaded joint numbers, identify the degrees of freedom associated with them, and use the latter to store the load in a $Q \times 1$ load vector R.

For N = 1 to NE
For NN = 1 to NLE
⎡If N = LE(NN)
⎣Call FE(NN, SF) (to find fixed end forces)
J1 = NI(N), J2 = NJ(N)
For K = 1 to 3
M1 = F(J1, K), M2 = F(J2, K)
If M1 ≠ 0 then R (M1) = −SF(K)
If M2 ≠ 0 then R(M2) = −SF(K)
FE is the routine that calculates the fixed end forces for uniformly distributed loads in terms of the load P(NN) and span length as
SF(2) = SF(4) = P(N)L/2
SF(3) = −SF(6) P(N)L²/12

⎡ For J = 1 to NLJ (total no. of loaded joints)
⎢ J1 = JL(J)
⎢ ⎡For K = 1 to 3 (no. of dof per joint)
⎢ ⎢M1 = F(J1, K)
⎣ ⎣If M1 ≠ 0 then R(M1) = R(M1) + PP(M1, K)

Solution of the equilibrium equations. Once the structure stiffness matrix KS and the load R vector are established, the equilibrium equations can be written in terms of the unknown joint displacements. The joint displacements r(I) are found by solving a system of simultaneous linear equations. The joint displacements are printed out along with joint numbers and the direction of the displacements. Any equation solver can be used. Since the structure stiffness matrix is symmetric, an eqution solver for symmetric matrices may be preferable. Routines for solving linear systems of equations are readily available. The following Gaussian elimination routine, for example, can be used as a subprogram for solving any linear system of equations. To call this routine KS and R are used for a and b. Note that upon solution the joint displacements are stored in the load vector R, since the latter is no longer needed.

Given the number of equations n, the matrix of coefficients, a(n, n) and the known vector b(n) (the right-hand side of the equations), the unknown values are placed in the vector x(n) upon the solution. In this routine the variables d and d_1 are used to store the intermediate values.

Gaussian elimination procedure for ax = b

```
For k = 1 to n − 1
   For i = k + 1 to n
   d = a(i, k)/a(k, k)
      For j = k to n
      a(i, j) = a(i, j) − d * a(k, j)
      b(i) = b(i) − d * b(k)
   x(n) = b(n)/a(n, n)
For k = n − 1 to 1 step −1
d₁ = 0
   For m = k + 1 to n
   d₁ = d₁ + a(k, m) * x(m)
   x(k) = (b(k) − d₁)/a(k, k)
```

Element-displacement and member-end force vectors. Before calculating the element force vectors, we must establish the element displacement vectors, in the global coordinates from the joint displacements in the global coordinates as follows.

```
For M = 1, NE (total no. of elements)
N1 = NI(M)                          (Identify the joints associated with the
N2 = NJ(M)                          element)
   For K = 1 to 3                   (Find the degree of freedom number for
                                    each joint)
      If (F(N1, K) ≠ 0,
      J1 = F(N1, K)
      D(K) = r(J1)                  (Set the values of the displacement vector
                                    equal to that of joint displacements)
      If (F(N2, K) ≠ 0,
      J2 = F(N2, K)
      D(K + 3) = r(J2)
   MM = SN(M), NN = MN(M)           (Find the section and material numbers
                                    for the element)
   A = AA(MM), I = IS(MM), E = EE(NN)
   X1 = X(N1), X2 = X(N2), Y1 = Y(N1), Y2 = Y(N2)
   Call the routine to obtain matrix B, Eq. (12.12b)
      For M1 = 1 to 6              (Find the force vector S = BD)
      S(M1) = 0.
         For N1 = 1 to 6
         S(M1) = S(M1) + B(M1, N1) * D(N1)
```

For NL = 1 to NLE (add fixed-end forces)
  If M = NL
  Call routine FE to find the fixed end forces SF for element LE(M)
  For L1 = 1 to 6
  S(L1) = S(L1) + SF(L1)

The force vector can be printed out after it is calculated for each element.

An alternative is to calculate the element stiffness matrix in the local coordinates, KE, and the transformations matrix, T, Eqs. (12.7a), and (12.8b). However, the above procedure using B is easier to program and more efficient to use.

Note that all the global stiffness matrices are symmetric and half of the terms can be found by symmetry after the terms in the first half are calculated.

OUTPUT MODULE

The output module provides numerical and graphic results of the calculations, including the joint displacements in the global coordinates and the member forces in the local coordinates. Graphic representations of the geometry and loads are useful for quick detection of errors in the input data.

19.3 EXAMPLES OF MICROCOMPUTER PROGRAMS

Microcomputers provide useful tools both for analyzing structures and for educational purposes. A large number of programs are available commercially, from listing in text books, via shareware, or from some academic institutions.

A number of programs have been developed by the author for the purpose of structural analysis and as teaching aids for the methods discussed in this book. These programs are interactive and include numerical and graphic outputs. Two programs for continuous beams and their influence lines (based on the singularity method) can handle rigid and elastic supports, variable cross-sections, and internal hinges. The input data are in the form of spreadsheets and can be recalled and the values can be modified on the screen. Two programs for two-dimensional trusses and frames allow the solution of those structures. The author has also developed macros for LOTUS that may be used with spreadsheets for automated solutions of frames and trusses. The steps in the solution as described in this text are shown on the screen. These programs can be obtained from the author. A more suitable device for instruction of structures as well as for solution of a variety of problems is a toolbox developed by the author for use with MATLAB, described under symbolic programs. The toolbox allows the solution of static, stability, and dynamic problems for a variety of structures, including two- and three-dimensional trusses and frames, arches, bow-girders, plate and solid finite elements, etc., and combinations of such elements.

The toolbox is divided into two parts. The first one consisting of interactive programs for application with data modification capabilities. The second part consists of a variety of modules suitable for teaching and learning the methods of structural analysis discussed in this book plus other problems such as those of structural dynamics. With these modules the problems are solved in the same manner as the hand solutions discussed in this book. However the matrices and operations are performed by the computer in response to simple commands issued by the user.

LABORATORY EXPERIMENTS

Program Modules for the Stiffness Method

The following assignments are to be done in a computer laboratory or by using a microcomputer in a number of sessions.

These program modules may be written in BASIC, FORTRAN, C, or any other language. After all the modules are completed, they can be combined into a single program. In this case the modules should either be written in the same language or in compatible languages. Each module may also be assigned in several parts. The modules may be written in the form of subprograms, with each subprogram performing a specific task. Each module should be tested before combining it with the rest.

a. Write a main program module (see section The Main Routine in the text) that calls the input module, the solution routine, the equation solver, and the force module.

b. Develop a program for input data (see the section Input Data) to read the input data for the joints, elements, sections, material properties, and loads. This subprogram should also calculate the fixity matrix to establish the degrees of freedom associated with the joints of each element.

c. Write a program module consisting of a set of routines (see Solution Routines in the text) for the solution process. These will include:

A subprogram for calculating the stiffness matrix in the local coordinate system.

A subprogram to calculate the transformation matrix.

A subprogram to find the element stiffness matrix in the global coordinate system. This may be done either directly or by using the stiffness matrix in the local coordinates and the transformation matrix.

d. Write a subprogram for the solution of a system of simultaneous equations. Use this program to calculate and print the joint displacements in the global coordinates.

The Gaussian elimination procedure given under Solution of equilibrium equations, or any other procedure may be used. A short routine may be added for reading the input and providing the output in order to use this module as a stand alone equation solver for other applications.

e. Write a program module that establishes the vector of member-end displacements and calculates and prints the member-end forces.

f. Optionally, a subprogram may be developed for plotting the nodes and elements with their numbers.

PROBLEMS

Problems 19.1 to 19.16 are to be done by a spreadsheet program. The copy command should be used to perform the repetitive operations.

19.1. Calculate the expressions for the moment and deflection of Example 5.3 for x from 0 to 30 ft with 1-ft increments. Graph the moment and deflection diagrams.

19.2. Set up a spreadsheet similar to the table in Example 6.3 to calculate the deflection of a truss from the given bar forces and properties.

19.3. Set up a spreadsheet to calculate the expressions for moment and deflection in Example 9.2 from 0 to 30 ft with 2-ft increments. Use the @IF function to set the values of step function equal to 0 or 1. Plot the moment and deflection curves.

19.4. Set up a spreadsheet to carry out the solution of Example 10.2 by moment distribution.

19.5. Use a spreadsheet to calculate the influence line of moment in Example 9.10 (for a fixed x and varying ξ). Plot the influence line.

19.6. Use a spreadsheet to calculate and plot the influence line of R_2 in Example 9.12.

19.7. Repeat Example 11.4, solution of the continuous beam, using a spreadsheet to calculate the stiffness matrices and fixed-end forces. Also calculate the member-end forces after the solution of the equilibrium equations is obtained. If the spreadsheet program allows the soluton of simultaneous equations (e.g. Lotus or Aseasy), then carry out the solution of the equilibrium equations with the spreadsheet as well.

19.8. Repeat Problem 19.7 for Example 12.2, to carry out the solution of the indeterminate truss by the stiffness method.

19.9. Repeat Problem 19.7 for Example 12.5 to carry out the solution of the two-dimensional frame problem.

19.10. Use a spreadsheet to calculate the thrust of a cable with a 60-ft span under three loads, 10, 20, and 10 kips applied at 20, 30, and 40 ft. The sag at the mid-span is 5 ft. Use Eq. (14.4) in Example 14.4. Plot the sag versus distance to get the deflected shape of the cable.

19.11. Repeat Example 15.4 using a spreadsheet to calculate the stiffness and transformation matrices and complete the solution.

19.12. For a hinged-end 10 ft long column, with an initial deflection 2 in at the center, calculate and plot the total deflection (Eq. (16.13)) while increasing the ratio of the axial to buckling load from zero to 100% by 5% increments.

19.13. Repeat Example 16.4 using a spreadsheet.

19.14. Repeat Example 17.7 using a spreadsheet.

19.15. Repeat Example 18.3 using a spreadsheet.

19.16. Repeat Example 18.6 using a spreadsheet.

Problems 19.17 to 19.24 are to be done by a symbolic manipulation program.

19.17. Carry out the integration by a symbolic manipulation routine to find κ in Example 6.1.

19.18. Derive the first row of the flexibility matrix for an arch (Eq. (15.2a)) using a symbolic manipulation program.

19.19. Calculate the terms 1,1 1,3 and 2,2 of the geometric stiffness matrix (Eq. (16.21b)) using a symbolic manipulation program.

19.20. Repeat Example 16.3 using a symbolic manipulation program.

19.21. Repeat Problem 16.5 using a symbolic manipulation program to find the buckling load in terms of E, I and l.

19.22. Repeat Problem 16.12 using a symbolic manipulation program.

19.23. Repeat Problem 16.13 using a symbolic manipulation program.

19.24. Repeat Problem 16.14 using a symbolic manipulation program.

Note: Programming problems for the stiffness method are given under Laboratory Experiments for this chapter.

BIBLIOGRAPHY

Au, Tung A., and P. Christiano: *Structural Analysis,* Prentice-Hall, Englewood Cliffs, NJ, 1987, 629 pp.

Berg, Glen V.: *Elements of Structural Dynamics,* Prentice-Hall, Englewood Cliffs, NJ, 1989, 268 pp.

Chajes, Alexander: *Structural Analysis,* 2nd ed., Prentice-Hall, Englewood Cliffs, NJ, 1981, 468 pp.

Cook, Robert D.: *Concepts and Applications of Finite Element Analysis,* 2nd ed., Wiley, New York, 1981, 537 pp.

Gerstle, Kurt: *Basic Structural Analysis,* Prentice-Hall, Englewood Cliffs, NJ, 1974, 498 pp.

Hibbler, Russel C.: *Structural Analysis,* 2nd ed., Macmillan, New York, 1990. 688 pp.

Hsieh, Yuan-Yu: *Elementary Theory of Structures,* 2nd ed., Prentice-Hall, Englewood Cliffs, NJ, 1982, 416 pp.

Kennedy, John B., and Murty K. S. Modugula: *Elastic Analysis of Structures Classical and Matrix Methods,* Harper & Row, New York, 1990, 800 pp.

Langhaar, Henry L.: *Energy Methods in Applied Mechanics,* Wiley, New York, 1962, 350 pp.

Laursen, Harold L.: *Structural Analysis,* 3rd ed., McGraw-Hill, New York, 1988, 475 pp.

Leet, Kenneth M., *Fundamentals of Structural Analysis,* Macmillan, New York, 1988, 565 pp.

Leonard, John W.: *Tension Structures,* McGraw-Hill, New York, 1988, 177 pp.

Martin, Harold: *Introduction to Matrix Methods of Structural Analysis,* McGraw-Hill, New York, 1966, 331 pp.

McCormack, Jack, and Rudolf E. Elling: *Structural Analysis: A Classical and Matrix Approach,* Harper & Row, New York, 1988, 608 pp.

Megson, T. H. G.: *Strength of Materials for Civil Engineers,* 2nd ed., Edward Arnold, London, 1987, 357 pp.

Melosh, Robert J.: *Structural Engineering Analysis by Finite Elements,* Prentice-Hall, Englewood Cliffs, NJ, 1990, 308 pp.

Norris, C. H., J. B. Wilbur, and S. Utku: *Elementary Structural Analysis,* McGraw-Hill, New York, 1976, 673 pp.

Popov, E. P.: *Introduction to Mechanics of Solids,* Prentice-Hall, Englewood Cliffs, NJ, 1968, 571 pp.

Reddy, J. N.: *Energy and Variational Methods in Applied Mechanics,* Wiley, New York, 1984, 545 pp.

Sack, Ronald L.: *Structural Analysis,* McGraw-Hill, New York, 1984, 652 pp.

Shames, Irving H.: *Introduction to Solid Mechanics,* 2nd ed., Prentice-Hall, Englewood Cliffs, NJ, 1989, 701 pp.

Smith, J. C.: *Structural Analysis,* Harper & Row, New York, 1988, 699 pp.

Timoshenko, S. P., and J. H. Gere: *Theory of Elastic Stability,* 2nd ed., McGraw-Hill, 1966, 541 pp.

Timoshenko, S. P., and D. H. Young: *Theory of Structures,* 2nd ed., McGraw-Hill, 1965, 629 pp.

Wang, Chu-Kia, and Charles G. Salmon: *Introductory Structural Analysis,* Prentice-Hall, Englewood Cliffs, NJ, 1984, 591 pp.

Weaver, William, and Paul R. Johnston: *Finite Element for Structural Analysis,* Prentice-Hall, Englewood Cliffs, NJ, 1984, 448 pp.

West, H. H.: *Analysis of Structures,* 2nd ed., Wiley, New York, 1989, 707 pp.

White, R. N., P. Gergely, and R. G. Sexsmith: *Structural Engineering,* Vols. 1 & 2, Introduction to Design Concepts and Analysis, 2nd eds., Wiley, New York, 1976, 670 pp., Vol. 3, Behavior of Members and Systems, Wiley, New York, 1974, 583 pp.

Zienkiewicz, O. C., and R. L. Taylor: *The Finite Element Method,* Vol. 1, Basic Formulation and Linear Problems, 4th ed., McGraw-Hill, New York, 1989, 648 pp.

ANSWERS
TO SELECTED
PROBLEMS

Chapter 1

1.2 (a) 10.7 lb/ft   (b) 1 160 lb/ft   (c) 2 870 Pa   (d) 23.5 psf
(e) −7.0 psf   (f) $F_1 = 750$ lb, $F_2 = 1\,500$ lb

1.5 $E = 3.43 \times 10^3$ ksi

1.7 $P_s = 0.84$ P, $P_c = 0.1$ P, $P_w = 0.06$ P; 8.4 in² concrete

Chapter 2

2.3 $R_{1x} = -120$ lb, $R_{1y} = -18$ lb, $R_2 = 18$ lb

2.6 $R_{1x} = 155.0$ kips, $R_{1y} = 153.4$ kips, $R_{2y} = 158.3$ kips

2.8 $R_{1x} = 0$, $R_{1y} = 57.2$ kips, $R_2 = 8.8$ kips

2.9 $R_{1x} = 9$ kips, $R_{1y} = 57.5$ kips, $R_2 = 25.75$ kips

2.11 $R_{1x} = 18$ kips, $R_{2x} = 18$ kips, $R_{1y} = 24$ kips, $R_{2y} = 24$ kips

2.13 $R_{1x} = -10$ kips, $R_{3x} = 10$ kips, $R_{1y} = 10$ kips, $R_{3y} = 0$

2.15 $R_{1x} = 94$ kips, $R_{2x} = 94$ kips, $R_{1y} = 98$ kips, $R_{2y} = 67$ kips

2.17 $R_{1x} = 0$, $R_{1y} = 0$, $R_3 = 5$ kips, $R_4 = 5$ kips, $R_6 = 0$

2.19 $R_{1x} = 56.67$ kN, $R_{2x} = -56.67$ kN, $R_{3x} = 0$, $R_{1y} = 170$ kN, $R_{3y} = 0$

2.21 $R_{1x} = -12$ kN, $R_{1y} = 9.4$ kN, $R_2 = 9.6$ kN

2.23 $R_{1x} = 0$, $R_{2x} = 0$, $R_{1y} = 15$ kips, $R_{2y} = -20$ kips, $R_3 = 40$ kips,
$M_1 = -75$ kip · ft

Chapter 3

3.1 unstable

3.3 internally and externally determinate

3.5 indeterminate

3.7 determinate

3.9 unstable

3.11 internally indeterminate

3.14 inspection shows it is unstable (3 bars (parts) and 3 joints are added to a stable part)

3.16 $F_{12} = 56$ kN (c), $F_{24} = 14$ kN (c), $F_{23} = 30$ kN (T), $F_{13} = 10$ kN (T), $F_{34} = 10$ kN (T)

3.19 $F_{12} = 6.75$ (T), $F_{13} = 6$ (T), $F_{14} = 3.75$ (T), $F_{24} = -11.25$ (c), $F_{34} = -4.5$ (c), $F_{35} = 6$ (T), $F_{45} = -7.5$ (c), (kips)

3.20 $F_{12} = 41.67$ (c), $F_{13} = 33.33$ (T), $F_{23} = 0$, $F_{24} = -53.33$ (c), $F_{25} = 25.0$ (T), $F_{35} = 33.33$ (T), $F_{45} = -15$, $F_{46} = -8.33$ (T), $F_{48} = -46.67$ (c) (kips) symmetric on right side

3.21 $F_{12} = -139.75$ (c), $F_{13} = 125$ (T), $F_{23} = -17.8$ (c), $F_{24} = -130.6$ (c), $F_{45} = -36.2$ (c), $F_{46} = -117.4$ (c), $F_{47} = 20$ (T), $F_{57} = 36.1$ (T), $F_{68} = -106.2$ (c), $F_{78} = 54.1$ (T) (kips)

3.24 $F_{12} = -16.40$ (c), $F_{14} = 12.67$ (T), $F_{23} = -14.91$ (c), $F_{24} = 2.83$ (T), $F_{34} = 3.33$ (T), $F_{35} = -14.91$ (c), $F_{45} = 6.6$ (T), $F_{46} = 9.68$ (T), $F_{56} = -19.38$ (c) (Forces in kips)

3.27 $F_{12} = -16.73$ (c), $F_{13} = -6.34$ (c), $F_{23} = 8.66$ (T), $F_{24} = -16.73$ (c), $F_{34} = -1.34$ (c) (kips) symmetry on other side

3.29 $F_{17} = 21.63$ (T), $F_{23} = -16$ (c), $F_{28} = 7.21$ (T), $F_{78} = 12$ (T) (kips)

3.32 $F_{63} = -54.28$ (c), $F_{67} = -2.29$ (c), $F_{6-10} = -36.92$ (c), $F_{10-7} = 34.36$ (T) (kips)

3.33 $F_{47} = -20$ (c), $F_{57} = -2.80$ (c), $F_{58} = 2.80$ (T), $F_{68} = 20$ (T) (kips)

3.36 $F_{23} = 0$, $F_{24} = 50$ kips (T)

3.40 $F_{23} = 235.7$ kips (T), $F_{24} = -119.26$ kips (c)

Chapter 4

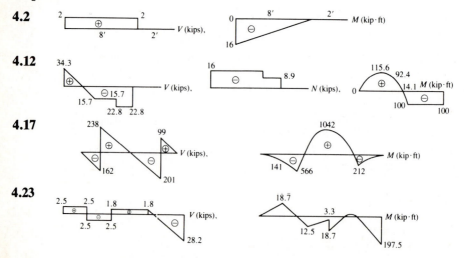

4.2

4.12

4.17

4.23

4.27

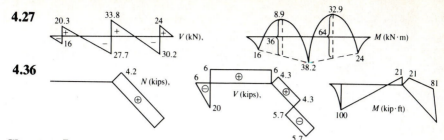

4.36

Chapter 5

5.3 $u = \dfrac{1972}{E}$

5.5 $v = 0.038\text{ in}$

5.8 $v_{max} = 0.00652\,\dfrac{w_0 l^4}{EI}$

5.11 $\theta_1 = -\dfrac{64}{9EI},\ \theta_2 = \dfrac{1472}{EI},\ v(16) = -\dfrac{4\,153}{EI}$

5.14 $v_3 = -\dfrac{p l^3}{64EI},\ \theta_3 = \dfrac{p l^2}{96EI}$

5.16 $v_1 = 0,\ \theta_1 = -\dfrac{720}{EI},\ v_2 = -\dfrac{8\,100}{EI},\ \theta_2 = -\dfrac{180}{EI},\ v_4 = \dfrac{22\,464}{EI},\ \theta_4 = \dfrac{2\,016}{EI}$

5.18 $v_2 = \dfrac{p l^3}{48EI},\ v_3 = \dfrac{5 p l^3}{384EI}$

5.20 $v_3 = -\dfrac{48\,600}{E},\ \theta_3 = 0;\ v_5 = \dfrac{86\,400}{E},\ \theta_5 = \dfrac{780}{E}$

5.23 (a) $v_3 = \dfrac{37.5}{EI}$, (b) $v_3 = 0.1917\text{ cm}$

5.25 $v = 0.993\text{ in (up)}$

5.27 $v_3 = M_3' = -\dfrac{4\,200}{EI}$

5.29 $v_{end} = \frac{19}{24} P l^3,\ M_{tmax} = Pl,\ M_{bmax} = \frac{9}{4} Pl$

5.31 $v_3 = \dfrac{l^3}{6EI}(8P_1 + 3P_2)$

5.33 $u = \dfrac{700}{3EI}$

Chapter 6

6.1 $v_3 = \dfrac{10\,000}{3EI}\text{ ft }(E\text{ in ksf, }I\text{ in ft}^4)$

6.3 $v_5 = 0.29$ in

6.6 $F_{12} = -42.4$ (c), $F_{13} = -70.7$ (c), $F_{23} = 48.5$ (T), $F_{24} = -11.5$ (c), $F_{25} = -26.1$ (c), $F_{34} = 2.16$ (T), $F_{35} = -51.5$ (c), $F_{45} = 18.5$ (T) (kips)

6.7 $F_{24} = 1.24$ P

6.9 $v_3 = \dfrac{1\,464.8}{EI}$

6.12   6.57 in

6.14 $v = \dfrac{P_0 l^3}{15EI}$

6.16 $v = \dfrac{4\,221}{EI}$

6.18 $v_3 = \dfrac{10\,133}{EI}$

6.20 $v_b = 1.78$ in, $v_s = 0.32$ in

6.22 $v_3 = \dfrac{pl^4}{389.64EI}$ (1.47% error), $M_0 = -\dfrac{pl^2}{19.74}$ (64.5% error),

$M_{1/2} = \dfrac{pl^2}{19.74}$ (17.8% error), $M_1 = -\dfrac{pl^2}{19.74}$,

$V_0 = \dfrac{pl}{3.14}$ (57% error), $V_{1/2} = 0$, $V_1 = \dfrac{pl}{3.14}$

Chapter 7

7.2 $R_2 = 37.5$ kips or $M_2 = -625$ kip · ft

7.3 $R_2 = \dfrac{5P}{18}$

7.7 $R = 0.48$ kN, $M = -8.43$ kN · m

7.8 $M_2 = -38.57$ kN · m, $M_3 = -200$ kN · m, $R_2 = 62.91$ kN, $R_3 = 136.93$ kN

7.10 $R_5 = 0.694$ kips

7.14 $M_2 = -282$ kip · ft

Chapter 8

8.1

 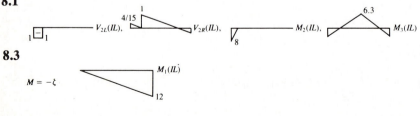

8.3

$M = -\zeta$

8.5

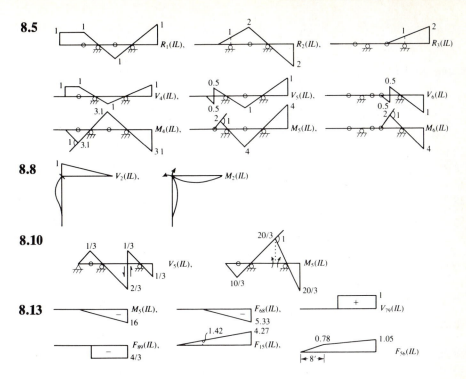

8.8

8.10

8.13

8.14 For load (d): $V_{max} = 15.275$ kips, $M_{max} = 155.75$ kip · ft
For load (e): $V_{max} = 21$ kips, $M_{max} = 210$ kip · ft

Chapter 9

9.1 $v_{mid} = \dfrac{19\,167}{EI}$

9.2 $v_0 = \dfrac{-7\,455}{EI}$ (up), $v_{mid} = \dfrac{16\,740}{EI}$ (down)

9.5 $v = \dfrac{34\,282}{3EI}$; $\theta = \dfrac{2\,381}{21EI}$

9.7 $v = \dfrac{-101\,920}{81EI}$ (up)

9.9 $v = \dfrac{5\,265}{4EI}$

9.11 $v_{mid} = \dfrac{147\,712}{15EI}$

9.14 $R_1 = 7.19$ kips, $R_2 = 20.62$ kips, $R_3 = 2.18$ kips

9.16 $M = -M_2 + R_2 x + R_3(x - 12)H(x - 12) - x^2$; $V = R_2 + R_3 H(x - 12) - 2x$,
$M_2 = 9$ kip · ft, $R_2 = 9.75$ kips, $R_3 = 26.25$ kips

9.17 $R_1 = 1 - \dfrac{\xi}{10}$; $V = -\dfrac{\xi}{10} + H(\xi - 10)$; $M = \tfrac{1}{2}\xi - (\xi - 5)H(\xi - 5)$

Chapter 10

In the design sign convention

10.4 $M_1 = -41.86$, $M_2 = -26.3$, $M_3 = -0.35$ (kip · ft)

10.6 $M_1 = -106.8$, $M_2 = -61.42$, $M_3 = 30.71$ (kip · ft)

10.8 $M_1 = -75$, $M_2 = -31.25$, $M_3 = -50$ (kip · ft)

10.11 $M_1 = -12.5$, $M_2 = -50.1$, $M_3 = -25.1$, $M_4 = 0$ (kip · ft)

10.12 $M_1 = -84.86$, $M_2 = -28.3$, $M_3 = 0$ (kN · m)

10.13 $M_{mid} = 279$ kip · in

10.15 $M_1 = 60$, $M_2 = -96$, $M_3 = 0$ (kN · m)

10.18 $M_{21} = -49.31$, $M_{62} = -3.77$, $M_{23} = -56.97$, $M_{32} = -35.36$, $M_{73} = 15.82$, $M_{34} = -31.63$, $M_4 = -40$ (kip · ft)

10.22 $M_{12} = -11.21$, $M_{21} = -38.07$, $M_{34} = 2.10$, $M_{23} = -38.07$, $M_{32} = -63.65$, $M_{34} = -63.65$ (kip · ft), Base shear: @ 1: 3.29, @ 4: 1.71 (kips)

10.24 $M_{12} = -47.07$, $M_{21} = M_{23} = -52.87$, $M_{32} = 0$ (kN · m)

10.26 $M_{12} = \dfrac{1.5EI}{468\,000}$, $M_{21} = M_{23} = -\dfrac{3EI}{468\,000}$

10.28 $M_{34} = 6$, $M_{43} = -3$ (kip · ft)

10.31 $M_{12} = -36.93$, $M_{21} = M_{23} = -38.5$, $M_{32} = M_{34} = 16.15$, $M_{43} = -88.42$ (kip · ft)

Chapter 11

In the analysis sign convention

11.1 $M_1 = 475$ kip · in, $M_2 = 50$ kip · in

11.6 $M_2 = -233.3$ kN · m

11.9 $r_1 = \dfrac{1.2343(10^5)}{EI}$, $r_2 = \dfrac{-1.6971(10^5)}{EI}$, $M_{12} = 2\,486$, $M_{21} = -429$

$M_{23} = 429$, $M_{32} = -1\,200$ (kip · ft)

11.11 $\theta = 0.0003$ rad, $v = 0.0353$ in, $V_1 = -1.25$ kips, $M_1 = -93.75$ kip · in

11.14 $\begin{bmatrix} P_1 \\ M_1 \end{bmatrix} = \dfrac{2EI}{l^3} \begin{bmatrix} 6 & -3l \\ -3l & 2l^2 \end{bmatrix} \begin{bmatrix} v_1 \\ \theta_1 \end{bmatrix}$

11.15 $\begin{bmatrix} V_1 \\ V_2 \end{bmatrix} = \dfrac{GAs}{l} \begin{bmatrix} 1 & -1 \\ -1 & 1 \end{bmatrix} \begin{bmatrix} v_1 \\ v_2 \end{bmatrix}$

Chapter 12

12.2 $\{u_2, v_2, u_3\} = \{0.0015 - 0.0031\ 0.0022\}$ (m)

$\begin{bmatrix} U_1 \\ U_2 \end{bmatrix}^{12} = \begin{bmatrix} 3.6 \\ -3.6 \end{bmatrix}$, $\begin{bmatrix} U_1 \\ U_2 \end{bmatrix}^{23} = \begin{bmatrix} 18.38 \\ -18.38 \end{bmatrix}$, $\begin{bmatrix} U_1 \\ U_2 \end{bmatrix}^{13} = \begin{bmatrix} 13.0 \\ -13.0 \end{bmatrix}$ (kN)

12.4 $\begin{bmatrix} u \\ v \end{bmatrix} = \begin{bmatrix} 0 \\ -0.1302 \end{bmatrix}$ (in), $\begin{bmatrix} U_1 \\ U_2 \end{bmatrix}^{21} = \begin{bmatrix} 51.60 \\ -51.60 \end{bmatrix}$,

$\begin{bmatrix} U_1 \\ U_2 \end{bmatrix}^{31} = \begin{bmatrix} -83.9 \\ 83.9 \end{bmatrix}$, $\begin{bmatrix} U_1 \\ U_2 \end{bmatrix}^{41} = \begin{bmatrix} 51.6 \\ -51.6 \end{bmatrix}$ (kips)

12.7 $\{u_2 \ v_2 \ u_3 \ u_4 \ v_4\} = \{0.0135 \ -0.2445 \ 0 \ -0.1875 \ -0.0135 \ -0.2445\}$

$\begin{bmatrix} U_1 \\ U_2 \end{bmatrix}^{12} = \begin{bmatrix} -3.75 \\ 3.75 \end{bmatrix}$, $\begin{bmatrix} U_1 \\ U_2 \end{bmatrix}^{13} = \begin{bmatrix} 18.0 \\ -18.0 \end{bmatrix}$, $\begin{bmatrix} U_1 \\ U_2 \end{bmatrix}^{23} = \begin{bmatrix} -6.25 \\ 6.25 \end{bmatrix}$,

$\begin{bmatrix} U_1 \\ U_2 \end{bmatrix}^{45} = \begin{bmatrix} -3.75 \\ 3.75 \end{bmatrix}$, $\begin{bmatrix} U_1 \\ U_2 \end{bmatrix}^{34} = \begin{bmatrix} -6.25 \\ 6.25 \end{bmatrix}$, $\begin{bmatrix} U_1 \\ U_2 \end{bmatrix}^{35} = \begin{bmatrix} 18 \\ -18 \end{bmatrix}$

12.10 $\{u_2 \ v_2 \ u_3 \ v_3 \ u_4\} = \{0.0604 \ 0.0125 \ 0.0479 \ -0.0125 \ 0.0125\}$

$\begin{bmatrix} U_1 \\ U_2 \end{bmatrix}^{12} = \begin{bmatrix} U_1 \\ U_2 \end{bmatrix}^{23} = \begin{bmatrix} U_1 \\ U_2 \end{bmatrix}^{34} = \begin{bmatrix} 125 \\ -125 \end{bmatrix}$, $\begin{bmatrix} U_1 \\ U_2 \end{bmatrix}^{14} = \begin{bmatrix} -125 \\ 125 \end{bmatrix}$,

$\begin{bmatrix} U_1 \\ U_2 \end{bmatrix}^{13} = \begin{bmatrix} -176.8 \\ 176.8 \end{bmatrix}$, $\begin{bmatrix} U_1 \\ U_2 \end{bmatrix}^{24} = \begin{bmatrix} 176.8 \\ -176.8 \end{bmatrix}$

12.14 $\{u_2 \ \theta_2 \ \theta_3\} = \{0.0012 \ -0.0003 \ -0.0003\}$
$\{V_1 \ M_1 \ V_2 \ M_2\} = \{5 \ 5.87 \ -5 \ 9.13\}^{21,34}$,
$\{-2.35 \ -5.87 \ 2.35 \ -5.87\}^{23}$

12.17 $\{\theta_1 \ u_2 \ \theta_2 \ \theta_3 \ \theta_4\} = \{0.0001 \ 0.0073 \ -0.0007 \ 0.0003 \ -0.0002\}$
$\{V_1 \ M_1 \ V_2 \ M_2\} = \{-2.15 \ -142.0 \ 2.15 \ -12.45\}^{21}$,
$\{10.70 \ 142.0 \ 10.5 \ -118.13\}^{23}$, $\{2.15 \ 118.13 \ -2.15 \ 36.32\}^{34}$,
$\{-0.20 \ 12.45 \ 0.20 \ -36.32\}^{14}$

12.20 $\{u_2 \ v_2 \ \theta_2 \ u_3 \ v_3 \ \theta_3\} = \{0 \ -0.112 \ -0.0021 \ 0 \ -0.1128 \ 0.0021\}$ (in, rad)
$\{U_1 \ V_1 \ M_1 \ U_2 \ V_2 \ M_2\} = \{0 \ 9.89 \ 469.64 \ 0 \ -5.09 \ 69.73\}^{12}$,
$\{-3.6 \ 3.6 \ -69.73 \ -3.6 \ 3.6 \ 69.73\}^{23}$,
$\{0 \ -5.09 \ -69.73 \ 0 \ 9.89 \ -469.65\}^{34}$ (kips, in)

12.25 $\{u_2 \ v_2 \ \theta_2^+ \ u_3 \ v_3 \ \theta_3^+ \ u_4 \ v_4 \ \theta_4 \ u_5 \ v_5 \ \theta_5^+\}$
$= \{0.0001 \ -0.0002 \ -0.0078 \ -0.0263 \ 0 \ 0 \ -0.0265$
$\quad 0 \ -0.0259 \ -0.0262 \ 0.0078\}$

$\begin{bmatrix} U_1 \\ U_2 \end{bmatrix} = \begin{bmatrix} 7.07 \\ -7.07 \end{bmatrix}^{12} , \begin{bmatrix} 21.21 \\ -21.21 \end{bmatrix}^{56} , \begin{bmatrix} 0 \\ 0 \end{bmatrix}^{13,36}$

$\{U_1 \ V_1 \ M_1 \ U_2 \ V_2 \ M_2\} = \{5 \ 5 \ 0 \ -5 \ -5 \ 25\}^{14}$,
$\{-20 \ 0 \ 0 \ 20 \ 0 \ 0\}^{34}$, $\{5 \ -15 \ -25 \ -5 \ 15 \ 0\}^{45}$
Note that the cables do not carry any load according to small deflection theory. (See Example 6.11.)

12.27 $\{\theta_1 \ u_2 \ v_2 \ \theta_2^- \ u_3 \ v_3 \ \theta_3^+ \ u_4 \ v_4 \ \theta_4 \ u_5 \ v_5 \ \theta_5^+ \ u_6 \ v_6 \ \theta_6^- \ u_7 \ v_7 \ \theta_7^+ \ u_8 \ \theta_8\}$
$= \{-0.0016 \ 0.0095 \ -0.2785 \ 0.0001 \ 0.1043 \ -0.1738 \ 0.0007$
$\quad -0.0190 \ -0.1109 \ 0 \ 0.019 \ 0.0748 \ 0 \ 0.0285 \ -0.2785$
$\quad -0.0001 \ -0.0063 \ -0.1738 \ -0.0007 \ 0.0380 \ 0.0016\}$

Force Vector: Beam Elements
$\{U_1 \ V_1 \ M_1 \ U_2 \ V_2 \ M_2\} = \{-20.66 \ 0.35 \ 11.85 \ 20.66 \ -0.35 \ 72.01\}^{12}$,
$\{-20.66 \ -0.61 \ -72.01 \ 20.66 \ 0.61 \ -73.99\}^{24}$,
$\{-20.66 \ 0.61 \ 73.99 \ 20.66 \ -0.61 \ 72.01\}^{46}$,
$\{-20.66 \ -0.35 \ -72.01 \ 20.66 \ 0.35 \ -11.85\}^{68}$,
$\{25.32 \ 0.16 \ -11.85 \ -25.32 \ -0.16 \ 57.77\}^{13}$,
$\{21.40 \ -0.56 \ -57.77 \ -21.40 \ 0.56 \ -83.47\}^{35}$,
$\{21.40 \ 0.56 \ 83.47 \ -21.40 \ -0.56 \ 57.77\}^{57}$,
$\{25.32 \ -0.16 \ -57.77 \ -25.32 \ 0.16 \ 11.85\}^{78}$. For truss elements
$\{U_1 \ U_2\} = \{-9.04 \ 9.04\}^{23}$, $\{-11.22 \ 11.22\}^{45}$, $\{-9.04 \ 9.04\}^{67}$ (kip, in)

Chapter 13

13.1 $\{V_1 \ M_1 \ V_2 \ M_2\} = \{17.19 \ 1350 \ 7.81 \ 0\}$ (kips, in) without settlement
$\{17.75 \ 1513 \ 7.25 \ 0\}$ with settlement

13.4 $\{M_1 \ M_2\} = \{-15.63 \ -31.25\}^{12}$, $\{0 \ 0\}^{23}$, $\{-18.75 \ -9.38\}^{24}$ (kN · m)

13.7 $\{U_1 \ V_1 \ M_1 \ U_2 \ V_2 \ M_2\} = \{8.69 \ 0 \ 285.28 \ -8.69 \ 0 \ -285.28\}^{12}$,
$\{0 \ 8.69 \ 285.28 \ 0 \ -8.69 \ 757.36\}^{23}$,
$\{0 \ -6.31 \ -757.36 \ 0 \ 6.31 \ 0\}^{34}$,
$\{0 \ 4.21 \ 0 \ 0 \ -4.21 \ 504.91\}^{56}$,
$\{0 \ -2.1 \ -504.91 \ 0 \ 2.1 \ 0\}^{67}$ (kip, in)

13.9 $\{v_2 \ \theta_2\} = \{-0.25 \ 0.0398\}$ (in, rad)
$\{V_1 \ M_1\} = \{41.29 \ 2990.9\}$ (kip, in)

13.11 (a) $K = 20.14$ kip/in, (b) $\{V_1 \ M_1 \ V_2 \ M_2\} = $
$\{8.06 \ 483.33 \ -8.06 \ 483.33\}$ (kip, in)

13.13 v_2 (top) $= -69.4$, v_3 (bottom) $= -69.43$ cm

13.16
$$\begin{bmatrix} 10\,855 & 0 & 0 & -10\,855 & 0 & 0 \\ 0 & 28 & 1\,102 & 0 & -28 & 1\,102 \\ 0 & 1\,102 & 48\,794 & 0 & -1\,102 & 39\,384 \\ -10\,855 & 0 & 0 & 10\,855 & 0 & 0 \\ 0 & -28 & -1\,102 & 0 & 1\,102 & -1\,102 \\ 0 & 1\,102 & 39\,384 & 0 & -1\,102 & 48\,794 \end{bmatrix}$$

13.18 with 3 elements, two in the first span 15 and 5 ft long
$\{V_1 \ M_1 \ V_2 \ M_2\} = \{0.06 \ 2.89 \ -0.06 \ 8.68\}^{14}$,
$\{0.06 \ -8.68 \ -0.06 \ 12.54\}^{42}$, $\{-0.05 \ -12.54 \ 0.05 \ 0\}^{23}$ (kip, in)

13.21 $\{U_1 \ V_1 \ M_1 \ U_2 \ V_2 \ M_2\} = \{9.99 \ 0.01 \ 2.15 \ -9.99 \ -0.10 \ 0\}^{12}$,
$\{14.13 \ 0 \ 0 \ -14.13 \ 0 \ 0\}^{23}$ (kip, in)

13.23 Member end moments $\mathbf{M} = \{0 \ -157.5 \ 157.5 \ 0 \ 0 \ -157.5 \ 157.5 \ 0\}$ (kip · m)

13.25 Bar forces $U_{12} = 56.57$, $U_{23} = 56.57$, $U_{34} = 16.57$
$U_{14} = 16.57$, $U_{13} = 23.43$, $U_{24} = -33.13$ kN

13.26 $\mathbf{M} = \{0 \ -48.49 \ 48.49 \ 139 \ -139 \ -48.49 \ 48.49 \ 0\}$ (kN · m)

Chapter 14

14.2 $v_2 = 26.32$ ft, $v_3 = 42.7$ ft, $v_4 = 31.54$ ft

14.4 $T = 3415.26$ kN

14.6 From horizontal of point 1 $h_2 = 5.39$ m, $h_3 = 3.78$ m, $T_{12} = 56.22$ kN, $T_{23} = 57.99$ kN, $T_{34} = 71.39$ kN, $T_{45} = 90.91$ kN

Chapter 15

15.3 $N = 545 \sin \phi + 995 \cos \phi - 900 \sin \phi \cos \phi$,
$V = -95 \sin \phi + 545 \cos \phi - 900 \sin \phi \cos \phi$,
$M = 29\,850(1 - \cos \phi) - 13\,500(1 - \cos \phi)^2 - 16\,350 \sin \phi$

15.5 At the right end $N = 35.6$, $V = 4.9$ kN

15.8 $\{U_1\ V_1\ M_1\ U_2\ V_2\ M_2\} = \{0.23\ 0.25\ 0.30\ -0.25\ 0.23\ -0.22\}$ (kip, in)

15.13 $\{35.10\ -7.32\ -29.90\ -35.10\ 7.32\ -32.30\}^{12}$
$\{13.89\ 13.89\ 32.26\ -13.89\ 13.89\ -32.26\}^{23}$
$\{35.10\ 7.32\ 32.26\ -35.10\ -7.33\ 29.90\}^{34}$ (kN, m)

15.15 $\mathbf{K} = \begin{bmatrix} \mathbf{K}_{11} & \mathbf{K}_{12} \\ \mathbf{K}_{21} & \mathbf{K}_{22} \end{bmatrix}$, $\mathbf{K}_{11} = \mathbf{F}_{11}^{-1}$, $\mathbf{K}_{22} = \mathbf{F}_{22}^{-1}$, $\mathbf{K}_{21} = \mathbf{Q}\mathbf{K}_{11}$, $\mathbf{K}_{12} = \mathbf{K}_{21}^T$

$f_{11} = \dfrac{\beta_1 l^3}{5\beta^2}$, $f_{12} = f_{21} = \dfrac{\alpha \beta_1 l^4}{20\beta^2}$, $f_{13} = f_{31} = \dfrac{\alpha l^3}{3\beta}$

$f_{22} = \dfrac{\beta_1^2 l^3}{30\beta^2}$, $f_{23} = f_{32} = -\dfrac{\beta_1 l^2}{6\beta}$, $f_{33} = l$

for $\mathbf{F}_{22}$ the terms that are different from those in $\mathbf{F}_{11}$ are
$f'_{12} = \alpha l^4(25 + \beta_1(93 + \beta_1(122 + \beta_1(58 - \beta_1(3 + 7\beta_1)))))/60(1 + \beta_1)^5$
$f'_{22} = l^3(10 + 2\beta_1)^3(7 - 2\beta_1) + \beta_1(41 + \beta_1)^3(-1 + \beta_1)/30(1 + \beta_1)^3$
$f'_{23} = l^2(3 + 2\beta_1^2) + \beta_1(8 - \beta_1^3)/6(1 + \beta_1)^{7/2}$

$\beta_1 = \alpha^2 l^2$, $\beta = \dfrac{1}{\sqrt{1 + \beta_1}}$, $\alpha = \dfrac{4h}{l^2}$,

$C_1 = \cos(\gamma + \phi_1)$, $S_1 = \sin(\gamma + \phi_1)$, $\phi_1 = \cos^{-1}\beta$

$\mathbf{T} = \begin{bmatrix} \mathbf{T}_1 & \mathbf{0} \\ \mathbf{0} & \mathbf{T}_2 \end{bmatrix}$, $\mathbf{T}_1 = \begin{bmatrix} C_1 & S_1 & 0 \\ -S_1 & C_1 & 0 \\ 0 & 0 & 1 \end{bmatrix}$, $\mathbf{T}_2 = \begin{bmatrix} C_1 & S_1 & 0 \\ S_1 & -C_1 & 0 \\ 0 & 0 & 1 \end{bmatrix}$,

$\gamma =$ angle of the chord with global X axis.

Chapter 16

16.2 $N_{cr} = \dfrac{\pi^2 EI}{(0.7l)^2}$

16.5 $N_{cr} = \dfrac{15.88 EI}{l^2}$

16.7 $N = 75$ kN with 6 elements (2 per span)

16.9 $N = 1008$ kips with 6 elements

16.11 $N = 691$ kips with 6 elements

16.15 $\mathbf{K} = \mathbf{K}_e + \mathbf{K}_s$, $\mathbf{K}_e =$ elastic stiffness matrix, same as that for beams,

$$
\mathbf{K}_s = kl \begin{bmatrix}
\dfrac{13}{25} & \dfrac{11l}{210} & \dfrac{9}{70} & \dfrac{-13}{420} \\[2ex]
 & \dfrac{l^2}{105} & \dfrac{13l}{420} & -\dfrac{l^2}{140} \\[2ex]
 & & \dfrac{13}{35} & -\dfrac{11l}{210} \\[2ex]
 & \text{Sym} & & \dfrac{l^2}{105}
\end{bmatrix}
$$

Chapter 17

17.3 $J = 3.2\,\mathrm{m}^4$

17.5 $\{T_1\ T_2\} = \{-225.9\ 225.9\}^{13}$, $\qquad \{-45.9\ 45.9\}^{34}$, $\qquad \{74.12\ -74.12\}^{42}$
(kip · in)

17.7 $\{V_1\ T_1\ M_1\ V_2\ T_2\ M_2\} = \{-7.7\ -166.3\ -1\,107.5\ 7.7\ 166.3\ 0\}^{12}$
$\{2.3\ 0\ 166.3\ -2.3\ 0\ 166.3\}^{23}$, $\quad \{-7.7\ 166.3\ 0\ 7.7\ 166.3\ -1\,107.5\}^{34}$
(kip, in)

17.9 $\{V_1\ T_1\ M_1\ V_2\ T_2\ M_2\} = \{-10.14\ 0\ 0\ 10.14\ 0\ -50.72\}^{12}$,
$\{-10.14\ 0\ -49.28\ 10.14\ 0\ -1.44\}^{23}$, $\{0\ 1.44\ 0\ 0\ -1.44\ 0\}^{34}$
$\{-0.29\ 0\ -1.44\ 0.29\ 0\ 0\}^{45}$ (kN, m)

17.11 $\{V_1\ T_1\ M_1\ V_2\ T_2\ M_2\} = \{9.98\ -0.07\ 99.54\ -9.98\ 0.07\ 0.25\}^{12}$
$\{-0.02\ -0.07\ -0.25\ 0.02\ 0.06\ -0.24\}^{23}$,
$\{-0.02\ -0.06\ 0.24\ 0.02\ 0.06\ -0.44\}^{34}$ (kN, m)

Chapter 18

18.1 $R_{1x} = -4.17$, $R_{1y} = -2.08$, $R_{1z} = -16.67$, $R_{2x} = 0$, $R_{2y} = -3.33$,
$R_{2z} = 13.33$, $R_{3x} = -10.83$, $R_{3y} = 5.41$, $R_{3z} = 43.33$ kips

18.4 $\{U_1\ U_2\} = \{-582.7\ 582.7\}$ (kips)

18.6 $\{U_1\ U_2\} = \{-10\ 10\}^{12}$, $\{8.47\ -8.47\}^{23}$, $\{3.63\ -3.63\}^{24,46}$,
$\{-10\ 10\}^{56}$, $\{8.47\ -8.47\}^{57}$, $\{-2.42\ 2.42\}^{26}$ (kips)

18.9 The non-zero forces are $\left\{ \begin{matrix} U_1 \\ U_2 \end{matrix} \right\} = \left\{ \begin{matrix} 56 \\ -56 \end{matrix} \right\}^{12}$, $\left\{ \begin{matrix} -24 \\ 24 \end{matrix} \right\}^{23}$, $\left\{ \begin{matrix} 124.71 \\ -124.71 \end{matrix} \right\}^{26}$,

$\left\{ \begin{matrix} -134.58 \\ 134.58 \end{matrix} \right\}^{16}$ (kips)

18.12 $\{U_1\ V_1\ W_1\ T_1\ M_{1y}\ M_{1z}\ U_2\ V_2\ W_2\ T_2\ M_{2y}\ M_{2z}\} =$
$\{0.9\ -22.13\ 0\ 20.85\ 0.57\ -677.69\ -0.90\ -22.87\ 0$
$-20.85\ 0.48\ 745.13\}^{12}$, $\{1.22\ -25.85\ 0\ 1.79\ 0\ -792.21$
$-1.22\ -19.15\ 0\ -1.79\ 0.24\ 189.45\}^{23}$, $\{2.40\ -19.23\ 0\ 10.42\ -0.49$
$-307.95\ -2.40\ -25.77\ 0\ -10.42\ -0.35\ 896.28\}^{24}$,
$\{67.95,\ -2.40\ -0.31\ -0.10\ 36.66\ -288.89\ -67.95\ 2.40\ 0.31$
$0.10\ 18.46\ -143.11\}^{26}$, $\{2.61\ -19.39\ 0\ -42.28\ -0.32\ -315.54$
$-2.61\ -25.61\ 0\ 42.28\ -0.45\ 876.14\}^{35}$,
$\{38.54\ -2.60\ 1.22\ 0\ -148.69\ -318.42$
$-38.54\ 2.61\ -1.22\ 0\ -74.06\ -157.68\}^{37}$ (kip, in)

INDEX

TA645.A73 1991
Arbabi, F. (Freydoon).
    Structural analysis and
behavior.

                        18752014

1-2-92

A000018752014

SUBJECT
TO RECALL **DATE DUE**

SEMEST DEC 0 1993		

Demco, Inc. 38-293